3 9082 13450 9978

D1072225

THE FATE OF ROME

The Princeton History of the Ancient World

THE FATE OF ROME

CLIMATE, DISEASE, AND THE END OF AN EMPIRE

KYLE HARPER

PRINCETON UNIVERSITY PRESS
PRINCETON AND OXFORD

Published by Princeton University Press, 41 William Street,
Princeton, New Jersey 08540
In the United Kingdom: Princeton University Press, 6 Oxford Street,
Woodstock, Oxfordshire OX20 1TR

press.princeton.edu

Jacket art courtesy of Shutterstock
Jacket design by Faceout Studio, Spencer Fuller

ISBN 978-0-691-16683-4

Library of Congress Control Number: 2017952241

British Library Cataloging-in-Publication Data is available

This book has been composed in Sabon Next LT Pro

Printed on acid-free paper. ∞

Printed in the United States of America

1 3 5 7 9 10 8 6 4 2

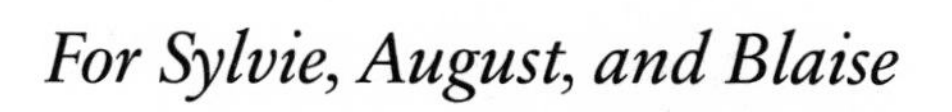

For Sylvie, August, and Blaise

In my beginning is my end. In succession
Houses rise and fall, crumble, are extended,
Are removed, destroyed, restored, or in their place
Is an open field, or a factory, or a by-pass.
Old stone to new building, old timber to new fires,
Old fires to ashes, and ashes to the earth
Which is already flesh, fur and faeces,
Bone of man and beast, cornstalk and leaf.

—T. S. Eliot, "East Coker"

CONTENTS

MAPS

Timeline

200 BC | 100 BC | AD 1 | AD 100 | AD 200

CLIMATE HISTORY

ROMAN CLIMATE OPTIMUM
200 BC – AD 150

LATE ROMA

DISEASE HISTORY

Antonine Plague 165

IMPERIAL HISTORY

Gibbon's "Happiest Era" 96–180

War with Parthia 161–6

Reign of Marcus Aureliu
161–80

Severan Dynas
193–235

HISTORICAL FIGURES

Aelius Aristides 117–81

Marcus Aurelius 121–80

Faustina 130–75

Galen of Pergamum 130–210

Septimius Severus 145-21

AD 300 | AD 400 | AD 500 | AD 600 | AD 700

RANSITIONAL PERIOD
50 - 450

LATE ANTIQUE LITTLE ICE AGE
450 - 700

ue of Cyprian 249–62

Justinianic Plague
first outbreak 541–43
Subsequent outbreaks to 749

Conversion of Constantine 312

Battle of Adrianople 378

Sack of Rome 410

Death of Attila the Hun 453

Last Western Emperor 476

ennium Games 248

rd-Century Crisis 250–70

Reign of Justinian 527–65

Reconquest of Africa 533–34

War with Persia 602–28

Reign of Heraclius 610–41

Hegira of Mohammad 622

Battle of Yarmouk 636

p the Arab 204–49

ian of Carthage 200-58

dius II 210–70

cletian 244–312

Constantine 272–337

Theodosius I 347–95

Stilicho 359–408

Claudian 370–404

Alaric 370–410

Attila 406–53

Theoderic 454–526

Justinian 482–565

Theodora 500–48

Procopius 500–54

John of Ephesus 507–88

Maurice 539–602

Gregory the Great 540–604

Mohammad 570–632

Heraclius 575–641

THE FATE
OF ROME

Prologue

NATURE'S TRIUMPH

Sometime early in the year AD 400, the emperor and his consul arrived in Rome. No one alive could remember a time when the emperors actually resided in the ancient capital. For over a century, the rulers of the empire had passed their days in towns closer to the northern frontier, where the legions held the line between, as the Romans thought of it, civilization and barbarism.

By now, an official imperial visit to the capital counted as a pretext for magnificent fanfare. For even without the emperors, Rome and its people remained potent symbols of the empire. Some 700,000 souls still called the city their home. They enjoyed all the amenities of a classical town, on an imperial scale. A proud inventory from the fourth century claimed that Rome had 28 libraries, 19 aqueducts, 2 circuses, 37 gates, 423 neighborhoods, 46,602 apartment blocks, 1,790 great houses, 290 granaries, 856 baths, 1,352 cisterns, 254 bakeries, 46 brothels, and 144 public latrines. It was, by any measure, an extraordinary place.[1]

The arrival of an emperor on the scene set in motion a sequence of carefully staged civic rituals, designed to assure the City of its preeminence within the empire and, at the same time, to assure the empire of its preeminence among all the principalities of the world. The people, as the proud stewards of the imperial tradition, were keen judges of this kind of ceremony. Rome, they were pleased to be reminded, was "a city greater than any the air encompasses on the earth, whose grandeur no eye can behold, whose charms no mind can measure."[2]

A grand imperial procession coiled its way to the forum. Here was where Cato and Gracchus, Cicero and Caesar, had made their political fortunes. The ghosts of history were welcome companions as the crowd gathered on this day to hear a speech of praise honoring the consul, Stilicho. Stilicho was a towering figure, a *generalissimo* at the zenith of his power. His imposing presence was affirmation that peace and order had returned to the empire. The show of confidence was reassuring. Just a generation before, in AD 378, at Adrianople, the legions of Rome suffered the worst defeat in their proud history. Ever since, the world had seemed to wobble on its axis. Goths entered the empire *en masse* and were an inscrutable mix of ally and enemy. The death of the emperor Theodosius I, in AD 395, revealed that the eastern and western halves of the empire had grown apart, as silently and consequentially as the drift of continents. Internal strife had menaced the African provinces and threatened the food supply. But, for the moment, the consul had calmed these rough waters. He had restored "the equipoise of the world."[3]

The poet who spoke in the consul's honor was named Claudian. An Egyptian by birth, whose native tongue was Greek, Claudian made himself one of the last true giants of classical Latin verse. His words betray the sincere awe the capital inspired in a visitor. Rome was the city that, "sprung from humble beginnings, has stretched to either pole, and from one small place extended its power so as to be co-terminous with the sun's light." She was the "mother of arms and of law." She had "fought a thousand battles" and extended "her sway o'er the earth." Rome alone "received the conquered into her bosom, and like a mother, not an empress, protected the human race with a common name, summoning those whom she has defeated to share her citizenship."[4]

This was not poetic fancy. In Claudian's time, proud Romans could be found from Syria to Spain, from the sands of Upper Egypt to the frostbit frontiers of northern Britain. Few empires in history have achieved *either* the geographical size *or* the integrative capacities of the Roman commonwealth. None have combined scale and unity like the Romans—not to mention longevity. No empire could peer back over so many centuries of unbroken greatness, advertised everywhere the eye wandered in the forum.

For nearly a millennium, the Romans had marked their years by the names of the consuls: thus Stilicho's name was "inscribed in the annals of the sky." In gratitude for this immortal honor, the consul was expected to entertain the people in traditional Roman style, which is to say with expensive and sanguinary games.

We know thanks to Claudian's speech that the people were presented an exotic menagerie worthy of an empire with global pretensions. Boars and bears were brought from Europe. Africa gave leopards and lions. From India came the tusks of elephants, though not the animal itself. Claudian imagines the boats crossing sea and river with their wild cargo. (And he includes an unexpected but wonderful detail: the sailors were terrified by the prospect of sharing their ship with an African lion.) When the hour came, the "glory of the woods" and "the marvels of the south" would be

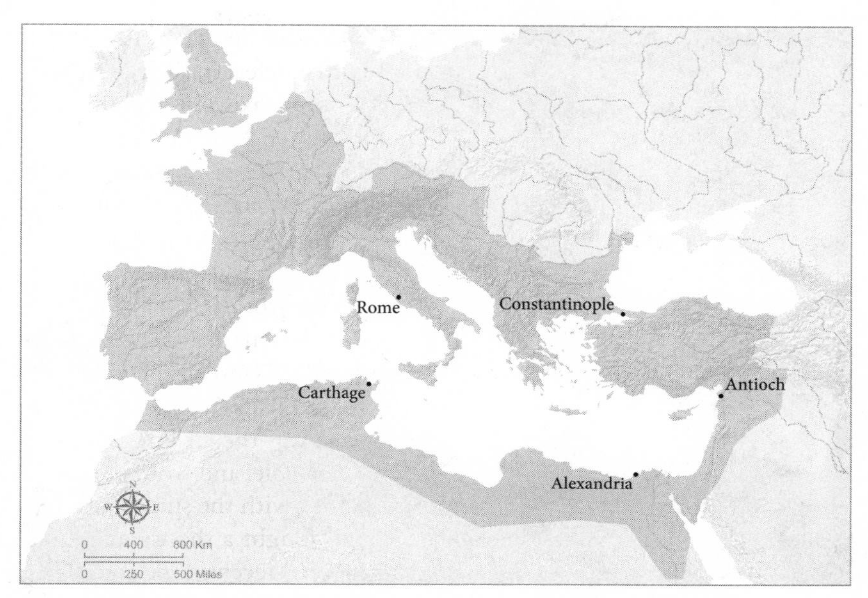

Map 1. The Roman Empire and Its Largest Cities in the Fourth Century

sportingly massacred. The blood-letting of nature's most ferocious beasts, in the confines of the arena, was a pointed expression of Rome's dominion over the earth and all her creatures. Such gory spectacles were a comforting familiarity, connecting the present inhabitants of Rome to the countless generations who had built and kept the empire.[5]

Claudian's speech pleased its hearers. The senate voted to honor him with a statue. But the confident notes of his oration were soon drowned out, first by brutal siege and then the unthinkable. On August 24 of 410, for the first time in eight hundred years, the Eternal City was sacked by an army of Goths, in what was the most dramatic single moment in the long train of events known as the fall of the Roman Empire. "In one city the earth itself perished."[6]

How could this happen? The answers we might give to such a question will depend very much on the resolution of our focus. On small scales, human choice looms large. The Romans' strategic decisions in the years leading up to the calamity have been endlessly second-guessed by arm-chair generals. On a larger canvas, we might identify structural flaws in the imperial machinery, such as the exhausting civil wars or the exorbitant pressures on the fiscal apparatus. If we zoom even further out, we might view the rise and fall of Rome as the unavoidable fate of all empires. Something along

Figure P.1. Relief of Caged Lions on Ship, Third Century. (DEA PICTURE LIBRARY / Getty Images)

those lines was the final verdict of the great English historian of Rome's fall, Edward Gibbon.

In his famous words, "The decline of Rome was the natural and inevitable effect of immoderate greatness. Prosperity ripened the principle of decay; the causes of destruction multiplied with the extent of conquest; and as soon as time or accident had removed the artificial supports, the stupendous fabric yielded to the pressure of its own weight." The ruin of Rome was but one example of the impermanence of all human creations. *Sic transit gloria mundi.*[7]

All of these answers can be true, simultaneously. But the argument put forth in these pages is that to understand the prolonged episode we know as the fall of the Roman Empire, we must look more closely at a great act of self-deception, right at the heart of the empire's triumphant ceremonies: the undue confidence, enacted in the bloody ritual of staged animal hunts, that the Romans had tamed the forces of wild nature. At scales that the Romans themselves could not have understood and scarcely imagined—from the microscopic to the global—the fall of their empire was the triumph of nature over human ambitions. The fate of Rome was played out by emperors and barbarians, senators and generals, soldiers and slaves. But it was equally decided by bacteria and viruses, volcanoes and solar cycles. Only in recent years have we come into possession of the scientific tools that allow us to glimpse, often fleetingly, the grand drama of environmental change in which the Romans were unwitting actors.

The great national epic of Rome's beginnings, the *Aeneid*, famously proclaims itself to be a song about "arms and a man." The story of Rome's end is also a human one. There were tense moments when human action decided the margin between triumph and defeat. And there were deeper, material dynamics–of agrarian production and tax collection, demographic struggle and social evolution–that determined the scope and success of Rome's power. But, in the very first scenes of the *Aeneid*, the hero is tossed on the spiteful winds of a violent storm, a plaything of the elemental forces of nature. What we have learned in recent years is making visible as never before the elemental forces that repeatedly tossed Rome's empire. The Romans built a giant, Mediterranean empire at a particular moment in the history of the climate epoch known as the Holocene—a moment suspended on the edge of tremendous natural climate change. Even more consequentially, the Romans built an interconnected, urbanized empire on the fringes of the tropics, with tendrils creeping across the known world. In an unintended conspiracy with nature, the Romans created a disease ecology that unleashed the latent power of pathogen evolution. The Romans were soon engulfed by the overwhelming force of what we would today call emerging infectious diseases. The end of Rome's empire, then, is a story in which humanity and the environment cannot be separated. Or, rather, it is one chapter in the still unfolding story of our relationship with the environment. The fate of Rome might serve to remind us that nature is cunning and capricious. The deep power of evolution can change the world in a mere moment. Surprise and paradox lurk in the heart of progress.

Here is an account of how one of history's most conspicuous civilizations found its dominion over nature less certain than it had ever dreamed.

Environment and Empire

The Shape of the Roman Empire

Rome's rise is a story with the capacity to astonish us, all the more so since the Romans were relative latecomers to the power politics of the Mediterranean. By established convention, Rome's ancient history is divided into three epochs: the monarchy, the republic, and the empire. The centuries of monarchy are lost in the fog of time, remembered only in fabulous origins myths that told later Romans how they came to be. Archaeologists have found the debris of at least transient human presence around Rome going back to the Bronze Age, in the second millennium BC. The Romans themselves dated their city's founding and the reign of their first king, Romulus, to the middle of the eighth century BC. Indeed, not far from where Claudian stood in the forum, beneath all the brick and marble, there had once been nothing more than a humble agglomeration of wooden huts. This hamlet could not have seemed especially propitious at the time.[1]

For centuries, Rome stood in the shadow of her Etruscan neighbors. The Etruscans in turn were outclassed by the political experiments underway to the east and south. The early classical Mediterranean belonged to the Greeks and Phoenicians. While Rome was still a village of letterless cattle rustlers, the Greeks were writing epic and lyric poetry, experimenting with democracy, and inventing drama, philosophy, and history as we know them. On nearer shores, the Punic peoples of Carthage built an ambitious empire, before the Romans knew how to rig a sail. Fifteen miles inland, along the soggy banks of the Tiber River, Rome was a backwater, a spectator to the creativity of the early classical world.[2]

Around 509 BC the Romans shuffled off their kings and inaugurated the republic. Now they gradually step into history. From the time they are known to us, Rome's political and religious institutions were a blend of the indigenous and the adopted. The Romans were unabashed borrowers. Even the first code of Roman law, the Twelve Tables, was proudly confessed to be plagiarized from Athens. The Roman republic belongs among the many citizenship-based political experiments of the classical Mediterranean. But the Romans put their own accents on the idea of a quasi-egalitarian polity. Exceptional religious piety. Radical ideologies of civic sacrifice. Fanatical militarism. Legal and cultural mechanisms to incorporate former enemies as allies and citizens. And though the Romans themselves came to believe that they were promised *imperium sine fine* by the gods, there was nothing ineluctable about Rome's destiny, no glaring geographical or technological secret of superiority. Only once in history did the city become the seat of a great empire.

Rome's rise coincided with a period of geopolitical disorder in the wider Mediterranean in the last centuries before Christ. Republican institutions and militaristic values allowed the Romans to concentrate unprecedented state violence, at an opportune moment of history. The legions destroyed their rivals one by one. The building of the empire was bloody business. The war machine whetted its own appetite. Soldiers were settled in rectilinear Roman colonies, imposed by brute force all over the Mediterranean. In the last century of this age of unbridled conquest, grand Shakespearean characters bestride the stage of history. Not by accident is western historical consciousness so disproportionately concentrated in these last few generations of the republic. The making of Rome's empire was not quite like anything that had happened before. Suddenly, levels of wealth and development lunged toward modernity, surpassing anything previously witnessed in the experience of our species. The teetering republican constitution generated profound reflections on the meaning of freedom, virtue, community. The acquisition of imperial power inspired enduring conversations about its proper exercise. Roman law helped to birth norms of governance, by which even the masters of empire might be held to account. But the scaling up of sheer power also fueled the cataclysmic civil violence that ushered in an age of autocracy. In the apt words of Mary Beard, "the empire created the emperors—not the other way round."[3]

By the time Augustus (r. 27 BC–AD 14) brought the last meaningful stretches of the shore under Roman dominion, it was no idle boast to call the Mediterranean "*mare nostrum*," our sea. To take full measure of the Roman accomplishment, and to understand the mechanics of ancient imperialism, we must know some basic facts about life in an ancient society. Life was slow, organic, fragile, and constrained. Time marched to the dull rhythms of foot and hoof. Waterways were the real circulatory system of the empire, but in the cold and stormy season the seas closed, and every town became an island. Energy was forbiddingly scarce. Human and animal muscle for force, timber and scrub for fuel. Life was lived close to the land.

Eight in ten people lived outside of cities. Even the towns had a more rural character than we might imagine, made lively by the bleats and brays—and pungent smells—of their four-legged inhabitants. Survival depended on the delivery of rain in a precarious environment. For the vast majority, cereals dominated the diet. "Give us this day our daily bread" was a sincere petition. Death always loomed. Life expectancy at birth was in the 20s, probably the mid-20s, in a world where infectious disease raged promiscuously. All of these invisible constraints were as real as gravity, defining the laws of motion in the world the Romans knew.[4]

These limits cast into relief the sheer spatial achievement of the Roman Empire. Without telecommunications or motorized transport, the Romans built an empire connecting vastly different parts of the globe. The empire's northern fingers reached across the 56th parallel, while the southern edges dipped below 24° N. "Of all the contiguous empires in premodern history, only those of the Mongols, Incas, and Russian czars matched or exceeded the north-south range of Roman rule." Few empires, and none so long-lived, grasped parts of the earth reaching from the upper mid-latitudes to the outskirts of the tropics.[5]

The northern and western parts of the empire were under the control of the Atlantic climate. At the ecological center of the empire was the Mediterranean. The delicate, moody features of the Mediterranean climate—arid summers and wet winters against a relatively temperate backdrop—make it a distinct type of climate. The dynamics of a giant, inland sea, combined with the knuckled texture of its inland terrains, pack extreme diversity into miniature scale. Along the empire's southern and eastern edges, the high pressure of the subtropical atmosphere won out, turning the land into pre-desert and then true desert. And Egypt, the breadbasket of the empire, plugged the Romans into wholly other climate regimes: the life-bringing Nile floods originated in Ethiopian highlands watered by the monsoons. The Romans ruled all this.[6]

The Romans could not impose their will on so vast a territory by violence alone. The maintenance of the empire required economies of force and constant bargaining with those inside Roman boundaries and beyond. Over the course of the empire's long life, the inner logic of imperial power, those economies and bargains, shifted shape many times.

Augustus gave order to the regime we recognize as the high Roman Empire. Augustus was a political genius, gifted with an uncannily long lifespan, who presided over the death throes of the republican constitution. During his reign, the campaigns of conquest, which had been fueled by elite competition for power in the late republican regime, started to slow. His reign

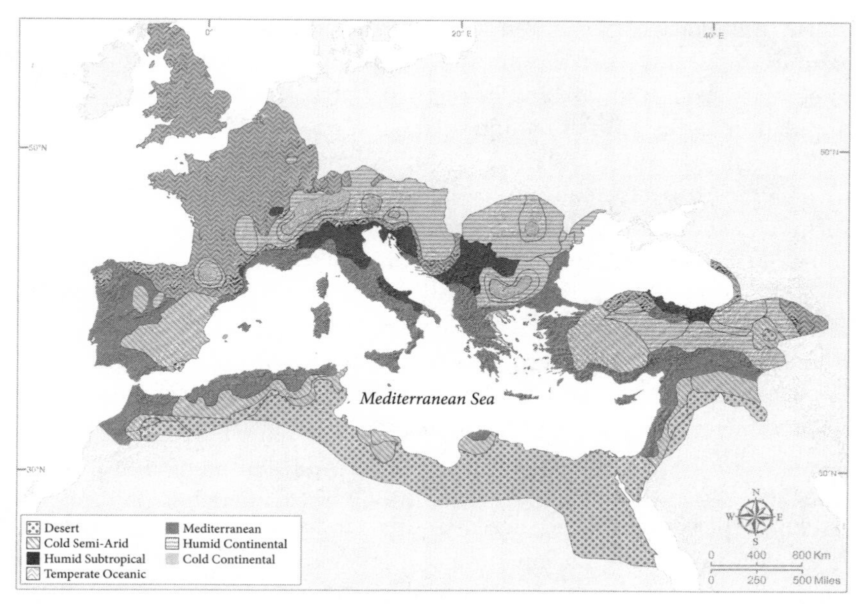

Map 2. Ecological Zones of the Roman Empire

was advertised as a time of peace. The gates to the Temple of Janus, which the Romans left open in times of war, had been closed twice in seven centuries. Augustus made a show of closing them three times. He demobilized the permanent citizen legions and replaced them with professional armies. The late republic had still been an age of gratuitous plunder. Slowly but surely, though, norms of governance and justice began to prevail in the conquered territories. Plunder was routinized, morphed into taxation. When resistance did flare, it was snuffed out with spectacular force, as in Judea and Britain. New citizens were made in the provinces, coming like a trickle at first, but subsequently faster and faster.

The grand and decisive imperial bargain, which defined the imperial regime in the first two centuries, was the implicit accord between the empire and "the cities." The Romans ruled through cities and their noble families. The Romans coaxed the civic aristocracies of the Mediterranean world into their imperial project. By leaving tax collection in the hands of the local gentry, and bestowing citizenship liberally, the Romans co-opted elites across three continents into the governing class and thereby managed to command a vast empire with only a few hundred high-ranking Roman officials. In retrospect, it is surprising how quickly the empire ceased to be a mechanism of naked extraction, and became a sort of commonwealth.[7]

The durability of the empire depended on the grand bargain. It was a gambit, and it worked. In the course of the *pax Romana*, as predation turned to governance, the empire and its many peoples flourished. It started with population. In the most uncomplicated sense, people multiplied. There had never been so many people. Cities spilled beyond their accustomed limits. The settled landscape thickened. New fields were cut from the forests. Farms crept up the hillsides. Everything organic seemed to thrive in the sunshine of the Roman Empire. Sometime around the first century of this era, the population of Rome itself probably topped one million inhabitants, the first city to do so, and the only western one until London circa 1800. At the peak in the middle of the second century, some seventy-five million people in all came under Roman sway, a quarter of the globe's total population.[8]

In a slow-moving society, such insistent growth—on this scale, over this arc of time—can easily spell doom. Land is the principal factor of production, and it is stubbornly finite. As the population soared, people should have been pushed onto ever more marginal land, harder and harder pressed to extract energy from the environment. Thomas Malthus well understood the intrinsic and paradoxical relationships between human societies and their food supplies. "The power of population is so superior to the power of the earth to produce subsistence for man, that premature death must in some shape or other visit the human race. The vices of mankind are active and able ministers of depopulation. They are the precursors in the great army of destruction, and often finish the dreadful work themselves. But should they fail in this war of extermination, sickly seasons, epidemics, pestilence, and plague advance in terrific array, and sweep off their thousands and tens of thousands. Should success be still incomplete, gigantic inevitable famine stalks in the rear, and with one mighty blow levels the population with the food of the world."[9]

Yet . . . the Romans manifestly did not succumb to mass-scale starvation. Herein is to be found the hidden logic of the empire's success. Far from steadily sinking into misery, the Romans achieved per capita economic growth, straight into the teeth of headlong demographic expansion. The empire was able to defy, or at least defer, the grim logic of Malthusian pressure.

In the modern world, we are accustomed to annual growth rates of 2–3 percent, on which our hopes and pension plans depend. It was not so in ancient times. By their nature, pre-industrial economies were on a tight energy leash, constrained in their ability to extract and exchange energy more efficiently on any sustainable basis. But premodern history was neither a slow, steady ascent toward modernity, nor the proverbial hockey stick—a flat-line of bleak subsistence until the singular energy breakthroughs of the Industrial Revolution. Rather, it was characterized by pulses of expansion and then disintegration. Jack Goldstone has proposed the term "efflorescence" for those phases of expansion, when background conditions conduce to real growth for some happy length of time. This growth can be extensive, as people multiply and more resources are turned to productive use, but as Malthus described, this kind of growth eventually runs out of room; more promisingly, growth can be intensive, when trade and technology are employed to extract energy more efficiently from the environment.[10]

The Roman Empire set the stage for an efflorescence of historic proportions. Already in the late republic, Italy experienced precocious leaps forward in social development. To a certain extent, the prosperity of Italy might be written off as the result of sheer takings, naked political rents seized as the fruits of conquest. But underneath this veneer of extracted wealth, real growth was afoot. This growth not only continued after the military expansion had reached its outer bounds—it now diffused throughout the conquered lands. The Romans did not merely rule territory, transferring some margin of surplus from periphery to center. The integration of the empire was catalytic. Slowly but steadily, Roman rule changed the face of the societies under its dominion. Commerce, markets, technology, urbanization: the empire and its many peoples seized the levers of development. For more than a century and a half, on a broad geographical scale, the empire writ large enjoyed both intensive and extensive growth. The Roman Empire both staved off Malthusian reckoning and earned uncalculated political capital.[11]

This prosperity was the condition and the consequence of the empire's grandeur. It was a charmed cycle. The stability of the empire was the enabling background of demographic and economic increase; people and prosperity were in turn the sinews of the empire's power. Soldiers were plentiful. Tax rates were modest, but collections were abundant. The emperors were munificent. The grand bargain with the civic elites paid out for both sides. There seemed to be enough wealth everywhere. The Roman armies enjoyed tactical, strategic, and logistical advantages over enemies on every front. The Romans had achieved a kind of favorable equilibrium, if more fragile than they knew. Gibbon's great *History of the Decline and Fall of the Roman Empire* launches from the sunny days of the second century. In his famous verdict, "If a man were called to fix the period in the history of the world, during which the condition of the human race was most happy and prosperous, he would, without hesitation, name that which elapsed from the death of Domitian [AD 96] to the accession of Commodus [AD 180]."[12]

The Romans had edged outward the very limits of what was possible in the organic conditions of a premodern society. It is no wonder that the fall of such a colossus, what Gibbon called "this awful revolution," has been the object of perennial fascination.

Our Fickle Planet

By AD 650, the Roman Empire was a shadow of its former self, reduced to a Byzantine rump state in Constantinople, Anatolia, and a few straggled possessions across the sea. Western Europe was broken into fractious Germanic kingdoms. Half the former empire was swiftly carved off by armies of believers from Arabia. The population of the Mediterranean basin, which once stood at seventy-five million people, had stabilized at maybe half that number. Rome was inhabited by some 20,000 souls. And its denizens were none the richer for it. By the seventh century, one measly trunk route still connected east and west across the sea. Currency systems were as fragmented as the political mosaic of the early middle ages. All but the crudest financial institutions had vanished. Everywhere apocalyptic fear reigned, in Christendom and formative Islam. The end of the world felt nigh.

These used to be called the Dark Ages. That label is best set aside. It is hopelessly redolent of Renaissance and Enlightenment prejudices. It altogether underestimates the impressive cultural vitality and enduring spiritual legacy of the entire period that has come to be known as "late antiquity." At the same time, we do not have to euphemize the realities of imperial disintegration, economic collapse, and societal simplification. These are brute facts in need of explanation, as objective as an electricity bill—and measured in similar units. In material terms, the fall of the Roman Empire saw the process of efflorescence run in reverse, toward lower levels of energy capture and exchange. What we are contemplating is a monumental episode of state failure and stagnation. In Ian Morris's valiant effort to create a universal scale of social development, the fall of the Roman Empire emerged as the single greatest regression, *in all of human history*.[13]

Explanations for the fall of Rome have never been lacking. There is a traffic jam of contending theories. A German classicist catalogued 210 hypotheses on offer. Some of these have held up to scrutiny better than others, and the two that enjoy pride of place as leading contenders for large-scale explanation emphasize the inherently unsustainable mechanics of the imperial system and the gathering external pressures along the frontiers of empire.

The first emperor, Augustus, established the constitutional framework of monarchy; rules of succession were purposefully indeterminate, and the accidents of Fortune played a perilously large role. As time progressed, contests for power and legitimacy played out as self-destructive wars for command of the armies. Concurrently, the ever-growing professional corps of imperial administrators displaced the webs of local elites in running the empire, making for a more bureaucratic and more brittle state. The mounting fiscal pressures progressively overheated the system.[14]

Meanwhile, the borders of empire stretched across northern Britain, along the Rhine and Danube and Euphrates, and past the edges of the Sahara. Beyond the march, jealous and hungry peoples dreamed of their own destiny. Time was their ally; the process we now call secondary state formation saw Rome's adversaries become more complex and formidable over the centuries. These threats relentlessly drained the resources of frontier zones and heartland alike. In tandem with dynastic strife, they were fatal to the fortunes of empire.

These familiar theories have much to recommend them, and they remain integral to the story presented in these pages. But in recent years, students of the past have been increasingly confronted by what might be called *natural* archives. Natural archives come in many forms. Ice cores, cave stones, lake deposits, and marine sediments preserve records of climate change, written in the language of geochemistry. Tree rings and glaciers are records of the environment's history. These physical proxies preserve the encoded record of the earth's past. Equally, evolutionary and biological history have left a trail for us to follow. Human bones, in their size and shape and scars, preserve a subtle record of health and disease. The isotope chemistry of bones and teeth can tell stories about diet and migration, biological biographies of the silent majority. And the greatest natural archive of all may be the long strands of nucleic acids we call genes. Genomic evidence can cast light on the history of our own species as well as the allies and adversaries with whom we have shared the planet. Living DNA is an organic record of evolutionary history. And the ability to extract and sequence ancient DNA from archaeological contexts is allowing us to reconstruct the tree of life into the deep past. Occasionally, it has let us finger some of history's microbial mass murderers with forensic identification as dramatic and final as any courtroom evidence. Technology is revolutionizing what we know about the evolutionary story of microbes and men.[15]

Most histories of Rome's fall have been built on the giant, tacit assumption that the environment was a stable, inert backdrop to the story. As a by-product of our own urgent need to understand the history of earth systems, and thanks to dizzying advances in our ability to retrieve data about the paleoclimate and genomic history, we know that this assumption is wrong. It is not only wrong—it is immodestly, unnervingly wrong. The earth has been and is a heaving platform for human affairs, as unstable as a ship's deck in a violent squall. Its physical and biological systems are a ceaselessly changing setting, and they have given us what John Brooke calls "a rough journey" for as long as we have been human.[16]

Our awareness of climate change is understandably preoccupied by the fact that greenhouse gas emissions are altering the earth's atmosphere at an alarming and unprecedented pace. But anthropogenic climate change is a recent problem—and frankly only part of the picture. Since long before humans started to load the atmosphere with chemicals that trap heat, the climate system has swayed and varied due to natural causes. For most of the two-hundred thousand years or so of human history, our forebears lived in the Pleistocene, an age of jagged climate oscillations. Small changes in the path of the earth, and slight variations in the tilt and spin of the earth around its axis, are constantly changing the amount and distribution of energy arriving from our nearest star. Across the Pleistocene, these mechanisms, known as orbital forcing, created icy interludes lasting millennia. Then, about 12,000 years ago, the ice broke, and the climate entered the warm and stable interglacial known as the Holocene. The Holocene was the necessary backdrop to the rise of agriculture and the growth of complex political orders. But it turns out the Holocene has been a time of sharp climate changes, dramatically important on human scales.[17]

While orbital mechanics still drive deep changes in the Holocene climate, solar energy varies in other consequential ways on shorter time-scales. The sun itself is an inconstant star. The eleven-year sunspot cycle is only the most familiar of an array of periodic variations in the solar dynamo; some drastically affect the earth's insolation. And our planet has played a role in natural climate change: volcanic eruptions spew reflective sulfate aerosols high into the atmosphere, screening the arrival of the sun's heat. Even in the friendly Holocene, then, orbital, solar, and volcanic forcing interacted with the inherently variable systems of the earth to make the climate far more volatile, and precarious, than we might have thought.[18]

The discovery of rapid climate change in the Holocene is a revelation. We are learning that the Romans were, in planetary perspective, lucky. The empire reached its maximal extent and prosperity in the folds of a late Holocene climate period called the Roman Climate Optimum (RCO). The RCO reveals itself as a phase of warm, wet, and stable climate across much of the Mediterranean heartland of empire. It was an inviting moment to

Table 1.1 Roman climate periods

Roman Climate Optimum	ca. 200 BC–AD 150
Roman Transitional Period	ca. AD 150–AD 450
Late Antique Little Ice Age	ca. AD 450–AD 700

make an agrarian empire out of a pyramid of political and economic bargains. Alongside trade and technology, the climate regime was a silent, cooperative force in the seemingly virtuous circle of empire and prosperity. As the Romans stretched their empire to its limits, they had no idea of the contingent and parlous environmental foundations of what they had built.

From the middle of the second century, the Romans' luck ran into short supply. The centuries that form the object of our inquiry witnessed one of the most dramatic sequences of climate change in the entire Holocene. First, a period of climate disorganization covering three centuries (AD 150–450) set in, which we will propose to call the Roman Transitional Period. At crucial junctures, climate instability pressed on the empire's reserves of strength and intervened dramatically in the course of events. Then, from the later fifth century, we sense the stirrings of a decisive reorganization that culminated in the Late Antique Little Ice Age. A spasm of volcanic activity in the AD 530s and 540s brought on the most frigid spell in the entire Late Holocene. Concurrently, the level of energy arriving from the sun slipped to its lowest point in several millennia. As we will see, the deterioration of the physical climate coincided with unprecedented biological catastrophe to overwhelm what was left of the Roman state.

This book will argue that the influence of the climate on Roman history was by turns subtle and overwhelming, alternatingly constructive and destructive. But climate change was always an *exogenous* factor, a true wild card transcending all the other rules of the game. From without, it reshaped the demographic and agrarian foundations of life, upon which the more elaborate structures of society and state depended. With good reason, the ancients revered the fearsome goddess *Fortuna*, out of a sense that the sovereign powers of this world were ultimately capricious.[19]

Nature wielded still another terrible device, capable of crashing in upon human societies like an army in the night: infectious disease. Biological change was even more forceful than the physical climate in deciding the fate of Rome. Of course, the two were not, and are not, unconnected. Climate change and infectious disease have been overlapping but not coterminous forces of nature. Sometimes climate change and pandemic disease were synergistic in their effects. At other times, they were more than temporally coincident, since perturbation in the physical climate can instigate ecological or evolutionary changes that spill over into disease events. In the course of the centuries we will consider, they often worked in concert to bear on the destiny of the Roman Empire.[20]

There is one truly categorical difference between climate change and infectious disease. The climate system, until recently, vibrated on its own tempo and terms, without human influence. By contrast, the story of infectious disease is far more intimately shaped by human interference. Human societies in effect create the ecologies within which deadly microbes live, move, and have their being. In many ways, an unintended and paradoxical consequence of the Roman Empire's ambitious social development was the lethal microbial environment that it fostered. Inadvertently, the Romans were complicit in building the disease ecologies that haunted their demographic regime.

To understand how the Romans lived and died, much less the fate of their empire, we must try to reconstruct the specific juncture of human civilization and disease history that the Romans encountered. The pathogens that have regulated human mortality are not an undifferentiated array of enemies. The biological particulars of germs are unruly and decisive facts of history. The history of germs has been dominated by the brilliant model devised in the 1970s and most famously expressed by William McNeill in his classic *Plagues and Peoples*. For McNeill, the connective thread of the story was the rise and then confluence of different Neolithic germ pools. Agriculture brought us into close contact with domesticated animals; cities created the population densities needed for germs to circulate; the expansion of trading networks led to the "convergence of the civilized disease pools," as pathogens that were endemic in one society leapt ravenously into virgin territories.[21]

In recent years the shine of the classic model has started to fade. The ground has quietly but decisively shifted around it. The 1970s were the peak of a triumphant moment in western medicine. One by one the scourges of the past fell before the advance of science. There was confident talk of a transition in which infectious disease would become a thing of the past . . . But the terrifying roster of emerging infectious diseases—HIV, Ebola, Lassa, West Nile, Nipah, SARS, MERS, and now Zika, to name only a few of several hundred—shows that nature's creative destruction is far from spent. And all of these emerging infectious diseases have something insidious in common: they arose from the wild, not from domesticated species. Pathogen evolution and zoonotic diseases from the wild now loom larger than before in the dynamics of emerging infectious diseases.[22]

These insights have yet to be applied in a complete and consistent way to the study of the past, but the implications are revolutionary for the way we think about the place of Roman civilization in the history of disease. We should try to imagine the Roman world, through and through, as an ecological context for microorganisms. To start with, the Roman Empire was precociously urbanized. The empire was a great buzzing switchboard of cities. The Roman city was a marvel of civil engineering, and no doubt toilets, sewers, and running water systems alleviated the most dread effects of waste disposal. But these environmental controls were poised against overwhelming forces, a thin and leaky tide-wall against an ocean of germs. The city crawled with rats and teemed with flies; small animals squawked in alleys and courtyards. There was no germ theory, little hand washing, and food could not be kept from contamination. The ancient city was an insalubrious home. Humble diseases spread by the fecal-oral route, inducing fatal diarrheas, were probably the number one killer in the Roman Empire.

Outside the cities, landscape transformation exposed the Romans to equally perilous threats. The Romans did not just modify landscapes; they imposed their will upon them. They slashed and burned forests. They moved rivers and drained basins and built roads through the most intractable swamps. Human encroachment on new environments is a dangerous game. It not only exposes us to unfamiliar parasites but can trigger cascading ecological change with unpredictable consequences. In the Roman Empire, the revenge exacted by nature was grim. The prime agent of reprisal was malaria. Spread by mosquito bite, malaria was an albatross on Roman civilization. The vaunted hills of Rome are knobs rising above a glorified swamp. The river valley, not to mention the pools and fountains throughout the city, were a haven for the mosquito vector and made the eternal city a malarial bog. Malaria was a vicious killer in town and country, anywhere the *Anopheles* mosquito could thrive.[23]

The Roman disease environment was also formed by the connectivity of the empire. The empire created an internal zone of trade and migration as had never existed. The roads and sea lanes of the empire moved not only peoples, ideas, goods—they moved germs. We can watch this pattern unravel at different rates of speed. It is possible to follow the diffusion of sluggish killers, such as tuberculosis and leprosy, which spread across the Roman Empire with a slow burn, like lava. When fast-moving infectious diseases finally hopped onto the great conveyor belt of Roman connectivity, the consequences were electric.

We will emphasize the paradoxical relationship between Roman social development and the disease ecology of the empire. Despite the benefits of peace and prosperity, the empire's inhabitants were unhealthy, even by premodern standards. One sign of their low level of biological well-being is their short stature. Someone like Julius Caesar, who was said to have been tall, may only have stood out in a society where men were, on average, a little under 5′ 5″. The burden of infectious disease weighed visibly on Roman health. But here is where we need to pay closer attention to the specificity of the Roman disease pool. If we look carefully at the patterns of mortality in space and time, we note a telling absence in the Roman world. There were not large-scale, interregional epidemic outbreaks. Most epidemics were spatially contained, local or regional affairs. The reasons for this absence lie in the intrinsic biological limits of the germs themselves. Microbes that depend on fecal-oral transmission, or hitchhiking inside arthropods, can only spread so far so fast. But starting in the second century, the combination of Roman imperial ecology and pathogen evolution created a new kind of storm, the pandemic.[24]

The centuries of later Roman history might be considered the age of pandemic disease. Three times the empire was rocked by mortality events with stunning geographical reach. In AD 165 an event known as the Antonine Plague, probably caused by smallpox, erupted. In AD 249, an uncertain pathogen swept the territories under Roman rule. And in AD 541, the first great pandemic of *Yersinia pestis*, the agent that causes bubonic plague, arrived and lingered for over two hundred years. The magnitude of these biological catastrophes is almost incomprehensible. The least of the three pandemics, by casualty count, was probably the mortality known as the Antonine Plague. We will argue that it carried off perhaps seven million victims. That is considerably lower than some estimates. But the bloodiest day of battle in imperial history was the rout of the Romans at Adrianople, when a desperate force of Gothic invaders overran the main body of the eastern field army. At most twenty thousand Roman lives were lost on that baleful day, and while it magnified the problem that these were soldiers, the lesson of the comparison is all the same: germs are far deadlier than Germans.

The great killers of the Roman Empire were spawns of nature. They were exotic, deadly intruders from beyond the empire. For this reason, a parochial history of the Roman Empire is a kind of tunnel vision. The story of Rome's rise and fall is entwined with global environmental history. In the Roman period, there was a quantum leap forward in global connectivity. Roman demand for silk and spices, slaves and ivory, fueled frenzied motion across borders. Merchants moved over the Sahara, along the Silk Roads, and above all across the Indian Ocean and into the Red Sea ports built by the power of empire. The exotic beasts brought to the slaughter in the Roman spectacles are like macroscopic tracers, illuminating for us the very routes that brought the Romans into contact with unimaginable new frontiers of disease. The most basic fact of global biodiversity is the latitudinal species gradient, the greater richness of all life in proximity to the equator. In temperate and polar regions, recurring ice ages have periodically scraped clean the experiments of evolution, and there is simply less energy and less biotic interaction in colder climes. The tropics are a "museum" of biodiversity, where time and higher levels of solar energy have conspired to weave imponderably dense tapestries of biological complexity. This pattern holds for microorganisms, including pathogenic ones. In the Roman Empire, human-built networks of connectivity sprawled insouciantly across zones of nature's making. The Romans helped build a world where sparks could light a conflagration on an intercontinental scale. Roman history is a critical chapter in the bigger, human story.[25]

There is an evolutionary history of germs that we are only beginning to see, but here we can make an earnest deposit by trying to see Roman history as one, perhaps unusually important, chapter in a much longer, global story of pathogen evolution. The Romans helped to create the microbial environment within which the random game of genetic mutation played out its cunning experiments. If the fate of the Roman Empire was shaped by the overwhelming force of pandemic disease, it was an uncanny mixture of structure and chance.

The urgent study of earth science and the genomic revolution are teaching us that climate change and emerging infectious diseases have been an integral part of the human story all along. The hard question has become not whether, but how, to insert the influences of the natural environment into the sequence of cause and effect.

A Human Story

The integration of knowledge from fields as disparate as the natural, social, and humanistic sciences is called consilience. *Integration* means that historians are far from passive recipients of new data from the sciences. Indeed, the interpretation presented in this book relies on our still advancing knowledge of those entirely human parts of the narrative. Centuries of ongoing humanistic scholarship have helped us understand the stresses and strains—the true nature and inner workings—of the Roman Empire at a level of detail that would make Gibbon jealous. This book tries to build on those insights, which are as fresh, ingenious, and surprising as the latest genomic study or paleoclimate archive.[26]

The question is how to explain the long sequence of momentous changes that rendered an empire that was integrated, populous, prosperous, and complex at one moment in time—in the age of Marcus Aurelius (AD 161–180)—into something unrecognizable five centuries later. It is an intertwined story of state failure and stagnation. The Roman Empire was built in a Malthusian world of energy constraints, but it was able to shove back those limits through a heady combination of trade and technical advance. The power of the empire was both a premise and an outcome of demographic expansion and economic growth. The state and social development went hand in hand. The rousing forces of climate change and infectious disease constantly acted upon this complex system, in a series of two-way relationships. Even in the case of the physical environment, where forces entirely beyond human control operated, the effects of climate change depended on the specific arrangements between an agrarian economy and the machinery of empire. And the history of infectious disease is always thoroughly dependent on ecologies constructed by human civilization.

We will not shy away from attributing great causal influence to natural forces, even as we strive to avoid flattening out the texture of events in reductionist fashion. Relationships between the environment and the social order were never tidy and linear. Even in the face of the sharpest challenges, the people we will meet in these pages surprise us with the depth of their response to adversity. The capacity to absorb and adapt to stress is measured in the term *resilience*. The empire might be construed as an organism with batteries of stored energy and layers of redundancy that permitted it to endure and recover from environmental shocks. Resilience is not infinite, however, and to look for it in ancient societies is also to be alert for the signs of persistent stress and the thresholds of endurance beyond which lie cascading change and systemic reorganization.[27]

The end of the Roman Empire, as contemplated here, was not a continuous decline leading to inevitable ruin, but a long, circuitous, and circumstantial story in which a resilient political formation endured and reorganized itself, until it fell apart, first in the west and then in the east. The pattern of change will always be presented as a highly circumstantial interplay between nature, demography, economy, politics, and even, we will argue, something so ethereal and quixotic as systems of belief, which were repeatedly unsettled and reconfigured in the course of these centuries. The charge of history is to interweave these threads of the story in the right way, with a healthy respect for the realm of freedom and contingency, and a strong dose of sympathy for the humans who made their lives under the circumstances they were given.

As we set out to explore a historical episode of this magnitude, it is worth declaring at the outset a few of the main contours of the narrative. It is a story with four decisive turns, when the pace of events gathered momentum and disruptive change trailed close behind. At each of the points of transformation in the transit between the high empire and the early middle ages, we will try to seek out the specific and intricate lines of connection between natural and human systems.

(1) The first was a multifaceted crisis during the age of Marcus Aurelius, triggered by a pandemic disease, that interrupted the economic and demographic expansion. In its aftermath, the empire did not fall or disintegrate, but instead recovered its previous form without the same commanding dominance as before.

(2) Then, in the middle of the third century, a concatenation of drought, pestilence, and political challenge led to the sudden disintegration of the empire. In what has been called the "first fall" of the Roman Empire, the bare survival of an integrated imperial system was an act of willful reconstitution, and a close-run thing. The empire was rebuilt, but in a new guise—with a new kind of emperor, a new kind of government, a new kind of money, and, soon to follow, a new kind of religious faith.

(3) This new empire then roared back. But in a decisive and dramatic period of two generations spanning the end of the fourth and beginning of the fifth centuries, the coherence of the empire was conclusively broken. The entire weight of the Eurasian steppe seemed to lean, in new and unsustainable ways, against the edifice of Roman power, and as it chanced to happen, the western half of the empire buckled. This cataclysm, which Stilicho had aimed to avert, probably ranks as the most familiar version of Rome's fall. In the course of the fifth century, the Roman Empire was dismembered—in the west. But it was not the grand finale of the Roman Empire.

(4) In the east, a resurgent Roman Empire enjoyed renewed power, prosperity, and population increase. This renaissance was violently halted by one of the worst environmental catastrophes in recorded history—the double blow of bubonic plague and a little ice age. Demographic shock played out in a slow motion failure of empire, culminating in the decisive territorial losses to the armies of Islam. Not only was the remnant of the Roman Empire reduced to a Byzantine rump state, but the survivors were left to inhabit a world with fewer people, less wealth, and perpetual strife among competing apocalyptic religions, including Christianity and Islam.

The rise and fall of Rome remind us that the story of human civilization is, through and through, an environmental drama. The flourishing of the empire in the halcyon days of the second century; the arrival of a new kind of virus from far beyond the Roman world; the rupture of the imperial grand bargain in the aftermath of pandemic; the meltdown of empire amid a concatenation of climate and health disasters in the third century; the empire's resurrection by a new kind of emperor; the fanning of massive people movements across Eurasia in the fourth century; the revitalization of eastern societies in late antiquity; the neutron bomb of bubonic plague; the insidious onset of a new age of ice; the final collapse of anything recognizable as the Roman Empire and the lightning conquests of the armies of *jihad*. If this book achieves its purpose, it will have become a little harder to hear these turns of the past as anything other than the contrapuntal motion of humanity and the natural environment, sometimes parallel and sometimes contrary, but as utterly inseparable as the sonorous lines of a baroque fugue.[28]

The pace at which our knowledge is growing is equal parts exhilarating and daunting. By the time the ink hits the pages of this book, scholarship will have sailed on. But that is a happy conundrum, and it is worth the risks if we can start to build a provisional map, inevitably to be filled in and corrected as discovery advances. It is time to reconsider the awesome, uncanny power of nature in the fate of a civilization that continues to surprise and captivate us, and we will need patience, as well as some imagination, to go back and pretend we do not know the ending. The place to begin is with Rome's greatest doctor, reared in the lap of peace and prosperity, who could little have imagined that dynamic cycles in our nearest star, or the chance mutation of a virus in a far-off forest, could rattle the foundations of the bustling empire that ruled the world where he was seeking his fortunes.

The Happiest Age

The Great Doctor and the Great City

The doctor Galen of Pergamum was born in the middle of the reign of the emperor Hadrian, in September of AD 129. While not a son of the very upper crust, Galen belonged to the *haute bourgeoisie*, for whom the empire meant prosperity and opportunity. Galen's birthplace, Pergamum, nestled just inland from the Aegean in the rising folds of Asia Minor, was the sort of town that flourished under Roman rule, and it was an auspicious breeding ground for a medical prodigy such as Galen. As a bastion of Greek tradition, Pergamum allowed Galen to acquire unequaled command of Greek medical literature, including the vast Hippocratic corpus. Pergamum's famous temple of the healing god Asclepius (the son of Apollo whose snake-twined rod has given medicine its most famous symbol) was a beacon for convalescents. In Galen's time, the temple, already over half a millennium old, was at its apogee. "All Asia" flocked to the shrine, and just five years before Galen's birth, it was honored by the presence of Hadrian himself.[1]

Galen's precocious talents earned him the esteemed post of physician to the gladiators in Pergamum. But the imperial peace afforded Galen even further horizons. He had travelled the eastern Mediterranean, crossing Cyprus and Syria and Palestine, scouring for local knowledge of drugs and remedies. He had studied in Alexandria, where the chance to see real human bones had left an impression: "the physicians there employ ocular demonstration in teaching osteology to students. For this reason, if for no other, try to visit Alexandria." The Roman Empire gave Galen, on any reckoning, an uncommonly wide experience of the medical arts. And inevitably, a man of his prodigious talents was drawn to try his fortunes in the great capital itself.[2]

Galen came to Rome in AD 162, in the first year of the joint rule of the emperors Marcus Aurelius and Lucius Verus. The doctor liked to quote a saying that "Rome was the epitome of the whole world." Rare complaints that Hippocrates (fl. 400 BC) had never seen were commonplace to Galen, "because of the large number in the city of the Romans." "Daily ten thousand people can be discovered suffering from jaundice, and ten thousand from dropsy." The metropolis was a laboratory of human suffering, and for an aspiring intellectual like Galen, it was a grand stage. His rise was dizzying.[3]

Soon after arriving, he saved a philosopher from a fever, "despite being scoffed at" for "presuming to cure an old man" in wintertime; his reputation grew. Flavius Boethus, a native of Syria who had held the empire's highest honor as a consul, was keen to watch Galen "demonstrate how speech and breath are produced." Before an enthralled audience, with a refined taste for spectacles, Galen vivisected a pig, switching its screams on and off by ligating nerves in a virtuoso performance. Galen cured Boethus' son and then his wife of grievous ailments; the powerful man gave Galen a small fortune in gold and, more importantly, his patronage. Galen moved in the most fashionable circles. One sensational success followed another. When the slave of a famous writer was injured, a deadly abscess formed under the sternum. Galen excised the infected tissue in a surgery that exposed the beating heart to open view; against Galen's own measured pessimism, the slave lived.[4]

Still in his mid-thirties, Galen had become a living legend. "Great was the name of Galen."[5]

None of this had readied the doctor for the mortality event that we have come to know as the Antonine Plague. In AD 166, during his fourth year in the capital, a pestilence from the east moved toward the city. Epidemics were not uncommon in Rome. At first the waves of fever and vomiting may have seemed only the familiar swell of a grim mortality season. It would have soon become apparent that something unusual was at hand.[6]

In his masterpiece, *The Method of Medicine*, Galen vividly described his treatment of a young man struck by the disease "when it first appeared." A slight cough grew violent, and the patient expelled dark scabs from the ulcerations in his throat. Soon the telltale symptom of the disease appeared: the black rash that wrapped the bodies of its victims from head to foot. Galen thought there were remedies to blunt the force of the disease, but these are a register of pure desperation: milk from mountain cattle, Armenian dirt, the urine of a boy. The mortality event he lived through stands as not only perhaps the first pandemic in human history, but also a moment of rupture in the story of the Roman Empire. It seemed to most that the god Apollo had exacted some dark new punishment. For Galen the scientist, it was simply "the great plague."[7]

The purpose of this chapter is to survey the empire that reared Galen, down to the eve of the pandemic. It was the age that Gibbon judged the "most happy and prosperous" in the history of the human race. Of course,

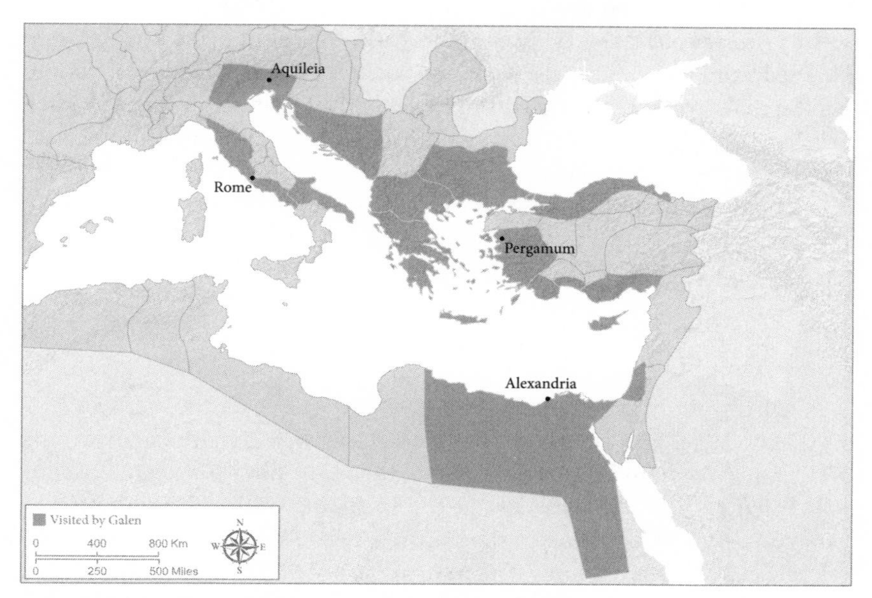

Map 3. Galen's World: Provinces Certainly Visited by Galen

in that appraisal, there was some remote attraction to the masters of the Roman world. But to pronounce the middle of the second century as the high point of Roman civilization is not an arbitrary or aesthetic judgment. In material terms, the Roman Empire set the stage for a stunning efflorescence, one of those periods in history when extensive and intensive growth conspired to edge forward social development. The empire itself was both a precondition of, and predicated on, this wave of development. The political frame of the empire and its social mechanics were interdependent.

At the same time, we will emphasize that the *pax Romana* was never the attainment of frictionless dominance; the strength of empire should be measured not by the absence of strains or challenges, but by the ability to withstand them. From this perspective, then, it becomes all the more necessary to seek out the reasons why the Antonine age has so often seemed like a bend in the course of history. Traditional answers, such as more formidable enemies across the frontiers and rising fiscal-political tensions, remain integral, but not adequate. Here we emphasize that the Roman efflorescence was built on a precarious and transient alignment of favorable climate conditions. And even more momentously, the structures of empire fashioned the ecological conditions for the arrival of an emerging infectious disease capable of unprecedented violence.

In a substantial sense, then, the trajectory of the empire's history was redirected from without, by the forces of nature. Of course, we do not have to believe that, spared these interruptions, the empire was set to endure in perpetuity. But the particular fate the empire did experience was so profoundly inseparable from the passing of the climate optimum and the shock of pandemic that, in any account of Rome's destiny, they merit a place squarely in the foreground.

The Dimensions of Empire

As Galen walked the streets of the imperial capital, among the many stones and statues competing to catch his eye, he might have noticed a column, which happens to survive, naming the thirty legions of Rome. Listed in geographical order, starting in the northwest corner of the empire and spiraling clockwise, the roster was a reassuring picture of Roman power. In the west, three legions guarded Britain, four the Rhine, and ten the Danubian provinces between the Alps and the Black Sea. In the east, eight legions were garrisoned from Cappadocia to Arabia, watching subject and foe alike. A mere two held the entire Roman position in Africa, one in Egypt and one in Numidia. One in Spain and two in the Alps rounded out the thirty legions. But even at this moment of equipoise, before the storms of war and pestilence, the empire was not a finished project. The Roman Empire was always poised between the primal will to conquer new peoples beyond its borders and the maintenance of security within the core zone of the empire. It never reached an entirely stable balance between these contradictory forces. Yet in the second century, across vast stretches of the tri-continental empire, an air of peace descended over the lands protected and patrolled by the force of Roman arms.[8]

In essence, the Roman Empire was a framework of military hegemony whose shape was determined by a blend of geographical facts and political technologies. The Roman Empire did not have natural or predestined frontiers. Even to think of clearly defined lines, such as mark territorial boundaries in modern states with advanced land survey, would be excessively precise. In the first place, the Romans ruled "peoples" or "nations." Appian, a Greek historian who served as a governor in the age of Hadrian, launched his history of Rome by describing the "boundaries of the *nations* which the Romans rule." He could sensibly enough point to the main geographical features at the edges of empire, such as the Rhine, the Danube, and the Euphrates, but he quickly noted that the Romans ruled peoples beyond these boundaries. The great legionary bases were set back within the frontiers, as reserves of strength, but also where they could act effectively as something between an imperial police force and a corps of engineers. The frontier zone was a dense network of smaller fortlets, watch-towers, and signal stations, sometimes stretching deep into unfriendly territory. The Quadi, a people beyond the Danube, were said to have revolted because they "could not bear the forts built to watch them."[9]

The Romans of the second century would not have recognized any grand plan to stop expansion and admire their finished work. With Augustus, territorial expansion slowed, but it did not grind to a halt. Aggression and diplomacy continued to enlarge the empire, sporadically. Even apparently defensive structures such as Hadrian's Wall were control systems, not expressions of sovereign territorial boundaries. Forward operations into Scotland continued intermittently for a century after the wall went up. Marcus Aurelius had serious designs to annex vast swaths of central Europe. And the efforts to control the regions beyond the Euphrates were a perpetual source of conflict.

The friction of expansion gradually drew lines of territorial hegemony that we call the limits of empire. These limits were derived from the features of the system the Romans had created, which required the coordination of military power, under Iron Age conditions of communication and transport, from an imperial center. The *political* coordination of the military machine was as important as its brute physical coordination. The emperor was the chief representative of the senatorial order, a narrow social group that maintained control of the armies by monopolizing the positions of high command as the birthright of their class. By the time of Marcus, in any given year, there were some 160 senators holding office somewhere in the empire, all coordinated from the nerve center of the capital.[10]

The Roman emperors had at least a crude sense of the "marginal costs of imperialism." "Holding power over the whole earth and sea, they choose to maintain their rule through prudence, rather than trying to carry their empire into the unknown, over miserable and profitless barbarians, some of whose embassies I have seen in Rome attempting to become subjects but being refused by the emperor on account of their worthlessness to him." The Romans had supposedly taken all the lands of the Celts, except where it was too cold or the soil too poor: "What the Celts had worth having belongs to Rome."[11]

The thirty legions amounted to some 160,000 men. The legions comprised the citizen army, in theory recruited exclusively from the ranks of Roman citizens, who often came from veteran colonies sprinkled across the empire. But the legions were just less than half the overall military. Complementing their strength were auxiliary units. Recruited from provincial populations, they were deeply integrated into the command structure and overall strategic design of the empire; long-term service was a well-trodden pathway to the privileges of citizenship. When we add the navy and irregular units, the Roman imperial war machine approached half a million men: "Not only was this the largest standing army that the world had yet known, it was also the best trained and best equipped."[12]

Maintaining the most powerful military force in history was not cheap. The defense budget was by far the largest item of state expense. The ordinary legionary of the second century was paid a stipend of 300 *denarii*, a healthy though not princely income; auxiliaries were probably paid 5/6 of this amount. Cavalry were paid higher, as of course were officers. Retirement bonuses and irregular donatives added still further costs. In all, the pay budget alone of the army in the second century was probably 150 million *denarii*, something like 2–3 percent of the entire GDP of the empire (about the share of present-day defense spending in the United States). In sheer size, the army and its budgets were historically massive.[13]

At the same time, as contemporaries recognized, the imperial framework established by Augustus represented a sharp and conscious departure from the extreme military mobilization of the Roman republic, which had been a whole society at arms. "During the days of the Republic when the senate appointed army commanders to their posts," wrote a third-century historian, "all Italians used to bear arms." In the empire, by contrast, the army was a professional force. Augustus "stationed mercenary troops on fixed rates of pay to act as a barricade for the Roman Empire." The Roman peace rested on the discipline, valor, and loyalty of a giant paid army. The fiscal machinery underneath the military hegemony formed the basic metabolic system of the empire.[14]

The dimensions of the Roman Empire were thus determined by the geophysical realities of coordinating such an army across three continents, the commitment to maintain class control over the military, and the cost of maintaining a force on this scale. At its height, Roman military dominance created prolonged stretches of peace, a bounty reaped by subject and citizen alike. At the heart of empire, the agonies of war could be put far out of mind. "Many provinces do not know where their garrison is; all men pay taxes to you with greater pleasure than some people would collect them from others." "The cities shine with radiance and grace, and the whole earth has been adorned like a pleasure garden; gone beyond land and sea is the smoke rising from the fields and the signal fires of friend and foe."

These fulsome praises come from a famous speech delivered by an unusually gifted, and then very young, Greek orator named Aelius Aristides before the emperor Antoninus Pius in AD 144. Whatever discount for flattery we wish to allow for a provincial on the make, his eloquent appraisal of what he called "the greatest empire and a surpassing power" cannot help but leave an indelible impression of life under imperial rule. "You have caused the word *Roman* to belong not to a city, but to be the name of a sort of common race." The positive verdict on the age, for Gibbon certainly, originated from such fawning tributes. Not all empires have evoked such jubilant praise from their subjects, and as we will soon see, there is ample material evidence that the seductions of empire were widely dispersed. Certainly, the loyalty of civic elites like Aristides was the glue of the empire.[15]

Aristides himself turned gravely ill in Rome and descended to the brink of death. He found his way to Pergamum, to recuperate at the shrine of Asclepius. As a young boy Galen saw the great orator, who spent years pursuing eccentric treatments suggested to him by the god. We will meet Aristides again, as the first known victim of the Antonine Plague.

Peoples and Prosperity

With some half a million soldiers on active duty, raw manpower was the main ingredient of Rome's military might. Mustering an army of this magnitude seems not to have been a dire strain during the high empire, certainly not in relation to what lay just ahead. In the words of Aristides, the empire "recruited only so many soldiers from each people as will neither be a burden for those who supply them nor by themselves will be sufficient to make up the complement of a single army of their own." The enticements of pay and privilege were sufficiently attractive, but at a more basic level, the ease of army recruiting was a benefit accrued from generous demographic increase. The Romans were not insensible of these connections. In the triumphal arch of Trajan at Benevento, for instance, the glorious victories of the army flow directly from the natural abundance—agricultural and human—granted to Rome by the gods.[16]

Contemporaries marveled at the fact that there were people everywhere. In his praise of Rome, Aristides wondered, "with so many occupied hills or urbanized pastures in the name of a single city, whose eyes could take it all in?" The debris of demographic expansion are evident in the archaeology of the provinces, from Syria to Spain, from Britain to Libya. The valleys were crowded, and expansion crept up the hillsides. Towns were hewn out of the lowlands, and cultivation was pushed beyond the limits of what had ever been known. Like a swell rising from the deep, the populations of three continents under Roman rule rose in a great, synchronized wave of growth that crested in the age of the Antonine emperors.[17]

Trying to reconstruct population levels in the ancient world is a crude business. It always has been. Already in the 1750s, David Hume and the Scottish divine Robert Wallace laid out arguments for wildly divergent views of the "populousness of ancient nations." The debate has not always been so genial (Hume helped correct the final manuscript of his adversary), but already in view were the outlines of a controversy that has continued down to the present day between "high counters" such as Wallace and "low counters" such as Hume. Even in recent times, credible voices have spoken in favor of peak numbers for the Roman imperial population ranging from ca. 44 million to 100 million.[18]

Where there *is* broad agreement is around the fact that the populations within the empire grew in the 150 years after the death of Augustus (AD 14) and reached their maximal extent on the cusp of the Antonine Plague. But absolute figures remain necessarily more speculative. Though the debate between Hume and Wallace carries on among modern scholars, the soundest arguments point us to believe that there were some 60 million inhabitants in the Roman Empire when Augustus died and closer to 75 million a century and a half later, when Galen first arrived in Rome.[19]

Population growth was the unintended outcome of countless, razor-thin changes in the narrow margins between life and death. Ancient populations were squeezed between powerful, countervailing pressures. Mortality was blindingly high. Life in the Roman Empire was short and uncertain. As we will see in the next chapter, even by the low standards of all underdeveloped societies, the actuarial tables of the Roman world were grim. Average life expectancy at birth fell somewhere between twenty and thirty years. The blunt force of infectious disease was, by far, the overwhelming determinant of a mortality regime that weighed heavily on Roman demography.

In environments of high mortality, the obligatory response is high fertility. The burden of fertility fell heavily on the bodies of women. They bore the biological brunt of the need to replenish the ranks. Roman law allowed girls to be married starting at age twelve. Most women married in their mid-teens. Marriage was effectively universal: there were no spinsters in the Roman world. The Romans praised the widow who remained unmarried—precisely because she was the oddity in a society where death always stalked and remarriage was expected. Marriage was, first and foremost, a covenant

Table 2.1. Population of Roman Empire ca. AD 165

Region	*Population (mil.)*	*Density (per km²)*
Italy (w/ islands)	14	45
Iberia	9	15
Gaul & Germany	12	18
Britain	2	13
Danubian provinces	6	9
Greek peninsula	3	19
Anatolia	10	15
Levant	6	43
Egypt	5	167
North Africa	8	19
TOTAL	**75**	**20**

for procreation. "Women are usually married for children and succession, and not for mere enjoyment."[20]

From Augustus onward, the state employed powerful inducements to high fertility in its natalist policies, penalizing childlessness and rewarding fecundity. Women who bore sufficient numbers of children were granted robust legal privileges. Contraception was primitive, at best. Natural fertility was the reality in the Roman world. The woman surviving to menopause bore something like six children, *on average*. The entire age structure of ancient societies was bottom-heavy, dominated by the very young. The streets of an ancient city would have had the sound of an unruly nursery. It can be reasonably if crudely hypothesized, then, that the main source of population growth in the Roman Empire was not a decline in mortality but, rather, elevated levels of fertility. This conclusion is broadly consistent with Malthusian theory, which predicts that higher levels of welfare are realized in higher levels of fertility: as more people lived further above the subsistence level, they were able to convert these slim economic advantages into demographic success.[21]

We should issue at least one immediate caution. The Roman demographic regime was not a fine-tuned machine. If it seems likely that the Roman population achieved growth of ~0.15 percent per annum between Augustus and Marcus Aurelius (a rate that would lift a population of 60 million to 75 million in a century and a half), this achievement was not the smooth progress of fertility rates regulated just above replacement levels. Population biology in the Roman world was volatile. Where infectious diseases rule the mortality regime, death is seething and unpredictable, marked by uneasy lulls and sharp interruptions. As a result, the populations of the

Roman Mediterranean were not stationary, in the short or long run. Rather, populations could experience whirring growth, interrupted by violent, staccato reversals. Average vital rates are most meaningful over wide stretches of space and long periods of time, precisely because they flatten out the wild oscillations of epidemic mortality.

The Romans lived and died with precarious and savage waves of infectious disease, not serene averages. So, the expansionary trend is only the coarse-grained view of what was really the pulsating sum of careening growth, set back by spasmodic irruptions of intense death. The Romans knew that life was evanescent, and that the winds of death could sweep back their hard-fought gains in a moment.

By the time Marcus Aurelius and Lucius Verus assumed the imperial office, they held sway over a quarter of humanity. Few empires, none of the Iron Age and none so enduring, achieved such a feat. The Han Empire of China was the Eurasian counterweight of the Romans. As we will see, in our period the effective distance between the two was shrinking: the geographical manual of Ptolemy, written in the middle of the second century, espoused definite opinions about overland distances to the capital of "Serica," and the great astronomer knew of navigators who reached the far east by sea. Han China is in many ways an appropriate *comparandum*, but even its population seems never to have matched the Roman imperial apex of ~75 million (in the east, that would wait for the full development of the rice economies and the construction of the great canal systems). There is a more telling contrast. A Chinese writer of the mid-second century lamented the press of peoples in core regions of the eastern empire. "In the central provinces and inner commanderies, cultivated land fills the borders to bursting and one cannot be alone. The population is in the millions and the land is completely used. People are numerous and land scarce." In the Roman context, such laments are notable for their absence.[22]

In the Roman Empire, population growth appears to have been accomplished without sending society spiraling downward in a cycle of diminishing returns. Contemporaries sang the song of prosperity, not the dirge of grinding impoverishment. For what it is worth (which may well be limited), the articulate classes of the Roman Empire were more preoccupied by general decadence than destabilizing squalor. Maybe our urbane elite was totally insensible to the daily life of the poor. But, it is harder to stare past famine, and we ought to be struck by the broad absence of true subsistence crisis in the Roman world. Food shortages were endemic in the Mediterranean, thanks to its naturally fickle ecology. Unlike the later middle ages, when violent spasms of acute hunger wracked the population, the Romans seem not to have been haunted by the threat of outright mass starvation. The absence of evidence is never probative, but it is suggestive.[23]

More important are the various indices reflecting high levels of production, consumption, and well-being in the Roman Empire. We lack proper economic statistics such as those gathered by modern states. So historians in search of Roman growth have often turned to archaeological proxies of economic performance. Shipwrecks, iron smelting, housing stock, public buildings, and even fish salting operations have all been cited as tracers of Roman productivity. They do in sum suggest robust economic performance in the late republic and high empire. And the broad evidence for meat consumption, implied from tens of thousands of sheep, pig, and cow bones, is difficult to square with any picture of a society emaciated because the population had badly overrun its resource base. It is telling that archaeologists are usually the biggest believers in Roman economic development.[24]

Still, it can be objected that these indices are crude and less than conclusive, particularly if we are interested in *per capita* measures. How can we be sure that the archaeological evidence for more *stuff* is not merely the effect of having more *people*? Perhaps the most telling answer can be retrieved from the abundant scraps of papyrus preserved from Roman Egypt. The arid climate of the Nile Valley means that, from this province alone, we chance to possess an extraordinary number of public and private documents. These, in turn, afford us the only chronologically resolved series of prices, wages, and rents from the Roman world. Precisely because Egypt was a region subject to net extraction by the imperial center, we can be certain that any patterns we observe are not due to plunder or political rents. The papyri suggest that, far from succumbing to diminishing returns on a massive scale, the Roman economy more than succeeded in absorbing population expansion, to achieve real growth on a *per capita* basis. Wage growth for truly unskilled laborers—diggers, donkey drivers, dung haulers—outpaced slowly rising prices and rents, right down to the advent of the Antonine Plague.[25]

The copious monumental ruins of the Roman Empire's many cities might also be considered an index of the real wealth of the societies under Roman rule. The extent and nature of ancient urbanism has been the object of spirited disagreement among modern historians. But the conclusion seems increasingly irresistible, that the Roman Empire fostered a truly exceptional level of urbanization. The empire was home to a galaxy of cities—over one thousand of them. At the top, the population of Rome probably surpassed one million residents. Its scale was artificially inflated by the political entitlements of ruling an empire, but only partly. It was also the nexus of the entire economy, a hub of useful activity. Moreover, the

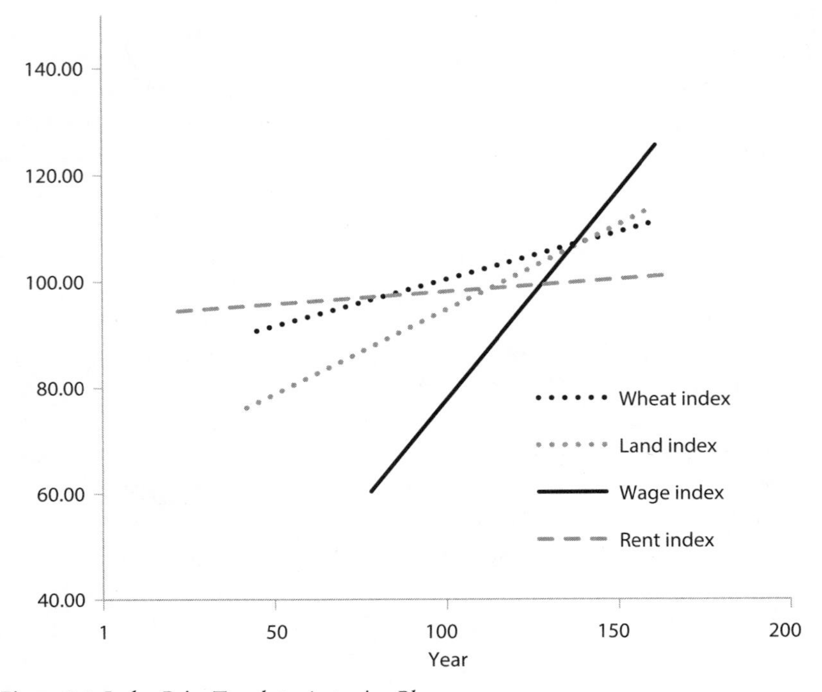

Figure 2.1. Index Price Trends to Antonine Plague

urban hierarchy was not overly top-heavy. Alexandria, Antioch, Carthage, and other metropoleis were surely several hundred thousand each (including, beyond the empire, the twin cities of Seleucia and Ctesiphon, the jewels of Parthia sited on the Tigris and serving as a hub of Persian Gulf exchange). Galen reckoned that Pergamum in his day had 120,000 inhabitants. There were perhaps dozens of cities approaching that size throughout the empire.

In the west, the arrival of the empire had catalyzed a construction boom, sometimes raising towns *ex nihilo*, sometimes simply overwriting their modest indigenous pasts. In the east, it was otherwise. Proud cities of fathomless antiquity could assimilate themselves to the empire's story or ignore it, as suited the circumstances; the emperors were usually happy to indulge and even patronize this civic pride. The cities of the Hellenic east experienced their heyday under Roman power, sprawling beyond their old confines and enjoying an age of unrivalled monumental construction. The case is very strong for imagining the towns of the Roman Empire not as parasitic consumers, gorged on political rents and entitlements, but as real nodes of value creation—with craft production, financial services, market activity, and knowledge exchange. In all, one in five of the empire's denizens may have lived in towns—a ratio inconceivable without significant levels of economic development. Here it is the simple fact that counts: for a long cycle, the Roman Empire nourished city life on a scale unlike anything that had come before, and unlike anything that would be repeated until the early modern period.[26]

The rewards of the imperial peace were thus widely dispersed. But this pattern hardly implies that the spoils were equitably shared. The distribution of wealth was highly unequal. Wealth and formal legal status formed the intricate architecture of a vaulting social hierarchy. At the bottom, legally, was a broad class of completely unfree persons. The Roman Empire was home to one of history's most extensive and complex slave systems—whose robust endurance is, incidentally, another oblique sign that overpopulation had not so reduced the price of free labor that servile labor was rendered unnecessary.

The humble and landless masses dominated quantitatively, but in town and country alike, markets and movement opened opportunities for the growth of a solid "middling" element. At the top levels of the pyramid, wealth was the benchmark of the formal aristocratic grades, such as town councilor, knight, or senator. Even though partible inheritance was the norm and there were institutional pressures militating toward the breakdown of truly massive estates, the largest private fortunes of the early empire were probably the largest that had ever been assembled in the history of humanity. There is no reason to doubt that the rich and middling elites reaped the primary benefits of Roman growth. And if elites did capture much of the economic growth, the higher wages earned by unskilled laborers only hint at the even more extraordinary achievement of the Roman economy.[27]

It is thus not quite the case that "the wealth of the Roman empire was simply a function of the enormous size of the population under its control." The greatest achievement of the Roman economy may simply have been that productivity growth was sufficient to absorb tens of millions of new working hands without sputtering from the glut of labor. That the economy reached some level of intensive growth on top of the blunt energy of more laborers is even more remarkable. This kind of intensive growth derives from two classical mechanisms: technology and trade. Technical development fosters what is called *Schumpeterian growth*, as new tools enhance the productivity of labor. Trade fosters *Smithian growth*, unleashing the forces of specialization and comparative advantage that were so important in classical economics. The two are complementary, allowing human labor to extract and harness energy more efficiently for productive uses. Although the Romans never threatened to break beyond the basic orbit of all preindustrial economies, trade and technology let them enjoy an extended phase of social development, one of premodern history's rare efflorescences.[28]

Archaeology is the best witness to the progress of technology and trade. It allows us to say that, in the Roman world, technical innovation was persistent if never quite revolutionary. Apart from some dramatic improvements in civil engineering, it is fair to say that "there was never really any such thing as Roman technology"—no characteristic breakthrough or package of innovations. Rather, the massive *diffusion* of technical advances across the wide empire, and large-scale capital accumulation and investment, amplified the gains of quiet ingenuity.[29]

Agriculture remained the primary sector; the spread of metal tools, better ploughs, new harrows, and a novel kind of reaper from Gaul accomplished real improvement. Agricultural processing experienced quantum leaps, with better screw presses, water-lifting machines, and salting vats in the vanguard. Water mills, it is now appreciated, were widely dispersed for the first time. "The large number of mills in ordinary civilian contexts—rural and urban—from all over the Empire shows that the water-mill quickly became an integral part of rural life even in drier areas of the Mediterranean lands." In this most obstinately slow-moving sector, the sum of technological improvement was not inconsiderable.[30]

Other sectors were slowly transformed. Manufacture, especially of ceramics, was not marked by radical technical discoveries, but organizational revolutions permitted mass production of a range of humble domestic goods. Mining and metallurgy seem to have been radically transformed under Roman rule; the easy availability of metal in turn had knock-on effects that it would be unwise to dismiss. To claim the Romans were effective construction scientists requires no special pleading. Transport technology was vastly improved. In the high empire, ships were larger and faster than ever before and for a long time after. "The size of Roman merchant ships was not exceeded until the fifteenth century, and the grain ships were not surpassed until the nineteenth." The lateen sail arrived in the Mediterranean in the early empire, possibly discovered through the Indian Ocean trade that boomed in the period. And it is likely that the massive port facilities built into Roman seacoasts made it safer than ever to try the hazardous shores of the Mediterranean. The sum and spread of these improvements amount to a quiet grassroots insurgency of technical advance.[31]

Trade was perhaps an even greater spur to growth. Commerce exploded under the *pax Romana*. The trade flowing to and from the imperial capital was a marvel to behold, as Aristides did not miss the chance to notice in his praise of Rome. "So many merchant ships arrive here, conveying every kind of goods from every people every hour and every day, so that the city is like a factory common to the whole earth." The author of the Biblical *Revelation*, a rather less friendly reviewer, agreed, imagining that upon the destruction of Rome, "The merchants of the earth shall weep and mourn over her; for no man buyeth their merchandise any more, the merchandise of gold, and silver, and precious stones, and of pearls, and fine linen, and purple, and silk, and scarlet, and all thyine wood, and all manner vessels of ivory, and all manner vessels of most precious wood, and of brass, and iron, and marble, And cinnamon, and odours, and ointments, and frankincense, and wine, and oil, and fine flour, and wheat, and beasts, and sheep, and horses, and chariots, and slaves, and souls of men."[32]

The city of Rome was a vortex of consumption, obviously, but trade networks spread like spiders' webs into all corners of the empire. Peace, law, and transportation infrastructure fostered the capillary penetration of markets everywhere. The clearing of piracy from the Mediterranean in the late republic may have been the single most critical precondition for the burst of commercial expansion that the Romans witnessed; risk of harm has often been the costliest impediment to seaborne exchange. The umbrella of Roman law further reduced transaction costs. The dependable enforcement of property rights and a shared currency regime encouraged entrepreneurs and merchants. Only lately have we come to appreciate the stunning advances in the Roman credit system. Roman banks and networks of commercial credit offered levels of financial intermediation not attained again until the most progressive corners of the seventeenth–eighteenth century global economy. Credit is the lubricant of commerce, and in the Roman Empire the gears of trade whirred. The empire by its nature systematically leveled barriers to trade.[33]

The result was a golden age of trade. Towns were the hubs of regional networks, which always retained pride of place within the landscape of trade. Most trade was local. Despite the quality of Roman roads, transport costs were steep, and conveyance by river or sea vastly cheaper than overland routes. Still, the scale of interregional trade is notable. Thanks to the indestructible nature of fired transport ceramics used to move liquid commodities, we can trace something of the intricacy and scale of the wine trade in the early empire. In a world with little taste for beer, with no tobacco or sugar, and without many other familiar stimulants, wine was the queen of commodities. It has been estimated that the city of Rome consumed 1.5 million hectoliters of wine each year: about 1/15 of the annual wine production of modern California.[34]

Trade and technology let the Romans stay ahead of the population crunch for a long cycle of development. All the same, there are no signs that the Romans threatened to induce accelerating breakaway growth, such as we take for granted in the modern world. The great liftoff only occurred when science was hitched to economic production and when fossil sources of energy, like coal, were exploited at scale. So, it does no discredit to the Romans to admit they had not transcended the basic mechanics of premodern economies. They were, simultaneously, precociously advanced and thoroughly preindustrial. We should not envision premodern economic development as a flat line of bleak subsistence until quickening growth from the Industrial Revolution onward. Rather, the experience of civilization has been one of consequential waves of rise and fall, consolidation and dissolution, with repercussions stretching far beyond a tiny elite squeezing rents from an underclass of indistinct peasants whose condition was more or less equally miserable from time immemorial. The Roman Empire was possibly the broadest and most powerful of these waves, prior to the ever-lifting crests of modernity.[35]

In short, the Romans accomplished real growth within the confines of a traditional organic economy, and this growth was not immaterial to the fortunes of the empire and its inhabitants. But loose ends remain, perhaps more visible now than before. There are not obvious signs that the Roman economy was already bumping against the hard limits of its potential. If the Roman economic system was neither careering toward its own demise, nor on the verge of endless growth, then why the turns that lay just ahead? There is something admittedly tidy about the theory that the cause of change came from *within* the system itself, that the comedown of the empire's economy was the inevitable revenge of overpopulation. To be sure, reprisals lurked somewhere down the road. But nature intervened first, without the provocation of a society that had overrun its carrying capacity.

History is full of these syncopated rhythms, with sudden, inexplicable beats that seem to come from nowhere to interrupt what only appeared to be the pattern. For long we have sought to explain the cycles of rise and fall in terms that were all too human, as though we were the only player in the band. But it increasingly appears that there has been another great instrument operating not far in the background, disposing conditions both amenable and unpropitious within which humans make their destiny. The climate has been an enabling as well as destabilizing force, and it looks to have been one, indispensable, player in the Roman efflorescence—and, subsequently, its unforeseen interruption.

The Roman Climate Optimum

Alexandria, perched on the shores of the Mediterranean at the western edge of the Nile delta, was one of those vitally radiant cities that flourished under Roman dominion. The capital of scientific inquiry (where Galen had studied real human bones), it was the home and headquarters of the great Ptolemy, who along with Galen was the most illustrious scientist of the Roman Empire. Like Galen, Ptolemy combined the accumulated learning of the ancient world with the hard-won advances of a rigorous empiricist. Like Galen, his theories would hold the field for a millennium. This most acute observer of the skies, however, reported patterns of local weather in Alexandria that have struck many subsequent readers as embarrassingly unlikely. On Ptolemy's testimony, it rained in Roman Alexandria every month of the year but August. Today there is about one day of rain from May to September, inclusive. This could not be a chance difference. Ptolemy's observations imply different atmospheric and hydrological circumstances in the southeast Mediterranean. It is but one, tantalizing, witness to the possibility that the climate of the Roman world was meaningfully dissimilar from our own.[36]

The Roman imperial project had an ally beyond anything the Romans could have imagined: the phase of Holocene climate that was the background

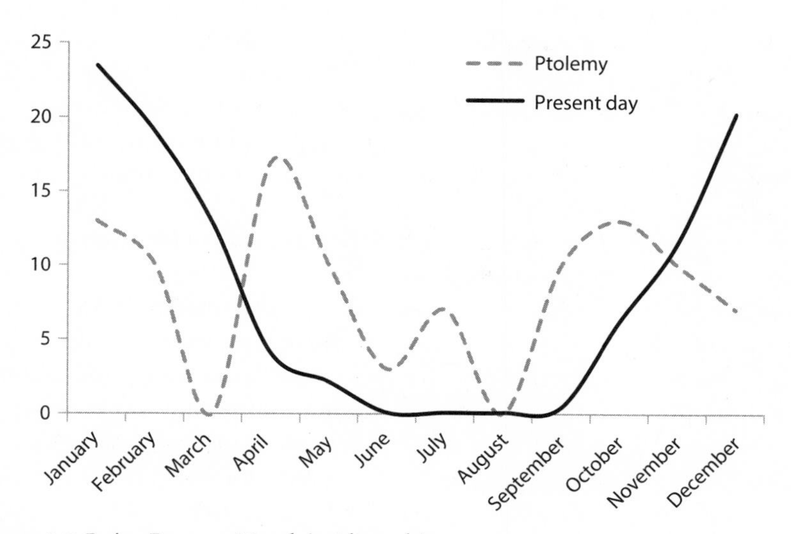

Figure 2.2. Rainy Days per Month in Alexandria

of their expansion. The last centuries BC and first centuries AD were favored by a warm, wet, and stable climate regime rightly known as the Roman Climate Optimum. The simultaneous efflorescence of the Roman Empire and China under the Han dynasty are one of history's many "strange parallels," synchronous pulses of growth and retraction on a global scale, that seem to require causal mechanisms on the same order of magnitude. Although it still lacks precise definition and remains imperfectly understood, the outlines of the Roman Climate Optimum insist that Rome flourished under hospitable environmental conditions. It deserves our exploration, not only because the climate can be such a powerful, constructive agent in an agrarian economy, but also because it underscores that Rome's already audacious experiment in growth rested on transient environmental foundations.[37]

In 1837 Louis Agassiz presented the term "Ice Age" to describe the imprint of radically variable past climates, traced in the geology of the Alps. Throughout the latter half of the twentieth century, his insights were resoundingly confirmed by sea sediments and ice cores that preserve deep archives of climate history. Our planet has been a wildly unstable place, its past full of surprises. The last Ice Age, far from a period of unbroken cold, was characterized by violent swings in the global climate system. The climate of the last hundred thousand years has been called a "flickering switch." Our hunter-forager ancestors survived through times that were not only much colder but also much more capricious. In a phase known as the Last Glacial Maximum, which started 25,000 years ago, the frost pushed humans south, so far that only stretches of southern Europe remained habitable. Chicago lay under the great Laurentide ice sheet.[38]

These savage oscillations were timed foremost to the rhythm of celestial mechanics, slight variations in the earth's rotation and orbit that affect the amount of energy received from the sun. The earth's tilt—that slight angle that puts each pole closer to the sun for half the year and causes seasons—actually oscillates between ca. 22° and 24.5° on a 41,000-year cycle. Moreover, the eccentricity of the earth's yearly journey around the sun—the precise bend of our elliptical pathway—changes, as our planet is tugged by the gravitational pull of other objects in the solar system. Most consequentially, the spin of the earth's rotation around its own axis slowly wobbles, like a top. Every 26,000 years, the earth's axis traces a cone in space, a movement called the precession of the axis. All of these orbital parameters overlap, variously amplifying each other and canceling one another out, massively altering the amount and spatial distribution of heat entering the earth's atmosphere. In the Pleistocene, the result of our planetary swaying and wobbling was, from a human perspective, chaos.[39]

Human civilization—agriculture, large state formations, writing, and so on—is a feature of the anomalous sliver of climate history known as the Holocene. The advent of this friendlier climate has been called "the end of the reign of chaos." Nearly 12,000 years ago, the ice broke. A favorable conjunction in orbital cycles led to abrupt and decisive warming. As the ice sheets melted, sea levels rose; as recently as 8,000 years ago you could stroll from Britain to the continent *à pied*. Relative to the Pleistocene, the Holocene has been warm and stable. But natural climate change did not cease with its arrival.

On a millennial scale, orbital forcing has still been driving long, profound shifts in the Holocene climate. After an early Holocene peak, the millennia of the Holocene have witnessed a sloping decrease in summer insolation in the Northern Hemisphere and a slow trend toward a cooler climate. The Middle Holocene (ca. 6250 BC–2250 BC) was a time of especially propitious climate. The Sahara was green. The Mediterranean was gentler, and miraculously fertile. It rained year round. Human expansion quickened across the Mediterranean, a grassroots dispersal without powerful kingdoms and empires overhead. The archaeologist Cyprian Broodbank called this happy age "how it might have been."[40]

From ca. 2250 BC, the Late Holocene started to take hold. The global climate was reorganized. There was a southward drift of what is called the Intertropical Convergence Zone, where the easterly trade winds converge around the equator. Desertification across the Sahara and Near East became sharper and irreversible. The monsoons weakened. There were more El Niños, and the pressure gradients in the North Atlantic diminished. Summers in the Northern Hemisphere became cooler. In the Mediterranean, the familiar seasonal alteration of dry and wet became increasingly pronounced. But, crucially, climate change progresses at multiple scales, concurrently. Against the backdrop of these millennial-scale patterns, there have been periods of decadal- to centennial-scale climate change. These shorter scale changes have variously reversed, scrambled, or accelerated the longer arc of late Holocene trends. Late Holocene climate change has been like a carousel, moving in different directions at different speeds simultaneously.[41]

The climate during the Holocene has also shifted on much shorter timescales. Orbital forcing, although gradual in its progress, can trigger abrupt changes, because of complex feedback and threshold mechanisms in the earth's systems. Smooth processes can produce jerky effects in the climate regime. Moreover, two additional forcing mechanisms have operated with particular influence on shorter time-scales during the Holocene: volcanism and solar variability. Volcanic eruptions spew clouds of sulfates into the atmosphere, reflecting radiation back into space. Even in the Pleistocene, mega-volcanoes made a mark, notably the Toba eruption of ~75,000 years ago that brought on a millennium of winter and is sometimes argued to have wiped out all but 10,000 of our ancestors. Solar variability is an equally powerful source of climatic instability. "In the galactic scheme of things, the Sun is a remarkably constant star." But, from the earth's perspective, our yellow dwarf is hardly changeless. Deep beneath the sun's visible surface, magnetic activity pulses. The eleven-year sunspot cycle is the most familiar manifestation. While solar luminosity varies only 0.1 percent on this cycle, its climatic effects are widely perceptible. Other, deeper cycles of solar variability have played a major role in Holocene climate change. In particular, a solar cycle with a periodicity of ~2300 years, known as the Hallstatt cycle, has driven deep shifts in the Holocene climate.[42]

These global forcing mechanisms leave us a long way from the local weather. The varying amount and distribution of energy reaching the earth impels change, but climate changes are actually expressed and experienced as moving patterns of temperature and precipitation. In general, changes in temperature tend to be more spatially coherent, true simultaneously across broader spans of the earth. Changes in precipitation are profoundly regional, because a larger and more sensitive array of mechanisms determine the timing, location, and intensity of rains.

In the lands ruled by Rome, variations of heat and moisture both mattered, and the consequences of climate change could be exquisitely local. The Roman Empire was, spatially, giant and unusually complex. Centered on the tightly knit Mediterranean core, it sprawled over three continents. Dura-Europus, a hub on the Euphrates absorbed into the Roman Empire, lay beyond the 40th meridian in the east; the empire's Iberian possessions stretched to 9° W. Hadrian's Wall lies above the 55th parallel, while along the southern fingers of the empire there were Roman cohorts stationed at Syene (24° N), and a Roman fort at Qasr Ibrim, lying at 22.6° N. Recently, evidence for a Roman detachment has been found on the Farasan Islands, overseeing Roman interests in the Red Sea, at 17° N! Because the equator receives more heat than the poles, meridional (north-south) gradients, not zonal (east-west) ones, decide climate differences. From an environmental perspective, the north-south scope of Roman imperialism was stunningly peculiar.[43]

It is not simply the raw square mileage of Roman territory that impresses, but also the specifics of the core region. The nexus of the empire was the Mediterranean Sea, a giant inland waterbody of 2.5 million km^2. The dynamics of the sea itself, in tandem with the crenellated landscapes that surround it, make the zone one of the most complex specimens of a climate

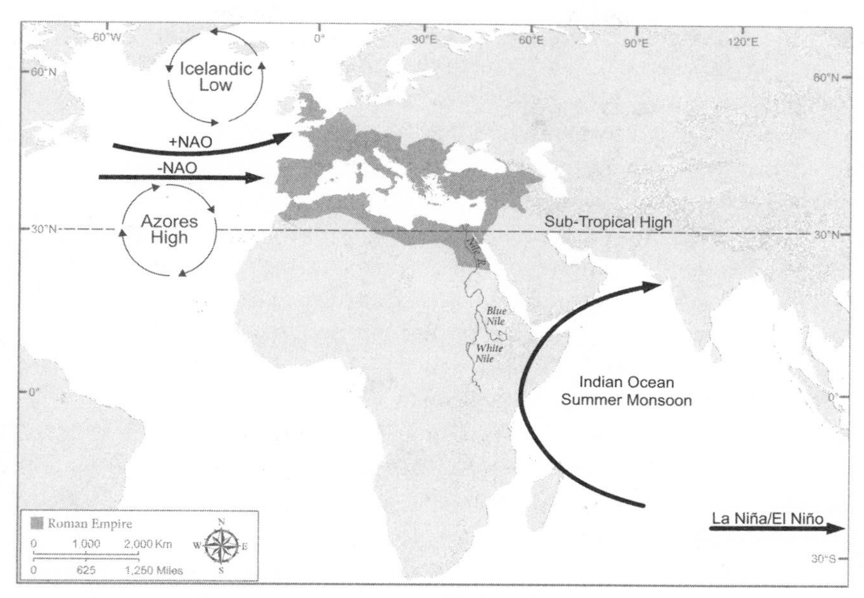

Map 4. Global Climate Mechanisms and the Roman Empire

regime in the world. Extremes of temperature and scarce water availability are a sensitive combination. Several interior zones of the Mediterranean that generate storms are notably finicky, capable of producing extreme precipitation. And what happens on the windward side of a mountain is often very different from the leeward. The Mediterranean region is a tessellation of microclimates. The predictable unpredictability of the Mediterranean has made it an intricate habitat. Strategies to mitigate risk, and the capillary integration of diverse landscapes, are essential to survival. Because of its spot on the globe and its unique local character, here resilience is a way of life. At the same time, an appreciation for the local flavor of Mediterranean environments should not lead us into believing that small-scale climate determinants were in any way autonomous from powerful regional and global controls. The western Mediterranean is under the more direct influence of atmospheric circulation patterns in the Atlantic, while the eastern Mediterranean is the plaything of several global mechanisms and lies exposed to the ridge of subtropical high pressure around 30° N that squelches precipitation in summer. Climate change, in short, is always experienced at the center of local, regional, and global dynamics.[44]

The problem of anthropogenic climate change has heightened the stakes for understanding the paleoclimate. Historians are the great, unintended beneficiaries of the ensuing scramble to scour the earth for natural archives, physical records that preserve clues about the history of the climate. Ice cores, tree rings, ocean sediments, lake varves, and mineral deposits in caves, known as speleothems, have furnished insights into the earth's past. In alliance with other indirect evidence, such as the shifting contours of glaciers and the archaeological distribution of pollens, these physical proxies provide a way to reconstruct the behavior of the climate in the distant past. And though it is now possible to understand the Roman climate in ways that were inconceivable only a decade or two ago, it is equally exciting that our knowledge continues to expand, at a pace that boggles the mind.[45]

Climate proxies speak in a cacophony of voices. The Roman Climate Optimum (sometimes called the "Roman Warm Period") is as widely recognized as it is poorly defined in time and nature. The chronological boundaries proposed here, ca. 200 BC—AD 150, are a coarse abstraction imposed on a range of evidence, but not arbitrarily. They allow us to describe a phase of late Holocene climate defined by global forcing patterns and a range of proxies displaying some coherence. Buoyed by high levels of insolation and weak volcanic activity, the RCO was a period of warm, wet, and stable climate across much of the vast Roman Empire.[46]

It starts with the sun. The sun was generous to the Romans. We are able to explore the sun's historic behavior thanks to physical tracers known as cosmogenic radionuclides. Cosmic rays—streams of high-energy radiation—whir throughout the galaxy. They constantly enter the earth's atmosphere, where they produce isotopes like Beryllium-10 or Carbon-14. Beryllium-10 atoms attach to aerosols and fall to the earth's surface within two to three years. The sun, however, interferes with the stream of cosmic rays passing toward earth; higher solar activity *depresses* the production of cosmogenic radionuclides. In consequence, the levels of Beryllium-10 being produced in the atmosphere—and falling to earth in precipitates where they are archived in ice sheets—vary in tune with solar activity. Cosmogenic radionuclides in ice cores are inversely related to solar activity and form a sensitive proxy of the changing amount of radiative energy reaching earth.[47]

These archives tell us that the RCO was a phase of high and stable solar activity. Between a grand solar minimum centered at 360 BC and another at 690 AD, solar radiation fluctuated within a modest band, reaching one peak at a grand maximum around AD 305.[48]

Meanwhile the volcanoes lay quiet. Of the twenty largest eruptions in the last two and a half millennia, none fall between the death of Julius Caesar and the year AD 169. Between the late republic and the age of Justinian

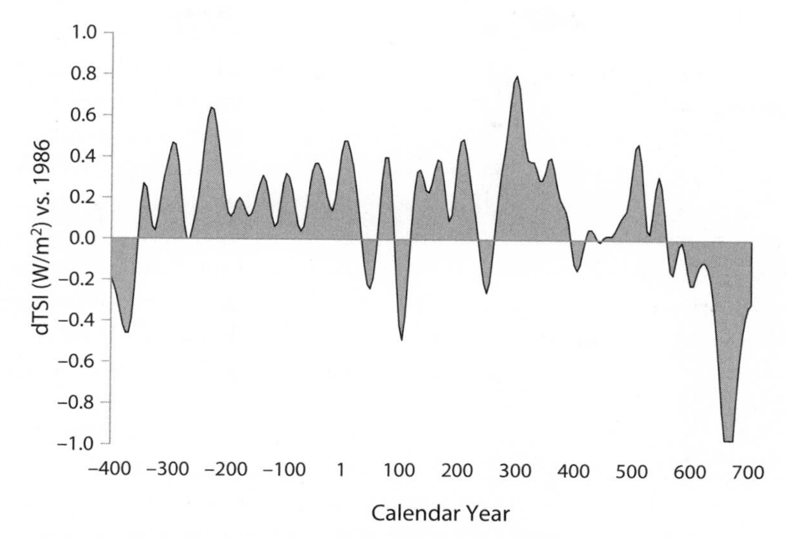

Figure 2.3. Total Solar Irradiance from [10]*Be (data from Steinhilber et al. 2009)*

(the 530s), there were no years of extreme post-volcanic cooling. Across the RCO, the stage was set for stability in the climate regime.[49]

Warmth ensued. The Romans themselves already thought so, as we learn from some of the very earliest human observations of climate change. The naturalist Pliny the Elder, writing in the first century, noted that beech trees, which used to grow only in the lowlands, had become a mountain plant. The cultivation of vines and olives advanced further north than ever before. And these botanical migrations were not due solely to human arts. Alpine glaciers tell the same story. Glaciers retreat and advance in complex rhythm with persistent changes in temperature and precipitation, giant movements that leave behind physical traces. Winter precipitation and, above all, summer temperatures control the balance between growth and melt, and individual glaciers have their own characteristic properties. Where the controls are understood, and where it is possible to date the growth or contraction of glaciers, glaciers are a frozen index of climate change. The signal of warmth in the Roman period is unambiguous. After a major glacial advance was finished by 500 BC, the ice retreated for hundreds of years, down to the first centuries AD. The Great Aletsch glacier may have reached or shrunk beyond its twentieth-century limits in the early imperial period. The Mer de Glace in the Mont Blanc Basin of the French Alps follows a similar pattern. Not until the third century AD was there a reversal, with sudden icy lunges down the slopes. The RCO was an age of melt in the Alps.[50]

Tree rings also testify to the warmth of the RCO. Tree growth can be controlled by temperature, precipitation, or a mix of both. The advantage of dendrochronology is its fine temporal resolution and high degree of statistical confidence. Continuous, overlapping series of regional tree growth can be established going back hundreds of years, allowing precise and robust reconstructions of the paleoclimate. Sadly the Mediterranean heartland has not been exceptionally forthcoming in providing old tree records, but a series of high-elevation trees in the Alps stretching back 2500 years shows a strong correlation with both local and more distant *Mediterranean* temperatures. The highest temperatures before the onset of modern warming were attained in the mid-first century, after which a very slow and uneven decline set in. In the first century, temperatures were in fact even higher than during the last 150 years.[51]

A final temperature proxy is retrievable from the caves of the Roman world. Year after year, the minerals in dripwaters accrete in caves to form stalagmites. The calcites in those cave formations are a mineral archive, the rock equivalent of tree rings, stretching back thousands of years. Those mineral rings include a small mixture of naturally occurring stable isotopes, such as $\delta^{18}O$, a heavy form of oxygen, or $\delta^{13}C$, a heavy isotope of carbon. The proportion of heavier isotopes in a sample is determined by the properties of the surrounding physical environment; in speleothems, heavy isotope ratios can reflect regional temperature, the source, amount, and seasonality of precipitation, and changes in the depositional processes at the site, which are sensitive to the local soil and vegetation cover. The temporal resolution can vary widely, from subannual to centennial. The karstic topography of the Mediterranean supplies an abundance of speleothem records, and the nearly unanimous consensus points to an age of exceptional warmth in the early empire.[52]

The precipitation records hold greater mysteries. There is less assurance that different regions will experience changes of the same timing, magnitude, or direction. The dynamics are more layered and more subtle. There are even, at times, stark trade-offs in the distribution of rain across Mediterranean regions. But in the RCO, the evidence for a period of greater humidity is startling in its consistency and breadth. The RCO was an era of rainfall in both the subtropical and midlatitude belts (in essence the southern and northern halves, respectively) of the Roman Empire. This pattern is striking and merits close examination. To aid us, an even wider array of proxies can be called upon, including physical evidence as well as human testimony in a variety of forms. They help us begin to piece together the strangely watery world of the early Roman empire.

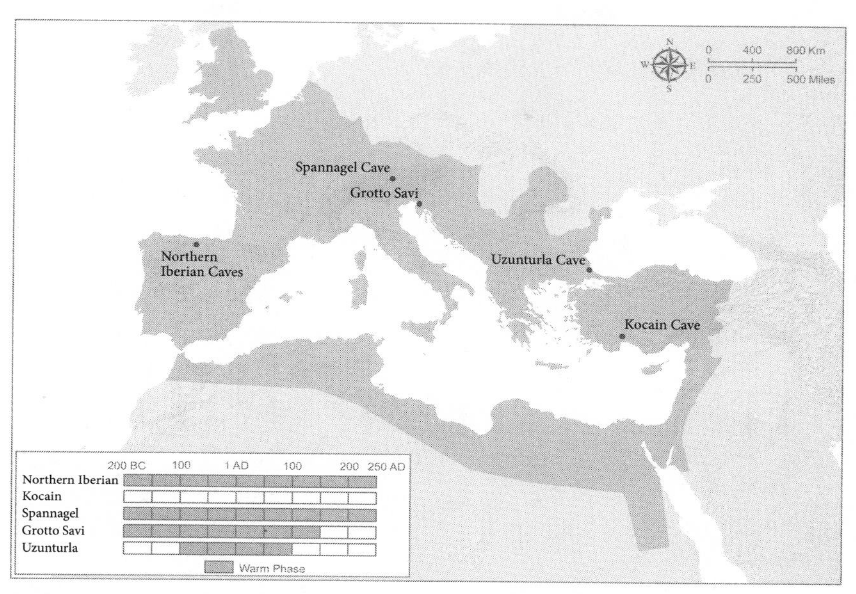

Map 5. Cave Temperature Records and the Roman Climate Optimum

In the northwestern Mediterranean, a wetter phase is so abundantly evident that in the specialist literature the centuries of the RCO are known as the "Iberian-Roman Humid Period." In the north-central Mediterranean, too, physical proxies point clearly to an age of humidity. Intriguing testimony to a different and wetter climate also comes from human observation in the city of Rome. Rome was "a fabulous artificial landscape," hewn into a swampy floodplain. The Tiber River was its soul, and despite the Romans' ingenious efforts to control it, on occasion the river jumped its banks to flood the city. Pliny the Younger described a flood in the reign of Trajan that, despite the spillway built by the emperor, floated the furniture of the aristocracy and the tools of the peasantry through the streets of Rome. Tiber floods are well documented but unevenly spaced across time. We are beholden to the written sources, so the distribution of flooding depends to some extent on the density of our evidence. But the pattern is unmistakable.[53]

Qualifications are in order. Flooding is an extreme phenomenon, not a measure of overall humidity. And the problem of disastrous flooding in the Roman Empire was exacerbated by the ravages inflicted on upland forests. The Roman Empire consumed fuel and materiel voraciously, denuding hillsides of the once dense sylvan texture that slowed and absorbed the

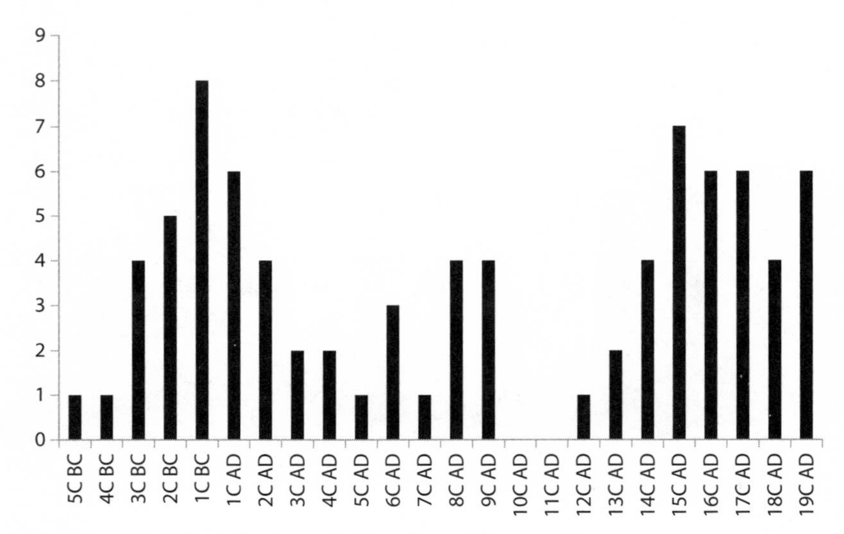

Figure 2.4. Tiber Floods by Century (data from Aldrete 2006)

rush of rainwaters. The distribution of floods remains striking still, and the comparison with the warm centuries of the Medieval Climate Anomaly is instructive: flooding was common in the Roman times, virtually absent in the central middle ages.[54]

The most startling pattern is the *seasonality* of Tiber inundation. The winter floods of medieval and modern times are as unsurprising as the sunrise. But the Roman pattern is nothing short of astonishing. The vast majority of floods struck in the spring to high summer. It is worth mentioning that the Roman poet Ovid implies the Equirria, horse races held in the middle of March, were regularly flooded. And there is simply no way to dismiss the fact that, in the entire medieval and modern sweep of time, Father Tiber did not overspill his banks in the summer, whereas in the Roman world he clearly did so. And this conclusion receives further confirmation from the weather calendar of the sagacious first-century Roman agronomist, Columella, who also assumes far more summer precipitation than is normal today. Like Ptolemy's Alexandria, Rome in the early empire seems to have known not just a climate dissimilar by tiny degrees. Some qualitative mechanisms of Mediterranean climate were subtly but decisively different in the first centuries.[55]

The southern arc of the Roman Empire sits even more exposed along the razor's edge of fatal aridity. But as we come to Roman North Africa and the Levant, it is necessary to pause and emphasize that climate change

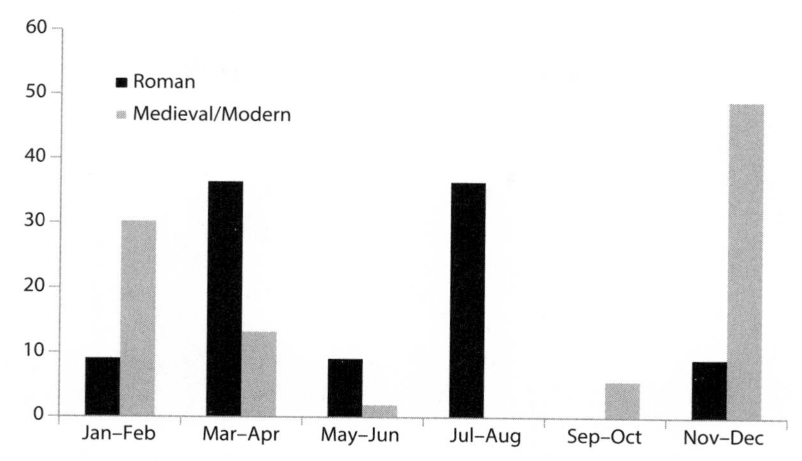

Figure 2.5. Seasonality of Tiber Floods (% of annual) (data from Aldrete 2006)

and human settlement do not move in perfect sync. Favorable climate was far from the only reason that exploitation of the landscape expanded in Roman times. The growth of population pushed people into marginal environments. But more than that, the thickening networks of exchange allowed farmers to edge daringly into zones of higher risk. Connectivity tempered the worst outcomes of dry years. Moreover, the growth of markets fueled entrepreneurial expansion, and Roman institutions deliberately incentivized the occupation of marginal land. The circulation of capital allowed a great burst of irrigation works across the semiarid landscape. The economic boom of Roman Africa was accomplished by the deployment of aqueducts, wells, cisterns, terraces, dams, reservoirs, and subterranean foggaras (long channels that transport groundwater from higher elevations to cultivated lowlands). Hydraulic technologies both indigenous and imperial criss-crossed the uplands and the valleys. By these devices water was assiduously collected and exploited in a semiarid region, where human occupation burgeoned as never before.[56]

At the same time, we should not discount the role of the climate as an ally and a nemesis. It has long been inferred from the literary evidence that the southern Mediterranean was wetter than it is in present times. Pliny the Elder reports elephants inhabiting forests in the Atlas mountains, on the southern fringes of the empire; their extinction in this region was probably a lethal combination of the ivory trade and long-term aridification. In Roman times, North Africa was the granary of Rome, noted for its exceptional fertility; now it is a major grain importer. Today the desert has crept over areas that were clearly under cultivation in the RCO. Opinion has swung about the importance of the physical climate in driving these changes. An older, rather deterministic theory gave way to a more subtle and open approach, in which human agency dominated the scene. But the continued accumulation of geophysical evidence for the role of the physical climate in late Holocene aridification is insistent, and a major inflection point seems to center around the end of the RCO, when the humid interlude ended and the creeping advance of the desert resumed.[57]

The most sensitive gauge of long-range precipitation changes in North Africa may lie beyond the Roman frontiers, among the southern neighbors of the empire. Recent work in the Fezzan, in southwestern Libya, has surprised us by revealing the extent and sophistication of the Garamantian kingdom. The Garamantian economy depended on trans-Saharan trade and settled agriculture. Agricultural practice here was revolutionized by foggaras. Extensive foggara networks let the Garamantian civilization flourish in the early centuries of the first millennium. Trade with Rome soared from the first century through the early fourth century. The archaeology traces the rise, and then fall, of a truly lost civilization.[58]

In the later part of this period, water scarcity proved an intractable and eventually overwhelming challenge. "It is even possible to trace the northward migration of terminal reservoirs as the water table dropped, the foggaras had to be deepened, and the result was the classic phenomenon of foggara-oasis migration as the outlets of the foggaras, and the agriculture and settlements around them, had to move downslope." Possibly the Garamantes overexploited a finite, fossil aquifer; almost certainly, the climate shifted around them. Tree rings from Saharan Cypresses suggest that aridification drove a long-term crisis around the desperate chase for water. Here was an ecologically vulnerable society, with relatively few layers of resilience, that can serve as an especially sensitive barometer of environmental stress. The Garamantes always lived on the edges of water scarcity. But progressive aridification following the RCO made the basis of subsistence ecologically impossible and brought the civilization to an utter end.[59]

Further east, in the Levant, copious attention has been lavished on the history of the region's water balance. The Levant has experienced sharp, centennial-scale oscillations against the backdrop of longer-term aridification. The shore levels of the Dead Sea, recovered from radiocarbon-dated sediments, provide a signal of regional precipitation. The lake was at a high stand from ca. 200 BC to 200 AD. Toward the end of this period the humidity started to break. These swings of fortune are vividly attested in the Talmud, which is full of second- and third-century rabbis who lived in a world where rain was uncertain and drought a devastating problem. "Said R. Eleazar b. Perata: From the day the Temple was destroyed (AD 70) the rains have become irregular in the world." It is tempting to put this down to gruff pessimism. But the rabbi may not have been altogether wrong. A speleothem from the nearby Soreq cave suggests that precipitation abruptly declined from ca. 100 AD. Clearly the third century was a period of water crisis, and the Dead Sea reached a low stand around 300. Again the RCO stands out as an era of humidity, on borrowed time.[60]

Across an unusually broad and diverse geographic range, then, warmth, precipitation, and stability characterized the RCO in the circum-Mediterranean. The RCO was a period when the longer-range effects of changes in orbital forcing, which drove cooling and drying across the entire late Holocene, were held in abeyance for a time, possibly by higher levels of solar activity. The RCO was a late manifestation of conditions that had prevailed in earlier millennia. We might think of it as the last dance of the mid-Holocene. The Mediterranean pattern, with its stark seasonal imbalances in precipitation, was not yet so completely expressed. Climate scientists are increasingly looking to the importance of changes in seasonality as a candidate to explain deep shifts in the Holocene climate. The RCO may have been the last phase of Holocene climate when the subtropics in this part of the globe received any meaningful summer precipitation. Ultimately the insistent direction of the late Holocene, masked for a few centuries, began to re-exert itself, unpredictably but with a vengeance.[61]

This drama was the work of nature. But if the final shift toward summer hyperaridity started in the later phases of the RCO, it enhances the possibility that the Romans had a modest role in accelerating climate change. Orbital, solar, and volcanic forcing are unmoved by human affairs, and the Romans did not pollute the atmosphere sufficiently to trigger climate change. But the Romans did fell forests in massive swaths. Woodland was cleared for agriculture, and the Roman economic machine consumed huge forests for fire and fuel. The Romans themselves witnessed this great deforestation and assumed it was an integral part of the civilizing process. "Day by day they press the forests to retreat up the mountain and to yield their place for cultivated land." The first-century poet Lucan equated the expansion of empire into Mauretania, for instance, with the arrival of the ax. Hadrian was concerned enough about the dwindling supply of long timber to claim certain Syrian forests as imperial property and exert control over their exploitation.[62]

In recent years, opinion has been tipping back toward the view that Roman deforestation was consequential. Deforestation matters, first and foremost, because it suggests that the Romans were bumping against some ecological limits. But it also influenced the climate. The loss of forest cover suppresses rainfall in the Mediterranean. Deforestation increases albedo (the energy reflected back off the earth's surface), so that more heat is reflected away from the ground. In consequence, there is less evaporation of soil moisture into the lower atmosphere. The effects are strong. Some climate models show that this sequence results in lower precipitation in Mediterranean regimes, *particularly during summer*. The case could be made that Roman deforestation interacted with natural patterns of late Holocene climate change to tip the circum-Mediterranean climate toward a regime with less summer precipitation. In this scenario, natural and anthropogenic causes interacted at the threshold between the RCO and the centuries of stress that loomed.[63]

The climate of Rome in the empire's halcyon days was a potent incubator of growth. It fueled the agricultural engine of the economy. The wheat harvest was sensitive to the timing and extent of both temperature and precipitation. Sustained temperature changes on the order of those experienced during the RCO let farmers carve entirely new landscapes of grain cultivation at higher elevations. Pliny the Elder praised the excellence of Italian wheat, and casually assumed that wheat grown "in the mountains" was not as impressive—but what is notable is the sheer fact that wheat *was* grown in the mountains. It has been estimated that, in hilly Italy, an extended rise of 1°C would have rendered, on conservative assumptions, an additional 5 million hectares of land suitable for arable cultivation; that is enough land to feed 3–4 million hungry bodies.[64]

The RCO would not only have extended the limits of cultivation, it also amplified the productivity of the land. Yields in Mediterranean agriculture respond positively to increasing temperature. A mild winter (during the season of germination and seedling growth) is more helpful than a scalding summer, but warmth is a gift to the farmer. And water is vital to the metabolism of plant growth. In the Mediterranean, rain is scarce and its arrival unpredictable. In the former territories of the Roman Empire, the wheat yield is highly sensitive to precipitation. In short, what we are learning of the RCO vindicates the honor of the Roman agricultural writers. They loved to relate marvelous tales of extraordinary yields, but what they presume to be *ordinary* yields have often seemed to compare too favorably with what we know of Italian agricultural productivity in the middle ages. The RCO was a boon to the fecundity of Mediterranean wheat farming.[65]

The RCO may have tempered the worst agricultural risks by offering more rain, more widely distributed, than in later periods. As the ubiquitous remains of irrigation technology throughout the Roman world attest, water management was a central preoccupation of farmers during the Roman period. The most dangerous threat is that rain in any given year will fall below the critical threshold of viability, around 200–250 mm for barley and 300 mm for wheat. The threat of total failure in any given year was palpably real. Based on modern data, Peter Garnsey has estimated that in parts of Greece the wheat crop might fail one year in four, barley one year in twenty. Thus, diversification, integration, and other forms of risk mitigation were indigenous throughout the Mediterranean to ensure bare survival. But regular rains during the RCO would have been a powerful ally in abating the dangers of weather-induced food crisis. Given the outsized influence of threshold effects, and the centrality of risk in Mediterranean agriculture, the conditions of the RCO were no small gift of security to the farmer living on the edges of subsistence.[66]

The quantity of rain and the length of the growing season are limiting factors for other Mediterranean staples, too. The Romans themselves recognized that it was possible to cultivate olives and grapes, sensitive to frost, in regions where the "relentless violence of winter" had once made their cultivation futile. The modern maps which claim to define the zone of "Mediterranean climate" by the limit of olive cultivation will mislead us, if we do not remember that those wavy boundaries will have undulated in historical time. There are, for example, heavy olive-crushing installations built at remote sites 500–700m above sea level in Roman Greece, high above the modern line of olive cultivation: either the peasants lumbered up the mountain with their harvest for processing, or these are the ruins of high-elevation agriculture that have been marooned by climate change. In sum, the conditions of the RCO rendered greater stretches of land pliable to the advance of human cultivation than in the centuries before or after.[67]

The climate was the enabling background of the Roman miracle. The RCO turned the lands ruled by Rome into a giant greenhouse. If we *only* count the marginal land rendered susceptible to arable farming in Italy by higher temperatures, on the most conservative estimates, it could account for more than all the growth achieved between Augustus and Marcus Aurelius. In such perspective, human toil can seem vain. The hard fatalism of the farmer is made of such stuff. The supreme sway of the climate is nothing if not humbling.

In historical perspective, we are only beginning to acknowledge the wavelike phases of growth and contraction that align with phases of climate history. The "Nature" that haunted the nightmares of Malthus turns out to be very real. But, it was not a fixed quantum. Rather, the physical environment of human civilization has been a capricious and inconstant foundation of human endeavor. We should not balk to attribute agency to nature in the shifting fortunes of civilization, nor does such a protagonist's part exclude the role of human agency and sheer chance. Trade, technology, and climate acted in unison to spur the Roman efflorescence. They were mutually reinforcing. Expanded, reliable, and fertile agrarian production inspired the specialization that is the heart of trade. Fecundity generated wealth that became technological capital.

The RCO catalyzed an experiment in growth unprecedented in its scale and ambition. But the Roman miracle was only as stable as the underlying conjunction, which depended so intimately on powers beyond human control.[68]

Resilience: Stress and Endurance in the Roman Empire

The emperor Hadrian was a ceaseless traveler. In the words of his ancient biographer, "Virtually no emperor traversed so many lands so swiftly." In AD 128, his journeys carried him through the African provinces. Hadrian was remembered as a hands-on prince, a reputation incidentally confirmed by an inscription from the legionary headquarters of Africa, recording a detailed speech the emperor gave after personally inspecting the exercises of the *legio III Augusta*. The imperial visit was long remembered, however, for another reason altogether.[69]

It seemed as if the arrival of the emperor brought a much-needed end to a fierce spell of drought. "When he came to Africa, it rained upon his arrival for the first time in five years, and on this account Hadrian was loved by the Africans." As it happens, the same drought is reflected in two contemporary inscriptions, erected at the behest of the very legionary commander praised by Hadrian in his address to the troops. A distant echo of its severity may appear in the price of wheat in Egypt: of ten attested wheat prices from Roman Egypt before the great pestilence, the very highest was in AD 128 (only four years prior, on the same estate, wheat had sold for 25 percent less). Whatever we wish to think of the emperor's numinous power over the skies, some clever historical sleuthing has revealed that his efforts included a quite practical measure, in the construction of a great aqueduct bringing water to Carthage. Stretching in total more than 120 km, it stands among the longest water supply devices ever built by the Romans.[70]

The widespread African droughts of the AD 120s may have been the first pangs of a secular aridity crisis that would grip the region over the next several centuries. The episode is also but one reminder, if one were needed, that the golden age of the empire was not a period of undisturbed tranquility. Sharp climatic variability is guaranteed in the Mediterranean, and the RCO at most moderated the excesses of year-to-year unpredictability. Acute epidemic crises were not unusual, at least at local-to-regional scales. Dynastic instability at home and geopolitical friction along the frontiers were virtually constant features of the Roman imperial enterprise. During the reign of Antoninus Pius, the crowning moment of the *pax Romana*, the rhetorical tutor of the prince-in-waiting, Marcus Aurelius, thought the empire resembled nothing so much as a windswept island beset by storms, pirates, and hostile fleets. Adversity was never absent from the Roman world, but at its height, the empire enjoyed a tremendous ability to maintain order amid the unrelenting turbulence.[71]

Resilience is the measure of a society's capacity to absorb shocks and to fund recovery from injury. Not every drought induces famine, and not every epidemic triggers collapse. But some do, and since the pattern of history is not pure contingency, we need mental tools to explain the links between such disturbances and their consequences. The resilience paradigm is such a mental tool, for it helps us conceive of the Roman Empire as an organism comprised of interdependent ecological (agricultural, demographic) and imperial (political, fiscal, military) systems. These systems, in turn, had functions whose success was imperiled by a range of risks, which human actors sought to mitigate or to manage by learned strategies of buffering, storage, and redundancy. The response to risk was costly, and so the ability to manage risk was not infinite; stress was inherent in the system; shifting threats or new shocks could add systemic stress to the regime.

The resilience paradigm allows us to see why the response of the system to an impulse was nonlinear; feedback mechanisms, critical thresholds, and changes operative on different timescales meant that one drought might have invisible effects, while another of just the same magnitude might seem to tip a society irreversibly toward catastrophe.[72]

The Roman Empire absorbed the countless, humble strategies of ecological resilience that made civilization in the Mediterranean possible. The Mediterranean climate demands versatility, and peasant wisdom accreted over the millennia to buffer farmers from the turbulence of nature. Strategies of diversification, storage, and integration evolved to reduce the danger of lean years. We have no keener observer of rustic lifeways in the ancient world than Galen. The doctor was professionally interested in the nutritional regime of country folk. He catalogued the exotic local grains that still characterized the agrarian fabric of many backwaters in the Roman Empire, where hardy varietals were often preferred. His sharp eye easily noticed un-Hellenic habits. "Barley is used for bread in many parts of the world." Even around Galen's Pergamum, the peasants made their bread from lesser grains "after taking their share of wheat to the cities." In times of true scarcity, the peasant had seeds of millet or panic ready; rough but reliable and fast, crisis crops were an insurance policy against hunger. So too were all forms of food storage, and Galen's writings preserve a trove of information about stockpiling acorns and drying and pickling legumes, fruits, and vegetables.[73]

The climate of the Mediterranean also fostered the evolution of *cultural* norms buffering against extreme risks. Traditional ideals of self-sufficiency, reciprocity, and patronage went hand in hand. The peasant fantasy of autarky was unrealistic, but it motivated a spirit of proud independence. Dio of Prusa, a Greek philosopher and politician writing just a few generations before Galen, described an encounter with a rustic family in his famous *Euboean Oration*. The family had married off a daughter to a rich man in a nearby village; when asked if they received help from the man, the peasant wife tartly denied it and insisted that they rather *gave* game meat, fruit, and vegetables to the daughter and her wealthy husband; they had borrowed some wheat for seed but repaid it immediately at the harvest. However romanticized, the story captures the "twin notions of self-sufficiency and reciprocity."[74]

And reciprocity between unequals shaded into patronage, a deeply established tradition in the stratified societies of the Roman Empire. In the contemporary letters of the wealthy Roman senator Pliny the Younger, we catch occasional glimpses of a benevolent patron at the highest levels showering aid and favors on his clients. The expectation of paternalistic generosity lay heavily on the rich, ensuring that less exalted members of society had an emergency lien on their stores of wealth. Of course, the rich charged for this insurance, in the form of respect and loyalty, and in the Roman Empire there was a constant need to monitor the fine line between clientage and dependence.[75]

These strategies of resilience, writ large, were engrained in the practices of the ancient city. Diversification and storage were adapted to scale. Urban food storage was the first line of redundancy. Under the Roman Empire, the monumental dimensions of storage facilities attest the political priority of food security. Moreover, cities grew organically along the waters, where they were not confined to dependence on a single hinterland. Cities stranded inland were most vulnerable to short-term climatic shocks. "Cities on the sea coast easily endure a shortage of this kind, importing by sea the things of which they are short. But we who live far from the sea profit nothing from our surplus, nor can we produce what we are short of, since we are able neither to export what we have nor import what we lack."[76]

When food crisis did unfold, the Roman government stood ready to intervene, sometimes through direct provision but more often simply by the suppression of unseemly venality. In AD 92–3, a cruel winter caused the price of grain to soar in Pisidia; the Roman governor, we know thanks to an inscription, condemned the injustice of profiteering and held the price of grain where it had been previously, "so that the mass of ordinary people should have some means of buying it." Often, intervention was private in nature. The classical cities had a strong ideological expectation that the wealthy would pour their resources into visible public goods; this culture of civic euergetism, so characteristic of the moral economy of the classical city, was nothing but the enlargement of the norms of reciprocity and patronage that cushioned individuals against the vagaries of the environment. We learn of a grandee from Roman Macedonia who held the office of high priest; at his own expense, he carried out road repair, held games and contests for the people, and sponsored beast hunts and gladiatorial combats; most telling of all, he sold grain below market price "in times of urgent necessity."[77]

The emperors improvised on these strategies, on a grand scale. The emperor Trajan would "divert and direct the abundance of the land now here, now there, as the moment and necessity demand. He would feed and protect a rescued nation across the sea as though it were some part of the Roman people and plebs." "Hadrian had seen many cities, more than any other emperor, and he 'took care of them all,' so to speak, giving water to some, harbors to others, grain to some, public works to others, money to some, and honors to others."[78]

The most familiar system of resilience was the food supply of Rome. The remnants of the monumental public granaries that stored the food supply of the metropolis are still breathtaking. It was said that the emperor Septimius Severus had cared so assiduously for the provision of Rome that upon his death there was enough grain stored to feed the city for seven years. The grain dole was the political entitlement of an imperial people, under the patronage of the emperor. The inhabitants of Rome had rights of first refusal on the emperor's generosity. An imperial letter of the second century, inscribed at Ephesus, promises that the eastern city could procure Egyptian grain, on the condition that the harvest was sufficient for Rome. "If, as we pray, the Nile provides us with a flood of the customary level and a bountiful harvest of wheat is produced among the Egyptians, then you will be among the first after the homeland." In the second century, some 200,000 citizens of Rome received 5 *modii* of wheat monthly; that amounts to 80,000 tons of wheat annually, just for the public dole. To feed the million mouths of the capital, the sea was criss-crossed by a flotilla of deep hulled grain ships. The signal boats in the vanguard of the Alexandrian fleet were a welcome sight, bringing joyous crowds to the shores of Italy to gawk at their arrival. What is most remarkable, though, is that the transport of grain to Rome was left in private hands; merchants were given modest subsidies for carrying grain to the city, but so much resilience was built into the grain market that, during the high empire, Rome could be fed without an elaborate system of requisition.[79]

The food system was sturdily built to withstand sudden, short-term shocks. The resilience of the food system serves to highlight by contrast the relatively meager infrastructure that existed to cushion the impact of demographic shocks. The next chapter will focus on the disease regime of Rome, but here is the place to underscore that the Romans were nearly helpless in the face of epidemic mortality. They had few tools at their disposal to mitigate the threats of infectious disease or to recover speedily from cutting losses. Ancient medicine was, frankly, probably more harmful than helpful. While basic nursing was no small advantage to the sick and ailing, the prescription of hot baths and cold plunges, and the common practice of bleeding patients, could only have added to the death rolls. Ordinary people turned to magic, which was ubiquitous. Certainly the Roman state possessed the technology to apply the kinds of quarantine that began to develop in the late middle ages, but the religious view of disease seemed to dominate public response: the Greeks and Romans reacted to mortality events with arcane sacrifices or the erection of apotropaic statues of Apollo to ward off disease. Even the rudiments of public health were conspicuously absent in the Roman Empire.

The mortality regime was sharp-edged, and in the absence of any effective means of redress, the response of ancient society was to tune fertility to high levels. Moreover, adoption was a mundane part of life, a realistic response to a mortality regime that always threatened family survival. The extensive practice of child exposure in the ancient world, which often resulted in death or cycled infants into the slave trade, might be seen as a somber release valve in a system fixedly set on high fertility. Finally, the ease of internal migration within the empire was a kind of demographic resilience; movement, mostly toward cities, skimmed the surplus of some areas to compensate the deficits of others. But, ultimately, the facts of biology were immoveable. Human societies of the late Iron Age had evolved few responses to buffer the effects of sharp mortality crisis. They could only claw back, slowly, from the setbacks of epidemic disease. When these bursts of mortality first became something more than a local disaster, the unprecedented shock sent the empire reeling.[80]

Just as the societies of the Roman world were built to withstand the pressures of ecological turbulence, so, too, the imperial system was designed to endure the slings and arrows of political misfortune. The regime established by the first emperor, Augustus, was perduring. Rome was ruled by a monarch in all but name, who administered a far-flung empire with the aid, first and foremost, of the senatorial aristocracy. It was an aristocracy of wealth, with property requirements for entry, and it was a competitive aristocracy of service. Low rates of intergenerational succession meant that most aristocrats "came from families that sent representatives into politics for only one generation."[81]

The emperor was the commander-in-chief, but senators jealously guarded the right to the high posts of legionary command and prestigious governorships. The imperial aristocracy was able to control the empire with a remarkably thin layer of administrators. This light skein was only successful because it was cast over a foundational layer of civic aristocracies across the empire. The cities have been called the "load-bearing" pillars of the empire, and their elites were afforded special inducements, including Roman citizenship and pathways into the imperial aristocracy. The low rates of central taxation left ample room for peculation by the civic aristocracy. The enormous success of the "grand bargain" between the military monarchy and the local elites allowed imperial society to absorb profound but gradual changes—like the provincialization of the aristocracy and bureaucracy—without jolting the social order.[82]

In the first century, an empire of conquest settled down into a symbolically unified territorial empire, with regular and rational, if heterogeneous, rates of taxation. The Roman army continued to mount large-scale campaigns of conquest from time to time, but most of its activity was defensive in nature and might be described as a mix of civil engineering and local surveillance. Through careful management, the political power of the army remained latent for most of the high empire. The coordination of the fiscal and military machinery of the state, across three continents, with Iron Age technologies of communication and travel, is one of the most intricate accomplishments of any premodern polity.[83]

The basic stability of the Augustan settlement belies the fact that the regime was under constant threat from within and without. As the ghost of

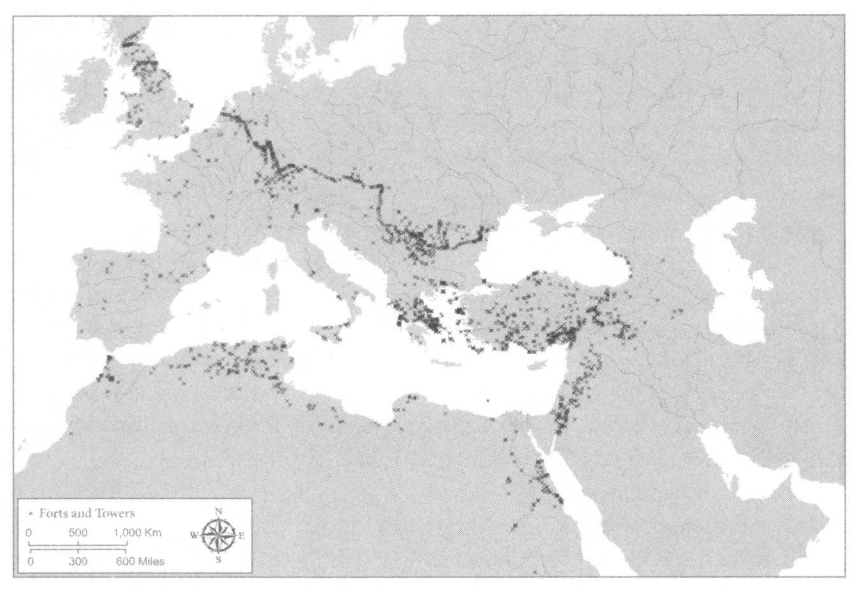

Map 6. Remnants of Roman Imperial Power (data from darmc.harvard.edu)

the republic faded from memory, the possibility of a revolutionary regime change became as remote as a fantastic dream. But the dynastic solution of Augustus was tenuous, and sane emperors took great pains to establish a smooth succession. Biology often failed, one way or another, and succession crisis was an irrepressible feature of the regime. A monogamous mating system, and a merciless mortality regime, left many emperors without a biological heir. By comparative standards, the reigns of Roman emperors were astonishingly short, so the uncertainty of the imperial dynastic system was a high-stakes problem. The lengthy reigns and series of imperial successions by adoption during Gibbon's happiest age were anomalous—a mix of dumb luck and a stable empire. Occasionally, as in AD 69 or 193 or 235–8, uncertainty boiled into outright civil war. But, whenever there was dynastic change, the new boss looked much like the old boss, just more provincial.

It is a testament to profound continuity that the historian Cassius Dio, writing in the early third century, could place a long speech in the mouth of the Augustan advisor Maecenas, which described the fundamentals of a regime that had endured across the entire span of time separating Augustus and Dio. The stability of the Augustan system is a tribute to the resilience of the aristocracy, the administration, the cities, and the imperial ideology that undergirded the regime.[84]

The imperial regime was expected, above all else, to remain victorious. *Victoria* was worshipped as a goddess of empire, symbolizing martial prowess as well as the safety that was guaranteed by Roman arms. The maintenance of imperial legitimacy and military hegemony was expensive. The entire state budget in the high empire was on the order of 250 million *denarii*, about two-thirds of which was consumed by the army (civilian salaries, the grain supply, public infrastructure, and patronage were among the other expensive line items); if GDP was some 5 billion *denarii*, then state expenditure was about 1/20 of GDP. The annual revenues of the state were amassed from a wide variety of land and head taxes, plus tolls, inheritance and manumission taxes, and extensive state-run mining operations.

The incidence of Roman taxation was, from one vantage, bearable. Because the fiscal system evolved piecemeal out of a protracted course of conquest and diplomacy, tax rates were heterogeneous, right down to the reforms of the later third century; averages are misleading, but a target rate in the range of 10 percent of annual agricultural production is not an unreasonable guess. In wheat equivalent, it has been observed, the Roman state was collecting more on a *per capita* basis than the English or French government was able to raise in the seventeenth century, although it was far short of the revolutionary rates achieved by the most modern eighteenth-century states.[85]

There was only a little cushion built into the fiscal machinery. In theory, the target rates would have allowed the treasury to collect a modest surplus each year. In reality, central collections were probably far below nominal goals. The lines of stress in the fiscal regime were never very hidden. Tax collection was a flashpoint of provincial resistance, and its successful execution depended on the collusion of local elites and their agents, like the "publicans" who are the emblem of villainy in the New Testament. Emperors were regularly in need of cash. (Vespasian famously taxed urine distributed from the public latrines and reassured his doubting son, Titus, that money could not stink: *pecunia non olet.*) Domitian (r. AD 81–96) gave the soldiers a raise equal to one-third of their annual salary—the only pay increase in the two centuries separating Augustus and Septimius Severus; his generosity strained the state's finances. In the second century, Hadrian had to cancel outstanding debts to the government, and only two generations later, in the aftermath of the pandemic, Marcus Aurelius did the same. Although these remissions were advertised as acts of generosity, they are in fact a signal that, even at the height of imperial prosperity, funding a tri-continental empire was not without its stresses.[86]

Rome's military dominance makes it easy to overestimate the reality of "peace." Edward Luttwak's *Grand Strategy of the Roman Empire* remains instructive in this regard. As the Roman Empire was transformed into a territorial empire, hegemony was maintained by an economy of force. The top strategic priority was the displacement of violence to the outer ring of provinces; but, in the course of time, the protection of these outer rings became a goal of statecraft.

The Roman frontier system epitomized the resilience of the empire; it was designed to bend but not break, to bide time for the vast logistical superiority of the empire to overwhelm Rome's adversaries. Even the most developed rival in the orbit of Rome would melt before the advance of the legionary columns. The Roman peace, then, was not the prolonged absence of war, but its dispersion outward along the edges of empire. Peace, insofar as it was ever a concrete goal of the Roman state, was elusive, always receding beyond the horizon. Even in the supposed heyday of peace, in the reign of Antoninus Pius, conflict within the borders and beyond was rife. In his reign, we know of a rebellion in Greece, an uprising of the Jews, extensive military operations in Britain, turmoil in Dacia, trouble in Africa, and insurrection in Spain. There was a major debasement of the currency ca. AD 155–57. It is revealing that Aelius Aristides, who authored the great paean to the Roman Empire, is also the probable author of another oration, which for long was believed to have been delivered amid the chaos of the third century; in fact, it is probably Antoninus Pius who is described as delivering the state through violent storms back to safe harbor.[87]

The ship at sea, battered by storms, was a prominent metaphor for the empire at its height. It reminds us, though, that the ship of state would not capsize under the weight of a single, giant wave. For even if the impending catastrophes, soon to toss the Roman Empire, were an order of magnitude greater than anything the empire had endured, their effects were subtle and registered ultimately in the long run. Even in the aftermath of the misfortunes that lurked just ahead, the empire was able to draw on its stores of resilience to right the ship of state.

This pattern certainly complicates the writing of Roman history. Many things were about to happen at once, both within the constitutional order of Rome and beyond, on the Danubian plains and the Iranian plateau. But the effect of the Antonine crisis was to end a certain trajectory, one of exuberant social development, that had made it possible to project a sense of stability and easy command in the ruling of the empire, even in the face of perpetual friction. Once the ground had shifted under the Romans, with the arrival of a less hospitable natural environment and the advent of a new, microscopic enemy more ferocious than any the Romans had ever faced, the storm clouds gathering on the distant horizon began to appear more imposing than usual.

The New Age

When Galen had first journeyed to Rome in AD 162, he would have felt sweeping in the other direction, along the roads and sea lanes, a massive military mobilization destined for the eastern provinces. Parthia was about to feel the full brunt of Roman power. Their king, Vologaeses IV, had taken the accession of Marcus Aurelius and Lucius Verus as an opportunity to test the fledgling emperors. Lucius was sent to Antioch, which would serve as headquarters for the greatest Roman military operation in over half a century. It was a war that would stir deep feelings of jubilation and then dread. The Romans came to believe that the Parthian campaign of Lucius Verus brought the pandemic down upon the empire. In truth, the war was both a display of Roman power at its absolute crest and a subtle turning of the tide.[88]

Lucius and Marcus were determined to flex the muscles of empire. The Romans might lose a battle, but they undoubtedly possessed what Luttwak called "escalation dominance." Nowhere is this more in evidence than during the Parthian campaign. Antioch was the command center; to enhance its connection to the supply lines of the imperial heartland, Roman engineers reshaped the landscape by building a canal that made the Orontes river more easily navigable. At least three legions were deployed from Europe to Asia, moved more than 3600 kilometers on Roman roads.

An equally impressive concentration of military expertise was summoned for the campaign. Unlike most of their aristocratic peers, both Marcus Aurelius and Lucius Verus lacked personal experience of field command. But the assembly of veteran leadership more than compensated. The war council included the most decorated senatorial commanders from around the empire, including C. Avidius Cassius, a senator of Syrian origin (and descendant of the Seleucid kings) who had distinguished himself in the reign of Hadrian. The war cabinet was reflective of the Roman imperial order: a senatorial elite, open to provincial talent, trained to command in a far-flung and sometimes recalcitrant empire.

Thus arrayed, the Roman imperial machine was inexorable. The war was a sanguinary rout. Again the Romans demonstrated their ability to project violence on an overwhelming scale. When aligned behind unified leadership, concentrated in a specific theater, and plugged securely into imperial supply lines, the Roman armies of the second century were an insuperable force, even against the empire's most formidable rival.[89]

Tidings of the victory were cheered in the capital. When Lucius returned to Rome in AD 166, the city witnessed the first official triumph it had seen in over half a century. But quickly the news from the east was dimmed in red. One of the heroes of the campaign, Avidius Cassius, had allowed his armies to surround Seleucia on the Tigris, a Hellenistic foundation deep inside Babylonia. At a global crossroads of trade, wealthy Seleucia was "the greatest of cities," a peer of the greatest towns of the empire; Seleucia had readily surrendered, but the Romans sacked the town anyway, claiming the inhabitants broke faith. Even by Roman standards, the violence was unnerving.

Amid the plunder, a Roman legionary chanced to unseal a chest inside a temple. The temple was a sanctuary of the god known as Long-Haired Apollo. Thence, the Romans believed, a pestilential vapor was unleashed that soon "polluted everything with contagion and death, from the frontiers of Persia all the way to the Rhine and to Gaul." This story became the official line on the arrival of an unfamiliar pestilence in the Roman Empire. In fact, the Parthian campaign and sack of Seleucia were largely incidental to the outbreak and course of the mortality event destined to become known by the family name of the emperors, the Antonine Plague. Its advent marked an epoch in both Roman and natural history.[90]

As the strange disease was snaking its way across the empire, Galen tried to cut short his Roman career. He made a narrow escape from the city, "like a runaway slave." He hastened overland to Brindisium and crossed on "the first boat to weigh anchor." Galen feared he would be detained by the emperors. His fears were soon to be realized. Lucius died, but Marcus summoned Galen to Aquileia, where he had set up winter camp in preparation to launch a military campaign in the north. Marcus and Galen were surrounded by a mortality event that was unlike anything either of them had experienced. The course of their lives was to be shaped by the unraveling of "the great plague." In one sense, the Antonine Plague was a creature of chance, the final unpredictable outcome of countless millennia of evolutionary experimentation. At the same time, the empire—its global connections and fast-moving networks of communication—had created the ecological conditions for the outbreak of history's first pandemic.[91]

3

Apollo's Revenge

Aristides and Empire: Rich but Sick

When the talented orator we met in the previous chapter, Aelius Aristides, delivered his "Roman Oration" before the emperor Antoninus Pius in AD 144, he was not in the finest fettle.

Aristides had come to Rome, much as Galen would a generation later, an aspiring young provincial ready to try his fortunes on the grandest stage. He had been preparing his entire life. A son of the gentry, Aristides had been tutored throughout his youth by a celebrity cast of rhetorical teachers. After his father's death, Aristides had cruised the Nile, the ultimate Grand Tour. He failed to discover its exotic headwaters, but acquired a stock of colorful experiences he could recycle for a lifetime. Shortly after, he ventured to the capital. He journeyed west by land, along the Via Egnatia, the great Roman highway cutting through the Balkans. On the way, he contracted a nagging cold that turned violent, worsened by the dreary weather and the swampy landscape. He struggled to eat, and breathing became laborious. "I was very worried about my teeth falling out, so that I was always holding up my hands to catch them." The fevers struck, and by the time he reached Rome, "there was not any hope even for my survival." When Aristides delivered the "Roman Oration," he lifted himself off what he thought was his deathbed.[1]

The account of his sickness in Rome is only the earliest episode in what amounts to the most intimate medical diary from the ancient world, the *Sacred Tales*. The text is a memorial to the healing god Asclepius, whom Aristides considered his savior. The illness in Rome began a lifelong descent into miserable health and dependence on the god. Aristides suffered intestinal disorders, migraines, consumption, catarrhs, tumors, seizures, and endless bouts of fever. Aristides often recuperated at the Temple of Asclepius in Pergamum (which, from one angle, looks as chic as a Beverly Hills rehab clinic). There he was treated by Galen's teacher Satyrus. In later years, Galen recalled the orator's frail constitution. Today, the ailments that Aristides chronicled are sometimes put down to "psychosomatic" causes, neurosis, or hypochondria. But that is unjust. Merely the *cures* taken by Aristides would have been the undoing of many a hale man. Already in Rome his treatments were dire. "The doctors made an incision, beginning from my chest all the way down to the bladder. And when the cupping instruments were applied, my breathing was completely stopped, and a pain, numbing and impossible to bear, passed through me and everything was smeared with blood, and I was violently purged." His life of therapy had just begun. Over the course of decades, Aristides pursued remedies ranging from sadistic to simply bizarre. There is no reason to doubt the physiological reality of his wretched health.[2]

Despite it all, Aristides managed to become the most renowned orator of his age. When Smyrna suffered an earthquake, Marcus Aurelius was reduced to tears by the mournful plea for help Aristides composed (and the emperor, fulfilling his part in this polite reciprocal business, offered the infusion of imperial aid that was *de rigueur* in the grand bargain between the emperors and the cities). The *Sacred Tales* were received with immediate and universal acclaim in antiquity, and the ancients did not regard Aristides as the eccentric that moderns often have. The therapies he pursued on the advice of gods and doctors were perfectly within the mainstream of second-century medical practice. Aristides may have suffered more than most, but in an age when disease was a lurking reality for all, his helplessness and search for salvation fascinated, because they were a melancholy bond of solidarity with the rest of humanity.[3]

In the engrossing record of the *Sacred Tales*, there is one case where we can say with confidence what ailed Aristides. Far from eccentric, this story draws him even closer to the history of his age. He was in the suburbs of Smyrna at the height of summer in AD 165 when a pestilence "infected nearly all my neighbors." Aristides' slaves caught the infection, then Aristides himself contracted the disease. "If anyone tried to move, he immediately lay dead before the front door. . . . Everything was filled with despair, and wailing, and groans, and every kind of difficulty."[4]

This passing notice is only a tiny piece of a larger puzzle, but it is the earliest definite attestation from the Mediterranean for the disease event known as the Antonine Plague. Aristides described the "terrible burning of a bilious mixture." He himself suffered a "persistent lesion" in his throat. He was at death's door but was spared: Aristides believed that a young boy who died at the exact moment his own fever broke had been a kind of grim substitute. It has been ingeniously suggested that his salvation from the disease prompted Aristides—the grateful patient of Asclepius and faithful devotee of Apollo—to write the *Sacred Tales*, a solemn gift to an empire suffering under the weight of the pandemic. Aristides can stand as a symbol for a society that was utterly helpless in the face of disease—and soon to be swept up in the drama of a biological event whose magnitude was unfamiliar even in a world ceaselessly rocked by the waves of epidemic mortality.[5]

In the AD 160s the Roman Empire intersected the evolutionary history of an emerging infectious disease. It was a fateful encounter, but it was not an ineluctable one. The plague was not the predictable boomerang of overextended growth, and we should not pose the Roman Empire as the victim of a Malthusian meltdown in which demographic expansion outstripped the capacities of the underlying resource base. But neither was the pestilence pure happenstance. The *ecological* conditions inherent in the empire loaded the dice in favor of this kind of event. To understand the role of disease in the Roman world, we must try to think of the empire as an environment for its invisible residents. The dense urban habitats, the unflinching transformation of landscapes, the strong networks of connectivity within—and especially beyond—the empire, all contributed to a unique microbial ecology.

This chapter tries to say what can be said about the biology of dying in the Roman Empire, putting the array of specific microbes that stalked the empire in the foreground. The Romans are perhaps the earliest civilization where such a hazardous exercise can even be attempted. Explorers of the Roman past have discovered some unexpected resources to aid us. For not only do we have prolific medical geniuses, like Galen, to guide the way, but also the testimony of stones, bones, and genomes. The formulaic evidence of grave inscriptions, the physical evidence of skeletons, and increasingly the molecular evidence of the pathogens themselves, all contribute to a more rounded picture of health and human biology in the Roman Empire. What emerges is inevitably more tantalizing than conclusive. But the Romans seem to have built their empire at a dangerous juncture in time, and we can just now begin to see the shadowy outlines of a new, evolutionary history of infectious disease in which the centuries of Roman civilization form an especially important passage.

Even by the standards of underdeveloped societies, the denizens of the empire were unhealthy. We might say that they were, like Aristides, rich but sick. The empire's fetid cities were petri dishes for low-level intestinal parasites. The empire's violence against the landscape called forth scourges like malaria. The empire's thick webs of connection let chronic diseases diffuse across the empire. But the really decisive moment came when an acute infectious disease transmitted directly between humans found its way into the empire. We will argue that what Galen called the "great pestilence" was in fact caused by smallpox. Certainly, it was a disease preternaturally well equipped to prey on the Roman Empire, hurtling itself across the roads and sea lanes that bound together the mosaic of cities and peoples under Roman rule. The Roman Empire prepared the way for the pandemic, building new gateways for germ migration into the empire and new highways for transmission within its territorial boundaries.

The Antonine Plague was unlike anything anyone had ever witnessed. The pandemic stirred a primitive religious dread among the peoples of the empire. There is something fitting in the blame that was ultimately laid upon the god Apollo, a protean god, moving easily across all boundaries, and closely associated, since the days of Augustus, with the image of the empire itself. The new pathogen was called forth by the nature of the empire and its global tendrils. The arrival of this pandemic disease marked the beginning of a new age.

Toward the Disease Ecology of the Roman Empire

Before the triumph of public health and antibiotic pharmaceuticals, infectious disease was public enemy number one for humankind. From banal *Staphylococcus* infections to the glamorous superkillers like smallpox and bubonic plague, infectious disease was the primary agent of human mortality. But the ensemble of deadly germs threatening humanity has not been stationary; it changes in historical time and varies over space. The Roman disease pool was an artifact of epoch and place. To imagine it is to take a germ's eye view of the world and to enter into the evolutionary journey of those microscopic organisms with whom we share the planet. It is important that we resist the temptation to view the Roman experience with pathogens as *just another act*, germs flitting on and off the stage as ever before. To do so is to miss entirely the critical place of the first millennium in the ongoing story of infectious diseases, and the circumstantial alignment of the Roman Empire with specific pathogens at a particular moment in time.[6]

The genomic revolution has put the history of human disease, at present, in a state of flux. The tumbling costs of genome sequencing, as well as new techniques for recovering degraded DNA from archaeological contexts, are beginning to let us peer back deeper into the past than ever before. Genomes establish evolutionary relationships and therefore allow us to reconstruct Darwin's "great Tree of Life, which fills with its dead and broken branches the crust of the earth, and covers the surface with its ever branching and beautiful ramifications." Systems of genetic relationship—known as phylogenetic trees—are providing us with maps of the microbial past, defining evolutionary relationships in ways that can help us to locate the history of an organism in time and space. When archaeological genomes can be recovered to boot, they not only pinpoint the existence of a given species in a particular spot in a specific layer of the past—they also help extend and enrich microbial phylogenies and therefore our understanding of pathogen evolution.[7]

These biological archives are only now starting to unsettle a tale that took hold a generation ago, before the triumph of molecular evidence. In this story, humans brought with them from the Paleolithic a baseline of "heirloom" germs and parasites that had been handed down from our hominid predecessors. These pathogens were old friends, well adapted to life with us, so much so that many of them were mere nuisances. As our hunter-gatherer forebears dispersed around the planet, they also acquired new parasites on the journey, "souvenirs" of their trek. Still, the pathogen load was altogether light. Then, the Neolithic Revolution was the big bang of violent infectious diseases. Density-dependent bugs could flourish now that our roaming ancestors settled down into towns, and diseases could make the leap from domesticated animals to the humans who lived cheek by jowl with them. In McNeill's *Plagues and Peoples*, the undisputed masterpiece of disease history in the premolecular era, the development of more advanced civilization brought on the "confluence of the civilized disease pools" of Eurasia. The discrete puddles of endemic killers in the early Neolithic flowed together with dread, genocidal effect as societies came into contact. Global connectivity was transformational, first in the Old World context and subsequently in its trans-oceanic vastness.[8]

This was a makeshift story, crafted from whatever scraps of epidemiology, geography, animal medicine, and, only in later periods, textual sources could be assembled into a coherent narrative. It was an ingenious construction, and it is remarkable how well its outlines have held up. Sometimes the molecular evidence has directly confirmed the intuitions of the earlier generation of historians. It turns out to be true, in some cases, that our more intimate relationship with domesticates became an important microbial bridge: measles, for instance, is a cattle disease that leapt to us (though not until the later Roman period, in fact). At other times, germ genealogy—like human genealogy—has been full of surprises. Tuberculosis, for instance, is the ancestor of bovine TB: we made the cows sick, and not *vice versa*. But the conceptual revolution goes deep, and its most startling discovery is the sheer ongoing dynamism and sinister creativity of evolution itself.[9]

The first humans lived in a very different landscape of germs, but there were some familiar adversaries. Certain families of viruses, like the Picornavirus family that includes pesky but dangerous enteroviruses and rhinoviruses (aka the common cold), are diverse, globally distributed, and common to a wide range of vertebrates, meaning that they have been with us before we *were* us. Other microbes that could sustain themselves in the environment or in animal reservoirs without depending on humans did not need to wait for civilization to cause us exquisite harm. African trypanosomiasis, or sleeping sickness, is a vector-borne disease transmitted by the tsetse fly and has been a scourge for humans from prehistoric times to present. And even rather limited human numbers could sustain chronic infectious diseases. Yaws, a tropical infection related to the bacterium that causes syphilis, is very ancient. Ongoing genetic study promises to shed new light across the landscape of disease faced down by our Paleolithic forebears.[10]

There must also be irretrievable chapters of the story, comprised of fleeting, explosive dead ends. As long as humans were dispersed in slow-moving bands thinly scattered across the horizon, then acute, lethal diseases burned themselves out; the pathogen would have infected so many, so quickly that the susceptible population crashed, before the germ could spill into other human groups. So, in addition to accumulating the low-virulence bugs that still cause us misery, our hunting and foraging ancestors would have been assailed by vicious evolutionary novelties from wild animal hosts—that quickly went extinct or retreated to nature. But, on the whole, Paleolithic people enjoyed a friendlier disease ecology.[11]

The Neolithic Revolution remains a decisive transition. It led to sedentary lifestyles, more monotonous diets, denser settlements, landscape transformation, and novel technologies of travel and communication. All of these bear consequences for both microbial ecology as well as the structure and distribution of human populations. In some instances, the consequences must have been almost immediate; diseases that had long been in the background readily thrived under new circumstances. Sanitation and density were the fundamental problems of living in town, and so we should look to the humble but effective dysenteries, typhoid and paratyphoid fevers, rhinoviruses, and other food- and fecal-borne parasites as the agents of mortality in ancient cities from the beginning of civilized history. The scourges of early urban living were not the charismatic great killers, but banal, workaday diarrheas, fevers, and colds.

Regardless of its continuing importance, the Neolithic Revolution no longer looks like the big bang in the history of infectious disease. The rise of agriculture has been demoted from its privileged place because we no longer need a singular moment when humanity drew into fatally closer contact with a more or less stationary background of potentially lethal germs. The experience of the twentieth century has been a harsh teacher: emerging infectious diseases are a constant menace. Farm animals are only a small part of the biological brew from which new pathogens emerge. The continuing power of the wild to generate new adversaries is evident in the roll call of recent scourges like Zika, Ebola, and AIDS. Nature, in short, is full of wild germ reservoirs and potential new adversaries, and genetic mutation is constantly spinning off dangerous molecular experiments. These treacherous evolutionary experiments are not evenly or randomly distributed around the globe. Even today, the burden of infectious disease lies heavily across the tropics. It has ever been thus. The latitudinal species gradient is the most widely observed pattern of biodiversity on the planet, and it is hardly limited to microorganisms. In the lower latitudes, spared the chill erasure of repeated ice ages, the evolutionary clock has simply been running longer. Moreover, there is just more energy arriving from the sun – and thus more life and greater complexity. The biogeography of infectious disease, then, does not follow the spatial distribution of plant and animal domestication; rather, it obeys the deeper principles of geographical ecology. As we will see, it appears that two of the three great Roman pandemics were imports from southern climes; the third – the bubonic plague – was probably a creature of the steppe, native to wild rodents. Infectious diseases can emerge almost anywhere, but the dice are loaded against certain parts of the globe.[12]

The crucial interface between humanity and new diseases is not the farmyard, but the entire array of birds, mammals, and other creatures who incubate the next potential human pathogen. So the sheer growth of human numbers, and the interconnection of once disparate human groups, has been a feeding frenzy for germs with the tools to infect humans. By colonizing almost every corner of the planet, we have widened the interface between ourselves and the zone of evolutionary experiment; by multiplying prolifically into the billions, we have improved the prospects of microbes looking to make a career as acute, lethal germs. And the connections we have progressively built between human societies not only link old germ pools, but more profoundly they have turned separate groups into a metapopulation for roving killers to explore. The main drama of disease history has been the constant emergence of untried germs from wild hosts, finding human groups linked in ever-larger pacts of mutually assured infection.[13]

Ecology and evolution drive the history of infectious diseases. The deep history of human infectious disease has been propelled not by the unintended side effects of domestication, per se, but rather by the explosive growth in the size and complexity of those populations that harnessed agriculture and nomadism, and subsequently the linkage of those populations with each other and with parts of the globe that are hot zones of evolutionary ferment. This picture is still murky but sharpening quickly. The genomic evidence increasingly points not to the *early* Neolithic but rather the more recent millennia as the scene of the real action. The Bronze Age, with its metal technologies and webs of connectivity, may turn out to have been more biologically volatile and interesting than we had imagined: plague has just been found in archaeological samples from across central Eurasia. The Iron Age, leading right down to the dawn of the classical world, looks to have seen major evolutionary moments in the history of major diseases, like tuberculosis.[14]

The history of disease and human civilization is a story full of paradox and unintended consequences.

Disease, Health, and Mortality in the Empire

The city of Rome was a wonder in its own time. A Talmudic passage captures the sense of immensity the capital could inspire in its visitors: "The great city of Rome has 365 streets, and in each street there are 365 palaces. Each palace has 365 stories, and each story contains enough to feed the whole world." The entire empire was an object of awe. "The power of Rome is invincible in all parts of the habitable earth." But the grandeur that was Rome may have been as much a boon to its invisible inhabitants as its human creators.[15]

The Roman Empire built a disease ecology whose ramifications its creators could not have begun to imagine. The empire nurtured urban concentrations whose density had never been seen before and would not be seen for centuries thereafter. The empire facilitated movement and connectivity within its unusually wide and diverse geographical regions. The scale of environmental transformation carried out under Roman rule represented the greatest surge of ecological change between the Neolithic and Industrial Revolutions. The commercial networks binding the Romans to peoples beyond the frontier, especially in Africa and Asia, appear stronger than we had ever imagined. And outside of human control, after a period of stability during the Roman Climate Optimum, a phase of raucous climate disorganization began in the later second century.

The passage of time has rendered the microscopic interlopers who stalked the Roman Empire nearly as invisible to us today as they were to the ancients, who remained oblivious of germs to the last. Only by indirect means of approach can we hope to salvage some sense of the disease regime and health profile of the denizens of the Roman Empire. It is a picture full of holes. And it is misleading ever to speak of *the* Roman disease ecology, in the singular. As we will see, the empire itself was a force of microbial unification, and the Roman period was a consequential phase of disease history, but within the sprawling geography of the empire were countless local germ ecologies, shades and variations of environmental context that made all the difference on a small scale. As we zoom in and out on the Romans, we cannot do justice to an empire whose germ ecosystem was more like a variegated and exquisitely irregular wetland landscape, than a single homogenous pool.

The ultimate measure of a society's health is average life expectancy. Life expectancy at birth has been the holy grail of Roman historical demography, and like the grail, in this quest the prize has always receded just beyond the horizon. We are still not sure how long the Romans lived. Our ignorance begins with the weighty question of infant mortality. The Romans seem to have weaned infants perilously early, depriving them of maternal immunities and exposing them to infectious agents in food and water. Up to 30 percent of live births in the Roman Empire may not have survived the first treacherous year, so that any claims about "average" life expectancy are unduly shaped by this highly uncertain beginning.[16]

The most promising avenue of approach has been found among the debris of imperial taxation, in the form of census returns preserved on papyri from one province, Egypt. These documents furnish a profile of the registered population's age distribution. These distributions, in turn, can be fitted to abstract mortality schedules known as Model Life Tables. Using this approach, it has been suggested that life expectancy at birth (e_0) was 27.3 for females and 26.2 for males in Roman Egypt. Of course, we cannot precisely gauge how effective the Roman state was in capturing the entire population, and some undercounting is certain. More unfortunately, the Model Life Tables are based on recent populations and do not match the conditions of life in the Roman world exactly. So, the census papyri are more suggestive than conclusive. In the end it has seemed safest to say that life expectancy at birth in the Roman Empire was between twenty and thirty years, and probably in the middle of that range, in this corner of the empire.[17]

One of the subpopulations that we know best—Roman emperors—also died on a schedule that suggests a harsh mortality regime. This small but

Figure 3.1. Gold Coin (Aureus) Celebrating the Fertility of the Empress: Fecunditas Augustae (American Numismatic Society)

revealing sample shows that Rome's rulers experienced the same unfriendly lifespans as their humblest subjects. "The potential benefits of ample nutrition were more than offset by constant exposure to an aggressive germ community." The rich might be buffered by plenteous food, more commodious habitats, and above all the ability to retreat to the countryside in the lethal summer months. But these protections proved feeble, as the private life of the emperor Marcus Aurelius poignantly reminds us. He and his wife Faustina were engaged in AD 138, when Faustina was eight, Marcus seventeen. In April of AD 145, when she was fifteen, they wed. Over the next twenty-five years, Faustina bore at least fourteen children. Only two—one girl and one boy—are certain to have survived their parents. In the letters of Marcus, we catch glimpses of the fevers and diarrheas that laid low so many little scions of the imperial line, and a Stoic father tried by misfortunes. It is little wonder that, as Galen's reputation soared, Marcus called upon him to be the personal physician of his son, Commodus.[18]

So much has been extracted from the written record. Where the documentary evidence abandons us, bones are starting to come to the rescue. Skeletons tell stories. Spines and joints can attest to debilitating chronic illnesses or the grinding stress of hard labor. The vaults of skulls and the orbits of the eyes can preserve hard evidence for a condition called porotic hyperostosis, a marker of physiological stress. Chemical analysis of stable isotopes can trace patterns of diet and migration. Teeth are a record of diet, nutrition, and health. They are permanently corroded by the monotony of carbohydrates, and the striations in their enamel can preserve memories of stress during developmental periods. In short, the biological burden borne by the inhabitants of the Roman Empire is still written on their bones.[19]

Like the documentary record, the skeletal record is full of uncertainty and hidden bias. But with care these risks can be mitigated, and the great promise of bioarchaeology lies in the sheer volume and distribution of the osteological remains from the Roman Empire. Sadly, science has not yet fully exploited the potential of the skeletal record in Roman archaeology. Old obstacles are only starting to be overcome. Inadequately standardized methodologies, restricted sharing of data and access to material, and strong interobserver differences have limited the conclusions that can be drawn. But there is some excellent work underway and thankfully more and more of it, not least from the island province of Britain.[20]

Maybe the most intriguing facet of the skeletal evidence lies simply in the length of the Roman bones we have. Height is a crude but valuable proxy for biological well-being. Achieved stature varies over time and space. Genes influence stature variation, but so do social and environmental factors that drive or impede growth. Height is a function of net nutrition: the nutritional intake of the body minus the metabolic expense of labor and disease during the developmental period. The body's growth curve is plastic but only for the first twenty years or so of life; the body can partly "catch-up" on growth after periods of deprivation or hardship, until growth ceases. Proteins are ideal building blocks of growth, so that meat consumption is a boost to achieved stature. Diet is thus primary. At the same time, infectious disease is a costly expense on the balance sheet of net nutrition. The immune system is metabolically voracious, and many diseases block the absorption of nutrients. A mother's health, too, has profound and long-term ramifications on the well-being of her offspring.[21]

In modern times, economic development has fired a global "growth spurt." Circa 1850, Dutch men were on average 164–5 cm; today they are 183 cm—the tallest in the world. In some parts of east Asia, the shift has been stunning. In 1950, Japanese men were 160 cm; today they are 173 cm. In the developed world, we are now about as tall as our genes allow, and in broad strokes, modernity has given humanity almost half a foot lift.[22]

In principle, the hundreds of thousands or more skeletons sitting in museum cabinets form a potential archive of stature history. In practice, determining height outcomes from bones has proven a daunting challenge, and we still lack a good, comprehensive study spanning different regions of the empire. Moreover, although height estimation is maybe more humanly interesting than bone length, converting bone measurements into stature introduces some pesky uncertainties. One way around these methodological

Table 3.1. Femur Lengths from Britain[a]

	Roman Femurs		*Anglo-Saxon Femurs*	
	Mean length	*Number*	*Mean length*	*Number*
Males	444.0	290	464.8	155
Females	412.9	231	429.22	130

Note: [a] Data of Gowland and Walther forthcoming.

challenges is to consider both height estimates *and* raw femur measurements. While femurs do not respond as sensitively as other bones to stress, they preserve well and are easy to measure.[23]

In Britain, the Roman conquest was a health catastrophe, while the fall of the empire was a biological blessing. The inhabitants of Roman Britain were diminutive, probably ca. 164 cm (5′ 4½″) on average for adult males, 154 cm (5′ ½″) for females. The best study now shows that average femur length in Roman Britain was 444 mm for men and 413 mm for women; in post-Roman Britain, it was 465 mm for men and 429 mm for women. Undoubtedly, early medieval people would have looked down on their Roman predecessors.[24]

In Italy, the entire Roman period presents a valley between Iron Age and early medieval peaks of stature. There is one meta-study that reports quite healthy stature outcomes for the Romans, but it is problematic. The underlying samples do not inspire confidence. More importantly, if we add even crude chronological dimensions to the same data and update the analysis with more recent finds, it is evident that Italians in the Roman imperial period were shorter than their Iron Age and early Republican ancestors.[25]

At present, only one study of stature in Roman Italy across time inspires confidence, and it demonstrates that the pre-Roman peoples of Italy were much taller than the Romans. Average male femur length declined from 454 mm to 446 mm. For Roman women, the loss was even greater. From a pre-Roman average of 420 mm, there was decline in the Roman period to 407 mm. In the middle ages, mean stature increased again, exceeding the Iron Age baselines. Male femurs were now 456 mm, whereas female femurs returned to 420 mm on average. Moreover the distal bones of the arm and the leg—the radius and the tibia—show even more marked losses in the Roman era, 3–4 percent, about twice the degree of change evident from the femurs. The authors posit that average Italian stature under the Roman Empire was around 164 cm (5′ 4½″) for men and 152 cm (just under 5′) for women.[26]

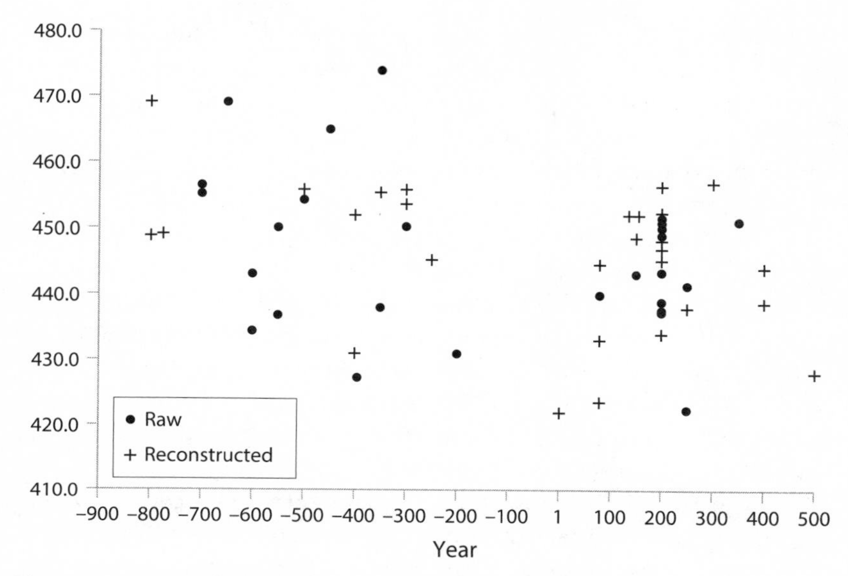

Figure 3.2. Mean Male Femur Length in Italy (mm) (see Appendix A)

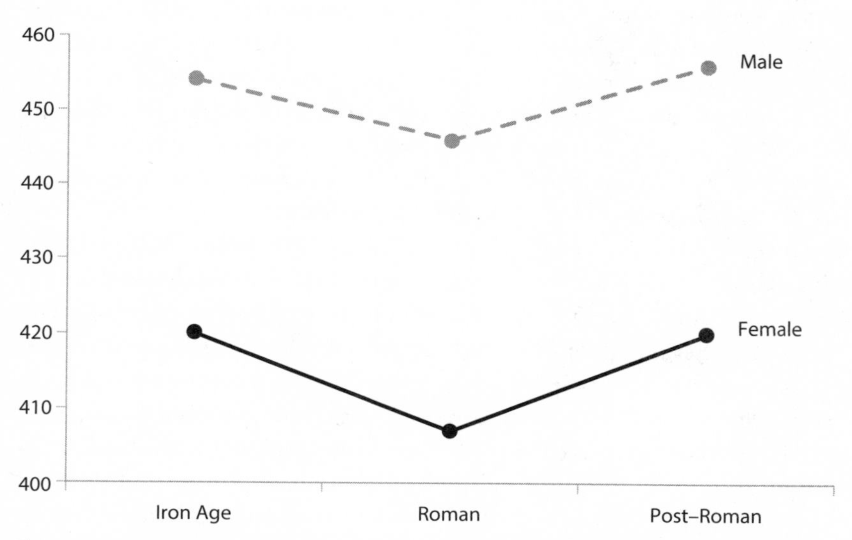

Figure 3.3. Average Femur Length in Italy (mm) (data from Giannecchini and Moggi-Cecchi 2008)

Why were the Romans short? Malnutrition would be a tidy answer, and it would be unwise to rule it out. But we should resist hastily drawing a line from the Romans' short stature to poor nutritional inputs and instead look to the disease burden as the culprit, at least in part. There are reasons to doubt that diet was the principal factor. For long, we relied on indecisive literary evidence from the upper classes to make educated inferences about Roman diet. Now, the chemical signatures of what the Romans ate are being traced in their bones. Stable isotopes of carbon and nitrogen occur naturally in the environment; due to the weight of their extra neutrons, heavy isotopes are cycled through nature down slightly different tracks. Nitrogen isotopes, for example, are a telltale sign of a creature's place in the food chain. The bone tissue of creatures toward the top of the pyramid is relatively enriched with heavy isotopes. Stable isotope ratios thus reflect the origins of the nutrients used to make human bones.[27]

Again we must be cautious, because the evidence is limited, and already it screams that there was no such thing as a "Roman diet"—only the aggregation of socially and regionally variable diets. But it turns out that many Romans, even poor ones, did not live on bread alone. Even bodies buried in the humblest way have shown some dietary enrichment from animal and especially marine protein. Most of the studies have focused on populations in and around the city of Rome itself, but evidence from Britain too suggests a diet that included meat and small amounts of seafood. The chemical makeup of human bones is consistent with the large number of animal bones found in Roman archaeological contexts and inferred as evidence for meat consumption. No doubt in a highly stratified society, many Romans hovered on the edges of subsistence. Certainly, there is important work still to be done, but so far, the bones do not obviously implicate malnutrition in the diminutive stature outcomes of the Romans.[28]

The conclusions to be drawn from Roman teeth point in the same direction, suggesting that disease played an outsized role in shaping Roman health. One major study compared teeth from two imperial era sites with an early medieval population. Neither epoch was an advertisement for oral hygiene, but their dental pathologies were unhappy in different ways. The early medieval teeth suffered more dental caries, lesions caused by an unbalanced diet of carbohydrates. The Roman teeth, by contrast, showed greater incidence of a growth defect called linear enamel hypoplasia (LEH). LEH occurs in childhood when the body is so stressed that the production of enamel is interrupted. Malnutrition or infectious disease—or the synergies between them—are to blame. Another study of seventy-seven rural laborers from an imperial-age cemetery in the Roman suburbs showed a very high frequency of enamel growth defects, but few other oral pathologies. This population was eating a diet in which meat was important and refined carbohydrates were marginal. More work is to be done, but so far the Roman dental record points to a population under ghastly physiological stress, with the disease burden a primary factor.[29]

It is worth drawing attention to the similar conclusions drawn from an unusually valuable sequence of burials from Roman Dorset, in southwest England. The Romans, down to around the second century, incinerated their dead. Hence there are often continuity gaps precisely in the late republican and early imperial strata. But an uninterrupted sequence of inhumation burials from Dorset gives us the rare chance to watch the empire come and go. The empire's arrival led to the hasty construction of the first town, in Roman style, with baths and aqueducts, drains, heating systems, and latrines. Despite the amenities, a reasonable answer to the question "What have the Romans ever done for us?" might have been, "Got us sick." Mortality rates went up. The very young and very old suffered most—precisely the segments with the weakest immunities. Men fared worse than women—and it bears noting that women have stronger natural immunity than men. Urbanization, social stratification, and mobility rendered the population more vulnerable to infectious disease. Similar patterns have been traced on the other side of England, at York, where the coming of the empire brought a more insalubrious environment, causing both a narrowing of the nutritional spectrum and greater exposure to infectious disease. Roman civilization was hazardous to provincial health.[30]

All of this evidence leads us to the conclusion that, not for the last time in history, a precocious leap forward in social development brought biological reversals. As the Golden Age Dutch were attaining the highest levels of income the world had ever known, their mean height stagnated. The lunge forward of the Industrial Revolution famously exacerbated health conditions and lowered average stature. In the United States, this cruel undertow of modernization is known as the Antebellum Paradox. Before income growth and public health could offset overcrowding and grinding labor regimes, men and women grew up to be shorter than their parents and grandparents.

In modernizing Britain, rickets, rheumatic fever, respiratory ailments, and diarrheas tragically stunted the growth of millions of people caught in the first violent waves of industrialization. The bodies of children bore the brunt of this brutality. Malthus had a crude notion of the effects of urban disease ecology. "There certainly seems to be something in great towns, and even moderate towns, peculiarly unfavourable to the early stages of life; and the part of the community on which the mortality principally falls seems to indicate that it arises more from the closeness and foulness of the air, which may be supposed to be unfavourable to the tender lungs of children."[31]

The Romans experienced their own version of this paradox, but neither breakaway technological growth nor novel institutions of public health came to the rescue. The Romans were helplessly caught in the vice grip of their own progress, with its confounding ecological repercussions. All signs point to an empire whose people were groaning under the weight of an exceptionally burdensome pathogen load, despite and in some ways because of the success of the Roman economy.

We wish we possessed nothing so much as "cause of death" statistics, such as began to appear at the end of the middle ages, to fill out this bleak picture of Roman health. We lack any direct indications of what microbial agents laid low the Romans, in what proportions. But we can try to imagine some of the specific health environments of the Roman Empire, and look for oblique clues to identify some of the most active agents of death in Roman times.

In the first place, the Romans were victims of their own strong preference for city life. Cities, thanks to their close quarters and their systems of supply, sewage, and sanitation, have distinct disease ecologies. Roman towns were magnets for migrants, seeking subsistence, opportunity, or excitement, and not a few were involuntarily transported for sale in the great slave markets that were standard in the Roman world. Without experience of local germ pools, migrants were immunologically vulnerable and surely perished in disproportionate numbers. Romans living in towns were victims of the urban graveyard effect, the differentially high mortality of cities. The very progress of development encouraged the growth of towns, but these in turn were health hazards.[32]

We should admit, however, that even in town the Romans had some intriguing forces working in their favor. Roman civil engineers brought fresh water gushing into the city. The aqueducts reaching from the cities into the uplands in so many parts of the empire delivered regular supplies of clean water, possibly the single most important health resource of all. The steady currents of water were not just for drinking and bathing—they also helped flush urban sewers. And the public toilets of the Roman Empire still impress. In the early empire, the emperors built magnificent public loos—they have been called a "hallmark" of Romanization—multi-seaters, some serving 50 or even 100 clients simultaneously. They consisted of tightly-spaced black holes along benches of marble, no lids. The most common decorative motif was the goddess *Fortuna*—a contemplative theme. In all, the impressive remains of aqueducts, sewers, and toilets have suggested to some modern historians that the Romans may have been spared the most squalid effects of premodern urbanism.[33]

There are strong grounds to be reserved in our optimism. The sewers of Rome, while massive, have not been held in esteem by modern experts. They were more storm culverts than waste disposal systems. The ingenuity of the monumental public toilet seems to have been summoned for the sake of imperial or civic vanity, rather than practical motives of hygiene. The more important system of *private* waste disposal registers as a befuddling missed opportunity. Domestic toilets were often not linked to sewer lines: the gaseous backflow, risk of flooding, and invitation to vermin outweighed the benefits. For the wealthy hilltop mansions, this situation may have been tolerable. For the rest of humanity, it meant being surrounded by the effluvia of the crowd. Internal cesspit toilets dominated in Roman houses, and chamber pots never went out of style. Private toilets were often built next to the kitchen. The Romans apparently used (and re-used) sponge sticks for the purposes we employ toilet paper. In the words of one classicist, "The hygienic implications of using such an implement are again at best dubious."

Every day, it has been estimated, the city of Rome alone produced over 100,000 pounds of human excrement, clumsily and incompletely removed from the city, to say nothing of the contribution from numberless animal residents. There was also a rollicking trade in human waste, valuable as fertilizer and fulling solution. It is altogether unsurprising to learn from an important recent study of Roman fecal remains that the Romans, in the imperial capital and beyond, were woefully infested by the telltale parasites of an unhygienic society, especially roundworm and tapeworm. Indeed, the spread of the empire only aggravated the incidence of intestinal worms. The environmental problems of urban life overwhelmed the inhabitants of the empire, just where we might have expected them to have pushed back the invisible tide.[34]

One surprising place that the Roman disease ecology has left its imprint, as well as clues to its nature, is in the seasonal patterns of death. In contemporary societies where infectious disease has been largely subdued, death comes in all seasons. But where infectious disease was a major cause of death, the grim reaper had uneven rhythms. Deadly microbes are environmentally sensitive. So are the vectors, like fleas or mosquitoes, that carry infectious agents. The patterns of mortality across the year can reveal the fingerprint of germs that are seasonally picky. Seasonal mortality is a crude forensic device. In the case of the Roman Empire, we are provisioned with a trove of data about seasonal mortality. When pagans died, they recorded the

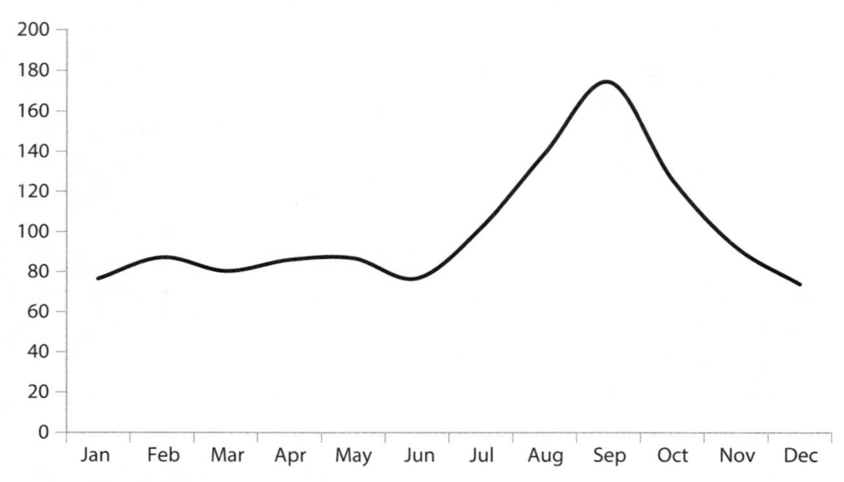

Figure 3.4. Seasonal Mortality in Ancient Rome

length of their earthly life on their tombstones. When Christians died, they recorded the *date* of death, considered the day of their rebirth into the afterlife. It inadvertently preserves a record of death's calendar in ancient Rome.

The Christian epigraphs from late antique (ca. AD 250–550) Rome preserve over 5,000 dates of death, a sample biased toward those who died between ten and forty years of age.[35]

The late summer and early fall were a time of surging death. The Romans knew that the dog days of summer were perilous. The sheer amplitude of variation is historically exceptional, suggesting an abnormally lethal disease pool in ancient Rome—an impression that is further underscored by the bias of our sample toward young adults, the hardiest element of a population. There were no meaningful differences between males and females in the seasonality of death, but there was age discrimination. Children, adults, and the elderly all succumbed heavily in the late summer and early fall, but the elderly suffered a distinct secondary peak in the winter, due to the vulnerability of older adults to winter respiratory infections. Most surprisingly, adults from 15–49 exhibit the greatest amplitude of all, with a massive spike centered on September. Just perhaps, many of these deceased were immigrants who lacked acquired immunities to the local disease pool and found the city an environment full of unfamiliar biological adversaries.[36]

We have already seen that the Roman emperors departed this earth on the same schedule as their subjects. The seasonal mortality data also argue that the Roman germ pool was no respecter of persons. Judging from the

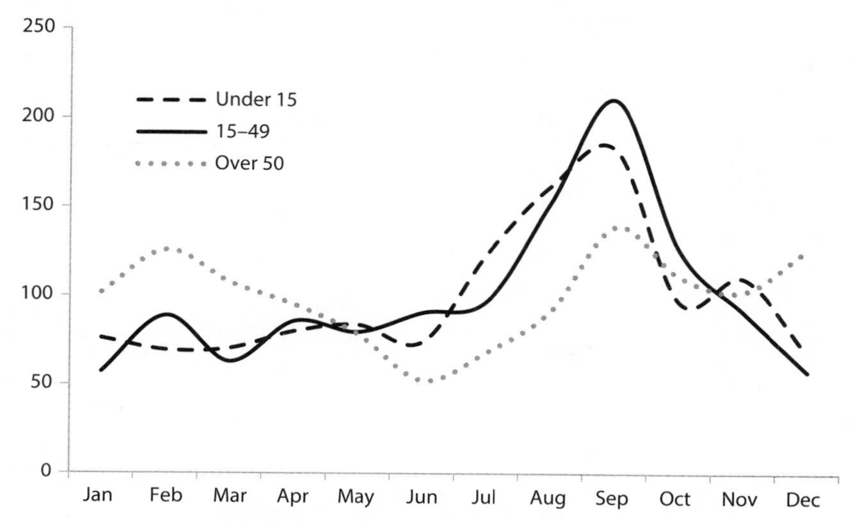

Figure 3.5. Seasonal Mortality in Rome, by Age

inscriptions carved into the elegant sarcophagi that held the mortal remains of the rich and famous, the summer-autumn wave of death was lethal for all. A sample drawn from the least impressive scratches, etched into the walls of the catacombs, evinces a similar pattern of death for these middling or lower-class denizens. This seasonal pattern was in fact noted by the doctor Galen, who certainly catered to the upper classes and who observed that autumn was lethal. He thought that the wild daily fluxes of autumn weather, with hot peaks and cold nights, tossed the body out of balance. "This irregularity of mixture is the factor that makes autumn most conducive to illness." In sum, the advantages of superior nutrition or more salubrious dwellings did not in the end keep the elites safe from the germ ecology of the city. They went the way of all flesh in the same manner as the least among them.[37]

The mortality crest starting in summer is a signature of stomach and intestinal diseases contracted through food and water. A range of acute diarrheal diseases must have been hyperendemic in Rome. All signs point to bacillary dysenteries and typhoid fevers. Bacillary dysenteries, especially *Shigellosis*, are spread via fecal-oral pathways in contaminated food and water. Flies can transport the bacteria, and inadequate personal hygiene aggravates its spread. *Shigellosis* comes on suddenly, bringing debilitating fever and bloody stool to its victims. Typhoid fevers like *Salmonella typhi* must also have been a profound menace. *Salmonella* bacteria are widespread in

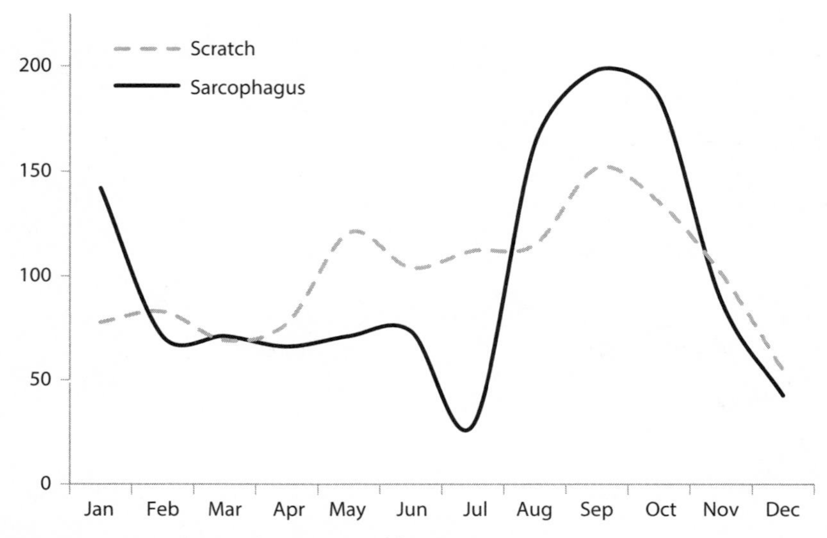

Figure 3.6. Seasonal Mortality in Rome, Class Proxy

nature and lurk in a range of animal reservoirs, although *S. typhi* is specific to humans. It too is spread via fecal-oral transmission, especially in water. Its onset is more insidious than *Shigellosis* but its outcomes were lethal in a society without medical controls. The interplay of summer heat and sanitation challenges instigated the mortality pulse in the hot months. Mighty Rome was overwhelmed by the most humble germs; however incongruous it might seem, diarrhea was probably the most deadly force in the empire.[38]

The death wave in ancient Rome continued into the fall. Here is a clue that puts us on the trail of a devastating killer stalking the Roman Mediterranean: malaria. Malaria is caused by the invasion of *Plasmodium* protozoa, single-cell parasites with complex life-cycles transmitted to humans by the *Anopheles* mosquito. Different species of *Plasmodium* protozoa may infect humans. *P. malariae* and *P. vivax* were ever-present dangers in Rome, but the story of mortality in the Roman world was unduly influenced by the most dangerous representative of the genus, *P. falciparum*, a virulent pathogen that was the agent of what the ancients called "semitertian fever," after the undulating pattern of fevers marked by acute intensification every other day. Even today, malaria causes high rates of morbidity and mortality, striking with peculiar violence the young or adults without any previous exposure. Where it is endemic, it has "an awesome power as a determinant of demographic patterns." Malaria was a pall over the city of Rome and core parts of its empire.[39]

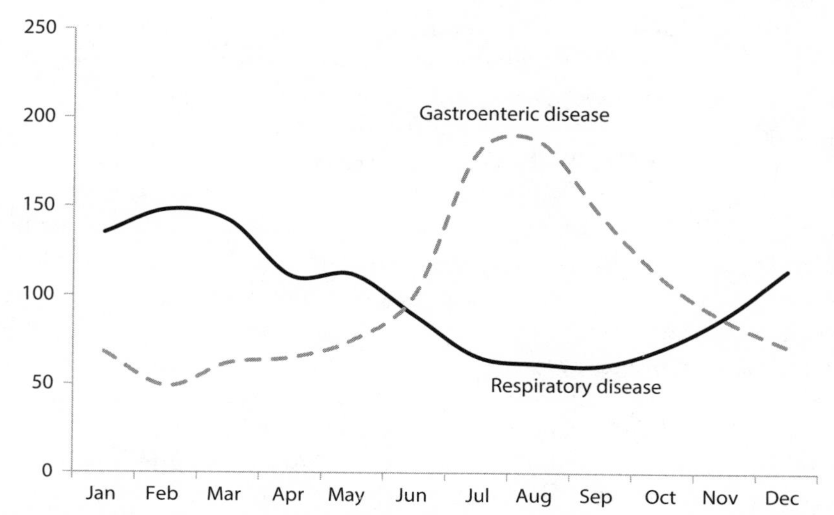

Figure 3.7. Cause Specific Mortality, Italian Towns, 1881–82 (data from Ferrari and Livi Bacci 1985)

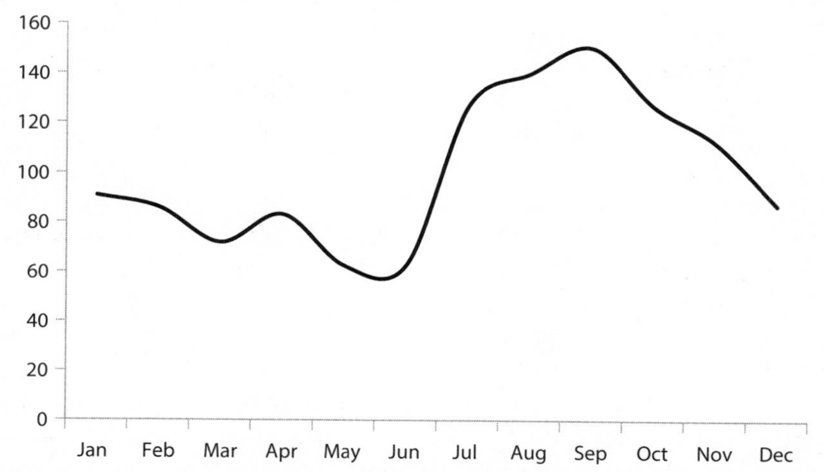

Figure 3.8. Seasonal Deaths by Malaria in Rome, 1874–76 (data from Rey and Sormani 1878)

The very name *malaria* means "bad air," and malaria is the ultimate ecological disease. *Plasmodium* is an ancient adversary, with origins in the African tropics, but genomic evidence now dramatically reveals that *P. falciparum* is a recent spinoff of a gorilla pathogen, perhaps less than 10,000 years old. By the time the Romans built their empire, malaria was no stranger. But the specific configurations of Rome's imperial ecology let malaria flourish. Its

DNA has just been sequenced from archaeological samples in two early imperial sites in southern Italy, providing solid confirmation of the pathogen's presence. Malaria was a wetland problem, endemic in central and southern Italy and similar regions. And in the very heart of the empire, in the city of Rome, documentary and written evidence lets us trace in unusual detail the ecology and effects of a specific endemic pathogen.[40]

Thanks to the work of Robert Sallares, we have a detailed biography of malaria and its special relationship with Rome. The ancient medical sources are an invaluable witness to the prevalence of malaria in the imperial capital. The greatest witness of all was none other than Galen. His careful work on intermittent fevers reflects the prominence of malarial diseases in second-century Rome. "We no longer need the word of Hippocrates or anyone else as witness that there is such a [semitertian] fever, since it is right in our sight every day, and especially in Rome. Just as other diseases are typical in other places, this evil abounds in this city." It occurred "most of all in Rome," whose inhabitants, Galen observed, were "most intimately familiar" with this malignant fever.[41]

The spatial dynamics of malaria are defined by the geographical contours of mosquito breeding. The Romans themselves were aware that swampy wetlands were pestilential places. Roman agricultural writers and architects alike have wise words about where and how to build houses that might deflect the deadly exhalations of the marshes. Rome itself was noted for its bad air. Its stagnant waters were spawning grounds for *Anopheles* eggs. The incidence of malaria is an eccentrically local question, and it is unsurprising that the patterns of seasonal mortality in other parts of the empire, also derived from ancient Christian tombstones, varied decidedly at times from the pattern in the capital city. In northern Italy, mortality peaked in summer without an autumn tail, while in southern Italy, where *Anopheles* thrives, the autumn spike points to the deadly work of *P. falciparum*.[42]

By its nature, Roman civilization seemed to unlock the pestilential potential of the landscape. The expansion of agriculture brought civilization deeper into habitats friendly to the mosquito. Deforestation facilitated the pooling of water and turned the forbidding forest into fields where mosquitoes more easily multiplied. Roman roads—like the Via Appia, paved by Trajan and cutting directly through the malarial Pontine marshes—"played a significant role in creating favourable new breeding habitats for *Anopheles* mosquitoes." Urban gardens and waterworks brought mosquitoes and humans into unbearably close quarters. The Romans were environmental engineers extraordinaire, and they knew it. "If anyone were to make a careful calculation of the abundance of water for the public, in baths, in pools, in canals, in urban houses, in gardens, and in suburban villas, and along the distance traveled the arches built, the mountains sliced, the valleys leveled, he would admit that nothing more marvelous exists in all the world." But the built environment catered to the proliferation of mosquitoes. The Roman Empire was an unintended experiment in mosquito breeding.[43]

Malaria is not just one disease among others. Because malaria is insidiously eager to team with other pathogens, its sinister reach went far beyond the dangers of primary infection. The effects of malaria include severe malnutrition, leaving its victims vulnerable to other infections. Galen knew the deadly quotidian form of malaria that particularly struck children; for those who survived it, the effects could last for decades in physical stunting and weakened immunity. Malaria clears the path for vitamin-deficiency disorders like rickets, and it can increase susceptibility to respiratory infections such as tuberculosis. Malarial environments seemed to accelerate the corruption of all life. "Why do men grow old slowly in places with fresh and pure air, while those in hollow and marshy places grow old rapidly?" But it could always kill quickly, too, and it is probable that immigrants were particularly vulnerable. Many wayfarers have succumbed to malaria in Rome. Centuries after Galen, the mother of Saint Augustine would contract malaria at Ostia, the port of Rome, and die after nine days of suffering.[44]

The frontiers of malaria are sensitive to both short and long-term changes in the climate. The ambient temperature influences the formation of the *Plasmodium* spores inside the mosquito, and the watery breeding habitats of *Anopheles* vary with moisture. The ancients were sensitive to these environmental influences. A text of the Roman period observed that a damp spring followed by a dry summer spelled a deadly autumn. The humid mid-Holocene was hospitable for the breeding preferences of the mosquito vector, and it may have been in these early millennia of fledgling civilization that malaria migrated into the circum-Mediterranean. In the territories ruled by Rome, at the border between the temperate latitudes and the subtropics, the epidemiology of malaria is exquisitely sensitive to climate fluctuations. But we must consider a malign possibility. If the RCO was indeed an especially wet period, it was a boon for the mosquitoes and the parasites they ferried.[45]

Malaria was endemic in Rome and other core regions. The right environmental motions could always flick the hair trigger that set the disease escalating toward *epidemic* heights. Galen knew the conventional wisdom: "when the entire year becomes wet or hot, there necessarily occurs a very great plague." In the early modern period, malaria flared into an epidemic every five to eight years in Rome and its environs. Malaria was surely one of the main drivers of epidemic mortality in ancient Rome. Death was not a steady drip in Rome. It came in seasons, and it came violently in epidemic years. The oscillations could be wild. The ancients were intimately familiar with the chaos of epidemic mortality and watched warily for signs of incipient pestilence. "If one person is sick it hardly throws the household into a panic, but where the creeping deaths show there is a pestilence, there is a din in the city and people flee, and they raise their fists at the gods themselves."[46]

The ancient sources counted the coming and going of deadly years. Modern historians have probably not taken enough notice of the fact that most ancient epidemics came from within and were regionally confined. The catalogue of all known plague years between 50 BC and the Antonine Plague is instructive. The list is not long, probably because plagues were such unexceptional events that many of them went without notice. The generic terminology of the Greek and Latin vocabulary for "plagues," *loimoi* and *lues*, betrays no understanding of the differential pathogenic origins of mortality events. Plagues were caused by miasma, polluted air, an angry god, or some indecipherable combination of divine wrath and environmental disturbance. Most of these could have been malaria, but it is hard to know; even the broader regional events could be pulses coordinated by climatic vibrations.[47]

For the period down to the coming of the pandemic, most of the plagues in the ancient histories were probably amplifications from within the seething pool of endemic diseases. Malaria and bacillary dysentery are prone to undulating variability across the years. It is telling that the naturalist Pliny the Elder believed the elderly were spared the onslaught of plague: it implies that their acquired immunities from previous infections buffered them when native diseases flashed into mass mortalities. The close association of epidemic years with short-term environmental disturbance, like floods, also betrays the guilt of the local disease pool, spun up by climate fluctuation into a surge of mortality. The Roman world was repeatedly the victim of its own boiling broth of microbes. It was not strafed from without by mobile, exotic pathogens.[48]

Dependent on vector-borne and environmental transmission, epidemics arising from diseases like malaria or dysentery were spatially bound. The webs of connectivity binding together the disparate regions of the empire were ready to facilitate microbial transport and transfer, but the communicable diseases that seem to have taken advantage first were not acute infections. Instead, chronic infections, like tuberculosis and leprosy, seized the opportunities provided by the empire's circulatory system. In fact, a combination

Table 3.2. All Known Epidemics, 50 BC–AD 165

Year	*Source*	*Event*
43 BC	Cassius Dio 45.17.8	Bad pestilence over "nearly all Italy"; certainly a dramatic year in climate history following massive volcanic eruption of 44 BC; Dio associates with Tiber flood; possibly malaria following floods.
23 BC	Cassius Dio 53.33.4	Unhealthy year in Rome; Tiber floods.
22 BC	Cassius Dio 54.1.3	Pestilence throughout Italy, associated with Tiber flood; Dio vaguely speculates "I suppose that the same thing happened in regions beyond too"; the context is that truly awful events led the Roman senate to believe they needed Augustus as consul or dictator.
65 AD	Tacitus, Ann. 16.13 Suetonius, Nero 39 Orosius, 7.7.10–11	Storm in Italy; terrible autumn plague in Rome swept away 30,000.
77 AD	Orosius 7.9	Plague in Rome in Vespasian's 9th year.
79/80 AD	Suetonius, Titus 8.3 Epit. de Caes. 10.13 Jerome, Chron. ann. 65 Cassius Dio 66.23.5	Eruption of Vesuvius scattered ashes far and wide; unprecedented plague in Rome, 10,000 died daily.
90 AD	Cassius Dio 67.11.6	People died from being smeared with needles, not only in Rome but virtually the whole world. (This obscure notice has defied clear understanding, and Dio does not actually claim there was an epidemic.)
117–138 AD	Hist. Aug. Hadrian 21.5	Famines, pestilence, earthquakes under Hadrian.
ca. 148 AD	Galen, Anat. Admin. 1.2 Galen, Ven. Art. Dissect. 7	Epidemic of "anthrax" in "many cities of Asia."

of textual, archaeological, and genomic evidence suggests that the Roman Empire played a major role in the biographies of both TB and leprosy.

TB is a devastating respiratory disease caused by the bacterium *Mycobacterium tuberculosis*. Long believed to be an ancient enemy, genomic evidence now suggests it may be as young as 5,000 years old. Transmitted directly between humans via airborne droplets, it loves dense and dirty cities. Its course can run from weeks to years, grinding down its victims with coughing and consumption. TB was a major cause of morbidity and mortality down to the twentieth century and remains a vicious global killer today. Like malaria, its presence can weigh heavily on any society where it takes hold.

TB was known to the earliest Greek medical authors, and it was certainly not a new problem in the Roman Empire. But we have recently learned that a major evolutionary moment in the history of the pathogen, leading to the most lethal modern lineages, occurred ca. 1800–3400 years ago. Those are still wide boundaries, hopefully to be refined with future work. But in the meantime, the skeletal record offers clues. Unlike most infectious diseases, TB leaves signature damage behind in its victims' bones, and thus can be tracked archaeologically. It is vanishingly rare in pre-Roman skeletons. Only one possible case has been found in Britain, for instance. Then, in the centuries of Roman dominance, TB becomes far more visible in the record. The empire has been called "a watershed moment for the spread of tuberculosis in Europe." The evolutionary histories of TB and the Roman Empire seem to have intersected in a fateful way. Possibly, the integration of far-flung towns aided the dispersal of one of history's great killers.[49]

The Roman Empire also accelerated the snail-paced diffusion of leprosy across Europe. Leprosy is a chronic infection caused by the bacteria *Mycobacterium leprae* and *M. lepromatosis*. Spread directly between humans, leprosy has a complex pathology but most characteristically destroys nerves and ravages the skin and bones, causing numbing and disfigurement, notably of the face. It is slow-acting, painful, and debilitating.

Leprosy may be truly ancient, hundreds of thousands of years old, although this remains an unsettled question. So far, the earliest known cases globally are from second-millennium BC India. It travelled from India to Egypt in the centuries just before Roman dominance, but Pliny the Elder and Plutarch in the late first and early second century both regarded it as a new disease. The physician Rufus of Ephesus was surprised that the great doctors of the past had failed to describe it. Leprosy begins to appear clearly in archaeological contexts from the Roman Empire. The DNA of *M. leprae* was recently recovered from the skeleton of a child aged four or five from a Roman imperial necropolis. The global genetic diversity of *M. leprae* shows that two of the main splits in the phylogeny of leprosy occurred around the time of the early Roman Empire. Again, genomic evidence argues for the importance of the early first millennium in the dispersal of a deadly bacterium.[50]

We should try to appreciate how circumstantial the Roman mortality regime was. The patterns of death in the empire were shaped by the needs, methods, and constraints of specific microbial organisms. These microbes had their own weapons and their own limits. Malaria was acute and lethal but bound by the geography and life-cycle of its mosquito vector. *Shigellosis* thrived in the dense, dirty cities but relied on the local pathways of the fecal-oral route. TB and leprosy, transmitted directly between humans, loved the endless vistas opened by Roman transportation networks, but they were slow migrants. The limits of these pathogens—and others, less visible in the record—were biologically self-imposed, not inherent in the disease ecology of the Roman Empire. For the right pathogen, the conditions of the empire might offer unfathomable opportunities.

In the short moral sketches of the great Plutarch—the author famous for his biographies of the illustrious Greeks and Romans—there is a set piece in which the question is posed whether there can be *new* diseases in the world. It was the kind of vaguely scientific conversation that became *au courant* among the lettered aristocracy of the Roman Empire. Plutarch has one speaker maintain that new diseases were possible. But he thought so only because there were still unexplored foods or fads that could insult the body in novel ways, such as the disconcerting fashion for hot baths. His disputant pressed the case that new diseases were in principle not even possible. The cosmos was closed and complete, and nature was not an inventor. The great doctors of the past stood in authoritative rebuttal of the idea. Then, in one of those pregnant moments of error, where the very bedrock of an ancient way of thought seems momentarily exposed to plain view, he insisted that "diseases do not have their own particular seeds." History is full of ironies, and it is a poignant one that even as Plutarch was composing this civilized disquisition, nature was distantly preparing the seeds of a new disease, an adversary uninhibited by most of the self-imposed limits that had bounded the familiar pathogens of the Roman world.[51]

The reassuring classical notion of unchanging nature would be rudely contradicted. The wild was preparing something new, something furious and vast.

The Romans and Global Networks

The naked-soled gerbil, *Gerbilliscus kempi*, is a rodent living in the belt of open savannah and dry forests that runs across Africa, in the middle lands between the Sahara desert and the wet tropics. These gerbils claim as their homeland a temperate expanse stretching from Guinea to southern Ethiopia. Many rodents serve as hosts for a genus of viruses known as the *Orthopoxviruses*. But the naked-soled gerbil is the only one known to harbor the species *Tatera poxvirus*, and this distinction makes our gerbil a rodent of unusual interest. The *Tatera poxvirus* is the closest relative of the camelpox virus. These two, in turn, are the nearest known relations of the species *Variola major*, better known as the smallpox virus.

These three species of virus emerged almost simultaneously from a common progenitor in a genetic divergence that separated them from an ancestral rodent *Orthopoxvirus*. Humans, camels, and naked-soled gerbils are each the only host of their poxvirus. Their biogeography places the evolutionary event that led to their divergence somewhere in Africa. The range of the naked-soled gerbil, in combination with the evolutionary history encoded in *Orthopoxvirus* genomes, points to Africa as the most likely birthplace of smallpox.[52]

The Roman Empire, in the centuries starting with the reign of Marcus Aurelius, was repeatedly the victim of biological events originating beyond its frontiers. The commercial links exposing Rome to the emerging infectious diseases of the world outside its borders were the most fateful constituent of the Roman disease ecology. A robust transfrontier trade was the corollary of Rome's historically precocious economic development. We are starting to appreciate the extent and vitality of the networks that ran in a great arc bending from the Red Sea to the Bay of Bengal, binding the Mediterranean to Arabia and Ethiopia, to India and the far east. While scholars long doubted the real import of this commercial network, the last decades have swung the pendulum. Archaeological excavation, the chance discovery of new documents, and a renewed appreciation for the vigor of Roman trade in general, have all opened our eyes to the real weight of the Indian Ocean exchange.

Our eyes are trained, by hindsight, to see the Atlantic as the waterway destined to bind global humanity together and fuel the surge of modern capitalism. In the first and second centuries, while the Atlantic was still an impassable barrier, the Indian Ocean looked poised to become the nexus of the globe. The coming of the Roman Empire was a catalyst. When the Romans annexed Egypt, it brought them to the borders of the Nubian kingdom of Meroe, the embryonic Axumite kingdom of Ethiopia, and the kingdoms along the eastern shore of Arabia. Augustus sailed a massive naval fleet down the Red Sea. Roman policy was assertive along the entire stretch of its southeastern border. The construction of Roman roads and canals connecting the Nile with the Red Sea was a boost to trade. The lucrative tolls charged on entering goods motivated the Romans to protect and to cultivate their mercantile networks. The protrusion of Roman power into the Red Sea has been vividly attested by the discovery of two Latin inscriptions, on the Farasan Islands off the coast of what today is the border between Saudi Arabia and Yemen. We learn that, in the very year Aristides delivered

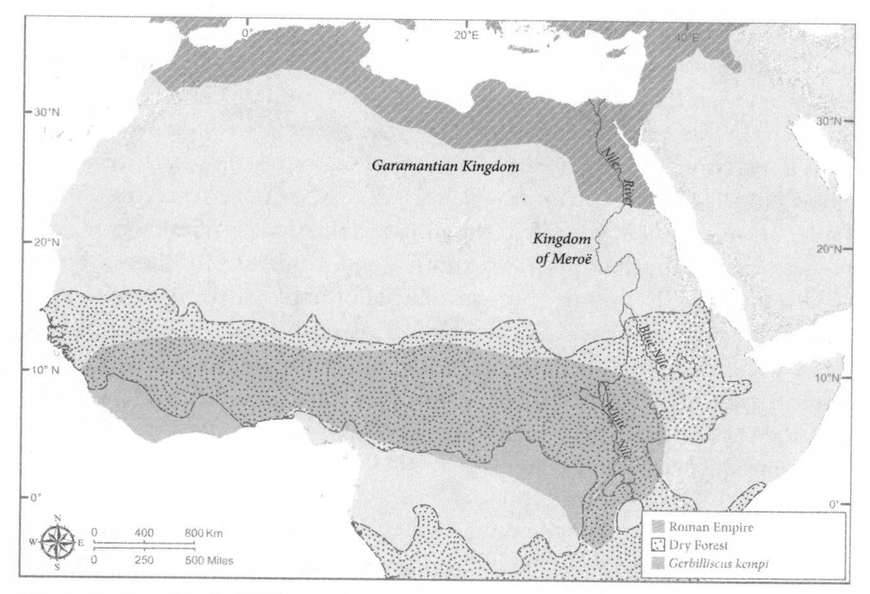

Map 7. Range of Naked-Sole Gerbil

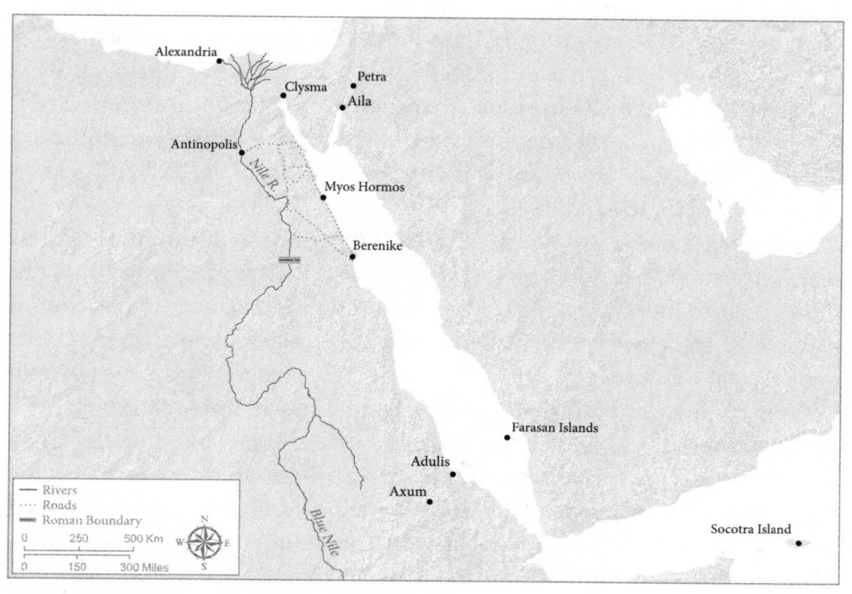

Map 8. The Romans and the Red Sea World

his oration in Rome, a detachment of the *legio II Traiana* had established a prefecture and built a fort, on an island 1000 km below the southernmost Roman port at Berenike in Egypt.[53]

Never before was the world so small. The geographer Strabo wrote that with the coming of the Romans the number of ships sailing each year from Myos Hormos to India had grown from 20 to 120 vessels per year. When Ptolemy wrote his *Geography* in Alexandria in the middle of the second century, he derived much of his information about the east from "those who are accustomed to sail to India," but with due allowance for the nature of these witnesses: "the merchant class generally . . . are only intent on their business, and have little interest in exploration, and often through their love of boasting they magnify distances." In his "Roman Oration," Aelius Aristides claimed that the cargoes from India and Yemen were so extensive that the Romans must have picked bare the orchards of these far-off lands. Aristides himself had traveled up the Nile to "Ethiopia," in search of the river's origins, all safely under the shadow of Roman power. What would have been an inconceivable adventure a few generations before had been turned into gentle tourism.[54]

Roman consumerism and the mobilization of capital were the sparks to the dry kindling of eastern trade. "Commercial exchange with India did not open up so much as explode." The commerce carried luxury goods like silk, spices, tortoise shell, ivory, gems, and exotic slaves. The *Circumnavigation of the Red Sea*, a text composed by a "businessman, not a man of letters," is a characteristic product of the age. It testifies to the commercial sophistication of the region stretching from East Africa to the Indian subcontinent. Written around AD 50 by a Greek merchant who knew the monsoon routes, it is a vivid and even chatty captain's view of the trading networks connecting the ports of Myos Hormos and Berenike, on the shores of Roman Egypt, to the far reaches of the Indian Ocean beyond.[55]

To emphasize the weight of luxuries in the trade is not to minimize its scale, diversity, or importance. A tariff listing 54 items subject to imperial duties in Alexandria suggests the range of high-value objects circulating in the eastern trade networks. Pliny the Elder estimated that the eastern trade drained 100 million *sesterces* each year from the empire. That is over 22,000 pounds of gold and about 1/6 of the empire's army budget. Pliny had a penchant for the catchy sum (with a twist of misogyny, blaming the frivolous tastes of Roman women), and his reports seemed exaggerated until a fragmentary papyrus appeared, preserving a contract between a commercial financier in Alexandria and a trader working the route from Egypt to Muziris in India. We learn that the return voyage of the ship (the "Hermapollon") carried ivory, nard, and other precious goods including some 544 tons

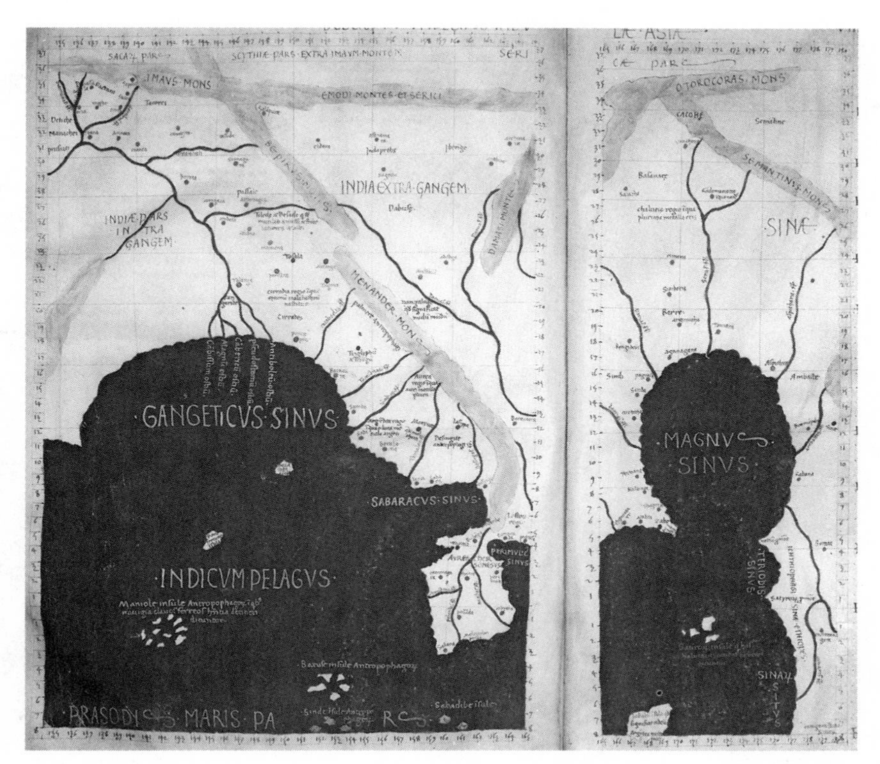

Figure 3.9. Ptolemy's World (15-Century Manuscript, Harley 7182, British Library; GRANGER)

of pepper. This single cargo was valued at 7 million *sesterces*: about the value of 23,000 tons of wheat or of 200 km^2 of Egyptian land.

Texts and documents underscore the primary importance of spices in the Indian Ocean commerce. There is no mistaking that the trade was fired by a taste for spices. The most famous Roman cookbook comes from this era, and it betrays what might seem to us a bit of overreliance on black pepper. In AD 92, the emperor Domitian built a spice quarter in Rome, at the very heart of the city where today the ruins of the Basilica of Maxentius and Constantine stand overlooking the forum. Pepper was not just an exotic luxury. A pound of it could be had for a few days' pay, and we find it on an order list for soldiers stationed on Hadrian's Wall. Not for the last time would consumer taste buds drive a global transformation with unforeseen ramifications.[56]

We are furnished with information principally from the Roman side, but we must keep in mind that indigenous sailors were agents of the trade too and that the flow of goods was multidirectional. Roman wares

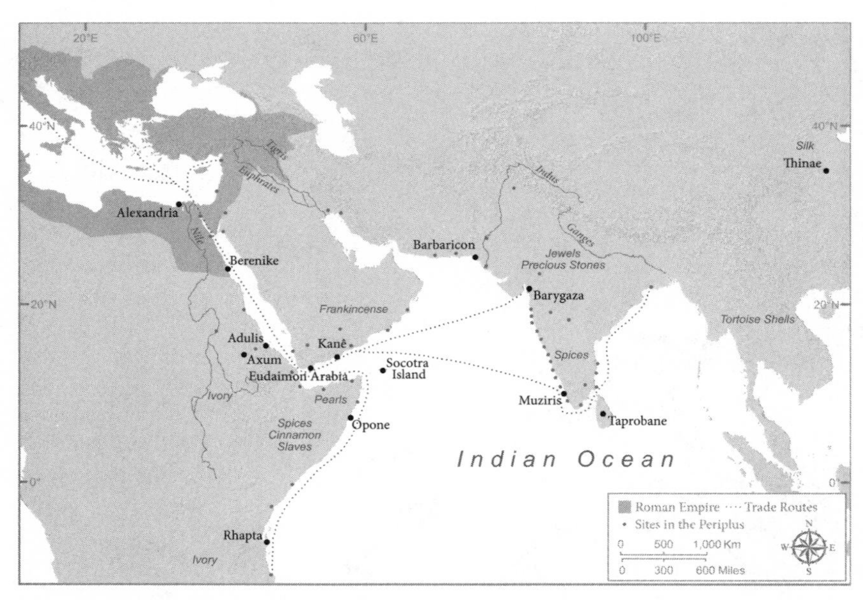

Map 9. The Romans and the Indian Ocean

as well as coins are found throughout the Indian subcontinent. Tamil poetry expresses local admiration for the "cool and fragrant wine" brought from western parts. The Indian poets describe the "beautiful large ships" of the westerners docking at Muziris, the very city whence the Hermapollon sailed; there they brought gold and returned "laden with pepper." There was certainly a permanent Roman trading colony here. The Peutinger Map, the most important map to survive from the Roman world, shows a Temple of Augustus at Muziris, a religious implantation by western traders, who carried goods and gods to the east and back. The trade along the coast of India was abutted by spurs curving far inland, through the Kushan Empire, to the Silk Roads and China beyond. The Chinese were the "silk people" to the Romans. Silk was a coveted commodity with a major market in the west, and during the early empire its exchange was conducted principally through the southern ocean route.[57]

It is a testament to the shrinking world that mutual awareness of Rome and China grew. The *Circumnavigation of the Red Sea* is the earliest western text to refer to the Han dynasty. By the second century the Chinese records are eminently aware of the *Da Qin*, the "great China"—in other words, Rome—far to the west. By the time Ptolemy wrote his *Geography*, Roman merchants had made it beyond the Malay peninsula. The annals of Chinese imperial history record the arrival of an embassy of Romans sent by

"Antun," Marcus Aurelius Antoninus. It is rightly suspected that this was no official embassy at all, but an adventurous trading party that had wandered into the Gulf of Thailand where they were seized by the Chinese emperor's forces. By the time these westerners were haled unprepared to the imperial court, the wares offered to the Chinese, namely elephant tusks, rhinoceros horn, and turtle shell, failed to impress. But, "this was the very first time there was communication." It was the same year that Lucius Verus and his army returned from the campaign in Parthia.[58]

East Africa was an integral part of this world. The author of the *Circumnavigation* described Adulis on its "deep bay," with roads stretching inland to the greater city of Axum, already the hub of the ivory trade and destined to become the major player along Rome's southern border. East Africa was one source for the exotic animals that tickled the Romans' imagination. It seems that the emperor Domitian managed to finagle the importation of a rhinoceros to Rome and widely celebrated it on his coinage. Beyond Axum, the Horn of Africa was under the strong hand of a king named Zoskales, "a stickler about his possessions and always holding out for getting more, but in other respects a fine person and well versed in reading and writing Greek." The merchant who wrote the *Circumnavigation* held considered opinions about what to buy and sell as far south along the African coast as Dar es Salaam.[59]

The Roman Empire threw open "all the gates of the inhabited world." The Greek orator Dio remarked that Roman Alexandria was "situated at the juncture, so to say, of the whole world and of the remote nations, like a market-place of a single city that brings everyone together in one place." There he could see, taking in the city's entertainments, not just "Aithiopians and Arabs, but even Bactrians, Scythians, Persians and a few Indians." This human mélange was a feature of the age. The recent discovery of graffiti inside a cave on Socotra, an island that lies 150 miles off the tip of the Horn of Africa, is a wonderfully unexpected window into this world. Over 200 scratchings from the Roman period record Indian, South Arabian, Aksumite, Palmyrene, Bactrian, and Greek traders rubbing shoulders. The island was then as now controlled from the Hadramawt in Yemen, but its eclectic graffiti are a testament to the kinetic energy generated by the Indian Ocean. By its physical position, Socotra was destined to be a kind of in-between place, and in the first centuries of this era it was a site of encounters in a corner of the world that was the nursery of incipient globalization.[60]

The merchants hugging the African shore and sailing the monsoon winds were also the agents of an invisible exchange. Where goods and gods go, so do germs. The real biological significance of the Indian Ocean system was not that it fused the "civilized disease pools of Eurasia," but rather that it formed a superconductor for emerging infectious diseases. The tropics are an evolutionary hothouse of disease. Central Africa is home to some of the richest vertebrate and microbial biodiversity on the planet. Consequently it has been and remains a dangerously productive zone of evolutionary experiment, the cradle of a disproportionate number of the pathogens capable of causing humans harm. The drama of disease history lies in the incessant collision of pathogen evolution and human connectivity. In the Roman Empire, those two forces came together in a particularly consequential way.[61]

The Great Pestilence

It is remarkable that we know as much as we do about the mortality event known as the Antonine Plague. At the same time, we necessarily see disease events that occurred nearly two thousand years in the past through a glass, darkly. And the mystery starts with the pandemic's port of entry into the Roman Empire.

The Romans believed that the mortality started with the sack of Seleucia. To be sure, Seleucia was a major entrepôt of the Persian Gulf, and Persian traders prowled the sea lanes of the Indian Ocean network. It is entirely likely that a spur of the pandemic ran through the gulf to Seleucia and diffused with the returning Roman armies. But the outbreak probably did not begin there.

The impious sack of Seleucia and the release of a noxious vapor from a temple of Apollo was a just-so story, forged in malice to blacken the memory of the co-emperor Lucius Verus and his general, Avidius Cassius. The Syrian general later tried to wrest control of the empire from Marcus Aurelius, and his name was stained in the annals of official historiography. The tale should never have enjoyed the credulity it has been granted. We have evidence for the disease inside the empire at least a year before the end of the Parthian campaign. The speeches of Aelius Aristides place the pestilence in western Asia Minor by 165. Furthermore, in the hilly interior of Asia Minor, in the hinterland of the ancient town of Hierapolis in Phrygia, a statue was erected in AD 165 to the god Apollo Alexikakos, the "Averter of Evil." This Apollo had an illustrious past: he had turned back the Plague of Athens, the most famous plague in Greek memory. In isolation, such a statue would hardly stand as proof positive of the epidemic, but it is another piece of circumstantial evidence for the progress of the disease into the empire, already before the return of the Roman armies.[62]

Once we are unattached to the origin story the Romans gave the pandemic, other clues about its itinerary seem more meaningful. The disease was almost certainly smuggled into the empire via the Red Sea axis. In the biography of the emperor Antoninus Pius (r. AD 138–161), we are provided the unusual report that there was a pestilence in Arabia during his reign. We might make little of it, but an inscription discovered in the region of Qaran, at the crossroads of the south Arabian kingdoms in ancient Yemen, pointedly verifies the mortality. Inscribed in Sabaic script around AD 160, the inscription refers to a pestilence that destroyed the town of Garw (*Bayt al Aḥraq*) and infected "the whole land" four years previously. The pestilence in Arabia in AD 156 cannot be identified as the agent of the Antonine Plague with certainty. But the coincidence is more than remarkable. If indeed the pestilence originated in Africa, it becomes likely that the plague in Arabia, heard in Rome, was a remote premonition of the storm on the horizon. A new microbe had escaped the continental interior and found its way to the promising webs of the Indian Ocean world.[63]

Once inside the Roman Empire, a germ with few self-imposed restraints, beyond its own consuming violence, was let loose. We have testimony of the plague's westward advance in the movements of Galen. He left Rome just as his renown had started to soar. His flight is a puzzle because he gave two different accounts of it. In an early work, he put down his return to Pergamum to obscure, hometown political circumstances: a bout of civil strife had ended back home. In his later tract *On My Own Books*, he admitted the "great plague" was the impetus for his departure. It is unclear whether he fled danger or rushed to the aid of his homeland. Regardless, it has gone unremarked how unusual it was to be able to *see* a pestilence coming, from without, sweeping across the Mediterranean. Galen escaped town just as or before the disease arrived in Rome. By mid-to-late AD 166, it was in the capital. The metropolis would have been a pathogen bomb, diffusing carriers of the disease across the western Mediterranean. The pandemic was raging among the troops in Aquileia by AD 168, advancing from one node of population to the next, unevenly diffusing in fractal spirals across the west. According to Jerome's chronicle, the army was devastated in AD 172.[64]

These are the thin, flitting shadows we see of the first wave of the disease, as it hastened across the empire from east to west. Otherwise, attestations of the pestilence are predictably haphazard. It wreaked havoc in the Nile delta, we happen to know from a carbonized document discovered in the region. A contemporary medical text, ascribed erroneously to Galen, claimed that the pestilence, "like some beast, foully destroyed not a few people, but even rampaged over whole cities and destroyed them." Gaul and Germany were hit. In Athens, Marcus Aurelius had to relax the entry requirements into the most exclusive club in town, letting even men of recent servile ancestry into the hallowed Areopagus; shockingly, the city could not find a chief magistrate in 167, 169, or 171; an orator from Athens, a few years later, wailed in a speech before the emperor, "Happy they who perished in the plague!" An inscription from Ostia, the port of Rome, records that an association of eastern traders had been severely reduced and struggled to pay their requisite dues. The plague reached deep inside Asia Minor and Egypt, and north beyond the Danube. Everywhere there might be evidence of the plague, it is to be found. The mortality event is the first truly deserving the label of a pandemic.[65]

The scope of the Antonine Plague astonished contemporary observers, who were accustomed to epidemics but not on such a spatial scale. The response to the crisis was, in the first place, religious. Plagues always stirred feelings of helpless, primitive dread, and the Antonine Plague touched deep chords of religious fear. From the mists of time, the god Apollo had been associated with pestilence; in Homer's epic poetry, he was the archer who sent arrows of plague. In the course of the outbreak, the rumor spread that a pestilential vapor had seeped from the temple of the Long-Haired Apollo in Seleucia. Some of the most remarkable testimony to the scale of the pestilence is to be found in the remains of the desperate, empire-wide attempt to placate the god whose anger was blamed for the catastrophe.

Ancient polytheism was a decentralized religion, its temples and priesthoods loosely embedded in the lives of the empire's towns and villages. The Roman Empire was an age of great piety but also great creativity in the worship of the gods, and its open roads fostered what has been called a "democratization" of religious authority. The fear of the pestilence was an easy opening for all sorts of enterprising soothsayers. One of the most memorable portraits in the writings of the wickedly funny satirist, Lucian, flays a contemporary mountebank named Alexander of Abonoteichus who sent oracles "to all the peoples" to ward off plague, including one oracle that invoked Apollo the Long-Haired. Alexander commanded certain sacred words to be inscribed on doorways as a charm against plague, but according to Lucian, those who followed his advice were especially cut down by the pestilence. It is important evidence that the dread of Apollo's revenge was real, and most of the empire's residents were probably closer to Alexander's credulous fear than Lucian's cool detachment.[66]

In fact, we have a remarkable number of inscriptions that reveal the wide influence of this sort of religious response. No fewer than *eleven* inscribed stones (ten in Latin, one Greek) have been found, in the far corners of the empire, with the short phrase "To the gods and goddesses, according to the interpretation of Apollo at Claros." C. P. Jones brilliantly deduced that these are in fact apotropaic inscriptions, all of them etched on plaques that were embedded in walls to ward off the dire pestilence. Further evidence continues to be brought to light. A pewter amulet from Roman London has just been published. It includes a longer version of the same apotropaic charm derived from the oracle of Apollo. In a convincing restoration of the amulet's text, Jones has shown that the god in fact prohibited *kissing*; kissing was an important form of social greeting in the classical Mediterranean, and if the disease was directly transmissible, the advice can only be considered medically sound.[67]

Long before the plague, Apollo had become one of the towering, syncretistic gods of the empire. His temples at Didyma and Claros were privileged centers of sacred communication, binding together the beliefs and practices of a far-flung and religiously heterogeneous patchwork of peoples. In the time of plague, towns across the greater Greek world dispatched embassies in desperate search of answers, and in at least *seven* different places, Apollo's prolix replies to these desperate embassies remain engraved in stone. "Woe! Woe! A powerful disaster leaps onto the plain, a pestilence hard to escape from, in one hand wielding a sword of vengeance, and in the other lifting up the deeply mournful images of mortals newly stricken. In all ways it distresses the new-born ground which is given over to Death—and every generation perishes—and headlong tormenting men it ravages them." Apollo ordered the town to purify its houses with ritual lustrations and to draw away the pestilence with fumigations. (There was solid precedent for the latter: half a millennium before, the famous doctor Hippocrates had ordered fumigation to repel a plague.)

In other cases, the oracle commanded libations and sacrifices to alleviate the grievous suffering. "You are not alone in being injured by the destructive miseries of a deadly plague, but many are the cities and peoples which are grieved at the wrathful displeasure of the gods." In several cases, the god ordered a statue of himself outside the city gates, drawing his bow, "which destroys diseases, as though shooting with his arrows from afar at the unfertile plague."

The outburst of Apolline religion generated by the Antonine Plague is utterly unlike anything else in the records of ancient epigraphy. What survives must be merely the very tip of a lost iceberg of religious dread. The religion of Apollo was hyperactive in this desperate moment, and although the apotropaic inscriptions are evidence for fear of plague rather than plague itself, the stones provide one sort of index to the broad reach of the Antonine pestilence.[68]

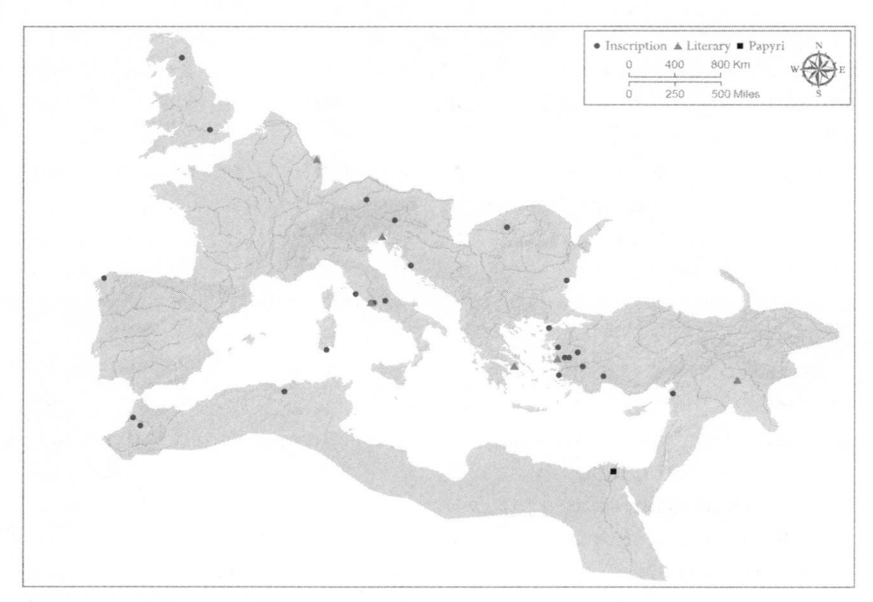

Map 10. Possible Indications of Antonine Plague

We inevitably wonder what pathogenic agent could account for such a vast mortality event. The question is not merely a matter of morbid curiosity. Pathogen biology determines the dynamics and dimensions of a disease event, and if we know the identity of the microbe behind the Antonine Plague, we can hope to fill in certain missing pieces of the puzzle. The only pathogen that has been under serious suspicion is the smallpox virus, and a virtual consensus around this identification has formed among Roman historians. Here we will advance the case that smallpox is in fact the best hypothesis, even stronger in some ways than has been recognized. But it is perilous to draw conclusions without positive molecular identification. Uncertainties remain and deserve to be spotlighted. And the story of smallpox, as told by genomic evidence, is in a state of upheaval. In the end, it may turn out that labeling the pathogen "smallpox" is a pardonable oversimplification of a rather more interesting and complex evolutionary reality.

Short of directly sequencing the microbe's genome from the archaeological remains of a victim, the identification of a historical pathogen depends on our knowledge of its pathology and epidemiology—its telltale behavior at the individual and population levels. Any identification, in turn, must at least be consistent with what is known of a suspect's phylogeny, its family history. As we seek the identity of the Antonine Plague, we are the recipients of at least one piece of unusual good fortune: antiquity's greatest doctor was on the scene. Galen treated "countless" victims during the "great pestilence," and although he did not compose a treatise specific to the disease, he has left behind scattered and occasionally detailed accounts of what he observed.

Cautions are immediately in order. Retrospective diagnosis is a risky business, even with an observer like Galen. We must remember that Galen was not writing for us, and in medicine, experience and observation are always filtered through terms and expectations that are prepared by the background culture. For all his brilliance, Galen's vision was blinkered by the constraints of his humoral theory: the belief that the body was a mix of four humors and that health was the balance of these humors. Galen had no concept of an infectious microbe, and, as with so many of his contemporaries, the possibility of an emerging disease seems not to have been available to him. For Galen, the Antonine Plague was always "the great" or the "longest lasting" plague, different in scale but not in kind from other epidemics. Galen lived in a world where a welter of infectious diseases was always present, simultaneously, and he was not trying to specify the symptoms caused by *this* pathogen. Pressing his texts into the service of retrospective diagnosis, then, is always a bit like trying to identify the ingredients of a stew from a review of its flavors.

Even through the lens of his humoral theory, though, Galen has left some perceptive clinical notes. To Galen, the etiology of the disease was an excess of the humor called black bile, literally a "melancholy," possibly an observation on the malaise of the plague's victims. In Galen's mind, its attendant symptoms were fever, a black pustular rash, conjunctival irritation, ulceration deep in the windpipe, and black or bloody stools. Those who had a "dry" constitution had the greatest chance of surviving an infection.[69]

Galen's lengthiest case history is recorded in the fifth book of his masterpiece, the *Method of Medicine*. It occurs in the midst of a longer discussion about how to heal wounds. In general, to be healed, wounds must be dried. Galen described a plague victim who had developed ulcers deep in his trachea and bronchial tree; Galen believed he had discovered a means whereby to dry the internal ulcers, thereby saving the patient. The man's whole body broke out with sores on the ninth day, "just as did almost all the others who were saved." He coughed up scabs. Galen made the patient lie on his back and hold a liquid drying agent in his mouth. He was restored to health. The recovering patient was desperate to be in Rome "where the plague was raging," but he could not rise until the twelfth day. This discussion prompted Galen's most important general reflections on the pathology of those suffering in the pestilence. Those who survived "seem to me to be dried and purged beforehand." Vomiting was thus a positive indication. Among those who were going to live, black extrusive pustules appeared close together over the whole body; in most cases there were "sores," and in all survivors there was "dryness."[70]

Galen believed that the fever putrefied the blood of the victims. "There was no need of drying medications for such *exanthemata* [extrusive pustules] for they spontaneously existed in the following manner: in some, in whom there was also ulceration, the surface fell off, which they call scabs, and henceforth what remained was already close to health, and after one or two days scarred over. In others, in whom there was not an ulceration, the *exanthem* was rough and itchy, and fell off like something scaly, and from this all patients became healthy." In his treatise *On Black Bile*, Galen described the black pustules covering the entire body that then dried and fell off like scales, sometimes many days after what Galen considered the turning point of the disease. These clinical observations describe the course of vesicular and then pustular lesions that scabbed off, leaving scarred but no longer pathological dermis in their place.[71]

A smallpox infection is the closest match to the disease Galen observed. It is worth reviewing in some detail the course of infection by the virus, *Variola major*, as it was observed by modern clinicians working around the globe in the decades leading up to the eradication of the disease. Smallpox was a directly transmitted disease. The virus was contracted by the inhalation of airborne droplets, expelled by an infectious person. Once smallpox virions invaded a new victim, the virus was exceptionally pathogenic: most infected people became sick at some level. The virus first multiplied in the mucosa, then in the lymph nodes and spleen, with bewildering speed; smallpox overran the initial immune response, and the body began scrambling to resist. This incubation phase could be relatively prolonged, 7–19 days, but usually around 12. During this false lull, the patient was not contagious. But neither was the victim yet immobilized, meaning the virus could travel far and fast.

Table 3.3. Galen's Pestilential Rash

Exanthēmata melana: black extrusive/pustular rash. The Greek implies bursting forth (etymologically "blooming," like a flower), pustules that rise out of the skin.

Helkos: wound, sore. For Galen, it implies a rupture in the continuity in the flesh (Galen K10.232). Galen repeatedly insisted that in the case of victims of the pestilence "all" sores were "dry and rough."

Ephelkis: scab. The natural hardening of a sore or wound.

Lemma: what is peeled off. The word is used of fish scales. For Galen, the victims whose rash did not turn into sores were scratched off like scales.

Epouloō: to scar over. Used frequently by Galen, to cicatrize.

Table 3.4 Progression of Smallpox Infection

Day	*Infectious*	*Pathology*
1	no	Asymptomatic
2		
3		
4		
5		
6		
7		
8		
9		
10		
11		
12		Fever, malaise, etc.
13		
14	Yes	
15		Macular rash
16		Papular rash
17		
18		
19		Vesicular rash
20		
21		Pustular rash
22		
23		
24		
25		Scabbing
26	Mildly	
27		
28		
29		
30		
31	No	Scarring
32		

The first symptoms were fever and malaise; they came on suddenly. The victim was soon infectious. There was some vomiting and diarrhea and back pain. In the most common course of the disease, the fever receded within a few days, just as the first harbingers of the skin pathology appeared. Painful lesions formed in the throat or mouth. A macular rash appeared on the face and the entire body, more densely over the face and extremities than the trunk. When the rash launched, it had an eventful two weeks or so, as the pox protruded from the skin and became vesicular. Then the bumps turned pustular, until after five days or so they began to scab. The patient was most infectious during the fever and onset of the rash but remained contagious until the scabs fell off. When the scabs fell, they left behind disfiguring scars. The whole course of infection was about 32 days.[72]

Such was the normal course of a smallpox infection. There were variations on the theme. A minority of cases in a normal outbreak could exhibit a hemorrhagic presentation. In an "early hemorrhagic" type of infection, the victim was overrun quickly, perhaps on the second day of the fever. Bleeding from various places on the body was visible, and while the skin became matted, the patient died before the characteristic rash could run its course. In a "late hemorrhagic" type of infection, bleeding manifested after the pustular lesions developed, appearing to seep through the skin. Hemorrhagic smallpox of both kinds tended to strike adults, and it was almost uniformly fatal.[73]

Smallpox could resemble other diseases presenting with pustular rashes, such as chickenpox or measles, especially in the early phases or mild cases. In the course of measles, a prodromal fever of 2–4 days accompanied by cough and conjunctivitis is followed by a rash spreading from the head to the rest of the body over about 8 days; unlike smallpox, the measles rash does not protrude from the skin, nor does it scar. In a chickenpox infection, fever and rash present simultaneously; skin lesions, shallower than in smallpox, spread over the body, in crops, and resolve quickly. The differential signs of smallpox are the deep pustules that protrude above the skin and feel deeply set within it, arising simultaneously over the body, in a lengthy course lasting two weeks, concentrating more in the extremities than the trunk, sometimes covering even the palms and soles.

Galen's observations are consistent with the symptoms of *Variola major*. In the course of smallpox infection, death comes around 10 days after symptoms appear, in line with Galen's belief that days 9–12 were critical. The fever was universal but not unusually intense, a credible description of smallpox. Galen implied that it was positive if the pustules were close together—not the opinion of modern clinicians, for whom confluent lesions were an ominous sign. But, here Galen was considering the subpopulation of survivors. The conclusions of a study by Littman and Littman remain sound: Galen had seen the hemorrhagic expression of smallpox. The sign of it for Galen was very black stool, a harbinger of the worst. Galen assumed that the excess of black bile experienced by all plague victims could express as the drying black rash or the passage of bloody stool. The former gave the patient reason to hope, while the latter meant the blood had been "completely cooked." Galen does not specify whether the pustules arose together or in crops, whether they concentrated on the middle of the body or in the extremities, or whether pustules appeared on the palms or soles; his description cannot allow definitive diagnosis. But the rash he described, from first appearance of protrusive lesions to scabbing and scarring, points to smallpox as the agent of the Antonine Plague, even across the cultural chasm that separates us from the ancient doctor.[74]

Smallpox is not a particularly old nemesis, and the genomic evidence is beginning to suggest that it experienced a brief but eventful existence. Molecular clock dating is a method of estimating how long ago an evolutionary event occurred: it provides a probability of how long it may have taken for a given level of genetic variation to have developed. One analysis placed the divergence of smallpox from its most recent common ancestor with *Tatera poxvirus* in Africa, only 2,000–4,000 years ago. Smallpox did not inhabit "civilized disease pools" of Asia time out of mind. And a new genomic study indeed suggests that smallpox underwent a major evolutionary event around the sixteenth century and that a more virulent form of the virus dispersed globally in this age of exploration and empire-building. The history of the disease between its origins and its modern career remains an open question.[75]

The earliest indications of smallpox in the literary evidence only date to the first millennium. Besides the Antonine Plague, there are possible allusions to smallpox epidemics in China by the fourth century; a striking account of a pestilence in Edessa in the late fifth century seems very likely to have been smallpox. Then, there are a number of descriptions of smallpox in medical texts from the sixth century on, ranging from an Alexandrian doctor named Aaron to classic texts of medieval Indian medicine such as the *Madhava nidanam* written by Madhava-kara in the early eighth century. In the late ninth century to the early tenth, the Persian physician Rhazes dedicated an extraordinary treatise to the differential diagnosis of smallpox and measles.[76]

A fuller picture is likely to emerge as more genomic data is recovered from archaeological samples. At present, one hypothesis is that *Variola* evolved from a rodent orthopoxvirus to become an obligate human pathogen, in Africa, sometime before the Antonine Plague. The biological agent of the second-century pestilence could represent an especially virulent lineage of *Variola* that went extinct, or an ancestral form of the virus that evolved into a milder medieval form of smallpox. And it still could have been caused by some other biological agent altogether, although there are no serious candidates at present. The genomic evidence will tell, in time. The deeper point is that the evolutionary history of human pathogens has been raucously turbulent in recent millennia.

Not many germs can accomplish what this pestilence did, in particular its transcontinental reach in the space of only a few years. The Antonine

Plague gives every sign of having been a highly contagious, directly transmitted infectious disease. While the ancients thought of pestilence as a *miasma*, a gaseous pollution moving through the atmosphere like a cloud of poison, we should not be misled into imagining the diffusion of the disease as a series of ever-widening concentric circles. To do so is to surrender all hope of retracing the transmission and population dynamics of the Antonine Plague. The pandemic more resembled a toxic and fissile pinball, shattering at each collision and diffusing outward in radiating streaks from each point of contact. The spread of the contagion was chaotic but structured by the possibilities and constraints inherent in the urban, interconnected empire built by the Romans. The pandemic moved from southeast to northwest, but it moved unpredictably, borne by human movements, not the winds. An arrow would be distorting, unless we imagine the fractal complexity underneath its overarching direction.[77]

The full measure of the challenge before us comes into view when we recall that modern historians have variously put the death toll of the Antonine Plague anywhere from 2 percent to over a third of the imperial population—a range from 1.5 to 25 million dead! Such are the ineluctable hazards of studying a disease event two thousand years in the past. We have no bills of mortality, and we must rely on what are really no more than chance glimpses of the effects of the disease from particular places at particular moments. These are crucially important, but they require caution, because the full impact of a mortality event is highly contingent on underlying social and ecological factors, and these varied widely even within the Roman Empire. The experience of the pestilence in a village, or in an army barrack, or in a metropolis, would have differed.[78]

Ultimately, the behavior of a pathogen at the population level depends heavily on its modes and means of transmission. The dynamics of an epidemic can be reduced to a small number of critical parameters: the *total contact rate*, the *transmission risk*, and the *case fatality rate*. The number of victims who contracted a disease was determined by the total contact rate times the transmission risk—the number of people an infectious patient came into contact with and the chances that those exposed were infected. In general, the transmission risk is almost purely biological. A virus like smallpox was highly contagious, but less so than extraordinarily communicable diseases like measles or flu. A report from rural Pakistan claimed that 70 percent of individuals living in a small household with an infected relative contracted smallpox, and 70 percent is a commonly used figure. Galen, who was chatty about his own medical history (he had the fevers four times in his youth), gave no indication that he was ever infected, despite treating hundreds of victims. So infection was likely but not universal, even among those exposed. While few people have passive or innate immunity to smallpox, survivors were conferred strong and lasting resistance.[79]

The most interesting variable was the total contact rate. The actual mechanisms by which a pathogen is transmitted have the greatest influence on the course of an epidemic. Smallpox, for instance, was an airborne pathogen. It departed from the original victim in a cough, sneeze, or saliva, and it entered the next victim via the nose or mouth. Because of its long incubation period, around 12 days, infected victims could carry the virus to new locations before they were immobilized. The patient was highly contagious for a total of about 12 days; the victim remained a potential source of infection for a few days longer, as the pustules scabbed. The virus was transmitted aerially, but it never flew very far—only three or four feet. We should be unsurprised by the reports, like those of Aelius Aristides, who described the pestilence tearing in succession through the members of his household. But, the strongest brake on the spread of smallpox is the 3–4 foot range of the virus and the immobility of victims during their infectious phase. Anything that could materially affect how many people come within the radius of danger will influence the dynamics of an infectious disease outbreak: from cultural norms of caring for the sick to large-scale transportation networks. The huge question of the Antonine Plague in the Roman Empire is a problem 3–4 feet in size, repeated millions of times.[80]

An array of structural facts aligned to enhance the contact rate in the empire and create a fertile environment for the spread of a contagious, directly transmissible pathogen. Thick and effective networks of transportation connected the empire by land and by sea. Still, the Roman world was an ancient society, and the time and expense of travel created friction against the diffusion of the pathogen. Urbanization fostered dense habitats, often with overcrowded housing units. But the vast majority of people lived in the countryside. The cultural conditions of the societies under Roman rule rendered them unexpectedly vulnerable to an acute infectious disease. The absence of germ theory (even though there was not total ignorance of *contagion*) meant there was no scientific reason to fear the infected, and the large-scale medical apparatus, based on home visits, circulated the disease throughout the cities. The new disease found in the Roman Empire a population with no prior social learning to buffer against an enemy like the one it now faced. Of course, the injunctions of Apollo against kissing, or the cautions of the ailing Marcus Aurelius around his son Commodus, suggest a rough-and-ready awareness of the communicability of the disease.

The case fatality rate of a disease depends on a combination of the pathogen's virulence and the population's biological status. Even a virus as lethal as epidemic smallpox had a case fatality rate in the range of 30–40 percent—so most who contracted it survived and became immune. Smallpox preyed on the very young (whose immune systems are developing) and the very old (whose immune systems are weak). The overall case fatality rate will depend on the age structure of the population struck by a virulent pathogen. Moreover, the pre-existing pathogen load can affect how lethal an outbreak is. In the New World, for example, "the pathogen load in the low, humid, and hot areas was heavier than elsewhere, interacting negatively with the new diseases imported from Europe." The Antonine Plague acted synergistically with the harsh disease environment to exacerbate mortality.[81]

Other factors softened the blow of the pestilence. The organized medical apparatus of the cities helped to ensure that the sick received care; while bloodletting and whatever "drying agents" doctors like Galen applied may have made matters worse, it is hard to overestimate the value of basic nursing which provides food and water to the suffering. It is often the difference between life and death. Galen noted that patients who could eat survived, while those who did not uniformly died. None of the literary sources for the Antonine Plague report social chaos in the midst of the pestilence; the integrity of the social order seems to have held together, except perhaps for a complex crisis in the Nile delta, where ecological change, social violence, fiscal debt, and pestilence led to utter social disintegration. In Rome, Marcus Aurelius was said to have provided for the burial of the poor at public expense; we are uninformed about any other city.[82]

After a long lull, the pestilence recurred in at least one major secondary episode. The pattern is in fact what might be expected of a directly transmitted virus that confers robust immunity on survivors. If a population is sufficiently large, the virus can quietly hide in corners of town, or it can

Table 3.5 The Epidemiological Factors of the Antonine Plague

Total Contact Rate	*Transmission Risk*	*Case Fatality Rate*
+ transportation networks	perhaps ~.70 for *Variola major*	+ age structure
+ population density		+ pathogen load
+ multifamily housing		– medical infrastructure
+ lack of germ theory		– lack of social breakdown
+ medical infrastructure		+ / – nutritional buffering
+ lack of social learning		

+ enhanced mortality – reduced mortality

continue to ricochet throughout other towns and villages before making a return. Once the proportion of susceptible hosts rises again, a new outbreak is possible. The first pulse of smallpox was felt in the empire by 165, and it whipsawed from region to region until at least 172. Its diffusion was shaped by the overlay of physical geography and human networks, in combination with the biological rhythms of the pathogen.

As the pestilence spread its tentacles throughout the Roman Empire, many of its offshoots would have quickly expired of the disease's own momentum. In that sense, the metropoleis like Rome and Alexandria were not just engines of germ circulation on the first wave, but their huge populations let the microbe lurk in small numbers, beneath our field of vision. Then, as birth and immigration replenished the ranks of susceptible hosts, the great cities became time bombs waiting to explode again, spewing the pathogen into their dispersed hinterlands once more. Thus it is unsurprising to find signs of the pestilence in Noricum in AD 182–3 and in Egypt in AD 178–9, known through the fortuitous survival of papyri and inscriptions. In Egypt, it is tempting to posit a rebound wave emitted from Alexandria around this time. In the west, a second major eruption is vividly attested at Rome in AD 191. Over 2000 died per day in the relapse. It horrified a populace that may have started to believe the worst was over.[83]

Most of our literary testimony looks out over the broad landscape of the deadly pandemic. In a few precious instances, we have the opportunity to zoom in for a more granular perspective. In one case, a carbonized papyrus roll from the Nile delta provides an up-close view of what has been called "demographic hemorrhage" in some twenty villages around the city of Mendes. Depopulation in these villages, which had fallen impossibly behind in their tax payments, may have started in the middle of the second century, driven by complex hydrological change in the river delta. But a text composed in AD 170 underscores the total depopulation of the villages that dotted this delta landscape. One village, Kerkenouphis, by AD 168–9, was said to have zero inhabitants because of a bandit uprising, tax flight, and "the pestilential situation." Here, the plague helped to push a marginal and distressed environment into total free-fall.[84]

The insidious mortality of the second wave is attested far up the river from the delta, at a village in the Fayyūm called Soknopaiou Nesos. Situated on the north shore of Lake Moeris on the very edge of the desert, the heart of this priestly village was the cult of the crocodile god and its temple; fishing, farming, and the caravan trade diversified its portfolio. In the winter of AD 178–79 the scourge revisited the village. Of 244 adult males alive in late AD 178, 59 died in January and another 19 in February of 179. The document gives us a snapshot of the death toll. It implies a mortality rate of 32 percent among the least vulnerable subpopulation, in a two-month span, during a secondary wave. If the case fatality rate was 50 percent, twice as many victims, 156 of the town's 244 men, could have contracted the disease. What this microcosm reveals is that the effective contact rates, in this densely settled corner of the empire, could be perilously high. A village like Soknopaiou Nesos was biologically connected to the world beyond, and once unleashed inside the settlement, the virus was able to race from one victim to the next.[85]

These two small case studies are valuable, but it would be a mistake to imagine them as representative samples. The delta villages were marginal settlements in a volatile environment, contending with a multipronged crisis. The Fayyūm village was exposed to Egypt's unusual population density and connected valley habitats. Both villages probably experienced something far graver than the average imperial settlement.

The military was struck hard by the disease. By AD 172, the chronicles reported, the army had been reduced to near extinction. The biography of Marcus Aurelius reported emergency conscriptions of slaves and gladiators and unusual levies of brigands. The army's straitened condition is documented by an inscription from a town in central Greece, normally exempt from legionary recruitment, sending more than eighty men ca. AD 170 into service, in what has been called a sign of a "serious manpower shortage" in the army. But the most remarkable index of the plague's demographic impact on the army has been inferred from a list of veterans discharged from the *legio VII Claudia*, after their twenty-five years of service in AD 195. Given reasonable assumptions about the annual intake and loss of a Roman legion, the sudden bulge of retirees in this year reveals that the legion had lost some 15–20 percent of its men, if not slightly more, in the initial wave of the pandemic, and hastily refilled its numbers in the years immediately following. While the barracks lifestyle may have accelerated the transmission of the pathogen, soldiers in the prime of life, with reliable systems of provision and care, should have died at significantly lower rates than other victims of the disease. Again, this sample is not representative so much as illustrative of what the killer was capable of once unleashed under certain conditions.[86]

Roman historians have sometimes argued that the severe demographic impact of the pandemic is registered in the sudden interruption in dated series of documents, including Egyptian papyri, building inscriptions, military discharge diplomas, and so on. It is a line of inquiry that has proven suggestive rather than conclusive, mostly because such gaps in the record indicate the presence of a crisis—not its cause. But a pestilence of such unaccustomed magnitude is far the best candidate for a trigger, and the specifically demographic roots of what quickly became a systemic crisis are independently confirmed in the long-range changes in real price levels.[87]

In the midst of the pestilence, imperial silver mining seems to have suddenly collapsed, sparking a short-term monetary crisis. In the provincial coinage of Egypt, there was debasement of the silver coinage starting in AD 164–65 and intensifying in AD 167–68. Then, from 170–71 until 179/80, there was a complete cessation of silver coinage from Alexandria, an extraordinary gap in provincial coin production. The unusual hiatus is paralleled in the civic mints in Palestine (from 166–67 to 175–76) and Syria (169 to 177), suggesting a much wider problem. The military mobilization against Parthia and the expense of the war machine had already stretched the imperial fiscal system, but the pestilence pushed it into a zone of critical danger. From the later 160s through the 170s, the monetary and fiscal infrastructure was teetering in response.

In Egypt we can follow the rapid changes in the price regime induced by demographic and monetary shocks. Nominal prices—prices expressed in the face unit of the currency, in this case the *drachma*—doubled. The coinage lost half of its purchasing power, evident in a range of commodity prices, including the most fundamental commodity, wheat.[88]

The economic impact of the pestilence was serious. Real land prices—the cost of land expressed in terms of wheat—plummeted. Suddenly, land

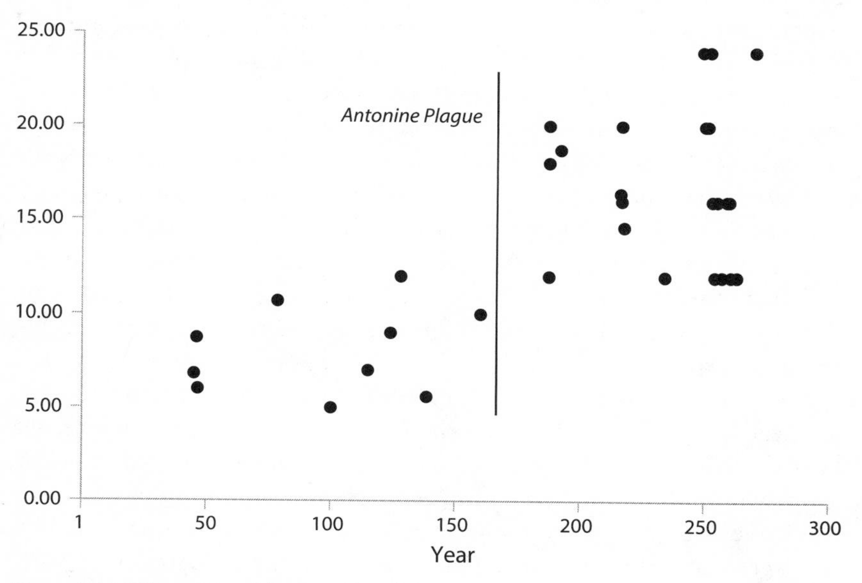

Figure 3.10. Wheat Prices (drachmai/artaba)

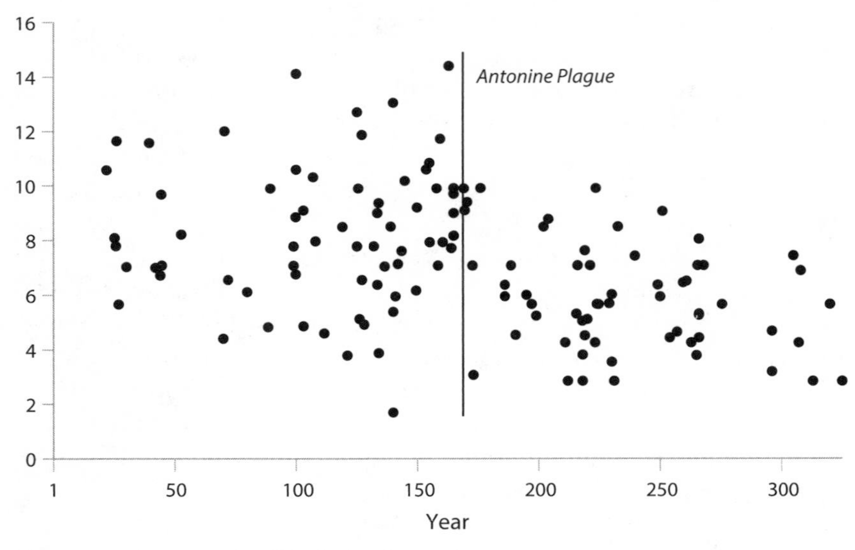

Figure 3.11. Rents in Kind (hl of wheat/hectare)

was less valuable, likely because demand for it had sharply contracted. The effect on real wages was a wash. While labor was presumably more scarce, and might have benefitted from the mortality shock in the form of higher wages, some damage to the economy—productivity losses from commercial recession or lower technical capital—prevented any detectible gains for ordinary workers. But real agricultural rents signal a deep shift in the relative weight of land and labor. The price that tenants had to pay to rent arable farmland was jolted downward and held at a new equilibrium for decades.[89]

In sum, all of the evidence—from the debris of an unparalleled religious response to the literary reports of a mortality event spanning the whole empire, from the glimpses into microcosms of the plague's violence to the widest-scale view of its economic effects—are consistent with the conclusion that the Antonine Plague was a mortality event on a scale the empire had never experienced before.

The need for a total death tally, a grisly summary of the pandemic's toll, has been irresistible. In the Antonine Plague, we have to reckon with the huge diversity *within* the geographical space bounded by Roman rule. Integrated coastal regions would have been most exposed to an empire-wide pandemic of a directly communicable disease. Vast swaths of the countryside were buffered by their own remoteness; Egyptian villagers fared worse than their counterparts in provinces with a more dispersed settlement pattern, such as characterized much of the west. The age structure of the empire surely meant that unimaginable numbers of infants and small children were carried off by the pestilence, a lost generation. The pre-existing pathogen load also would have exacerbated mortality.

Most efforts to gauge the total mortality of the Antonine Plague have fallen somewhere between 10–20 percent. The only epidemiological model of the pestilence, based on the assumption that the agent was smallpox, yielded an estimated mortality of 22–24 percent for the empire as a whole. We might imagine the possibility of very high contact rates and death rates for the core parts of the empire and, simultaneously, very strong buffers in the considerable hinterland and periphery of the imperial territory. The army mortality rate of ~15–20 percent might have been toward the high end of what was possible in the innermost heart of the empire, closely connected around the sea. Even if the lower end of that range could be transferred to Rome, it would imply that a minimum of 300,000 inhabitants of the capital contracted the disease, half of whom perished. It is easy to imagine how such devastation could account for the resounding horror that echoes in all of our sources. In the end, the zone of complete ignorance, especially the penetration of the disease in the countryside, looms too large for comfort. For what it is worth, a guess around 10 percent, if we truly intend the imperial population as a whole, seems prudent, maybe twice that in the areas most devastated by the pandemic. If the virus did carry off 7 to 8 of the empire's 75 million souls, it was, in absolute terms, the worst disease event in human history up to that time.[90]

In the course of history, many shooting-star pathogens must have leapt from forest or field only to annihilate themselves in a paroxysm of violence, burning through all susceptible hosts in some small tribe or village until extinguished. This evolutionary dead end might have been the fate of the microbe that caused the Antonine Plague, had it not ricocheted onto the stage at the moment in history just when the networks to conduct it to the wider world existed as never before. In that sense, the course of Roman history was redirected by the chance conjunction of microbial evolution and human society.[91]

Resilience and the New Equilibrium

The Antonine Plague marks a turning point, the end of a certain trajectory in the development of Roman state and society. But we should resist the temptation to treat the event as a fatal blow, consigning the imperial project to eventual ruin. Even if the empire suffered aggregate mortality as high as 20 percent, it would have reduced the imperial population back to levels seen late in the reign of Augustus. On the one hand, the undoing of a century and a half of robust growth in the blink of an eye was a staggering blow. On the other hand, the Augustan empire was not sparsely populated. And the Antonine Plague did not wreck the inner logic of the Roman demographic regime. Here is perhaps the most important difference between the Roman Empire and the New World populations ravaged by epidemic diseases, including smallpox. The context of colonization, slavery, and resource extraction incapacitated these reeling societies; the real impact of the microbial expansion was felt in the long run. "The long-term impact of the new diseases was the more negative the more 'damaged' the demographic system became and the less able it was to rebound after a shock."[92]

Precisely this kind of disintegration did not occur in the aftermath of the Antonine Plague. If anything, the survivors of the pestilence ramped up fertility to maximum levels in the decades after the plague. After the secondary pulse of the outbreak, there were no major epidemic events known before the Plague of Cyprian in AD 249. If the agent was smallpox, there is no evidence that it became endemic in the largest cities of the empire. The population grew again in the generations after the pandemic, though it never re-attained its earlier peak. Even the village of Soknopaiou Nesos seems to have rebounded. The Antonine Plague did not send the empire into a demographic tailspin from which it could not recover.

But the shock of this mortality event, an order of magnitude greater than anything the empire had ever experienced, stressed the capacities of the imperial system. The immediate political test was profound. The fiscal crisis confronted the empire with acute challenges; by AD 168, Marcus was auctioning the treasures of the palace to raise funds. Basic agrarian rhythms were disrupted. Galen reported "continuous famine for not a few years among many of the nations subject to Rome." Hungry town-dwellers descended on the countryside, and "in accordance with their universal practice of collecting a sufficient supply of wheat to last a whole year," stripped the fields, leaving the rustics to scavenge and survive on twigs and grass. It is noteworthy that this, our most vivid testimony to large-scale famine in the collective experience of the high Roman Empire, comes in the wake of the pandemic. But, in general, the fabric of empire did not come unwound.[93]

The effects of the pandemic were altogether more subtle. If the population was reduced to something near its Augustan level, the interceding time had wrought changes in the political and moral economies of empire. Not least of these were the weightier responsibilities of governance. A hegemonic empire of conquest had settled into a territorial empire, gradually assimilating its diverse peoples within a common polity that commanded their loyalty. The citizens and subjects of empire had demands in return, of peace and order. They looked to their government with expectation; we chance to know of one governor in Egypt, after the pestilence, who received 1804 petitions from his provincials, in one three-day assize. By the reign of Marcus Aurelius, the grand bargain between the empire and the civic elites around the empire had proven successful but never entirely stable; the provincial aristocracy insinuated itself in the highest ranks of imperial society, and the empire needed their service in a wider range of roles than ever. Their wealth, and their service, demanded a place and a prominence that Augustus could not have foreseen. In the reign of Marcus, the exigencies of war and plague, and the tolerant attitude of that philosophical emperor, opened the door for provincials with talent as never before. The pandemic hastened the provincialization of the empire.[94]

Beyond the frontiers proceeded greater changes still. Proximity to the empire fueled secondary state formation in the barbarian realms, and the rise of more formidable enemies across the Danubian border represented a deep geopolitical shift. The removal of three legions to the east for the Parthian campaign was a calculated risk. The plan was for Lucius to settle affairs in the east, then turn to the northern problem. While Lucius was conducting operations, Marcus was at Rome, already raising two new legions for action in the north. It all proved to be poorly timed. Lucius' victorious armies straggled home under the cloud of pestilence. Meanwhile the storm of plague broke in the west. The northern expedition was delayed by a year. News from the front was bleak: the Marcomanni and Quadi demanded quarter in the empire, or war. When Marcus and Lucius did set out for a northern campaign, the army was ravaged by pestilence in the winter camp at Aquileia. Galen, as the doctor had feared, was summoned by the emperor. Lucius himself succumbed.[95]

The northern wars of Marcus Aurelius are often considered a turning point in the fortunes of the empire. Something was now different. Even the "escalation dominance" of the Romans seemed to falter. Strike forces of invading barbarian troops pierced deep into the empire, both across the Alps and down the Balkans. Marcus spent most of his last decade in a grinding and indecisive campaign, interrupted by the attempted usurpation of the throne by Avidius Cassius. The Syrian senator who sacked Seleucia turned disloyal for reasons that remain obscure. The rebellion was snuffed out, but it was a distraction from the frontier operations. It was also a premonition of future rebellions.

The Stoic emperor spent his last years on the Danube, where he claimed victories that seem hollow in their consequence. Simply maintaining the empire had quietly become exhausting, and a margin of resilience had been lost. The expansion of the Roman Empire was premised on the growth that had made it possible. The plague was a shock to the system. The loss of population was immediately felt in the military recruitment crisis, but in the longer term it subtly shifted pressures deep beneath the surface. Recruitment was more difficult, and consequently the inducements had to be more lucrative. The provincials had earned their way to prominence through service in the name of empire, and the reckoning was soon at hand.[96]

The senator and historian Cassius Dio, one of those provincials who would rise to the commanding heights of the empire in the aftermath of the crisis, reflected on the ambiguous legacy of Marcus and his times. Marcus "did not meet with the good fortune that he deserved, for he was not strong in body and was involved in a multitude of troubles throughout practically his entire reign. But for my part, I admire him all the more for this very reason, that amid unusual and extraordinary difficulties he both survived himself and preserved the empire." It still seems a fair and considered verdict on the accomplishments of a man whose lot was to struggle against the shifting of the tides.

Despite the good offices of Marcus, the miraculous efflorescence of the *pax Romana* was clipped in its bloom. The empire survived, but the chill gusts of a new age can be felt already in the emperor's Stoic reflections, as they are known to us from his remarkable diary. "As soon as a man has prepared the dead for burial, it is then his turn to be buried, all in a mere moment. So in the end, always keep your sight fixed on how ephemeral and worthless human affairs truly are. That which oozes in the body one day, becomes cadaver and ashes tomorrow. . . . Therefore stand high, like a rock. Beaten ceaselessly by the waves, it holds strong and calms the swell of the waters around it."[97]

The Roman Empire was a survivor. But the age of pandemics had arrived, and in future encounters with new germs, the empire was not to prove entirely equal to the challenges that nature was laying in store.

The Old Age of the World

A Millennium of Empire

On the 21st of April in AD 248, the city of Rome celebrated its 1000th birthday. For three days and three nights, the haze of burnt offerings and the sound of sacred hymns filled the streets. A veritable zoo, of the most extravagant creatures from around the world, was offered to the people, and massacred: thirty-two elephants, ten elk, ten tigers, sixty lions, thirty leopards, six hippopotami, ten giraffes, one rhinoceros (hard to come by, but incomparably fascinating), and countless other wild beasts, not to mention a thousand pairs of gladiators. These *ludi saeculares*, the traditional "century games" that Rome held to mark centennial anniversaries, summoned forth a host of archaic memories, "skilfully adapted to inspire the superstitious mind with deep and solemn reverence," in the words of Gibbon. The celebration still carried shadowy associations with the underworld and the diversion of pestilence. Despite the deliberate primitivism of the rites, the *ludi saeculares* could be credited, like so much else, as a creative rediscovery of the imperial founder, Augustus. The *ludi saeculares* were in every sense an imperial affair, a stage-crafted display of the awesome power that Rome enjoyed, uninterrupted for centuries on end. Little did contemporaries know they were witnessing a sort of valediction, the last secular games Rome would ever see.[1]

It is easy, at our distance, to imagine that there was some measure of denial in such an exuberant celebration of the Roman millennium—that the inhabitants of Rome were enjoying the ancient equivalent of cocktails on the deck of the Titanic. But we must not be blinded by hindsight. Rome in AD 248 offered much to inspire a sense of familiarity and confidence. Just a generation before, the "navel of the city," the *umbilicus urbis*, had been grandly refurbished, a monument affirming that Rome was the center of the world. The *pomerium*, the urban boundary, remained a construct of the imagination in an unwalled city that sprawled into its hilly countryside. The coins, including the very issues minted in AD 248 to honor the games, maintained their ponderous texture of true silver, so that to hold one of

Figure 4.1. Silver Coin (Antoninianus) of Emperor Philip Celebrating Millennium Games (American Numismatic Society)

them even today is to feel the combination of precious metal and public trust that steadied the value of the imperial money. We have a flavor of the confidently patriotic prayers whispered at the games: "For the security and eternity of the Empire, you should frequent, with all due worship and veneration of the immortal gods, the most sacred shrines for the rendering and giving of thanks, so that the immortal gods may pass on to future generations what our ancestors have built up." The secular games were an omnibus act of religious piety, mobilizing all of the city's most archaic reserves of cultic energy in an effusion of thanksgiving and supplication for the eternal empire.[2]

The emperor presiding over the spectacle on this occasion was Marcus Julius Philippus, or Philip the Arab. Hailing from the southern reaches of Syria, he was not a conspicuous outsider. The steady integration of the provinces had long since effaced the distinction between rulers and subjects. His reign began in a storm of confusion, amid a failed invasion of Rome's eastern neighbor that took the life of his predecessor; but Philip had skillfully extricated the Roman army, at a dear price, and headed for Rome, leaving the eastern provinces safely under the protectorate of his brother. Philip's reign started with an impressive show of energy: administrative reforms were attempted in Egypt, and a great burst of road improvements have been detected in places so removed as Mauretania and Britain. A satisfying victory was achieved against the northern barbarians, and, in AD 248, he was able to return to Rome to celebrate the millennial anniversary. As Philip clearly recognized, the City herself demanded obeisance as the focal point of power, at the nexus of people, army, and senate. In Rome, still, campaigns were planned, careers plotted, fortunes decided.[3]

The Rome of Philip would have felt familiar to Augustus. And yet, just one generation on, we find ourselves in a truly alien world. The serene confidence of the empire had been rudely shaken. Hulking stone fortifications, the Aurelian Walls, went up round a city where distance and mystique had so recently seemed protection enough. The silver had vanished from coins that were now not much more than crude wafers, spewed in desperate superabundance from the mints. A truly new kind of man—the Danubian soldier with little time or awe for the *urbs* itself—had irreversibly wrested control of the state from the moneyed senatorial aristocracy. Careers were made and unmade in the barracks of northern garrison towns, rather than in the old capital. Beneath the imperial city itself, in the maze of burial caverns known as the catacombs, there is evidence that the obscure cult of Christianity was, for the first time, making uncanny strides toward becoming more than a marginal curiosity. In short, in the space of a single generation, the lineaments of an entirely new age, the period we now call late antiquity, had come into view.

This generation of headlong change is cloaked in obscurity. The murder of Philip in AD 249 touched off a spiral of dissolution that would engulf the entire imperial order. Historians know these times as the "crisis of the third century." The empire seemed to pass under a maleficent star. All at once, aggressive enemies on the eastern and northern frontiers pushed into the empire; the teetering dynastic system was exposed, while in quick succession one usurper after another spilled civil blood in pursuit of the crown. Fiscal crisis was the inevitable consequence of war and intrigue.

With the advantage of hindsight, historians have had no trouble finding the roots of this crisis. The collection of causes gives the crisis of the third century an air of inevitability; it seems overdetermined. The last thing we might seem to need is another cause to add to the crowded queue. But, to introduce environmental crisis into the story is only to be faithful to the insistent evidence for the agency of climate change and pandemic disease. It might also inject a healthful sense of the circumstantiality of the crisis, which was not just the inevitable release of long accumulated pressure. The concatenation of very specific and sudden blows to the Roman Empire in the 240s and 250s forced the system beyond the threshold of resilience. A withering drought and a pandemic disease event to rival the Antonine Plague lashed the empire with a force that was an order of magnitude greater than the combined menace of Gothic and Persian intrusions. The collapse of frontiers, dynasties, and fiscal order was as much the *consequence* as the cause of the crisis. The edifice of empire buckled along the seams of structural fragility, but the blows from without provided the fresh destructive force.[4]

The language of "crisis" derives from Greek medical terminology. The crisis is the turning point of an acute illness, when the patient succumbs or recovers. It is an apt metaphor for the empire in the middle of the third century. It provokes us to remember that, by ca. AD 260, there was no guarantee of Rome's future. The frontier network had utterly failed; great chunks of the empire, both east and west, had cleaved themselves off under breakaway rulers; basic routines of governance vanished. The centrifugal force might well have prevailed.

Yet the patient recovered. Under the forceful leadership of a series of Danubian military officers, most of the empire was reassembled. But here the metaphor of crisis is stretched to its limits. The healed patient was not quite the same in the aftermath. The empire that reemerged was based on a new equilibrium, with new tensions and new harmonies of state and society. It required more than a generation of trial and learning to calibrate, but what emerged from the rubble of crisis has been rightly described as a "new empire." Whereas the Antonine crisis had sapped the empire's batteries of stored energy but left the foundations intact, the crisis of the third century was transformational. It should be called the first fall of the Roman Empire, and even in this dimly lit corner of the Roman past, we can see that the environment was a protagonist in turning imperial fortunes.[5]

If the purpose of the *ludi saeculares* was to invoke divine favor and ward off pestilence, the rites soon proved a stupendous failure. It was a point that was surely not lost on contemporaries.

The Long Antonine Age: The Severan Empire

The marriage of Marcus Aurelius and his wife Faustina was, even by Roman standards, prolific. But of their fourteen children, only one male descendant, Commodus, who had been placed under the medical supervision of Galen, survived his parents. He was enough. The lucky run of emperors without a male heir came to an end, and immediately the empire reverted to the biological principle of succession. The seventeenth emperor of Rome, Commodus was the first who had been born to the purple, reared from the cradle as the prince.

During his twelve years of rule, the empire found its footing after the trauma of war and pestilence. But Commodus lacked the civility of his father, and relations with the senate turned from sour to deadly. In AD 190–91, epidemic disease returned to the city with a vengeance, in concert with a gripping food shortage that spread from Egypt to Rome. Recriminations flew. The senate blamed the malfeasance of the emperor's cronies. A conspiracy was cautiously hatched; under the emperor's nose, safe men were appointed to critical posts; on New Year's Eve of AD 192, Commodus was strangled in the palace. The dynasty was toppled.[6]

The eventual winner of the imperial sweepstakes was a middling senator of modest physical stature and unexceptional accomplishment named Septimius Severus. His was a very Roman story. He was born in the middle of the reign of Antoninus Pius, in AD 145, just a year after Aelius Aristides delivered his hymn to the greatness of Rome. His hometown was Lepcis Magna, a Punic town on the Mediterranean coast that was practically a model of Romanization. The first Latin inscription dates to 8 BC. A temple of the Punic deity Milk'ashtart was reconsecrated as a temple of "Roma and Augustus." The accouterments of a Greco-Roman town came quickly: amphitheater, porticos, baths, aqueduct, arches. In the later first century, Lepcis was granted the status of a *municipium*, a town whose elected magistrates automatically became Roman citizens. Under Trajan, Lepcis became a *colonia*, all its citizens now citizens of Rome. Even in a city that boasted tremendous olive oil wealth, the ancestors of Septimius Severus stood out, vaulting to the highest echelons of Roman society. They paved the way for Septimius to follow a senatorial career, serving the empire from Syria to Gaul. When the coup took down Commodus, Septimius had been posted as governor of the militarized province of Upper Pannonia. The situation in Rome spiraled out of control, and Septimius was hailed as emperor by his troops.[7]

Although he himself was a great believer in astrology, there was nothing particularly foreordained about his success. Yet Septimius Severus was to prove one of Rome's most influential dynastic builders.

The dynasty he built would endure for more than four decades. It is important to see it in the right profile. Septimius soon styled himself a son of the Antonine dynasty. While this was an audacious fiction, the advertisement of Antonine heritage aptly expressed the fact that his empire was more an extension of the previous age than a premonition of darker times over the horizon. Historians have lately cut the crisis of the third century down to size, to a delimited period stretching from the middle of the 240s to the middle of the 270s. The rehabilitation of the Severan dynasty is an inseparable adjunct of this shorter, sharper crisis. The negative judgment of the ancient historians contemporary with the dynasty long colored modern opinion. Cassius Dio considered the end of Marcus' reign the end of a golden age and the beginning of an age of "iron and rust." But pessimism was absolutely *de rigueur* in Roman historiography (things were always getting worse), and Dio reflects the exquisite distaste of the senatorial order for the later representatives of the Severan dynasty, in which women played a prominent role. The deep veins of misogyny and strained relations between emperor and senate should not darken the achievements of a manifestly accomplished imperial dynasty.[8]

Septimius Severus was a wealthy senator from a coastal Mediterranean hub. He was not by any stretch an army man. His military credentials at the time of his accession were modest at best, far less impressive than other dynasty builders like Augustus, Vespasian, or Trajan. Septimius had to build his military resume on the go, washing away the distasteful memory of bitter civil war with a hasty but successful invasion of Parthia and a massive campaign to finish the conquest of northern Britain. Septimius had the army to thank for his power, and he harbored no illusions on this score. His advice to his sons, "get along, enrich the troops, and care little about everyone else," betrays his practical outlook. After the death of Commodus, the real "secret of the empire" had been revealed, that the army could be used as instrument of blunt force to seize the mantle of power. But, in the case of Septimius, the instrument was still wielded by a man of the senatorial order, a commander drawn from the ranks of the civilian class. And the commander, in line with the best of Roman traditions, would reward his loyal base in turn.[9]

The triumph of Septimius was an undisguised boon for the provincials. The sons and grandsons of Roman colonists strewn around the western Mediterranean had risen inexorably from the later first century. But with the Severans, we observe the entry of a fully provincial elite into the senate and the palace. The wars under Marcus, in combination with the demographic upheaval of the pandemic, had accelerated the entry of talented provincials into the upper ranks of the imperial order. An entire brigade of talented and wealthy Africans "stormed the heights" under the Antonines. Septimius followed in their stead, and the dynasty he built unleashed the full potential of the provinces.[10]

Fittingly, when his first wife—an obscure, hometown girl—passed away, Septimius, then governing Gaul, proposed to a daughter of the Syrian aristocracy named Julia Domna. The offer of engagement traveled a mere 4400 kilometers from Lugdunum to Emesa! This match made of empire became the core of a Libyan-Syrian dynasty that brought a distinctive style and openness to imperial culture. Septimius oversaw the full integration of Egypt into the mainstream of imperial society—a proper town council for

Alexandria and the entry of Egyptians into the senate. Septimius was not abashed to show his Libyan origins, and it was a heyday for North Africa. Early in his life, Septimius had a dream in which he looked down from a mountain on the whole world and saw it singing in harmony. Septimius was an active dreamer, but this one captures something of what his dynasty accomplished.[11]

The crowning moment was left to his son, Caracalla. In AD 212, at a stroke, he granted citizenship to all free inhabitants of the empire. The "Antonine Constitution" erased the already tenuous distinction between imperial rulers and colonized subjects. Universal enfranchisement belatedly affirmed that the Roman Empire had become a territorial state. It was a watershed. Mere moments after its enactment, we happen to find the denizens of a remote village, tucked in the mountainous folds of southern Macedonia, trying to sort out what their new status meant for customary relations between patrons and their freed slaves. A little later, we find women on the fringes of the Syrian desert asserting their rights to property ownership . . . by invoking the legislation of the emperor Augustus. Over the course of the third century, the diffusion of Roman law picked up pace as the new citizens learned to bend Roman law to their purposes. By the century's end, a traditional handbook for orators discouraged speakers from trying to flatter a city by praising its laws, "since the laws of the Romans are used by all."[12]

Not by accident was the Severan period the apex of classical Roman law. The greater portion of Justinian's *Digest* is comprised of excerpts from Severan jurists. The most conservative of all intellectual disciplines found its finest exponents in a series of officials from the eastern edges of empire. The jurists Papinian and Ulpian were both Syrians, and both served the Severan administration in the highest capacities. The spread of citizenship was matched by a higher degree of professionalism in the practice of law, and in the case of Ulpian we can say that some of his greatest writing was called forth by the need to equip governors for the challenge of responding to the new citizens. The law school in Beirut was established, destined quickly to become the epicenter of legal life and learning. Nothing more eloquently testifies to the decentering of imperial culture in the Severan age than the provincial contribution to Roman jurisprudence.[13]

The talent of the provinces found an outlet in the growing ranks of the imperial administration. The early Roman Empire was characterized by a "deficit of officials"; the central administration was a wispy cover, tossed over the sturdy civic foundations of public life. The expansion of the central imperial offices was an inevitable and organic process that unfolded in tandem with Romanization and the diffusion of market-based institutions.

Under the Severans, the pace quickened. The second aristocratic order, the equestrian, was energetically broadened; in the third century, there were still gentleman equestrians, but an increasing number of civil and military offices at the equestrian grade swelled the ranks of imperial knights. There is no need to see the senatorial and equestrian orders in conflict or tension in the Severan age. Throughout the reign of Septimius, senators "virtually monopolized the senior administrative posts and army commands." The Severan empire respectfully guarded the exalted place of the senate in running the empire, but the professional ranks of imperial service were now broader, and more representative of the vast territories under Roman rule.[14]

The most important political change in the age of the Severans was a subtle shift of power to the army. Augustus had successfully deweaponized the army as a political instrument, but the events that brought Septimius to the helm flashed its true potential. The consequences were felt in the purse. Early in his reign, Septimius gave the troops a 100 percent raise. The average legionary saw his pay increase from 300 to 600 *denarii* per year. The appreciation was long overdue. The soldiers had not seen a pay hike since AD 83–84, in the reign of Domitian. If the Egyptian evidence is broadly indicative, the years after the Antonine Plague had witnessed a doubling in nominal prices, so that the raise under Septimius was equivalent to a belated cost-of-living adjustment.[15]

But the raise may also signal something even more subtle and profound. The Roman state had always managed to field an army of nearly half a million men at arms with a light touch. The higher salary is only one sign that military recruitment was destined to become a more serious chore in the years to come. But it was not yet a crisis: Septimius succeeded in enrolling three new legions without manifest strain, and enlistment remained voluntary. Septimius did concede active-duty servicemen the right to marry, breaking a centuries-old tradition in which enforced bachelorhood was part and parcel of the discipline of a professional army. The right to marry was surely no small inducement to service, and it slowly changed the complexion of the military. In sum, Septimius' concessions to the troops were part power politics, part overdue adjustment, and part recruitment strategy.[16]

The fruits of Severan success were abundant. A bloom of cultural efflorescence, more inclusive than ever before, unfolded. The influx of provincial talent was a jolt to Severan culture. The ancient capital remained the focal point of imperial patronage. The building program of Septimius in Rome was ambitious, swaggering into dialogue with the constructions of the emperor Augustus. The arch of Septimius required rebuilding the *umbilicus urbis*, adjacent to the golden milestone of Augustus, where all roads symbolically converged. The grand Temple of Peace, destroyed by a calamitous fire under Commodus (much to the regret of Galen, who lost writings and precious drugs in the disaster), was rebuilt with élan; giant columns of red Aswan granite imposed on the viewer from the outside, while inside the extraordinary marble map, known as the *Forma Urbis Romae*, spread some 60 × 40 feet, showing every corner of the city with the intention of overwhelming the eyes. Septimius erected the Septizodium, a massive façade honoring the seven planetary gods, where the Via Appia met the Palatine Hill in the heart of the city. Caracalla sponsored monumental baths, and the last of the Severans, Alexander, constructed the final aqueduct of Rome. Great watermills and giant granaries went up around the city.[17]

At the time no one knew they were enjoying the last great burst of monumental public building in the classical Mediterranean; it was followed by an abrupt hiatus, before the cycle of church construction in late antiquity resumed the spirit of monumentality in a new guise. The building boom is but one sign that the Severan period was an age of economic and demographic recovery.

It was during these decades that the sour churchman Tertullian could declare, "it is clear to behold that the world itself is more intensely cultivated and built up than in olden times. All places are now crossed by roads, all are known, all are open for business. The most pleasant estates have obliterated what were once notorious wastelands. The deep forest yields to the ploughed field. Wild beasts flee before our herds. The desert is sown, and rocky fields are planted. The marshes have been drained, and there are now more great cities than there were once mere houses. None now fear the lonely isle or dread their craggy shore. Everywhere there are houses, everywhere people, everywhere the city, everywhere life! And the greatest testimony of all is the abundance of the human race." We might doubt these rosy observations, if they had been offered in a spirit of flattery. Tertullian had graver purposes: the talented polemicist needed to find credible proof against the doctrine of the transmigration of souls, and the unprecedented number of humans walking the earth seemed a glaring obstacle to the doctrine's logic![18]

The demographic recovery proceeded without the interruption of major epidemics. While the smallpox virus could have become endemic in the larger cities of the empire, there are no reports of the disease between the recurrence in Rome in AD 190–91 and scattered references in later centuries. The absence of evidence is never conclusive, but on balance the silence suggests that the pandemic burned itself out or hid in corners where its impact was limited. The retreat prepared the way for population rebound.

It has been the impression among papyrologists that the population of Egypt expanded again, though it never reached its pre-Antonine peaks. The village wasted by the pestilence, Soknopaiou Nesos, was clearly hanging on during the Severan period, and it is documented at least down to AD 239. The village of Karanis revived in the early third century and then virtually disappeared in the middle of the century, before another revival toward its end. Other cases follow the pattern. Oxyrhynchus, one of the best documented towns of Roman Egypt, is estimated to have been home to 11,901 souls in AD 199 and then to have grown to ca. 21,000 by AD 235: while the rate of growth implied by these numbers is too high, the direction of change is at least indicative. Broadly, the literary, papyri, and archaeological records agree that the Severan age was a period of demographic resurgence.[19]

Under the Severans the empire recovered its balance. If there was a corrosive agent in the new order, it was the brute revelation of the army's power. The genie could not be put back in the bottle. The son and successor of Septimius, Caracalla, after disposing of his brother, threw himself behind the soldiers. He increased the pay of the ordinary legionary by 50 percent again, to 900 *denarii* per year. While Septimius had debased the silver coinage early in his reign, the repercussions had been minimal. The fiscal exigencies, or sheer pride, of Caracalla required a more radical sleight of hand. He experimented with a new silver coin, the *antoninianus*, valued at two *denarii* but containing only 80 percent of the silver of two *denarii*. Yet the introduction of the new coin seems not to have provoked trouble. The state rigorously maintained that the public coinage embodied a face value established by fiat, not by the market value of the precious metal content. Remarkably, it worked. The *denarii*, with higher silver content, were not driven out of circulation, and there is no evidence for nominal inflation. The coinage was increasingly a fiduciary currency. Only with the benefit of hindsight does it seem that the Romans had built a pier, swaying out over the abyss.[20]

With the exception of a brief interlude on the death of Caracalla, the Severan dynasty ruled until AD 235. Its last representative, Alexander Severus, was cut down by his own men on campaign along the Rhine. The claimant was a man named Maximinus. An equestrian from the military gentry of the lower Danube, he was the first true outsider to hold the imperial throne. Maximinus would be remembered as a savage. He seems to have malingered in the north on campaign, despite the fact that the senate had confirmed his rule. He sent dispatches of his victories to the capital but also installed paintings of his campaigns outside the senate house. Judging from the silver content of his coins, he was able to maintain, despite the expense of his military operations, the financial equilibrium of the later Severans.

But in his disregard for the power politics of Rome, he was too far ahead of his time.

In the spring of AD 238, his regime folded. It was a textbook legitimation crisis. The revolt started in distant North Africa, where the locals refused to bear the crushing fiscal expectations of his agents. A rather bungling senatorial coup still managed to topple his regime. The career of Maximinus shows that sometimes the first act of history is the farce. Maximinus was a harbinger, but the age of the barracks emperors was not yet at hand.[21]

The Old Age of the World: Climate Change in the Third Century

It is irresistible, in retrospect, to see the career of Maximinus as a prelude. But that presumes too much about the next act. In AD 238, the senate resumed control of affairs, and soon the thirteen-year-old Gordian III was alone in power. He was capably advised by remnants of the Severan elite. He set off for the east to answer Persian aggression in northern Mesopotamia, and by AD 242, exactly eighty years after Lucius Verus, he arrived with a massive entourage at Antioch. Within two years, after a botched campaign, Gordian III was dead, deep behind enemy lines. Philip was hailed as emperor and hurriedly extricated the army, for an indemnity of 500,000 *aurei* (gold coins). The situation was not desperate. He "calmly" worked his way to Rome, stopping in cities throughout the east, Asia Minor, the Balkans, "much in the manner of princes who had ruled a more quiescent empire." He arrived in the capital and took up residence in the palace. In a short time, Philip proved an active administrator. A denizen of the imperial city in his reign might be forgiven for believing it was business as usual. But, within a year of the exuberant celebration of Rome's millennium, the fabric of the empire started to come unraveled.[22]

The Roman Empire had seen dynastic instability before. It had suffered humbling losses and survived years of dearth. But what was poised to transpire, starting in the later 240s, was without precedent: a comprehensive breakdown of the frontier system, the total demise of an ancient monetary regime, more-than-transient rival emperors inside Roman territories. The next years would see cascading change that shattered all centralized institutional control of circumstances. The crisis was "so extreme in itself that the Empire's survival is almost surprising." It is true that the margin of resilience had been eroded by the gradual progress of time and circumstance.

But contemporaries were aware of the sudden, wrenching, environmental background of the crisis, and to the congested list of causes behind the crisis we ought to add the shocks of climate perturbation and pandemic disease.[23]

Christians in this time of trouble would coin the idea that they lived in the "old age of the world." It was a metaphor they came to elaborate in a war of ideas. For in the midst of crisis, an untimely public spat erupted about the nature of the gods. The emperors soon fixed blame for the crisis on the failure of the Christians to worship the gods properly. The Christians protested that, in reality, the earth itself was simply passing into senescence. We would do well to take this polemic seriously, on its own terms, for it was articulated in a very specific key by highly trained rhetoricians. Less than a generation after Tertullian feasted his eyes on the ebullient vitality of civilization in Roman Africa, another Carthaginian, Cyprian, had come to believe it was obvious that "the world has grown old and does not stand in the vigor whereby it once stood, nor do the strength and liveliness that once availed it still abide. . . . In winter there is not such an abundance of rains to nourish the seeds. The summer sun burns less bright over the fields of grain. The temperance of spring is no longer for rejoicing, and the ripening fruit does not hang from autumn trees."[24]

Scholars have rummaged the libraries of ancient philosophy looking for the ancestry of Cyprian's metaphor. But we have somehow not taken seriously the most direct source of the metaphor's potency, the biological assumptions about aging. For the ancients, to age was to become cold and dry. The young were hot and moist, fervid with energy. These concepts were expressed with clarity in ancient conversations about diet. The young, for example, had to take care with wine, which threatened to overheat their already ardent systems. The excess of heat loosened their self-control, and its disinhibiting qualities made wine, in the words of a second-century novel, a kind of "sex fuel." But for the elderly, the warm wash of wine was invigorating. It slowed the desiccation of the body. Galen wrote often of the "dry nature of old people's bodies. The very reason that each part becomes dry is that it is unable to receive the same degree of nourishment because of the weakness of the heat." To grow old was a prolonged evaporation, leading ultimately to chill death. "Since death is the extinguishing of the innate heat, old age is, as it were, its fading away."

This view of aging is precisely what Cyprian had in mind, when he claimed that the world had grown grey. "The falling rays of the setting sun are not so bright or brilliantly fiery. . . . The fountain that once overflowed from abundant springs, now forsaken by old age, scarcely yields a drop." For

Cyprian, the world itself had become cold and dry. The world was a pale old man, leaning into the grave.[25]

The natural archives prove our human witnesses faithful. The smiling days of the Roman Climate Optimum came tripping to an end in the later second century. The break was not sharp. The RCO quietly faded away, and what replaced it was the Late Roman Transition, a period of indecision and disorganization, of sharper variability, lasting some three centuries. The changes were global in scale. Solar variability was the main external forcing mechanism. The sun weakened on the Romans. The beryllium isotope record shows a precipitous drop in insolation in the AD 240s. Cooling followed. In the Alps, after centuries of melt, the ice of the Great Aletsch started creeping down the mountain. So did the Mer de Glace glacier in the Mont Blanc Basin. Records as far apart as Spain, Austria, and Thrace show a coordinated bout of cooling. Cyprian was probably right to sense the chill winds of a cooler age in the middle of the third century.[26]

The outstanding feature of the RCO had been anomalous humidity across the Mediterranean. In the RCO, the long march of the Holocene toward greater aridity had taken a pause. But when the RCO broke, the effects of a longer cycle of aridification were unmasked.

In the short term, the AD 240s stand out as a moment of piercing drought in the southern rim of the Mediterranean. Drought parched Cyprian's North Africa. The bishop's public defense of Christianity was pitched to a society that had just survived a wrenching spell of aridity. Christians were inevitably blamed "if the rains fall from above but rarely, if the land is given over to dust and becomes desolate, if the barren earth sprouts hardly a few pale and thirsty blades of grass . . . if the drought causes the spring to cease." The failure of the skies left the cities short of food, but Cyprian acidly criticized the storehouses of the rich, who sought to profit in the crisis. The entire crisis was an evangelical moment, an invitation to the security of a faith that promised life beyond the present distress. "If the vine fails, the olive tree cheats us, and the burning field withers with crops dying in the drought, what is that to the Christians?" The desiccating landscape was the background of Cyprian's entire performance as a Christian spokesman.[27]

At the same time, drought struck in Palestine. Abutting the desert, the agricultural belt of the Levant always awaited the coming of the rains with pious suspense. In the rabbinic texts of the second and third centuries, precipitation is virtually a miracle. The hardness of the land was deeply embedded in the contemporary worldview; since the destruction of the Temple in AD 70, there had been dryness in the land. The monuments of rabbinic literature might not be the safest place to search for unbiased climatological records, but the memories of drought surrounding the sages of the AD 230s–240s are insistent, and we may posit a historical substratum to the legends of the rabbis. Ḥanina bar Ḥama was major rabbinic figure, a protégé of the great Judah I, who played a leading role in the school at Sepphoris and lived to ripe old age (died ~AD 250). In the stories attached to him, drought is an overbearing problem. In one episode, the rains for a time failed both in the Galilee and to the south in Judea. A rabbi in the south made it rain by instituting a public fast, while the drought in Sepphoris endured because "their hearts are hard." Eventually the waters came, but the memories of an epochal drought, and its long-awaited alleviation, clung to the memory of this leading rabbi.[28]

In straitened circumstances, the empire could rely on Egypt. The green ribbon of the Nile valley was miraculously fertile. This was the empire's great insurance policy. The valley's unique ecology hedged the empire against the petty vagaries of the Mediterranean climate. The Nile River drains two main branches. Its steady baseline flow discharges the White Nile, whose headwaters lie in equatorial Africa. The annual inundation—the surfeit of water and silt that rise above the baseline flow—is the handiwork of the Blue Nile. Some 90 percent of the Nile's floodwaters originate in monsoon rains that fall in East Africa in the summer; the Blue Nile gathers the runoff of the highlands in Ethiopia and carries it downstream, where it joins the regular flow of the White Nile at Khartoum. The result is the greatest natural irrigation pump in the world, harnessed by human civilization millennia before the coming of the Romans. The life-bringing waters and fertile silt rendered Egyptian agriculture exceptionally productive. Egypt was the breadbasket of Rome, and a boon to much of the empire.[29]

The yearly rise and fall of the river was a sacred rhythm, anticipated with hopeful prayers. As the ancients knew all too well, the divine gift of the flood was not constant. In the course of a lifetime, any priest or peasant watched good years and bad years pass by. What even their carefully trained eyes could not have noticed were the imperceptibly slow but ultimately decisive cycles of change beneath these annual variations.

In the very long term, over the millennia of the later Holocene, the Nile's discharge has gradually declined, as the monsoon belt has shifted southward and pulled the Intertropical Convergence Zone with it. Against the backdrop of this broader secular shift, on shorter timescales lasting decades or centuries, the Nile flood has been alternatingly dependable or erratic. Like the crests and troughs of a business cycle, the Nile flood has had lengthy mood swings that could affect the course of civilization along the valley

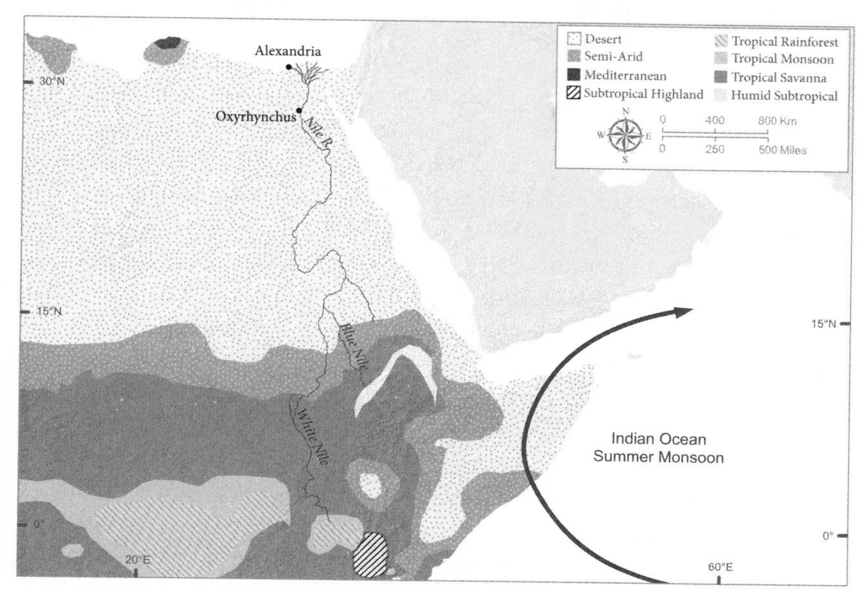

Map 11. Nile Hydrology and Climate Mechanisms

and beyond. For the period after AD 641, these phases can be followed in the world's oldest, continuous human record of climate: the Nilometer readings preserved by Arabic chronicles. In earlier periods, the record is patchy and indirect. But the evidence we do have argues that the centuries of Roman rule witnessed profound changes in the Nile's behavior.[30]

The Nile records again suggest that the Roman Empire's builders had benefitted from impeccable timing. Michael McCormick and I assembled a database of flood quality in the centuries of the early empire, based on earlier collations of the papyri data (often indirect and uncertain) for good and bad floods in the Roman period. The Nile record parts into two distinct phases, one running from the annexation by Augustus down to ca. AD 155, the second from AD 156 to the end of the third century. The earlier period was marked by more dependable inundations and a higher proportion of excellent floods; the later period saw a disproportionate number of the worst floods.

Moreover, in precisely the same years that witnessed the phase shift, the AD 150s, for the first time a new kind of document, the "declaration of unflooded land," appears in the papyri. Its origins are obscure, but these declarations may well have been a response to the onset of a more erratic regime of Nile flooding.[31]

The physical evidence for Nile variability is, alas, more indirect. There is a strong connection between the Nile inundation and the mode of global climate variability known as the El Niño–Southern Oscillation (ENSO). In El Niño years, the waters of the eastern Pacific are warmed, and the monsoon rains far to the west are suppressed; a strong El Niño is correlated with weak Nile floods. Today El Niños occur every 3–5 years, but ENSO periodicity has varied over time.

Unfortunately, high-resolution ENSO records going back to the first millennium remain rare and uncertain. But one sedimentation record from Ecuador suggests that during the Roman Climate Optimum, ENSO events were very rare (once every 20 years or so). The quiescent ENSO meant an active and reliable flood regime in Egypt, and it marks yet another way that the RCO exhibited features resembling the mid-Holocene. Then, in the centuries of the Roman Transitional Period, ENSO events became extremely common—every third year or so. The unusual good fortune of the Romans ran out, long after they had come to depend on levels of Egyptian productivity that assumed anomalously favorable conditions.[32]

What is not in any doubt is that, just when the Romans most needed a buffer against bad fortune, the Nile abandoned them spectacularly.

In AD 244, the waters failed to rise. In AD 245 or 246, the floods were weak again. By March of AD 246, before the harvest, public officials in Oxyrhynchus were taking emergency measures otherwise unparalleled in the record. There was a command to register all private stocks of grain, within twenty-four hours, under threat of drastic penalties. The state carried out compulsory purchases, at shockingly high prices, 24 *drachmai* per *artaba*. Normally the government set prices that were favorable to itself, but 24 *drachmai* was dear: about twice what we might expect for the period, implying acute desperation to acquire grain even at a high cost. Two years later, in AD 248, the shortage was still a gripping problem. A papyrus of that year refers to the "present emergency" and a scramble to fill the offices handling the public food supply. In another papyrus of AD 248, an individual refused to fulfill the obligatory office of food supply, surrendering all his belongings to dodge it. At this same moment, the bishop of Alexandria claimed that the riverbed was as parched as the desert—which, if it is not just a rhetorical figure, actually points to the simultaneous failure of the White and Blue Niles. In all, this amounts to the severest environmental crisis detectible at any point in the seven centuries of Roman Egypt.[33]

The climatic turbulence came at an inauspicious time. Much has been made of the payoff to secure the Roman army's retreat from Persia: 500,000 *aurei*. That was an exorbitant ransom. But we can crudely estimate the

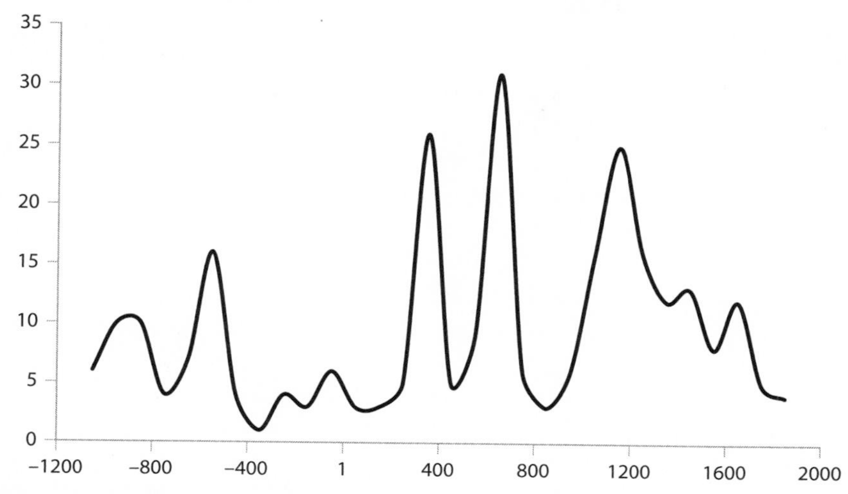

Figure 4.2. El Niño Events per Century (data from Moy et al. 2002)

impact of a provincial-scale drought in Egypt, if only to attune our imaginations to the possibilities. The wheat crop on a plot of land depended on any number of factors, including the quality of the land sown. But the flood was the silent partner in the farming business. On one well-known third-century estate, wheat yields on a series of arable plots within the same region ranged, in the space of a few years, from 7 to 16.6 *artabas* (the unit of dry measure, equivalent to 38.8 liters) per *aroura* (the unit of land, equivalent to .2756 hectares). Based on an average of ~12 *artabas* per *aroura*, the annual gross production of Egypt has been estimated at 83 million *artabas*. If a year with a poor flood reduced yields by only 10 percent, which seems a conservative estimate, the total economic cost to the province was 8.3 million *artabas*, at contemporary prices equal to 1 million *aurei* or twice the payment to the Persian king, Shapur.

The Roman state extracted at least 4–8 million *artabas* of wheat from Egypt each year; if a drought cost the state only 20 percent of its annual tax revenue from Egypt, the value would be 96,000–192,000 *aurei*. In fact, the damage could have been multiples of this: when the Nile failed in medieval Egypt, gruesome starvation often followed. A run of consecutive poor floods was exponentially worse, as the margins of resilience wore thin. While we cannot be precise or certain, it is reasonable to conclude that drought was at least as implicated in the start of the crisis as the sunk costs of the failed invasion.[34]

The challenge for us is to suspend our knowledge of what comes next. The entire generation leading up to crisis was not a prelude to the inevitable.

The Severan and post-Severan emperors had achieved a kind of narrow equilibrium, but the concatenation of geopolitical and environmental shocks were a dangerous threat to the new order. The droughts of the AD 240s alone would have pushed the imperial system to the brink of what it could manage. But nature had still another unhappy twist waiting for the Romans. Not for the last time, spasms in the global climate system were closely followed by the advent of an unfamiliar infectious disease. The sweeping violence of a new pandemic was, ultimately, more than the structures of the empire could bear. Just a few years after the jubilant celebrations of Rome's eternity, the empire found its continued existence entirely uncertain.

The Plague of Cyprian: The Forgotten Pandemic

Cyprian was born in the boom years of Roman Carthage, in the reign of Septimius Severus, to a family of modest prosperity. He received a liberal education and became a teacher of rhetoric. That is the sum of our knowledge about the early life of a man destined to become the most important figure of the western church in the third century.

The meager biographical details do not help us to understand why, around AD 245–46, Cyprian made the highly eccentric decision to become a Christian. In the early third century, there were probably no more than a few hundred thousand Christians, lightly scattered across the empire. The pagan gods still unquestionably ruled in the hearths and temples of the Roman Empire. We should not miss what a stroke of good fortune it was for the Christian movement in Carthage to gain a literate, much less a truly educated, entrant. It was a coup. No time was wasted making the most of it, and by AD 248 Cyprian found himself the bishop of Carthage. The ten years of his episcopate, down to his martyrdom in AD 258, would prove to be among the most consequential in the history of the church, thanks in large measure to the pestilence which historical memory has attached to his name.[35]

The bishop's writings furnish the most vivid surviving testimony to the epidemic, and his legacy was soon associated with the event in Christian chronicles. From there, the plague went down in history connected to the name of Cyprian. It is a name that has often misled. The established view, represented in the solid tomes of the *Cambridge Ancient History*, describes the plague as "one which affected Africa in the mid-third century." Because William McNeill noticed the Plague of Cyprian in his history of infectious disease, it still finds mention in general histories of disease. But the Plague of Cyprian has fallen into complete oblivion among scholars of antiquity. In the most authoritative recent surveys of the period, it fails to garner even a passing remark.[36]

This neglect has many causes, including changing fashions that have tried to question the severity of the third-century crisis. But more subtly, the neglect originates in a failure to appreciate how exceptional true pandemic events have been. The simple fact of a mortality event attested contemporaneously at far ends of the empire merits close investigation. The Plague of Cyprian was not an episode in the life of third-century Carthage; it was a transcontinental disease event of rare magnitude.

The Plague of Cyprian struck in a period of history when basic facts are sometimes known barely or not at all. Yet the one fact that virtually all of our sources do agree upon is that a great pestilence defined the age. Inscriptions, papyri, archaeological remains, and textual sources collectively insist on the high stakes of the pandemic. In a recent study, I was able to count at least seven eye-witnesses, and a further six independent lines of transmission, whose testimony we can trace back to the experience of the pestilence. What is starkly lacking, however, is a Galen. The dumb luck of having a great and prolific doctor to guide us has run out. But, now, for the first time, we have Christian testimony. The church experienced a growth spurt during the generation of the plague, and the mortality left a deep impression in Christian memory. The pagan and Christian sources not only confirm one another. Their different tone and timbre give us a richer sense of the plague than we would otherwise possess.[37]

The pestilence came from Ethiopia and migrated north and west across the empire. So the chronicles tell us, and we might suspect slavish emulation of the plague account in Thucydides, the model literary description of a plague, familiar to every educated Greek. But two telling clues corroborate the possibility that again a microbial agent had invaded the empire from the southeast. First, archaeologists have discovered a mass grave adjacent to a body-disposal operation at the site of ancient Thebes, in Upper Egypt. Lime was mixed on site, to be poured over bodies that were then hastily incinerated. The disposal site dates to the middle of the third century, and the utter uniqueness of the corpse-burning and mass disposal enterprise argues that something about the disease had startled the inhabitants into extreme measures. The more decisive evidence for the pandemic's southern origin is provided by the bishop of Alexandria, who places the disease in the Egyptian metropolis by at least AD 249. The first dateable evidence for the pandemic in the west comes from AD 251, at Rome. The chronology affirms an eastern point of entry and vindicates the chronicles.[38]

The Plague of Cyprian raged for years. The chronicles report a plague lasting fifteen years, but it is unclear exactly which fifteen-year span they mean. There may have been a second wave sometime around AD 260. The emperor Claudius II in AD 270 was supposed to have been killed by a pestilence, but whether his death truly belongs to the same pandemic is entirely obscure. The sources insist upon a prolonged event, as the mortality coiled its way around the empire, with at least two pulses in the city of Rome. One of the later chronicles actually preserves the significant detail that some cities were struck twice. It is unfortunately impossible to be more precise. The Plague of Cyprian is in the background of imperial history from ca. AD 249 to AD 262, possibly with even later effects around AD 270.[39]

The geographic scope of the pestilence was vast. "There was almost no province of Rome, no city, no house, which was not attacked and emptied by this general pestilence." It "blighted the face of the whole earth." The plague of Cyprian is attested everywhere we have sources. It hit the largest cities like Alexandria, Antioch, Rome, and Carthage. It attacked the "cities of Greece" but also more remote urban places like Neocaesarea in Pontus and Oxyrhynchus in Egypt. According to one report, the Plague of Cyprian raced through town and countryside alike; it "afflicted cities and villages and destroyed whatever was left of mankind: no plague in previous times wrought such destruction of human life." The Plague of Cyprian was an empire-wide event.[40]

The lack of a medical witness like Galen is partly compensated by the vivid account of the disease in Cyprian's sermon on the mortality. The preacher sought to console an audience encircled by unfathomable suffering. It took no mercy on his Christians.

"The pain in the eyes, the attack of the fevers, and the ailment of all the limbs are the same among us and among the others, so long as we share the common flesh of this age." Cyprian tried to ennoble the victims of the disease, likening their strength in pain and death to the heroic intransigence of the martyrs. Cyprian conjured the symptoms for his hearers. "These are adduced as proof of faith: that, as the strength of the body is dissolved, the bowels dissipate in a flow; that a fire that begins in the inmost depths burns up into wounds in the throat; that the intestines are shaken with continuous vomiting; that the eyes are set on fire from the force of the blood; that the infection of the deadly putrefaction cuts off the feet or other extremities of some; and that as weakness prevails through the failures and losses of the bodies, the gait is crippled or the hearing is blocked or the vision is blinded."[41]

Cyprian's account is central to our understanding of the disease. The pathology included fatigue, bloody stool, fever, esophageal lesions, vomiting,

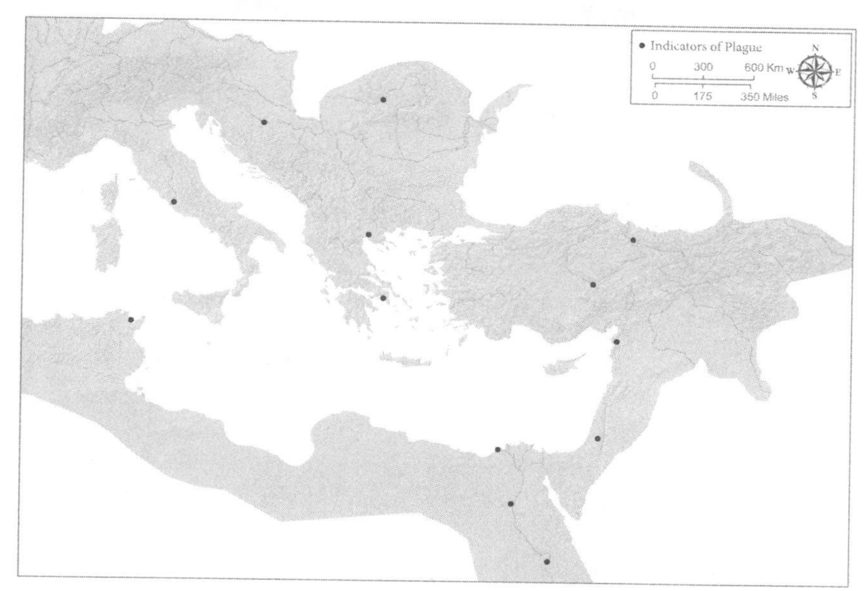

Map 12. Indications of Plague of Cyprian

conjunctival hemorrhaging, and severe infection in the extremities; debilitation, loss of hearing, and blindness followed in the aftermath. We can complement this record with more isolated and frankly uncertain hints from other witnesses. According to Cyprian's biographer, the disease was characterized by acute onset: "carrying off day by day with abrupt attack numberless people, every one from his own house." At a greater distance from the events, a folk tradition about the Plague of Cyprian from northern Asia Minor insisted on the sheer speed of the attack. "The affliction fell abruptly upon the people, penetrating faster than they expected, feeding on their houses like fire, so that the temples were filled with those laid low by the disease who had fled there in the hope of a cure." The same tradition remembered the insatiable thirst suffered by the victims of the disease (and here at last may be a merely ornamental emulation of Thucydides). "The springs and streams and cisterns were full of those burning with thirst because of the weakness brought on by the disease. But the water was too weak to quench the flame from deep within, leaving those once afflicted with the disease feeling just the same after the water as before."[42]

The course of the infection and illness was terrifying. This impression is confirmed by another North African eyewitness, a Christian not far removed from the circle of Cyprian, who insisted on the sheer unfamiliarity of the disease. "Do we not see the rites of death every day? Are we not witnessing strange forms of dying? Do we not behold disasters from some previously unknown kind of plague brought on by furious and prolonged diseases? And the massacre of wasted cities?" The pestilence, he argued, was a manifest encouragement to martyrdom, since those who died the glorious death were spared the "common fate of others amidst the bloody destruction of ravaging diseases." The Plague of Cyprian was not just another turn through the periodic cycle of epidemic mortality. It was something qualitatively new—and the evocation of its "bloody" destruction may not be empty rhetoric, if hemorrhagic symptoms are implied.[43]

The disease was of exotic origin and moved from southeast to northwest. It spread, over the course of two or three years, from Alexandria to other major coastal centers. The pandemic struck far and wide, in settlements large and small, deep into the interior of empire. It seemed "unusually relentless." It reversed the ordinary seasonality of death in the Roman Empire, starting in the autumn and abating in the following summer. The pestilence was indiscriminate; it struck regardless of age, sex, or station. The disease invaded "every house."[44]

One account predictably blamed the "corrupted air" that spread over the empire. But another chronicle tradition, going back to a good contemporary historian in Athens, recorded that the "disease was transmitted through the clothes or simply by sight." The observation is notable; in a culture without even a rudimentary sense of germs, the comment betrays a pretheoretical sense of contagion. The concern that the disease could be transmitted by clothing or eyesight suggests at least a dim awareness of an infectious origin. And it just might provide a further hint that the disease affected the eyes. The ancients harbored plenty of eccentric notions about the powers of eyesight, among them that it was tactile, ejecting a flow of particulates from the eye of the looker. The bloody eyes of Cyprian's victims may have presented a terrifying visage, in a culture where the eyes had the power to reach out and touch.[45]

The death toll was grim. We have an intriguingly specific report from the bishop of Alexandria, who claimed that "this immense city no longer contains as big a number of inhabitants, from infant children to those of extreme age, as it used to support of those described as hale old men. As for those from forty to seventy, they were then so much more numerous that their total is not reached now, though we have counted and registered as entitled to the public food ration all from fourteen to eighty; and those who look the youngest are now reckoned as equal in age to the oldest men of our earlier generation." The reckoning implies that the city's population had declined by ~62 percent (from something like 500,000 to 190,000). Not all of these need be dead of plague. Some may have fled in the chaos. And we can always suspect overheated rhetoric. But the number of citizens on the public grain dole is a tantalizingly credible detail, and all other witnesses agreed on the scale of the mortality. An Athenian historian claimed that 5,000 died each day. Witness after witness—dramatically if imprecisely—testified that depopulation was invariably the sequel of the pestilence. "The human race is wasted by the desolation of pestilence."[46]

These haphazard clues do not equip us well to identify the pathogenic agent of the Plague of Cyprian. But the range of suspects capable of causing a disease event of this scope is not large, and some possible agents can be almost certainly exculpated. Bubonic plague does not fit the pathology, seasonality, or population-level dynamics. Cholera, typhus, and measles are remote possibilities, but each poses insuperable problems. Smallpox must be a serious candidate. The two-generation lapse between the episode under Commodus and the Plague of Cyprian means that effectively the entire population would have been susceptible again. The hemorrhagic form of the disease might also account for some of the features described by Cyprian.

But in all the case for smallpox is weak. A North African author claimed it was an unprecedented disease (though whether he would have had any memory of previous smallpox epidemics is of course questionable). None of our sources describe the full-body rash that is the distinctive feature of smallpox. In the church history of Eusebius, written in the early fourth century, an outbreak more like smallpox was recounted in AD 312–13. Eusebius both called this a "different illness" than the Plague of Cyprian and also distinctly described the pustular rash. The exotic origins of the third-century event, again from beyond the Roman Empire, do not suggest the eruption of a now-endemic pathogen. Finally, the putrescent limbs and permanent debilitation of the Plague of Cyprian are not a fit for smallpox. None of these clues are conclusive, but collectively they militate against the identification of smallpox.[47]

Any identification must be highly speculative. We would offer two candidates for consideration. The first is pandemic influenza. The influenza virus has been responsible for some of the worst pandemics in human history, including the "Spanish Flu" epidemic that carried off some 50,000,000 souls at the end of World War I. The lack of clear evidence for influenza from the ancient world is puzzling, because the flu is old and it was undoubtedly not a stranger in the ancient world. Influenza is a highly contagious acute respiratory disease that comes in many forms. Most types are relatively mild, causing familiar coldlike symptoms. Other rare types of influenza are more menacing. Zoonotic forms of the disease, especially those native in wild aquatic birds, can be pathogenic to other animals, including pigs, domestic fowl, and humans; when these strains evolve the capacity to spread directly between humans, the results are catastrophic. There have been four global outbreaks in the last century, and avian influenza (which includes some dreaded strains such as H5N1) remains a terrifying threat today.[48]

Pathogenic zoonotic influenzas are viciously lethal. They induce an overheated immune response which is as dangerous as the viral pneumonia itself; hence, the young and healthy are paradoxically put at risk by the vigor of their immune response. The lack of any respiratory symptoms in the account of the Plague of Cyprian is a strike against the identification. But it is worth reading some observations of the 1918 pandemic. "Blood poured from noses, ears, eye sockets; some victims lay in agony; delirium took away others while living. . . . The mucosal membranes in the nose, pharynx, and throat became inflamed. The conjunctiva, the delicate membrane that lines the eyelids, becomes inflamed. Victims suffer headache, body aches, fever, often complete exhaustion, cough. . . . Often pain, terrific pain. . . . Cyanosis. . . . Then there was blood, blood pouring from the body. To see blood trickle, and in some cases spurt, from someone's nose, mouth, even from the ears or around the eyes, had to terrify. . . . From 5 to 15 percent of all men hospitalized suffered from epistaxis—bleeding from the nose." Pandemic influenza might indeed account for the horrifying experience of the Plague of Cyprian.[49]

The winter seasonality of the Plague of Cyprian points to a germ that thrived on close interpersonal contact and direct transmission. The position of the Roman Empire astride some of the major flyways of migratory birds, and the intense cultivation of pigs and domestic fowl such as chickens and ducks, put the Romans at risk. Climate perturbations can subtly redirect the migratory routes of wild waterfowl, and the strong oscillations of the AD 240s could well have provided the environmental nudge for an unfamiliar zoonotic pathogen to find its way into new territory. The flu is a possible agent of the pestilence.

A second and more probable identification of the Plague of Cyprian is a viral hemorrhagic fever. The pestilence manifested itself as an acute-onset disease with burning fever and severe gastrointestinal disorder, and its symptoms included conjunctival bleeding, bloody stool, esophageal lesions, and tissue death in the extremities. These signs fit the course of an infection caused by a virus that induces a fulminant hemorrhagic fever. Viral hemorrhagic fevers are zoonotic diseases caused by various families of

RNA viruses. Flaviviruses cause diseases like Yellow Fever and Dengue Fever, which have some resemblance to the symptoms described by Cyprian. But Flaviviruses are spread by mosquitoes, and the geographic reach, speed of diffusion, and winter seasonality of the Plague of Cyprian rule out a mosquito-borne virus.[50]

Other families of viral hemorrhagic fevers are borne by rodents or transmitted directly between humans. Arenaviruses, like Lassa Fever, are spread by rodents. Old World arenaviruses are endemic in reservoirs in Africa, and it is plausible that the Plague of Cyprian was caused by such an agent. However, great rodent-borne pandemics will probably have to wait for the Justinianic Plague. The distinctive biology of the plague bacterium and its intricate interspecies dynamics make bubonic plague capable of continental-scale pandemics. The speed of travel and scale of the outbreak during the Plague of Cyprian would be unlikely for an arenavirus.

The speed of diffusion points to direct human-to-human transmission. The belief that caring for the sick and handling the dead were fraught with danger underscores the possibility of a contagion spread between humans. Only one family of hemorrhagic viruses seems to provide a best match for both the pathology and epidemiology of the Plague of Cyprian: filoviruses, whose most notorious representative is the Ebola Virus.[51]

Filoviruses are millions of years old. Fragments of their genetic material are anciently embedded in mammalian genomes, and for millions of years they have infected bats, insectivores, and rodents. Yet filoviruses, like Ebola Virus and Marburg Virus, were only recognized in the second half of the twentieth century during a series of small-scale outbreaks. The Ebola epidemic of 2014 brought further attention to the family. The natural host of the Ebola Virus remains unconfirmed, although bats are suspected. Ebola Virus grabs public attention because of its ghastly clinical course and extreme case fatality rates.

To cause an epidemic, the Ebola Virus must first leap from its host species to a human; this probably occurs when humans come into contact with infected bats or apes. Once infected, after a brief incubation period (on average 4–10 days, sometimes longer), victims suffer intense fever and a disease that breaks down multiple systems simultaneously, including gastrointestinal and vascular involvement. Conjunctival injection and severe hemorrhagic symptoms could well account for the disturbing reports of Cyprian. Tissue necrosis and permanent disfigurement of the limbs might reflect Cyprian's description of extremities turning putrid and becoming irreversibly disabled. Case fatality rates, even with modern treatment, are grotesquely high: 50–70 percent. Death usually comes between days 6 and 16;

survivors are thought to possess immunity. The Ebola Virus is transmitted by bodily fluids, but *not* aerial droplets; it spreads easily within households. Caregivers are at special risk, and cadavers remain a potent source of infection. The observance of traditional burial rites has been a problematic risk factor even in recent outbreaks.[52]

Retrospective diagnosis from anguished reports of nonmedical personnel across nearly two thousand years is never going to offer great confidence. But the hemorrhagic symptoms, the shocked sensibilities, and the insistence on the novelty of the disease all fit a filovirus. An agent like Ebola Virus could diffuse as quickly as the Plague of Cyprian, but because of its reliance on body fluids for transmission, it could exhibit the slow burning, "unusually relentless" dynamics that so struck contemporary observers. The obsession with deadly corpses in the third-century pandemic strikes a profound chord, given the recent experience of the Ebola Virus. The uncertainty lies in our profound ignorance about the deep history of pathogens like Ebola that never became endemic in human populations. As historians, we understandably default to the familiar suspects. But our broadening awareness of the incessant force of emerging disease, at the frontier between human society and wild nature, suggests a place for significant disease events in the past, like the Plague of Cyprian, caused by zoonotic diseases that wreaked havoc and then retreated back to their animal hosts.

The Roman Empire was once more the victim of a pest from outside the endemic pool of native diseases. The global climate turbulence of the AD 240s, which clearly affected the monsoon systems, stirred ecological changes that may have led to the eruption of the Plague of Cyprian. For over a decade, it wound its way through the empire, diffusing swiftly but burning

Table 4.1. The Plague of Cyprian

Pathology	*Epidemiology*
Acute onset fever	Exotic origins, east-to-west
Weakness	Empire-wide within 2 years
Bloody diarrhea	"Relentless," enduring, 15 years
Esophageal hemorrhage	Dangerous to caregivers
Continual vomiting	Corpses contagious
Conjunctival bleeding	Directly transmissible, by sight
Putrescence in the limbs	Struck households
Permanent disability	Indiscriminate
Loss of hearing, sight	Urban and rural
	Winter peak
	High mortality
	Follows severe drought

slowly. The pandemic struck soldiers and civilians, city dwellers and villagers alike. Pagan and Christian authors, with very different outlooks and very different motivations, writing at far ends of the empire, uniformly agreed that this pestilence was unlike anything the empire had faced before.

In the Antonine Plague, the fibers of the imperial structure were frayed but not pulled asunder. By the time of the Plague of Cyprian's appearance in AD 249 there was much that was different. The empire's stores of reserve energy were depleted. Perhaps this microbial enemy was just more sinister. In this event, the center could not hold. There is much that must remain uncertain about the Plague of Cyprian, but not this: in its immediate wake, anarchy was loosed on the world.

The Blood-Dimmed Tide

At the century games, choirs of boys and girls sang hymns boasting of the empire's unquestioned supremacy. In AD 248, the empire worked. There was one emperor, in Rome, the city whose people remained the symbolic focus of the empire. Philip's legitimacy was affirmed by senate and army. Even in years of dearth, this legitimacy allowed him to control the machinery of an empire stretching from Britain to Egypt, Syria to Spain. Each year, the cycle of tax collection brought in enough grain to feed the people and army; between money collected and the empire's silver mines in central Europe, the emperor could pay the soldiers strung along the vast frontier. The money paid to soldiers had real value; the *denarius* went as far as it had under Septimius. The empire obeyed one man. But the stupendous fabric was about to come undone. Philip's later coinage shows signs of unprecedented stress. The army revolted on the northern frontier, and soon Decius, the man sent to quell the rebellion, installed himself on the throne. The empire had passed a point of no return.[53]

The demise of Philip inaugurated two decades of chaos. Between the millennial celebration in AD 248 and the accession of the soldier-emperor Claudius II in AD 268, the history of Rome is a confusing tangle of violent failures. The structural integrity of the imperial machine burst apart. The frontier system crumpled. The collapse of legitimacy invited one usurper after another to try for the throne. The empire fragmented, and only the dramatic success of later emperors in putting the pieces back together prevented this moment from being the final act of Roman imperial history. A thoroughgoing fiscal crisis made it impossible to collect taxes and maintain the currency with any credibility. This failure violated what the Romans recognized as the fundamental axiom of empire: "an empire requires soldiers, and soldiers require money." As the currency regime dissolved, the infrastructure of the private Roman economy started to crumble. The fire fed on itself. An accelerating spiral of disorder engulfed the empire.[54]

By design, the Roman frontier system was defensible, not impenetrable. But almost simultaneously, in the early AD 250s, the defensive network imploded along all of the main fronts. A later historian summarized the vastness of the failure. "The Alemanni, having devastated the Gauls, penetrated into Italy. Dacia, which had been adjoined by Trajan beyond the Danube, was then lost. Greece, Macedonia, Pontus, and Asia were destroyed by Goths. Pannonia was plundered by Sarmatians and Quadi. Germans advanced all the way to the Spains and subjugated the noble Tarraco. The Parthians [i.e. the Persians], having occupied Mesopotamia, began to lay claim to Syria." The military crisis was marked by the concatenation of attacks in multiple theaters and barbarian incursions into parts of the soft interior normally insulated from the violence of the imperial periphery. The smell of blood seemed to draw attacks like never before.[55]

In the words of an oracle, "the universe will be cast into chaos with the destruction of mankind in pestilence and war." The relationship between pestilence and frontier insecurity was obvious to contemporaries. Sober sources drew a causal link between the demographic damage of the pandemic and military adversity. In one case, the advances of the Persian King Shapur I were directly motivated by his awareness that the Roman army was depleted by the mortality. The barracks were auspicious for the spread of a virus transmitted directly from one victim to the next. Germs were the first, invisible wave of attack in the great invasions.[56]

The frontiers buckled in the early AD 250s. The first to break was the Danubian front, where Carpi and Goths invaded in AD 250. In the summer of AD 251, the emperor Decius and his army were slaughtered at the Battle of Abritus by the able Gothic king Cniva. The Romans lost control of the entire Danube line.

Next the Euphrates frontier fell. In AD 252, Shapur I went on the offensive in the east. It was a lightning campaign unlike anything the eastern provinces had suffered. Syria was overrun, and Persian armies plundered the interior of Asia Minor. At the same time, new tribes of Goths took to the sea and rampaged from the Black Sea to the Aegean. Helpless cities as far as Ephesus were devastated.

In the mid-250s, the Rhine system disintegrated. Franks and Alemanni raided the wealthy provinces of Gaul from around AD 256; for virtually a generation, this territory was the victim of large-scale looting. When the emperor Gallienus tried to respond with a northern operation, the heart of the empire was exposed, and by AD 260 an invasion from the upper Danube reached the outskirts of Rome itself. That same year, Gallienus learned that his father and co-emperor Valerian had been shamefully captured alive by Shapur I. The great victory monument carved into the rock cliffs at Naqš-i Rustam celebrates the humiliation of the Romans. On every front—including obscure violence in Africa and Egypt—the Roman Empire was gravely wounded.[57]

Simultaneous pressure along both major frontiers was always a formula for catastrophe. Now, too, the foes were more formidable. The Persians were ably led. The Gothic confederation represented the peril of more advanced social formations beyond the northern frontier. There had been a slow "technology convergence" between the Romans and their Germanic neighbors. The evolution of more sophisticated enemies weighed invisibly on the entire edifice of the Roman Empire. But, once the pestilence hollowed out the Roman frontier shield, the structural weaknesses of the imperial system were exposed to hungry and ambitious peoples on the far side of its borders, with ancient grudges against the belligerent empire. There should be no doubting the causal importance of the pandemic in the military crisis. It exposed the latent threat and allowed the frontier system to be overwhelmed by the violent tide.[58]

We hear of hastily assembled popular militias defending cities deep within the empire and of walls hurriedly built. An altar to the goddess Victory was erected in AD 260 in Augsburg, celebrating the success of the provincial army alongside "the people." Together, this makeshift militia routed the invading barbarians on return to Germany and freed "many thousands of Italian captives." Even the "people of Rome," long coddled by their privileges, were armed to repel the invaders in AD 260.

By the AD 260s, there were functionally three Roman Empires, one in the Gauls, one in the east ruled from Palmyra, and the central core state controlled by Gallienus. This last was eventually reduced to the defense of Italy and the Balkan routes leading to Italy, and we know that in the later AD 260s even the towns of Greece, such as Athens, were effectively reliant on what self-help could be scraped together from resources to hand. Strategic annexes like Dacia and the entire territory between the Rhine, Danube, and Main Rivers—known as the *agri decumates*—were evacuated, and lost, forever. The Roman Empire was unbundled, and it is no wonder that Gallienus, who was able to maintain his office in his shrinking core region until AD 268, cuts a pathetic figure in the collective historical memory of the Romans.[59]

The ebbing of the state's power is mirrored in the coinage. For better or worse, it represents the closest thing we have to a running commentary on the status of the empire. In the AD 250s and 260s, the silver content of the currency fell precipitously. The ancient denominations, the *sesterces* and *denarii*, were unceremoniously melted down; soon these august coins simply ceased to exist, entirely replaced by the *antoninianus*, a revolution quite as imponderable as the disappearance of the dollar would seem to us. Then, in the space of less than two decades, the *antoninianus* was progressively debased until it was a billon coin, a base metal token with an imperceptible wash of silver. The momentum of a currency crisis accelerated, as private holders must have sought to hold on to good metal, pulling it from circulation. Indeed, no other era of Roman history is so productive of coin hoards.

We have glimpses of the gathering currency crisis from Egypt. The coinage maintained its fiduciary value, for a time. But in a papyrus of AD 260, we find a governor forcing bankers to accept "the divine currency of the Augusti." It is telling—both that the bankers tried not to accept it and that the governor could force them to do so. In the generation of pestilence and debasement, there were wild gyrations in the price level of goods and services, of nearly 100 percent. This instability appears modest only in light of what was to come. At the end of the crisis, during the reign of the restorer Aurelian, the effort to put the pieces back together failed. The fiduciary value of the coinage collapsed. Prices leapt tenfold, and a century of galloping

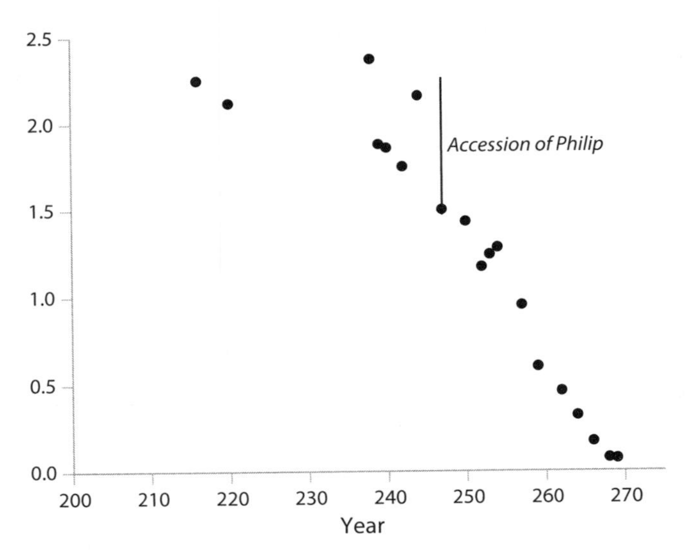

Figure 4.3. Silver (g) per Antoninianus (data: see note 60)

inflation was on the horizon. A thousand-year epoch of silver money was doomed to end.[60]

Humbling military losses, imperial fragmentation, and the inability to pay the troops in hard currency finally undid Gallienus. What is surprising is that his rule lasted so long. It is a testament to the deep reserves of resilience and the power of ideological legitimacy at the core of the Roman Empire. It may also reflect the sheer inability of any alternative to gather strength in the chaos of the long-lasting pestilence. But in AD 268, Gallienus was assassinated in Milan. The coup was orchestrated by a Danubian military officer named Claudius. Claudius II was not just another in a long and blurry line of claimants to the throne. His ascent signaled the arrival of a wholly new kind of emperor and marks not so much the end of crisis as the beginning of a new age. The ground had been cleared by the shock of drought and pestilence, war and fiscal collapse. At last, the age of the barracks emperors was at hand.

Restoration and Revolution

The generation that elapsed between the death of Philip and the ascent of Claudius II was an age of endings in Roman history. Places and villages quietly disappear from the record. The census records in Egypt stop in the 250s. The last of the ancient private endowments vanish. Inveterate habits of public epigraphy simply halt. The grandeur of the civic temples is dimmed. We can even trace the sudden demise of individual ateliers, as the disintegration of economic life and the flow of capital and investment were abruptly snapped. So many of the fibers that once imperceptibly held together the classical order find their ending in this period.

This ground-clearing was both the precondition and the consequence of the political revolution that swept Claudius II into power. The line of emperors that begins with Claudius II liked to advertise their work as a kind of "restoration." But the imperial system that coalesced in the aftermath of pestilence and crisis had a new inner logic. It was a revolution founded on the twin principles that defined the new equilibrium: the imperial machinery would be controlled by military emperors of Danubian extraction, and their soldiers would be rewarded in honest gold. Order was restored around these sturdy premises of the new state.[61]

Ironically, the blueblooded prince Gallienus prepared the way for the rise of the soldier emperors. Of impeccable senatorial ancestry, his wealthy

Figure 4.4. Gold Coin (Aureus) of Claudius II Celebrating the Loyalty of the Army (American Numismatic Society)

family had ancient roots in Etruria. His father rose in the service of the Severan regime and attained the consulship. Socially and geographically, the rule of a man like Gallienus hewed to traditions going back to the very foundations of the imperial office. But in his reign, control of the legions was wrested from the senatorial class.

According to a later source, Gallienus "feared that the imperial power would be transferred to the best of the nobility through his sloth," so he became "the first to prohibit the senators from undertaking a military career or entering the army." Whatever the motives, from just this moment, it indeed becomes impossible to find senators commanding Roman forces. The high position of legionary commander, *legatus legionis*, had been the lynchpin of senatorial control over the army. The replacement of senators by professional soldiers in the high commands dispelled a uniquely Roman aristocratic ethos and broke an ancient sociopolitical order stretching back centuries to the late republic. Plague and war again pulled down one cadre of elites and allowed the rise of another, but this time the reconstitution was more radical, and it was a pattern destined to endure.[62]

If Gallienus hoped to preclude usurpation, his policy was sorely miscalculated. For centuries, legionary command had been the staging ground for imperial pretenders. Only now, it would be professional soldiers rather than well-bred generals who could rally the troops behind their cause. The ascension of Claudius II, who, notably, had commanded the crack unit of imperial cavalry, was the immediate fulfillment of this possibility. The death of Gallienus marked the end of a certain kind of emperor.

But as revolutionary as the social background of Claudius II was, his geographical background was equally consequential. He hailed from Upper Moesia or Lower Pannonia. This corridor of the Danubian plain was anciently filled with Roman veteran colonies. Over the centuries, when legionaries laid down their arms, they mixed with local populations; sons of soldiers followed their fathers patriotically into service; the Danubian frontier hardened one and all to the realities of war. A military culture grew up. The region produced few senators, but many decorated officers. Decade after decade, these officers loyally served their commissioned superiors, but with the empire in shambles and their homeland overrun, they seized the mantle for themselves.[63]

The life of Claudius II was cut short by the plague. His revolution survived him. Once the Danubian military officers had seized control of the machinery of empire, they refused to relinquish it. Walter Scheidel has brilliantly shown that, down to the reign of Phocas (AD 610), nearly three-quarters of Roman emperors originated from a region constituting 2 percent of the empire's territory. The Theodosian dynasty is virtually the only aberration from the pattern, and it is the exception that proves the rule. The Theodosian dynasty was born in the absolute "perfect storm," in a moment of sudden desperation after the massacre of the officer corps at the Battle of Adrianople (AD 378). From AD 268, the moneyed Mediterranean aristocracy was displaced by a cadre of professional soldiers hailing from a small, northern corner of the frontier. The region was what Ronald Syme called a "zone of energy," at the critical overland juncture where the eastern and western halves of the empire met. The Roman Empire was taken over not just by the military elite of any frontier, but, elites specifically from *this* place.[64]

Great empires are often swallowed by their own periphery. That is not what happened to the Romans in the third century. The Roman Empire was restored by an internal frontier zone. The barracks emperors identified as Romans. Ancient Roman blood ran in their veins. They show a streak of impatient traditionalism, for instance, in the application of Roman law. The ethos of the Danubian emperors led them to protect the empire as a whole; Aurelian, the immediate successor of Claudius II, dedicated his energies to the reconquest of the eastern and northwestern provinces of the empire. There was no egregiously conspicuous enrichment of the ancestral homeland in the centuries of Danubian rule. The people of Rome, not of Sirmium or Naissus, remained the beneficiaries of outsized political entitlements. But the work of restoration required bold strokes. The city of Rome was respected as the symbolic center of the empire, but the barracks emperors were not hesitant to set up palace in garrison towns closer to the

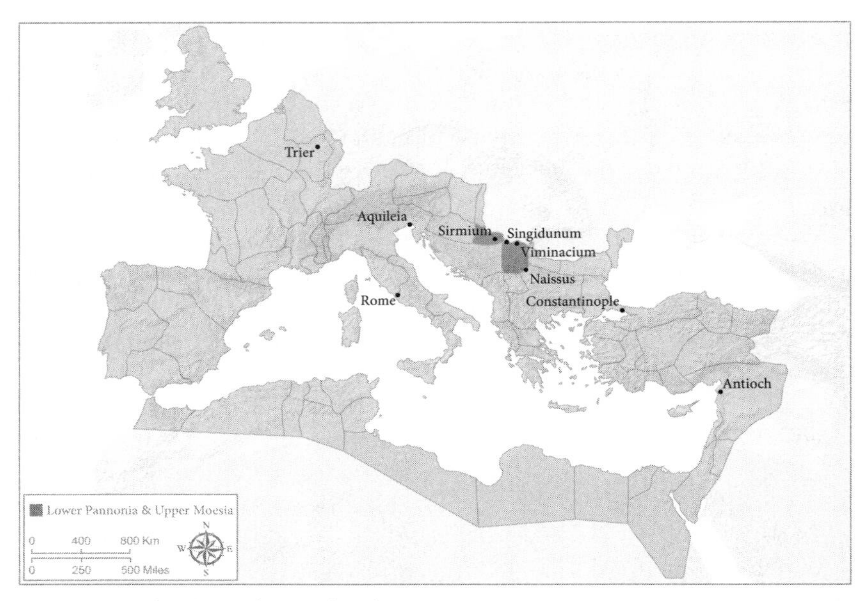

Map 13. Two Provinces That Produced Most Late Roman Emperors

scene of the action. The administrative apparatus would be unrepentantly overhauled. Constitutional niceties were set aside for the higher cause of reassembling the empire.[65]

Where the later emperors were clearly not impartial was in the favor bestowed upon the army, especially its officer corps. Claudius II rewarded the loyalty of the soldiers who elevated him . . . in gold. A perceptive scholar of ancient coinage has suggested that this moment was the beginning of late antiquity. The act was born of necessity, with the silver currency in disarray. But it proved unforgettable.

Henceforth emperors paid accession bonuses in gold. The implications were not subtle: where present, the emperor personally handed out the gold, and loyalty oaths were sworn. These bonuses were regularized, and the soldiers would receive one every five years, lest they learn to regret an emperor's longevity. In the course of time, the soldiers' regular stipends, denominated in silver currency, became worthless, and the donatives functioned as a salary. Great victories continued to deserve bonuses too. We gain a sense of the possibilities from a treasure discovered at Arras in northern France in 1922. A clay pot belonging to a military officer held precious jewels, silver objects, and 472 coins, including 25 gold medallions, earned during a military career that seems to have stretched from ca. AD 285–310.

Figure 4.5. Medallion of Constantius I, Arras Treasure (Bibliothèque nationale de France)

One of the gold medallions weighed 53 grams, celebrating the reconquest of Britain by Constantius I, father of Constantine, who is acclaimed as the "restorer of eternal light." Diligence and loyalty were handsomely repaid.[66]

The politics of gold would redefine state and society from the inside out. The age of the barracks emperors was to be the age of gold.

The spiritual repercussions of the crisis are inevitably more elusive, less mechanical, but in the long run they were even more consequential. Massive mortality events provoke unpredictable religious responses. Fervor and despair change the atmospheric pressure of spiritual life. The Antonine Plague called forth an empire-wide turn to the most archaic layers of Apollo worship. The Plague of Justinian, as we will see, pushed Mediterranean cultures toward a sharp apocalyptic mood. Later, in the Black Death, the persecution of the Jews and the flagellant movement were direct reactions to the plague, while a more abstract cultural fascination with death has seemed linked to the harrowing experience of mass mortality in the later middle ages.

The crisis of the third century was a moment of truth for the traditional civic religions of the ancient world. It also opened the door to the uncanny growth of a marginal religious movement known as Christianity. Within the space of a generation, the confident archaism on full display at the millennial games of Philip had yielded to a religious landscape where high-pitched voices of dissent were more audible than ever before.

Already in its incipient phases, the crisis sparked religious conflict. Spontaneous acts of prayer and sacrifice were a proper reaction to the accession of a new ruler. But sometime toward the end of the year AD 249, the emperor Decius *required* all citizens to partake in an act of sacrifice and

Figure 4.6. Silver Coin (Antoninianus) of AD 251–53 Showing Apollo the Healer (American Numismatic Society)

deployed the machinery of the empire to enforce the order. It may be more than coincidence that, as the pestilence raged in Alexandria and appeared westward bound, the emperor devised a scheme of universal supplication. To the ancient mind, plague was an instrument of divine anger. The Antonine Plague had provoked spectacular acts of religious supplication at the civic level, fired by the great oracular temples of the god Apollo. Apollo was soon at work in the Plague of Cyprian, too. The emperors started minting a new image on the currency, invoking "Apollo the Healer." Religious solutions were desperately sought in Rome. "The peace of the gods was sought by inspection of the Sibylline books, and a sacrifice was made to Jupiter the Healer as they had commanded." The plague unleashed an urgent combination of fear and piety. Whether or not the disease triggered the initial orders of Decius to sacrifice, the Plague of Cyprian was soon implicated in the religious upheaval of the age.[67]

Scholars have become wary of calling the religious policy of Decius a "persecution." That is, perhaps, too one-sided a view. The desire to extirpate Christianity was not the entire impetus for the policy. The empire-wide order of Decius to sacrifice might be imagined as a scaled-up version of the civic responses that the Antonine Plague had once provoked. But now, in an age of universal citizenship, the response to the crisis was all-encompassing, and compliance was not voluntary. None of this is incompatible with the possibility that suppressing Christianity was a conscious goal of Decius from the beginning. After all, the Christians' refusal to sacrifice was not only an act of defiance; it imperiled the protection of the gods in the face of the enveloping disaster.[68]

Christians were being scapegoated. The religious polemics between pagans and Christians called forth Cyprian's defense of the faith; especially in his apologetic masterwork, the *Ad Demetrianum*, his principal agenda was to exonerate Christians of guilt for drought, pestilence, and war. We are lacking the prosecution's side of the conversation, but we catch muffled echoes of it a generation later, in the bitter words of the pagan philosopher Porphyry. He blamed the insolence of the Christians for the health catastrophes of the age: "And they marvel that the sickness has befallen the city for so many years, while Asclepius and the other gods are no longer dwellers among us. For no one has seen any succor for the people while Jesus is being honored." It was an attitude that may well have prevailed in the AD 250s.[69]

Decius set up a religious dragnet. Citizens had to prove their loyalty with an act of pagan sacrifice. The individual certificates of sacrifice survive in abundance in the papyri of Egypt. Christian refusal to participate led to even more intense response from the central government, now explicitly aimed at the growing church. Valerian implemented measures that were unequivocally aimed at hunting out Christians. Looking back, the Christian church saw this whole episode as one great trial, the culmination of centuries of imperial effort to repress the faith. But this obscures the circumstances of the persecution, and it misrepresents how tiny the Christian movement remained.

We have only the most impressionistic sense of Christianity's expansion. Down to AD 200, Christians are virtually invisible in the documentary record. If not for later events, the Christians of the first two centuries would be hardly a footnote to history. In the later second century, it has been estimated that there were on the order of 100,000 Christians. By AD 300, there had been staggering change. The clearest sign is the sudden spread of Christian personal names. It has recently been estimated that an astounding 15–20 percent of the population may have already been Christian in Egypt. Precision is specious, but even on the most cautious set of assumptions, the unavoidable conclusion is that the third century witnessed the explosive transformation of Christianity into a mass phenomenon.[70]

The Jesus movement was propelled by missionary zeal from the start. But the dynamics of something so intimate as "conversion" must be sought in the specific conditions of each generation. The forces of attraction that drew small bands of urban eccentrics to the faith in the second century were not what catalyzed the mass movement of the third century. And even within the third century, the rate of change was not constant. The combination of pestilence and persecution seems to have hastened the spread of Christianity. That was the memory of one Christian community, at Neocaesarea in Pontus. In the folk legends that attached to the local hero of the faith, Gregory the Wonderworker, the plague was a pivotal episode in the Christianization of the community. The mass mortality painfully showed up the inefficacy of the ancestral gods and put on exhibit the virtues of the Christian faith. However stylized the tale may be, it preserves a kernel of historical recollection about the plague's role in the religious transformation of the community.

Christianity's sharpest advantage was its inexhaustible ability to forge kinship-like networks among perfect strangers based on an ethic of sacrificial love. The church boasted of being a "new *ethnos*," a new nation, with all the implications of shared heritage and mutual obligation. Christian ethics turned the chaos of pestilence into a mission field. The vivid promise of the resurrection encouraged the faithful against the fear of death. Cyprian, in the heat of persecution and plague, pleaded with his flock to show love to the enemy. The compassion was conspicuous and consequential. Basic nursing of the sick can have massive effects on case fatality rates; with Ebola, for instance, the provision of water and food may drastically reduce the incidence of death. The Christian ethic was a blaring advertisement for the faith. The church was a safe harbor in the storm.[71]

Once the fire of crisis was burned out, its ashes left behind a fertile field for Christian expansion. Gallienus called a halt to the persecution in AD 260; a peace lasting over forty years fell upon the church. The famous church historian, Eusebius, triumphantly described these days of unhindered growth. "How does one describe those multitudes worshipping and the throngs pressing together in every city and the brilliant assemblies gathered in prayer? Indeed because of these crowds the old buildings no more sufficed for them, and spacious churches were built from the very ground up in all the cities."

Christians moved confidently, in high circles. They are more visible than ever before. In Oxyrhynchus, the city in Egypt whose trash heap has yielded such a trove of papyri, the church becomes more than a shadow in these years. The first papyrus naming a Christian was recorded in AD 256. Shortly thereafter, we can follow the rise of the Christian community through a cleric, Papa Sotas, who may have been the town's first bishop, certainly its earliest known. His career is documented in no fewer than five papyri, which show him writing letters of recommendation, soliciting funds for the church, and moving freely around the eastern Mediterranean—in short, acting like a late antique bishop. In Oxyrhynchus, the church emerges abruptly, from virtual invisibility to a mood of swaggering confidence.[72]

Meanwhile in Rome the honeycomb of burial caverns we know as the catacombs expanded unhesitatingly. A few burial chambers went back to the late second or early third century; these soon became the hubs of sprawling complexes radiating outward. The third quarter of the third century marked the takeoff, when suddenly the Christian presence underground became something more than a handful of discrete burials. Now, long passageways lined with humble burial notches cut into the walls curled into the lamplit distance. The catacombs were not the romantic hideaway of an outlaw cult, nor the top-down design of ambitious popes. Rather, the catacombs were the continuation in death of the communal bonds that vivified the church above ground, sustained by expansive networks of patronage, a strong but complex sense of identity, and sharp beliefs about the afterlife. There was fluttering energy in this period, with insouciant social mingling in a diverse community that did not yet encompass the superwealthy. The shrines of the martyrs were not yet strongly organized. This was a shadow society, one that weathered the challenge of pestilence and persecution, and emerged on the other side ready for dazzling growth.[73]

If we knew nothing of Christianity, we would nonetheless describe the third century as an age of inversion within traditional polytheism. The ancient religions floundered. The grand tradition of temple building came sputtering to a standstill. The second century had been an age of exuberant religious construction. Hadrian completed the great temple of Olympian Zeus in Athens, left unfinished since the sixth century . . . *before Christ*. The temples were the gleaming "eyes" of a city. By the middle of the third century, they were tumbling into disrepair. In Egypt, the last temple inscription dates to the reign of Decius. Then, deafening silence. By the end of the century, temples that had recently been the incubators of the most ancient religious lore of humankind were turned into military barns. Rites of imponderable antiquity simply vanished. The old registrations of temple personnel and property ceased from AD 259. The collapse is truly startling. Perhaps it was more pronounced in Egypt, where municipal institutions were of more recent vintage, than in other parts of the empire, but the truth is that diligent efforts have yielded relatively meager evidence for the vitality of temple life elsewhere too. By any measure, the crisis of the third century was an unrestrained catastrophe for the traditional civic cults.[74]

It is important to ask *why* it was so. There was no such thing as a coherent "paganism," except in the mind of Christian polemicists. Ancient polytheism was diffuse. It was an ensemble of loosely interconnected religions, immanent in nature and ingrained in the life of the family and the city. The polytheism that flourished in the Roman Empire was built into the vaulting social hierarchies of the ancient city. We meet the authentic paganism of the high empire not in high theological speculation but in the street life of the cities. A famous example is known from Ephesus, where a wealthy Ephesian citizen and Roman knight named C. Vibius Salutaris established an endowment in honor of the goddess Artemis. The interest from the endowment, maintained by the temple, funded magnificent religious pageants celebrating the long history of the Ephesians; effusive gifts of cash were given to the citizens along archaic tribal lines; blood sacrifices were made to the goddess. These religious endowments were utterly wiped out in the financial chaos. The old patterns of civic patronage were destabilized. The ancient gods did not lose out in a crisis of faith. They were embedded within an order whose foundation itself cracked.[75]

The superstructure fell, but ancient polytheism hardly died out. The particulates of natural religion were everywhere. A traveler walking down a Roman road would see "an altar garlanded with flowers, a leaf-shaded grotto, an oak loaded with horns, a beech crowned with animal-skins, a sacred hillock within an enclosure, a tree-trunk with an image carved in it, a turf altar moistened with a libation, or a stone smeared with oil." No crisis could wash out the tenaciously rooted ground cover of folk polytheism. In the third century, the Christians remained surrounded by the sounds and smells of seething polytheism. But when the loftier expressions of public religious life faltered, the Christians seized the moment. The church inserted its voice obtrusively into the public conversation, in a way that even in the Severan period would have seemed almost impossible. The church was ready to talk terms with the empire. By the turn of the fourth century, the Christian community had become a force to be reckoned with. The barracks emperors vacillated between policies of eradication and cooptation, until the most successful among them pledged himself fully, and somewhat unexpectedly, as the protector and patron of the faithful. It was an age of bold strokes.[76]

The Road to Recovery

The emperor Aurelian (AD 270–75) reconquered the secessionist territories. He built walls around the city of Rome and attempted a thoroughgoing reform of the coinage. He insisted on the worship of *Sol invictus*, the unconquered Sun god, something of an outsider in the Mediterranean pantheon, but easily enough domesticated. He paraded the queen of Palmyra,

Zenobia, through the streets of Rome in the glorious rites of a Roman triumph and proclaimed himself the "restorer of the world."

In reality, his reign was a heady mix of old and new. The restorative work of the barracks emperors was carried out in the name of tradition. The success of their project has led modern historians to question even the reality of crisis. But we should not take it for granted that the Roman Empire would be reassembled into a unitary state with a pan-Mediterranean geographic framework. Han China did not survive its parallel crisis intact. The Roman Empire was given a second life, a fact which should cause us to marvel at the achievement of restoration, not to doubt the gravity of the crisis.[77]

The empire's fortunes reached a low tide in the AD 260s. It was the demographic bottom too. Here the work of recovery was much slower. The Plague of Cyprian and the broader crisis were disorienting. Interior regions accustomed to peace were brutally violated; old social hierarchies buckled. Throughout the west, rural settlement patterns reveal a rift. Life returned, but slowly, and to a different, more wary rhythm. The cities were never quite the same; even the healthiest late antique cities were smaller than they had formerly been, and in aggregate, even after the recovery, there were simply fewer major towns. The old days when army recruitment could be handled with a light touch were forever gone. Late antique statecraft would have a harder edge, by necessity. But the project of restoration laid the groundwork for another century and a half of imperial integration and economic resurgence.

The long fourth century was, in its way, a new golden age, less brilliant than the Antonine efflorescence in material terms, but extraordinary by any other standard. Yet somewhere within the new equilibrium lurked the seeds of divergence between the eastern and western halves of the empire. The project of restoration led, eventually, to the establishment of a second Rome, at Constantinople. The foundation of the new capital was a stroke of genius that would shift the geopolitical balance, more profoundly than anyone could have imagined. When the hand of global climate change set off a chain reaction of people movements and refugee crises that realigned the pressures bearing on the edges of Roman territory, it would break the empire along the lines of stress that had slowly developed. Only half the empire would survive the next fall.[78]

Fortune's Rapid Wheel

The Reach of Empire

Among the minor works of Claudian, the poet whom we last met celebrating the consulship of Stilicho, is an attractive composition called "The Old Man of Verona." It celebrates an unnamed farmer whose slow and innocent life was never tossed by the upheavals of Fortune. He lived blissfully beyond the sweep of time. The old man would die in the same humble cottage where he was born. Never as a wandering stranger had he tasted the waters of foreign rivers. He reckoned the years by "the turning harvests, not the consul's name." He remembered the time "when that mighty oak was only a little acorn." For him, neighboring Verona was as foreign as "sun-drenched India," and Lake Garda was as remote as the shores of the Red Sea. Yet, the farmer's small horizons were the stuff of his happiness. "Let another seek out the farthest edges of the western lands. The seeker may have more of an adventure—but the old man has more of a life."[1]

This is a charming idyll. But it is not impossible that Claudian, wearily trudging across the valley, had truly encountered such a peasant, living honestly, right in the shadow of high politics. It might well have moved Claudian personally. The old man's rooted existence stood in pointed contrast to the poet's own experience of life. Claudian was an Egyptian poet who had ventured west to become a cultural sensation at court as well as the hired mouthpiece of the empire's most powerful figure, the *generalissimo* Stilicho. If such an old farmer really did exist, the poem holds a special poignancy. It is usually dated around AD 400. In the very next year, an army of Visigoths led by Alaric tore through the Po Valley. Stilicho met them in a bloody encounter at Pollentia and chased them back east, through the plains of northern Italy. The tranquility of the countryside was shattered. The decisive confrontation, in fact, took place at Verona, where Stilicho's troops repulsed the invaders. It was his crowning achievement as a battlefield commander.[2]

Thereafter things went wrong, swiftly. On the last day of AD 406, the Rhine frontier crumbled. Events gathered pace. In 408, Stilicho's regime fell in a *coup d'état*, and the general was soon executed. The empire had lost all semblance of situational command in the west. The Gothic leader, Alaric, who had an eye for the main chance, surrounded Rome. In August of AD 410, the eternal city was sacked. The violation of the ancient capital was damaging enough, but the symbolic reverberations were even more profound. "The frame of the fragile world" had collapsed. Rome did not fall in a day, but still the sack of the city stands as a pivotal moment in a pivotal generation, when the central imperial power lost control of the western provinces. This time, the losses were to prove irreversible. Over the course of the fifth century, the western Roman Empire fell apart. No one, in any corner near or far, was untouched by an event of this magnitude.[3]

For historians, explaining the rapid disintegration of the empire has proved an enduring challenge. "Few things are more difficult in late-antique history than to know why, *in the western half of the empire*, the Roman military and the Roman government failed." If anything, the scope of the problem has only become even more daunting in recent years, as we have increasingly come to appreciate the robust recovery from the crisis of the third century. The empire roared back, and it is harder than ever to lay the blame for its demise on a progressive decay from within or a spiral of inevitable dissolution.

The Roman Empire in the later fourth century was the most powerful state on the globe and one of the most powerful that had ever existed. The emperor Theodosius I (AD 379–395) ruled over an empire larger than that of Augustus. Its sheer fiscal power remained historically exceptional—on a par with the most formidable of seventeenth-century polities. In some of the empire's territories, including most of the eastern provinces, the demographic and economic resurgence was nearly miraculous. Even in the west, the fall of the empire *caused* the decline, and not vice versa. There were structural weaknesses and human blunders, as ever, but it is no easy calculation to make these add up to an event as momentous as the disappearance of central imperial power in the west.[4]

To untangle the sequence of developments leading down to the failure of the empire, we must attune ourselves to the different rhythms of change in this period. The political restoration was a revolutionary and ongoing project. When, in the midst of the third-century crisis, the grand bargain between the narrow senatorial elite and the cities was replaced by a military autocracy, it opened the space for an age of experiments. There was more structural change in the administration of the empire in the hundred years between Diocletian (r. AD 284–305) and Theodosius (r. 379–395) than there had been in the first three centuries of empire, combined. The imperial system was radically centralized. There had been fewer than a thousand salaried officials in the employ of the early empire; in late antiquity, there were something like 35,000. The ramifications of this sweeping experiment were still being worked out, when external pressures intervened to test the top-heavy regime.[5]

We have also come to appreciate that late antiquity was an age of opposites, and none so consequential as the tension between society's dynamism and the state's desire for rigidity. The military emperors ruled without the same measure of patience or subtlety that had once been demanded of the monarch—the posture of restraint and respect that the earlier emperors had called *civilitas*. Constitutional inhibitions were few. The law codes that have furnished such a rich record of the period reflect an ambitious vision of the state's control. Often, the state imagined it could bind whole classes of individuals to their status or occupation, fixing fast all social relations. But the political restoration had unleashed a vibrant economic renaissance. The stabilization of the currency, in particular, let markets quickly regenerate. The state drew energy from the vitality of the private sphere, but its dreams of controlling this energy for its own ends were harder to achieve.

We will pause at some length over the dynamism of fourth-century society, because it throws into sharp relief what came next. The vaulting hierarchies of a highly articulated and wealthy society imploded, leaving in their place a poorer and simpler order in the west. Here is where the image of Claudian's independent peasant is most beguiling. The fall of the empire was not the substitution of one power for another, rumored in the distance; it was the end of an ancient order of state and society, one whose pervasive effects reached into the farthest corners of life, everywhere the Roman writ ran.

The rhythms of environmental change behind this drama were intricate. Compared to the pandemics and climatic turbulence the empire had experienced, the long fourth century was an interlude of peace. The role of the environment was subtle, but not insignificant. The climate was warmer. In many regions, the new growth germinated in the sunshine of a warmer climate. But the days of the Roman Climate Optimum did not return. The climate was now a more undependable ally. Moisture trade-offs between regions were more pronounced, in a climatic regime dominated by the Atlantic pressure gradients.

The demographic history of this period, too, is more subtle than before. While the empire was spared a major disease event, late Roman societies were beset by frequent spasms of epidemic mortality. The supergerms were on hiatus, but the dreadful array of native pathogens continued to make the empire an insalubrious environment. In the late empire, instability in the climate system and the tumult of war repeatedly stirred mortality events from within the local disease pool.

The real impact of environmental change in the fourth century may have been felt to the east. The Atlantic regime that ruled the climate of the empire in this period also brought savage aridity to the Eurasian steppe. An age of migrations was launched from the heart of Asia. We know far less than we would like about the inner drama of nomadic state and society in this consequential age. What is obviously new, however, is the sudden prominence of the steppe peoples in the affairs of the Roman Empire. The arrival of the Huns on the western fringes of the steppe overturned the Gothic order that had held for more than a century. Suddenly Goths pushed across Roman frontiers, and the application of pressure unexpectedly overwhelmed the structures of empire.

We need not go in for monocausal explanations. The coming of the Huns did not, by itself, spell the doom of the western empire. In the end the Huns conquered very little, and the effect of their entrance onto the scene must be measured within the particular circumstances that they encountered—the ongoing recovery, the ceaseless political experimentation, and the silent rift between east and west. But neither was the nomadic horde a mere feather that happened to push the empire beyond the threshold of resilience. The entire Asian steppe, for the first time in history, shifted its weight, hurling its most advanced state formations against the west. It was a trial that only half the empire found a means to survive.

The most sharp-eyed observer of the fourth century, the historian Ammianus Marcellinus, opened the last book of his Roman history and introduced his famous account of the Huns with the image of "Fortune's rapid wheel, which is always interchanging adversity and prosperity." Through century after century of empire, the Romans had endured countless adversities. But the challenges that coalesced in the later fourth and early fifth centuries proved insuperable. We can appreciate that their alignment, in both the human and natural dimensions, was quite as capricious as Ammianus ever imagined.

The New Imperial Equilibrium

The first order of business, for the barracks emperors who seized the mantle of power in the depths of crisis, was preserving the empire. They were willing to change capitals, currencies, even gods, to the end of restoring stability. But gradually it became necessary to regularize the new order.

Diocletian, a Danubian soldier who rose to the supreme power without any ancestral claims, proved in the two decades of his rule (AD 284–305) to be an ardent reformer. His chief innovation was the tetrarchy, the division of the emperor's role among four colleagues. The tetrarchy was an ingenious attempt to suppress civil war, while allocating the business of a sprawling empire among a team of four rulers. Diocletian laid down the groundwork of a new regime. His reforms stabilized and refined emergency measures that had arisen in the heat of crisis. Diocletian had "very little use for senators," continuing to prefer talent over ancestry and wealth. The provinces were "chopped into slices," so imperial governors could exert more direct control over their territories. Diocletian dissociated the civil and military offices, which had been indissolubly fused. At the top, the imperial court grew in scale and in pomp, with the emperor himself increasingly secluded and wrapped in majestic ceremony. The very words of the later emperors became "sacred."[6]

The primary task of Roman statecraft remained, as ever, paying the army. Diocletian made his own task harder by growing the beast he had to feed. The swelling of the army ranks appalled contemporaries. It was claimed that Diocletian doubled the size of the army. In truth, Diocletian's army was probably not much larger than its early imperial predecessor, at 400,000–500,000 men, but the return to this level from the depths of crisis, and after the loss of manpower in the Plague of Cyprian, was a steep and exhausting climb. Diocletian was active in fortifying the frontier, energetically repairing roads and military installations across the empire. His career at arms must be judged a resounding success. He pacified the north and re-asserted Roman dominance over Persia, extending Roman power to include an archipelago of heavily walled towns along a line stretching past the Euphrates into Mesopotamia. The problem of paying for it all exercised Diocletian's administrative genius. He resolutely dispensed with the archaic patchwork of local tax systems and replaced it with a single unified tax machine, based on standardized fiscal measures. Officials crisscrossed the whole empire carrying out a great new census. Even Italy unceremoniously lost its privileges and submitted to being taxed like any old province.[7]

There was no choice but to continue granting the soldiers donatives in gold. But Diocletian remained unwavering in his commitment to the regular stipend, paid in the old denominations. The money of Diocletian was centered on the *denarius* as the unit of account. Its purchasing power continued to sink, so Diocletian tried to stem the tide of inflation. He reformed the currency and implemented stern price controls. His famous Price Edict epitomizes the newly interventionist style of late Roman statecraft. It is evident from the law's preface that the soldiers were foremost in mind. "Prices go up not fourfold, not eightfold, but to such extortionate levels that it outstrips the ability of human language to describe these values and sales. And meanwhile one purchase wipes out the salary and bonus alike of the soldier, who hands over as loathsome profits for these plunderers the entire tax, paid by all the world in order to support the army." Diocletian decreed price limits for some 1200 commodities (from farm tools to freight charges, textiles to slaves, Gallic sandals to male lions). The Price Edict is a sidelight on the extent of economic specialization, even amid the inflation. But as his Christian enemies gleefully observed, and documentary evidence confirms, the policy was an abject failure.[8]

Diocletian's reforms opened the door for Constantine. Constantine was the son of a military officer, born at Naissus (modern Niš in Serbia), the town where Claudius II scored a decisive victory over the Goths. Constantine would fabricate a dynastic link with Claudius II, whose coup brought the Danubian emperors permanently into power. But Constantine's first order of business was to subvert the tetrarchic system, claiming his father's territories in AD 306, defeating his western rival Maxentius in AD 312, and finally extinguishing all remnants of Diocletian's system by taking the east in AD 324. Constantine was a polarizing figure already in his lifetime. He was a reformer and a regime builder who stands at the origins of the late antique state. His long reign (AD 306–337) allowed him to build a supportive network of allies and clients indebted to his rule. But he also built an order, a systemic structure of power that long outlasted his reign. The only proper comparison for Constantine is the first emperor, Augustus himself, a long-ruling figure whose regime became the template for a new equilibrium after decades of violent instability. It was not a comparison that was lost on the man himself or his fawning contemporaries.[9]

By the time of Constantine's ascent, the military class was firmly in control. It was now time for *détente*: Constantine's regime was able to reconcile the new elite with the remnants of the senatorial order. Constantine showed a renewed preference for the senatorial grade, reassigning top offices such as provincial governorships to men of senatorial rank. But he also redefined the senatorial order from within. He created a second senate for his new capital, Constantinople, that progressively grew in status to equal the Roman senate. Even more consequentially, he began the process of inflating the senatorial title, creating new pathways to achieve senatorial status. The senatorial order grew deliriously in numbers. This growth came at the expense of municipal aristocracies, as the updraft of wealth, prestige, and talent put new pressures on the traditional town councils. By granting senatorial rank as the reward for imperial service, Constantine set in motion the basic dynamics of the late Roman aristocracy. Constantine reorganized the entire system of rank and honors, vigorously centralizing the economy of honor and orienting it around the emperor's person.[10]

Like Augustus, Constantine solidified his new order with a rigorously conservative social policy. He eagerly protected veterans and peasants, the hardy foundation of the empire's power. Governors were told to take care, "so that the multitude of the lower classes may not be subjected to the wantonness and subordinated to the interest of the more powerful." Constantine's laws reinforced social hierarchy. He aimed to keep slaves and freedmen in their place. The reforms of Constantine reveal a thorough abhorrence of social mixing. Constantine tightened the famous adultery legislation of Augustus, and he strengthened the marriage prohibitions that kept the honorable elite distinct from the untouchable classes at the bottom of the social scale. He banned property transfers to illegitimate children (a small decency which Roman manners had discreetly tolerated) and restricted divorce (which the Romans had practiced quite liberally). Centuries of tradition and legal subtlety were no impediment to the emperor's will. In an age of brisk upheaval and regime change, Constantine's laws set the tone for late antiquity.[11]

The age of bold strokes let Constantine carry out these experiments. The most famous of all was his religious conversion. There is no reason to doubt the sincerity of the emperor's religious motives. Christianity was an idiosyncratic choice, and not an obviously calculated one. The church had grown, despite the renewed persecution under Diocletian, but Christians remained a peripheral group. Constantine's religious beliefs were, in the short term, a liability.

But his faith brought him the allegiance of a devoted and organized bloc, and Constantine turned his patronage of the church to advantage. He wasted no time intervening in burning ecclesiastical disputes and earnestly sought to establish doctrinal harmony. He was open-handed in his benefactions toward the church, and like any emperor he funded monumental building projects for his God of choice. He severed the lifeblood of funding for the old gods, surreptitiously looted the temples, and put blood sacrifices on their way to extinction. At the apex of the social pyramid, the emperor's tastes set a tone, even in matters so intimate and inscrutable as the worship of the gods. Constantine was the empire's patron in chief, and his favoritism rippled outward in expanding circles of influence. For Christianity, Constantine's uncanny choice was *the* watershed, the moment of irreversible acceleration.[12]

The foundation of a new Rome was equally idiosyncratic. For decades emperors had shuttled along a string of frontier-facing towns, cities like

York and Trier, Sirmium and Naissus, Nicaea and Antioch. When Diocletian decided to celebrate the twentieth anniversary of his rule in Rome, it may have been his *first time* to see the city. Rome remained the sentimental, symbolic, and ceremonial capital, but it had long been a truism that "Rome is where the emperor is."

The choice to establish a formal counterweight to Rome in the east was, nevertheless, an impolitic leap. The geographic choice was a stroke of brilliance. The military center of gravity was located in the Danubian provinces. Situated along the high road connecting west and east, Constantinople had ready access to the lands of the march. Fortified by Constantine and his successors, the city would prove virtually impregnable. A city on the sea, its hinterland was the entire wealthy arc of Hellenized provinces from Asia Minor to Egypt. While later emperors elaborated on the original plan, Constantine's ambitions for the city that would bear his name were capacious from the start. Here again he set in motion the forces that would define the coming centuries. Constantinople was a city of destiny.[13]

Constantine's reign set a pattern for late antiquity. It did not bring to an end the age of reform and experimentation, but now, for the first time since the breakdown of the Augustan order in the middle of the third century, there was an essentially stable set of relationships between the army, the aristocracy, and the imperial administration. In one final similarity with the first founder of the Roman Empire, by the time Constantine died, there were few living who could remember the old ways. His reign was the longest since that of Augustus. After three decades of rule, in May of AD 337, Constantine gave up the ghost. His body was carried in a gold coffin and brought to Constantinople, where he received an uncertain mix of traditional and Christian obsequies and was laid to rest in the middle of a memorial to the twelve apostles. His eulogy was prophetic. "The lot of imperial rule fell to him even after his death. For he administered the entire world in a kind of new life, ruling the empire in his name as the Victorious One, the Greatest One, the Augustus." The ghostly presence of Constantine would hang over the new order for centuries to come.[14]

The Enabling Environment

Environmental change cooperated with human initiative in the rebuilding of the late empire. The charmed conditions of the Roman Climate Optimum would never return; the last wisps of mid-Holocene-like weather, warm and humid everywhere, were irretrievably a thing of the past. The end of this era had been tumultuous. Global and regional instability peaked in the middle of the third century, coincident with the extreme droughts that seemed like the death rattle of the earth itself. But if the third century was the "old age of the world," the long fourth century was unexpectedly a new lease on youth.

The climate stabilized. After AD 266, there was not a major volcanic event for more than a century and a half. Solar output escalated, reaching its maximum across the entire Roman period around AD 300 and then maintaining high levels through the fifth century. The fourth century was a time of distinct warming. The Alpine glaciers were in full retreat by the middle of the fourth century. The fast-reacting Mer de Glace glacier in the Mont Blanc Basin melted to 1990s levels by the end of the century. Average temperature levels do not seem to have matched the highs of the early empire, but the sun smiled on the age of restoration.[15]

As the RCO receded, a phase of climate history that is more recognizable as the late Holocene became visible. Large-scale climate patterns were now under the dominant sway of the North Atlantic. Atmospheric pressure gradients in the North Atlantic have an outsized influence on the fate of societies stretching from western Europe deep into interior Asia. Two centers of opposing circulation in the Atlantic interact to shape the direction of the westerly storm tracks. The Azores High is a zone of permanently high atmospheric pressure to the west of the Mediterranean; the high pressure creates anticyclonic circulation, spinning air clockwise and blocking rainfall. To the north, the Icelandic Low is a zone of resident low pressure centered in the northern Atlantic; it creates cyclones and spins air counterclockwise, over western Europe. The fluctuation in pressure differences between these two zones is known as the North Atlantic Oscillation. The NAO is one of the truly great climate mechanisms of the globe.[16]

The strength of the NAO in the winter is high stakes. When pressure differences over the Atlantic are pronounced—in other words, the positive mode of the NAO index—they generate powerful cyclonic activity and spin the westerlies poleward; Britain and northern Europe get drenched. When pressure differences are relatively modest, weaker storm tracks dribble into the western Mediterranean, favoring the water balance of the south over the north. For instance, a frequently positive NAO in 2015–2016 contributed to record-setting rains in Britain and anomalous drought in parts of the western Mediterranean. Like a global-scale yard sprinkler, the swiveling NAO directs the spray of the storm tracks over the middle latitudes of the northern hemisphere.[17]

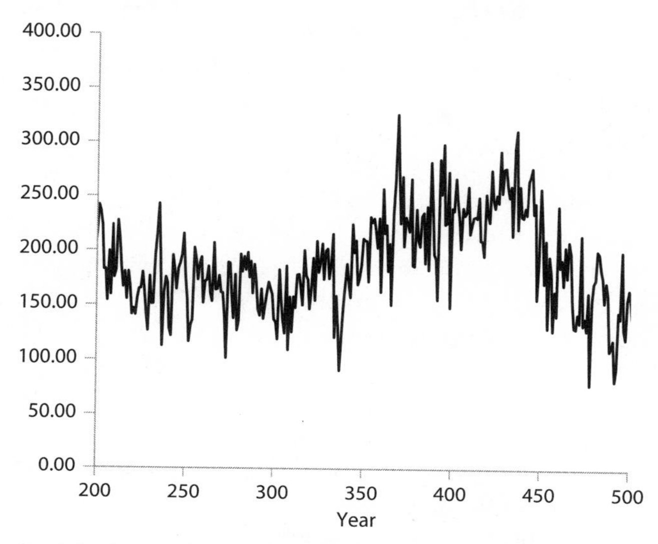

Figure 5.1. Precipitation Totals (mm) France/Germany (data from Büntgen et al. 2011)

The history of the NAO is recoverable from natural archives. The British Isles are directly exposed to Atlantic climate mechanisms, and a sensitive record of the NAO going back 3000 years has been found in the caves of Scotland, in stalagmites whose annual growth rate is sensitive to the phase of the NAO. A period of persistently positive NAO starts in the late third century. The fourth century stands out. Between the Bronze Age and the present day, the fourth century saw levels matched only during the Medieval Climate Anomaly. Other pieces of the puzzle then fall into place. In lake records from Spain, there are strong signs of aridity beginning some time in the fourth century. In northern and central Europe, by contrast, the rains were more abundant. The precipitation record derived from French and German oaks reflects high and rising levels of rainfall across the fourth and first half of the fifth centuries, as storm tracks moved over central and northern Europe.[18]

In the central Mediterranean, the effects of a positive NAO are unpredictable. Great troughs in the storm tracks can bring rains to Italy or miss the peninsula altogether. The northern parts of Italy may have been the recipient of continental rainfall, even as the south was dried by the decline of winter storm activity. Northern Italy rebounded in the fourth century, in part because of the strong presence of the imperial administration, but also perhaps because of reliable rains. In central and southern Italy, by contrast, recovery from the crisis of the third century was weak: the countryside of Campania was "an emptied landscape, an agrarian slum, with scattered cottages perched in the ruins of what had once been centers of thriving agrobusiness." Italy may have been poised right along the knife's edge of abundance and misfortune. Under the control of the positive NAO, the moisture regime of the Mediterranean was a flickering switch.[19]

The climate mechanisms in the eastern territories of the empire were more layered. The NAO still exerts influence, but the eastern Mediterranean climate sits at a truly global crossroads, tugged by the monsoonal systems of the tropics, affected by atmospheric pressure over Asia, and distantly modulated by the El Niño Southern Oscillation. And although *temperature* patterns in the eastern Mediterranean can be coherent across wide areas, precipitation is more dependent on local factors and thus more finicky. In late antiquity, the eastern Mediterranean regions seem to have experienced moisture trade-offs, with sharply divergent patterns in Anatolia and the Levant. In Israel, the fourth century inaugurated two centuries of life-bringing humidity, before more arid times resumed. In Asia Minor, nearly the opposite: the fourth century was modestly drier, while afterward a more humid period began.[20]

The fourth-century climate, then, was favorable but fluttering. The Mediterranean was perched along the uncertain, shifting edge of the winter storm tracks. Major droughts and famines are much more frequently recorded in the written records of the late empire. But we must hedge this observation with any number of qualifications. The resurgent population would have meant more mouths to feed once again. Droughts and famines were more commonly *attested*. But thanks to the triumph of Christianity, the kind and scope of evidence we have from late antiquity is radically different. We have more sermons, letters, and saints' lives. Many of them are from out-of-the-way places, places that were invisible in the earlier period. And the informants we have are altogether more garrulous when it comes to the hardships of ordinary life. Christian leaders made a living by helping the poor. We cannot necessarily say that there was more drought and famine, just because we know of more droughts and famines.[21]

The prime example of a climate-induced crisis in the late antique record is a food shortage that gripped Cappadocia in AD 368–9. We see the entire affair only through the eyes of the bishop, Basil of Caesarea, an avant garde figure in the church. Basil brought all of his rhetorical and administrative genius to bear on the crisis. Through his eyes, we watch as this inland society steeled itself against the oncoming shortage. It rumbled over the horizon, slowly. For Basil, the food crisis was a teachable moment, pulling back the veil on the stark social cleavages in Roman society. He takes us into the hovel of a poor father, forced to decide which of his children to trade for sustenance. "How can I put before your eyes the suffering of the poor? . . . He turns his eyes at last to his children, to take them to the market and find a way to put off death. . . . Imagine his deliberations. 'Which will I sell first? Which will the grain merchant like best?' . . . He goes, and with ten thousand tears, sells his most beloved son."[22]

What we cannot know is how many times this scene had played out before, without a Basil to report the bitter details. We should be cautious with stories like these, but the physical evidence for aridity in this region should also check us from too breezily dismissing the whole affair as the artful exaggeration of an ambitious bishop. The natural archives, and the general atmospheric regime of the fourth century, provide a realistic context for just this sort of acute crisis in Anatolia.

Basil's famine was, to all appearances, a local phenomenon. But if we comb the evidence carefully and compare it with the record of the high empire, we also find that the fourth century presents us with accounts of spatially widespread food crises, of a kind that are hard to find in the earlier days of empire. The most notorious of these struck in the mid-380s. In AD 383, "the hopes of all the provinces were betrayed by a miserable harvest." At the same time, the Nile flood was weak. This conjuncture was a recipe for emergency. A "general famine" ensued. We are richly informed about this episode, because it was perfectly timed to become a matter of religious polemic in the highest circles. This famine prompted an unusually eloquent dispute between the pagan senator, Symmachus, and the Christian bishop of Milan, Ambrose, over the removal of the Altar of Victory from the senate house in Rome. In the ongoing tug-of-war between pagans and Christians, the Altar of Victory had become a special totem. The pointed exchange between the senator and the bishop gives us a rare, aristocrat's-eye view of a truly extensive famine.[23]

For the pagan Symmachus, the unusual severity of this famine was caused by the anger of the gods. Bad harvests were normal in the course of events, easily overcome because "provinces come to the relief of each other, good harvests here supplying the deficiencies of bad harvests there." But the current famine far exceeded the ordinary "vicissitudes of the harvest-seasons," and the "general scarcity" was an unmistakable omen of the gods' displeasure. The rural poor were "kept alive by eating the twigs of forest trees." The city of Rome took emergency steps, expelling foreigners (though not the exotic dancing girls, who were allowed to remain) to protect its precious stores. For the Christian bishop Ambrose, the crisis was overblown. The northern provinces had enjoyed good harvests. And, he pressed, "Can we really believe that the Nile omitted to overflow its banks in the usual way, because it wanted to avenge losses of the priests of the city of Rome?" His efforts to block the altar's restoration carried the day.[24]

As with Basil's famine, this episode is recorded thanks to fortuitous circumstances, and it may simply be that these kinds of events were more common in an earlier age than the sources let on. But we should not lose sight of the real climate factors behind the inter-regional food crisis of the AD 380s. The abundance of the harvests in the north, coincident with the drought conditions in the south, is certainly plausible. And the ill-timed failure of the Nile is unexpectedly confirmed in a papyrus, in which a military recruit complains of the ravages of a famine in Upper Egypt. We know of other major food crises that rattled the entire empire in late antiquity, including a vicious sequence of drought and famine in the early AD 450s, that are too insistent for us to dismiss as purely an artifact of what we chance to know. It is altogether credible that the background climate of these centuries fostered short-term climate crises on a larger scale than before.[25]

The physical climate during the imperial renaissance was favorable but fickle. This pattern was mirrored in the biological history of the fourth century. Late Roman society, even in the period between the pandemics, still groaned under an oppressive mortality regime. The disease ecology of the early empire endured. The empire remained densely urbanized and thickly connected. Health outcomes in late antiquity were bleak. The Romans were still short. In fact, many of the skeletons that bioarchaeologists have attributed to the "Roman Empire" actually belong to fourth-century contexts, when the practice of inhumation rather than cremation had become virtually universal. As before, the cost of fighting infectious disease drained the body's resources and quietly depressed the stature levels achieved by the Romans. There was no crisis of the third century for germs. They had not let up on the Romans.

The seasonal patterns of mortality are a signature of the heavy endemic disease load. Between the conversion of Constantine and the sack of Rome in AD 410, we have thousands of Christian tombstones from the imperial city recording the day of the believer's departure from this life (and the sharp drop-off after AD 410 is itself a sign of disruption in the old capital). In aggregate, it is our single richest dossier of information on the seasonal rhythms of the grim reaper. The dog days of summer were deadly, as a wave of lethal gastrointestinal bugs overwhelmed the city. Mortality spiked in July but only peaked in August and September. The autumn crest surely

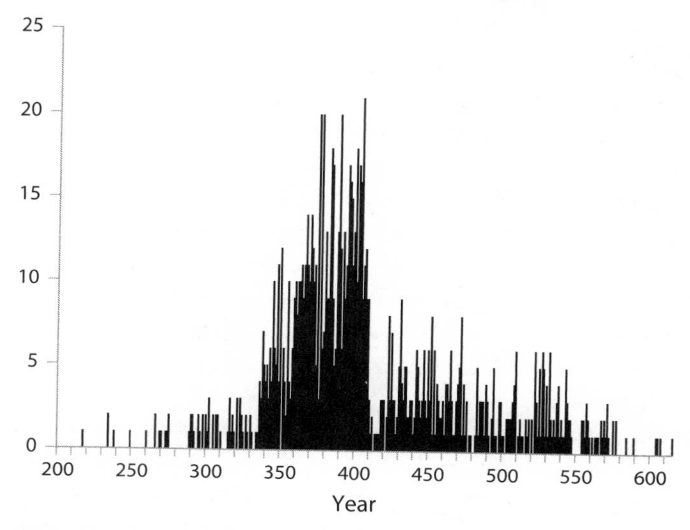

Figure 5.2. Number of Date-of-Death Inscriptions by Year

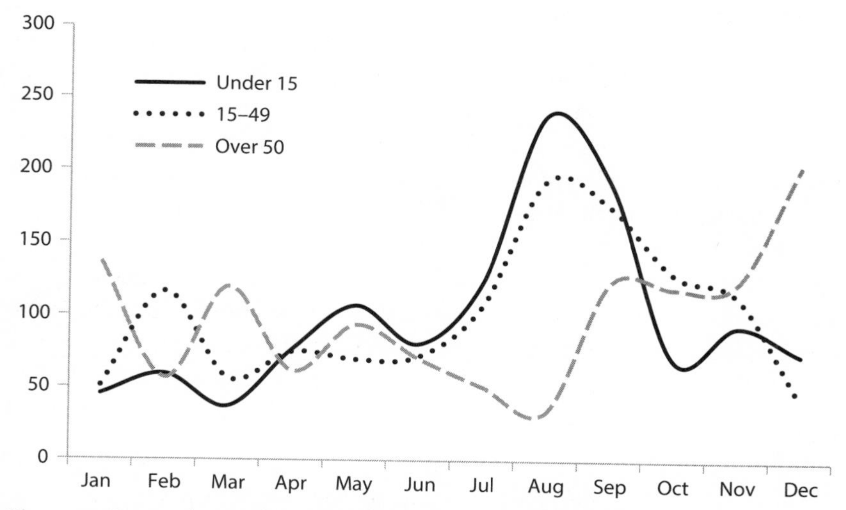

Figure 5.3. Seasonal Mortality in Rome to AD 410: Children, Adults, Elderly

points to the continuing prevalence of malaria. For the elderly, however, winter remained by far the most perilous season; the winter respiratory infections preyed on those who survived into their later years.[26]

The germs of Rome were vicious. But the headline from the long fourth century may be the absence of a catastrophic mortality event. In a thorough catalogue of the sources, Dionysius Stathakopoulos has identified fourteen epidemics in the fourth century and another eighteen in the fifth century. These tallies are rather higher than the number of outbreaks we can register in the early empire. What we see now spread before our eyes is a little more of the normal background of epidemic mortality across the empire. And the truly striking fact is the absence of interregional mortality events. The exception that proves the rule is a serious outbreak of a deadly pustular disease, perhaps smallpox, in parts of the eastern empire in AD 312–13. A drought was followed by famine and pestilence. Victims were struck with a fiery rash covering the entire body, and many of them were left blind. But the geographic reach of this event was anomalous. Most epidemics were highly constrained.[27]

Epidemics could grip cities or regions, but their infectious agents were not typically germs that could spread easily over vast distances. Disturbances like war and famine regularly triggered mortality events on a local scale. Siege warfare and armies on the move were always biological hazards. Sieges caused miserable crowding. They threatened access to food and endangered the supply of clean water. Armies on the march brought soldiers into contact with unfamiliar germs. Repeatedly in late antiquity, invading columns of foreign troops were repelled by the invisible shield of local germs. War and mortality crisis went hand in hand.

The twittering climate regime of late antiquity also had an intimate relationship with the pulses of epidemic mortality. Food shortage was a corollary of disease outbreak. Anomalous weather events might trigger explosive breeding of disease vectors. A devastating famine in Italy in AD 450–51 was coincident with a wave of malaria, for instance. Food crisis fanned desperate migrants in search of survival, overwhelming the normal environmental controls embedded in urban order. Food shortages forced the hungry to

Box 5.1.
Twin Calamities: How Climate Events Trigger Epidemics

→ Vector or host movement/breeding (e.g. mosquitos, rats)
→ Subsistence migration, crowding
→ Broken environmental controls (waste, corpse disposal)
→ Malnutrition
  → Consumption of poisonous substances
  → Weakened immune resistance

resort to consuming inedible or even poisonous food, all while depleting the power of their immune systems to resist infection.[28]

Ancient Mediterranean societies protected themselves as they could with buffers against the stress of environmental variability. In the later empire, our vivid sources provide us the chance to watch as towns tried, and sometimes failed, to blunt the impact of nature's violence. When systems of control snapped, catastrophe could ensue.

Our most acute report of local breakdown is the narrative of a famine and pestilence that swept Edessa and its hinterland. In March of AD 500, a plague of locusts destroyed the crops in the field. By April, the price of grain skyrocketed to about eight times the normal price. An alarmed populace quickly sowed a crop of millet, an insurance crop. It too faltered. People began to sell their possessions, but the bottom fell out of the market. Starving migrants poured into the city. Pestilence – very probably smallpox – followed. Imperial relief came too late. The poor "wandered through the streets, colonnades, and squares begging for a scrap of bread, but no one had any spare bread in his house." In desperation, the poor started to boil and eat the remnants of flesh from dead carcasses. They turned to vetches and droppings from vines. "They slept in the colonnades and streets, howling night and day from the pangs of hunger." When the December frosts arrived, the "sleep of death" laid low those exposed to the elements. The heaps of corpses were all the church could handle. The migrants were worst affected, but by spring no one was spared. "Many of the rich died, who had not suffered from hunger." The loss of environmental control collapsed even the buffers that subtly insulated the wealthy from the worst hazards of contagion.[29]

It may be that Edessa, on the eastern edges of the empire, was too remote for easy rescue. But there is no doubting that the episode, for all its brutality, was regionally confined. Opportunistic bacteria and viruses seized such moments of disorder and weakness. Their success did not hinge on their overwhelming capacity for transmission, and they did not threaten to spark a conflagration reaching beyond the region. Until the great climate spasms of the AD 530s, and the arrival of a ferocious new pathogen that soon followed, the world of late antiquity was granted a time of reprieve from the most savage microbes. The men and women who lived and died in these centuries had to contend with the old familiar complaints. These were choppy waters. But for a season, the Romans were spared environmental catastrophes on an imperial scale. As we will see, peoples far beyond the frontier were not so fortunate, and the consequences were ultimately to rattle the empire itself.

The Vaulting Structure

Sometime in the reign of Constantine, a man named John was born in the far southeastern corner of the empire, at Lykopolis in Egypt. Lykopolis lay on the west bank of the Nile, along a lonely stretch of the river some 400 miles upstream from Alexandria. It was a week's sail from the Mediterranean, at a hard clip. A monastic writer named Palladius visited Lykopolis in the late fourth century, and the journey took him eighteen days "partly by foot and partly by ship on the river." But that was during the season of the flood, "when many fall sick, which indeed I did myself." The purpose of his voyage was to find the monk John, who lived alone in the hard, sun-swept hills above the town. John had become a religious celebrity, and a sighting of the holy man was as exotic and thrilling as an encounter with any wild creature.[30]

John was of undistinguished ancestry. Around the age of twenty-five, he renounced the world and trained in the monastic communities that were just emerging in Egypt. He walled himself in a cave high above the town and spent thirty years in isolation, excepting the regular delivery of his victuals. The gifts of healing and clairvoyance came to him (including the eminently practical ability to predict whether the annual inundation of the Nile would be abundant or meager). In his last years, John would receive visitors—all male, through a window in his cell, on Saturdays and Sundays. His legend spread on the wind, to the ends of empire. The emperor Theodosius I looked to John as a "personal oracle," and at least twice dispatched an imperial messenger to Lykopolis on the eve of military campaigns to gather the monk's premonitions.[31]

John's miraculous gifts were the stuff that fired the imagination of the fourth century. But the arid sands of Egypt have fortuitously preserved a small dossier of contemporary letters, in which we see the real depth of the recluse's entanglements in the world around him. In one letter, John intervened on behalf of a villager named Psois, who desperately employed the monk's help in nothing other than dodging the draft. The villager had mortgaged his two children to borrow eight gold coins, lent in turn to John to be used as an instrument of persuasion (what might impolitely be called a bribe). The effort had failed. Psois then amputated his finger, a gruesome but conventional means of disqualifying oneself for military service. It was risky. In AD 367, a law ordered that any conscripts found to have cut off a finger were to be burned alive. In AD 381, however, Theodosius I declared that "if any person by the disgraceful amputation of his fingers should evade the use of arms, he shall not escape that service which he seeks to avoid, but he shall be sealed by tattoo, and he shall perform military service imposed as a labor, since he declined it as an honor." Our shirker, Psois, seems to have been caught unawares by this very law of Theodosius, and it has been suggested that his draft dodging belongs in the direct aftermath of its promulgation in AD 381.[32]

We will probably never know whether John was finally able to come to the rescue of the unhappy Psois, but the colorful episode is an instructive example of how the machinery of the state shaped the intimate details of life, even in this remote corner of empire. We should not underestimate the sheer scale of life in the fourth century. At the same time, the draft-dodging villager also reminds us—as does the stream of laws in the codes—that military recruitment was a persistent problem, though not a purely demographic one. Conscription pitted the finite power of the imperial state, and its agents on the ground, against an uncanny array of forces. It would be a mistake to draw a straight line from the demographic nadir of the later third century to the military crisis of the later fourth century. There was too much history in between, and in fact unruly dynamism, rather than decay or decadence, posed the greater challenge to statecraft in the late empire.

The bracing reforms of Diocletian and Constantine, and the environmental background of the fourth century, set the scene for the Roman Empire's comeback act. The imperial recovery started with the demographic turnaround in the later third century. But the ongoing monetary crisis was an invisible drag on renewed takeoff. The silver currency remained in free fall. Diocletian tried to save the old money regime by brute force, dictating maximum prices and decreeing a market value for gold. He requisitioned the precious metal in huge purchases that brought the gold stock rushing into the empire's coffers at artificially low prices. His monetary policy failed, and galloping inflation continued into the fourth century.[33]

Monetary instability choked credit markets and stifled exchange. But in the reign of Constantine, a solution started to take shape: a true gold economy. Constantine let gold circulate freely at its market price. He also reduced the size of the gold coin, the *solidus*, to 1/72 of a Roman pound. These reforms paved the way for a full-blown gold system. They were stabilized by the creation of new taxes in gold money that assured the imperial state of revenue in precious metals. Constantine's reign was an economic watershed. Under Constantine and his sons, the gold *solidus* became the functional basis of a new economy. By the AD 340s, there were vastly more *solidi* in circulation, as the melted treasures of the old temples and gold from a new source of supply entered the market. By the 350s, the *solidus*

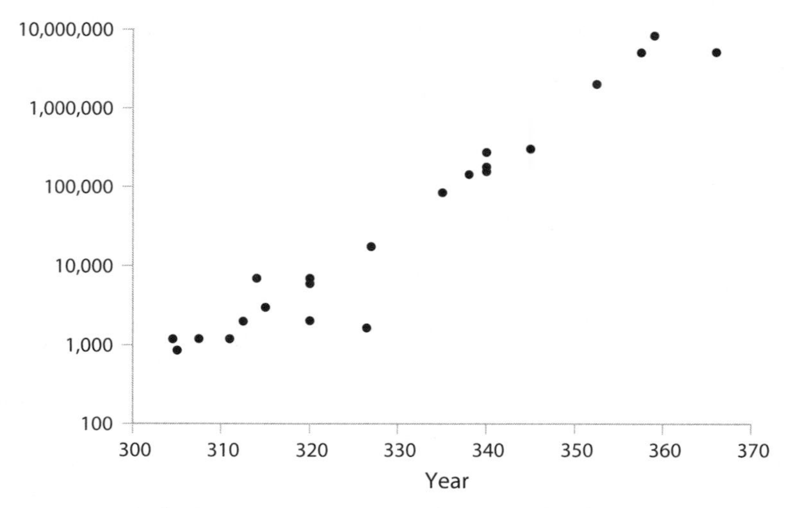

Figure 5.4. Nominal Wheat Prices, AD 300-375, denarii/artaba of wheat

Figure 5.5. Gold Coin (Solidus) of Constantine I (American Numismatic Society)

even started to replace the old *denarius* as the unit of account in common transactions. We should not underestimate what a flip of the imagination this required. For a thousand years, silver *was* money. Now, life would be re-monetized, with gold at the center.[34]

The state collected taxes in gold and paid its huge roster of officials in gold. The fiscal machinery was the pump of the economy's circulatory system. But in late antiquity the market economy quickly revived, and the real story of the age is the "particular late Roman fusion of market and fiscal forces." This fusion is reflected in the careers of those social climbers who took indiscriminate advantage of private markets and public emoluments alike. We know of one man, named Heliodorus, who made a fortune as a fish-sauce merchant. He invested his profits in land and slaves and went to law school. He served the emperor and in reward was given landed estates across Macedonia and Greece, "gold, silver, an abundance of slaves, and herds of horses and cattle." Such biographies throw light on the overlapping networks of capital and imperial patronage that gave such vitality to fourth-century society.[35]

The monetary restoration revived the financial sector. The great banks of the Roman Empire had all but vanished with the crash of the silver money regime, but in the fourth century they came back to life. The evidence for credit and banking in the fourth century exceeds that for any period of Roman history. Nothing from antiquity can match in vividness the priest John Chrysostom's portrait of the deposit bankers of Antioch at work in his day. Credit markets fueled capital investment and underwrote mercantile ventures. They turned the wheels of commerce. "The merchant who wants to get rich prepares a ship, hires sailors, summons a captain, and does all the other things necessary to set sail, and borrows money and tries the sea, and passes into foreign lands." The revival of money and credit awakened commercial networks across the Mediterranean. Saint Augustine, in his port town of Hippo, evoked the allure of the trading life. "'Sailing and trading,' another says, 'that's great! It's great to know many provinces, make money everywhere, not be beholden in town to some mighty man, to always travel in foreign lands and nourish the mind on a variety of business and nations, and then to come home, rich with the profits!'"[36]

Late antique trading networks grew on the memory of earlier centuries of commerce, but they were not beholden to the past. New regional circuits of exchange evolved, less dominated by Italian demand. Egypt and Palestine entered the wine trade in earnest from the third and fourth centuries. The archaeological distribution of one kind of ceramic pottery, known as African Red Slip Ware, is truly astonishing and traces the rise of Africa to a position of prominence in the long-distance networks connecting the entire empire. The lure of profits knit together the Roman world, making it a giant free trade zone worked by savvy professional merchants. "Like the merchant who conducts his trade and knows how to make a profit in his business not just by one route or in one manner, but who watches carefully all about him, with quick wit and alertly: if he should fail to make a profit, he turns to another deal—for his whole purpose is to make money and grow his business." A text known as the *Description of the Whole World and Its*

Figure 5.6. Mosaic Showing Beachside Trade, Hadrumentum, North Africa (Bardo Museum) (Leemage / Getty Images)

Peoples is the fourth century's contribution to the genre of rough-and-ready trader's geography. Written by an eastern merchant, it is a "practical guide to the best buys of the fourth-century empire's different shores." It hints at the scale of commercial integration in this period.[37]

The capitalism of the sea greased the wheels of social mobility in late antiquity. Far from an age of stagnant social relations, the recovery flung open the doors of opportunity. The possibilities were felt far and wide.

We happen to have the tombstone of a one-time peasant from a modest backwater in Tunisia. His grave told his story with rather unabashed self-regard. He had been "born into a poor dwelling and of a poor father, who had no property or household." Under the "raging sun," he reaped harvest after harvest and became a "gang leader instead of a laborer." "This effort and my frugal lifestyle brought success and made me master of a household and gained me a house, and my home itself lacks nothing." He was appointed to the town senate, all despite "starting out as a country boy." His case sketches the lines of possibility in an open society, even far from the centers of energy.

Closer to the nodes of power and wealth, the opportunities loomed still larger. The new, eastern senate in Constantinople filled its ranks against the backdrop of social disequilibria. To the horror of the old guard, the sons of coppersmiths, sausage-makers, fullers, and bath attendants suddenly wore the robes of senators. As usual in such circumstances, the marriage market helped to soften the edges of rapid upheaval. The biography of Saint Augustine is a case in point. His meteoric ascent from the dusty North African backcountry was completed by an engagement to a daughter of impeccable ancestry, ultimately left unconsummated by his sharp conversion to the religious life.[38]

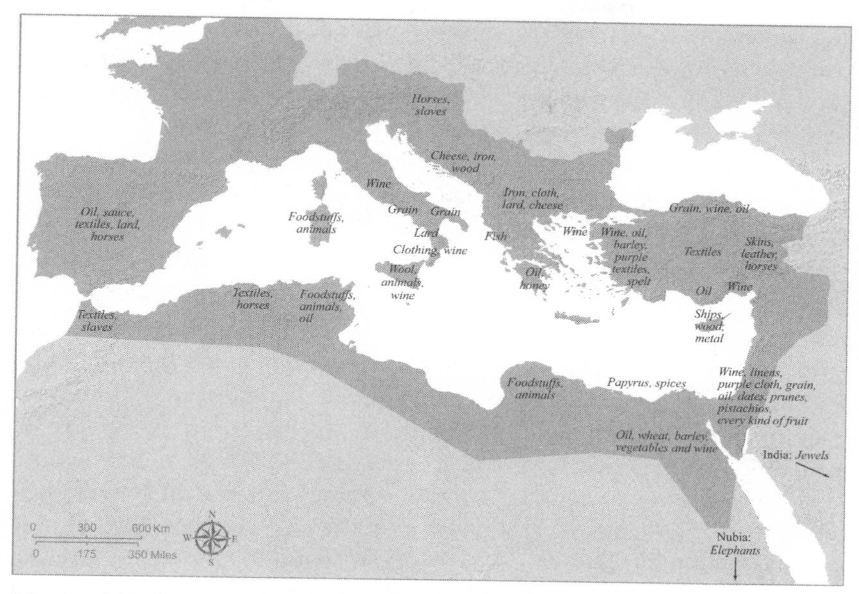

Map 14. A Trader's-Eye View of the Roman Empire: The Expositio

Under the shelter of the imperial recovery, a restless society stirred. Intricate layers of economic and legal stratification structured the social order. Most insidiously, the empire was still home to a genuine slave society. In fact, the slave system traces the physiognomy of late antique society with particular clarity. After the dislocations of the third century, the slave system experienced a kind of brutal resurgence behind the revival of the Mediterranean market economy. Slaves were everywhere. Their sweat and toil was the foundation of many aristocratic fortunes. An aristocratic woman named Melania the Younger, from one of the most blue-blooded lines in Rome, owned over 8000 slaves. One of her estates in southern Italy alone had 2400 unfree workers. The pious Melania freed thousands of her slaves, but even after renouncing the material world, she was trailed still by seventy-five slave-girls and eunuchs! Her case is exceptional but telling. Slavery is a manifestation of the hidden power of markets behind social relations—markets for commodities, markets for honor, markets for human bodies.[39]

Slave-ownership on Melania's scale was rare. More consequential were the elites, late antiquity's 1 percent, who owned "multitudes," "herds," "swarms," "armies," or simply "innumerable" slaves, both in their households and in the fields. We encounter these rich slave-owners any time we are afforded a glimpse of the lifestyle or economic foundations of the well-to-do in the fourth century. Occasionally we glimpse the fusion of public and private circuits of wealth in the patterns of slave-holding. We have a speech that praised a retired military officer who was virtuous but "not wealthy": "this man for a long time commanded many soldiers, but he was barely able to buy one farm, and even it was nothing to praise. He had eleven slaves, twelve mules, three horses, four Laconian dogs, but he terrified the souls of barbarians."[40]

Maybe most telling of all is the ubiquity of slave-ownership among the indistinct middling classes. "Even the household of the poor man is like a city. For in it there are also rulers. For instance, the man rules his wife, the wife rules the slaves, the slaves rule their own wives, and again the men and women rule the children." To own a slave was a standard of minimum respectability. In the fourth century, priests, doctors, painters, prostitutes, petty military officers, actors, inn-keepers, and fig-sellers are found owning slaves. Many *slaves* owned slaves. Even assistant professors in Antioch had a few slaves. The same pattern prevailed in the countryside, where all over the empire we find working peasants with households that included slaves. In the papyri from rural Egypt in late antiquity, "the ownership of a small number of slaves—one to four—was not remarkable. The economic importance of slavery in such households was not marginal."[41]

The scale of economic stratification was truly staggering. The top senatorial families of late antiquity owned stupendous wealth. According to the breathless report of a Greek observer, each of the great senatorial houses in Rome was like a city in its own right, with fora, temples, fountains, baths, and even hippodromes inside. Houses of the top rank had incomes of 384,000 *solidi*, while those of the next rank earned over 72,000 *solidi* per year. These incomes are the equivalent of something like the production of 80,000 family farms, per year. The saintly Melania inherited family estates strewn across the western Mediterranean: in Italy, Sicily, Spain, Gaul, Britain, and Africa. One of her ranches in Africa required two bishops. When her two young children died, she decided to liquidate this trust fund, built over generations. It was a scandalous breach of aristocratic duty. It also crashed the land market: she had trouble moving her Roman mansion. In the words of Chris Wickham, "The western senatorial elite . . . could boast both ancestry . . . and gigantic wealth, possibly, in the case of its leaders, greater in relative terms than any other aristocracy ever."[42]

The economic elite in this age accumulated private fortunes whose scale and geographic dispersal would not be matched again until the age of trans-Atlantic colonialism. But the dominant social process of the late empire was not the drastic concentration of wealth in a few hands. Far from being lop-sided, late antique society was dominated by middling persons, respectable but fragile, locked in networks of patronage. The late antique city was a hub of production, exchange, and services. It bustled with professionals, merchants, and craftsmen of unpretentious means. Many of them clung to slender patrimonies. Only rarely do we have the opportunity to see the combined weight of this unspectacular prosperity. But the fragments of land registers from Egypt provide solid proof that modest land-owners and small independent proprietors were numerically dominant. Wealth was stratified but not concentrated.[43]

The largest social bloc, as ever, remained the silent majority of laboring peasants. Many of them were landless farmers. Only sometimes in our sources can we hear the plaintive cries of workers hard squeezed by ambitious landlords. Their condition was parlous but not hopeless. The state, far from cravenly beholden to landed interests, wanted to protect its faithful tax base. Constantine, with a watchful eye on the "inflow of tax revenues," passed laws protecting the "multitude of the lower classes," in fiscal assessment, debt collection, and even tenancy contracts. The empire's shadow was inescapable. Despite the dreamy idyll of the poet Claudian, there was no such thing as a pristine peasantry untouched by time. Archaeology gives the lie to any such image. Late Roman peasants ate from plates spun in specialized manufacturies and slept under industrially produced rooftiles; we regularly find coins sprinkled on their farms. They were embedded in circuits of market and fiscal exchange. When Synesius, bishop of a town in Cyrenaica, wished to play up the isolation of the "rustic people" in the highlands of North Africa, he claimed that "there are people amongst us who suppose that Agamemnon, the son of Atreus, is still king, the great king who went against Troy." But, he confessed, "men know well that there is always an emperor living, for we are reminded of this every year by those who collect taxes."[44]

Beneath the tax-paying farmers, squeezed between private landlords and public tax collectors, were the truly poor. They haunted the shadows of late ancient society. Those who depended on their own labor hovered under the threat of poverty. They could be pushed into the ranks of what social scientists call the "conjunctural poor," when the natural turbulence of climate and disease overwhelmed their fragile stores of resilience. During a famine that swept Syria in AD 384–85, Antioch found its streets filled with hungry refugees, who had been unable to find even grass to eat and suddenly massed in town to scavenge. "Structural poverty" was also an abiding presence. In the countryside, the structural poor flitted on the edges of subsistence. Saint Martin, in Gaul, remarked on a "swineherd, shivering with cold and almost naked in his garment of skins." In town, they huddled at the gates for alms or around the public baths for warmth. These were the poor whose groans floated through the air of any late Roman town. They were naked and homeless. "Their roof is the sky. For shelter they use porticos, alleys, and the deserted corners of the town. They hide in the cracks of walls like owls. Their clothing consists of wretched rags. Their harvest depends on human pity."[45]

The destitute are more visible than ever before in late antiquity. This visibility is the direct result of a massive effort by Christian leaders to mobilize sympathy in their favor. It means that we are suddenly given the unexpurgated version of what ancient society was like—the unpleasant parts are now there to see. Bishops sought to make the "groans and gnashing" of the poor impossible to ignore. "Shall we neglect them? Walk on by?" The result was a new model of human solidarity, spun with magnificent rhetoric, that suddenly shines light into the unseen corners of the ancient city. Here the structural poor were gnawed by disease and disability. "You see a man who is transformed by his grievous afflictions into the form of an animal. His hands have been made into hooves or claws, leaving footprints on the man-made streets. Who can recognize that they are the prints a man has made having passed along the way?" "The sick man who is poor is doubly poor. For the poor who are in good health go from door to door, approaching the homes of the rich or setting up camp at the crossroads and there hailing all who pass by. But those trampled by illness, shut up in their narrow rooms and narrow nooks, are only able, like Daniel in his cistern, to wait for you, devout and charitable."[46]

We see the world of late antiquity primarily through the prism of the cities. Urban life rebounded from the late third century, but not entirely on the same terms as before. Some urban spaces were never revived, and others were repaired but transformed. Large-scale building resumed, but now churches were threaded into the fabric of urban life. In general, as political actors, the cities lost some of their former independence. The central government swept their sources of revenue and exerted a magnetic pull on the treasure and talent of their elites. But, in a sprawling empire, the cities necessarily continued to play a coordinating role in the administration of the empire, and they flourished as hubs of exchange and production in the late empire.

All of this was, as usual, magnified in the case of Rome. As a city, it was always a little artificial—propped up by the political rents and entitlements of ruling an empire. In the fourth century, after the clouds of crisis had cleared, the old capital enjoyed a kind of Indian summer. The city had long since

Figure 5.7. Rome, As Represented in the Notitia Dignitatum *(Sixteenth-Century Printing, University of Oklahoma History of Science Collections)*

lost its real political clout. Diocletian made one, uneasy, visit to the capital. Constantine, in thirty years, saw it all of three times. The year AD 348, a century after Philip the Arab's secular games, was allowed to pass without fanfare, "so little concern is there these days for the city of Rome." But in fact the city had lost little of its luster. When the emperor Constantius II entered the city in AD 357, he was memorably overawed ("on every side which his eyes rested he was dazzled by the array of marvelous sights"). Rome remained the symbolic center of the empire and a focus of tremendous wealth. Its *plebs* continued to enjoy unparalleled entitlements. By the reign of Aurelian (r. AD 270–75), the people were given baked bread rather than grain. Olive oil was given daily to the registered populace. A massive supply chain guaranteed wine to the people at a fraction of the market price. Pork, too, had been added to the free distributions, and no fewer than 120,000 recipients were on the pork dole in late antiquity. The imperial food subsidies artificially inflated the population of the eternal city. The best estimates put Rome's fourth-century population at ca. 700,000 inhabitants.[47]

In the east, New Rome grew at a pace that outran its builders' most ambitious dreams. Constantinople's population expanded ten-fold in less than a century, from some 30,000 to 300,000 residents. Grain that had once been earmarked for Rome was now destined for the eastern capital, and so many ships covered the sea between Alexandria and Constantinople that it was like a long artificial strip of "dry land." A magnificent system of waterworks went up, supplying the city with aqueducts that rivaled Rome. The city was bursting at the seams, and the walls were rebuilt repeatedly. Monumental civil engineering projects were carried out almost continuously down to Justinian. Constantinople was a creature of politics, its populace deliberately enlarged to suit the empire's pride. But, as with old Rome, we should not see it as a mere sponge. The city was a nexus of commerce, finance, and industry. It quickly became the true epicenter of Greek culture.[48]

Other great metropoleis, like Antioch, Carthage, and Alexandria, flourished without the same fabricated political supports as the two capitals. Alexandria still held out its claim to be the "greatest of the cities of the inhabited world." Its inventory boasted 2,393 temples, 47,790 houses, 1,561 baths, and 935 taverns. These were the super-cities, but just beneath them, the landscape was dotted with great towns in the range of 50,000–100,000, places like Ephesus, Jerusalem, Caesarea, Sardis, Thessalonike, Apamea, Trier, or Milan. True cities, in the 25,000–50,000 range, places like Hermopolis or Hippo or Scythopolis or Bordeaux, were more common still. All of these, and many even smaller, would have had the familiar texture of the classical city, with public baths, colonnades, fora, and other amenities. Now, too, basilicas and shrines elbowed their way into prime locations. Even if the cities had become more dependent on a centralized imperial state, the old habits of local patronage were not moribund.[49]

Urban vitality was rooted in the countryside. In the east, the fourth century was the beginning of a miraculous efflorescence of rural life. We will explore it in more detail in later chapters, but an unbroken cycle of growth is evident down to the sixth century. In the west, the rural resurgence was patchy. Massive expense was devoted to the late Roman fortifications that stretch along the Rhine and Danube. But the security situation seems to have depressed the border zones permanently. Many safer regions in the west, such as Britain, coastal Spain, northern Italy, and southern Gaul, saw robust settlement in the countryside. There was a "villa boom" in large parts of the late antique west. Most sites were modestly prosperous and clearly equipped to be little engines of agricultural production. But the boom was not universal. Parts of inland Spain and peninsular Italy limped along, never achieving demographic recovery. The varying fortunes of rural landscapes in the west were determined by the interplay of climate change, market integration, and the prospects of local security.[50]

The population grew, but the margins of abundance had been thinned. Even after the crisis had passed, the old, easy ways of military recruitment could not be resumed. The late antique state was heavy-handed. Diocletian and Constantine required the sons of soldiers and veterans to follow their fathers into the military life; army service became virtually a heritable

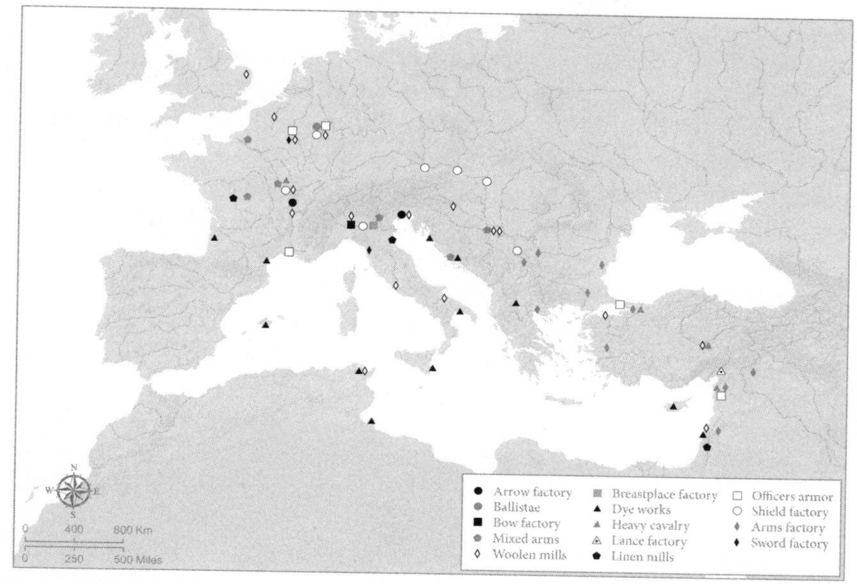

Map 15. The Imperial Machinery of Army Logistics

status. A combination of harsh violence and lucrative enticement was used to replenish the ranks. Standards were discreetly slackened: 5′ 7″ became the minimum height, in theory. Notoriously, barbarian units were enrolled to fill the gaps. But it would be simplistic to ascribe the challenges of military recruitment to "manpower shortage" *tout court*. The fourth-century state had to contend with at least one truly novel alternative to military service: the allure of the religious life for men who might have heeded the call to arms. "The huge army of clergy and monks were for the most part idle mouths." By the end of the fourth century, their total number was perhaps half the size of the actual army, a not inconsiderable drain on the manpower reserves of the empire. The civil service was also an attractive, and safe, career. The vexing issue of military recruitment in the fourth century was not directly a demographic problem.[51]

The raw military power wielded by the fourth-century Roman state was still extraordinary. Its scale of coordination was astonishing. The Roman army fielded half a million men, including 70,000 specialized troops, recruited and trained to ancient standards of discipline. The army was supplied and equipped by the most extensive logistical system the world had ever seen. The provision of weapons, armor, uniforms, animals, and food depended on the imperial machine that Diocletian and Constantine had built. The Roman soldier carried arms manufactured in over three dozen specialized imperial factories spaced across three continents.[52]

Officers wore bronze armor, embellished with silver and gold, made at five different plants. Roman archers would have used bows made in Pavia and arrows made in Mâcon. The foot soldier was dressed in a uniform (shirt, tunic, and cloak) made at imperial textile mills and finished at separate dye-works. He wore boots made at a specialized manufactory. When a Roman cavalryman of the later fourth century rode into battle, he was mounted on a mare or gelding that had been bred on imperial stud farms in Cappadocia, Thrace, or Spain. The troops were fed by a lumbering convoy system that carried provisions across continents in mind-boggling bulk. The emperor Constantius II ordered 3 million bushels of wheat to be stored in the depots of the Gallic frontier and another 3 million bushels in the Alps, before moving his field army to the west. "When an army of northern barbarians undertook a campaign, its leaders did not think in terms of millions of bushels of wheat."[53]

An unbiased observer in the later fourth century would have noted the Roman army's numerical, tactical, and logistical superiority on all fronts. But within the space of a few generations, the Roman imperial army in the west would cease to exist. The former territories of the west would be carved into successor kingdoms. The failure of empire was one of the greatest strategic implosions in history. As we have come to appreciate the reality of the empire's recovery in the fourth century, it has actually become harder to explain this failure. The collapse of the western empire was not in any simple sense the delayed consequence of unresolved tensions left in the aftermath of the third-century crisis. The renaissance of Roman power was interrupted by forces from beyond the empire. The sequence of events that set the collapse in motion lay far to the east, in the uncharted expanses of central Asia. The steppe was about to intrude into western history and put crushing pressure along the northern borders of the empire.

The New Geopolitics: The Mediterranean vs. Central Asia

The Eurasian steppe is a giant contiguous ecological zone stretching from the plains of Hungary to the eastern fringes of Mongolia. Its climate is continental and tends to extremes, with oppressive summers and savage winters. The steppe is too dry to support trees. But it is moist enough to avoid being

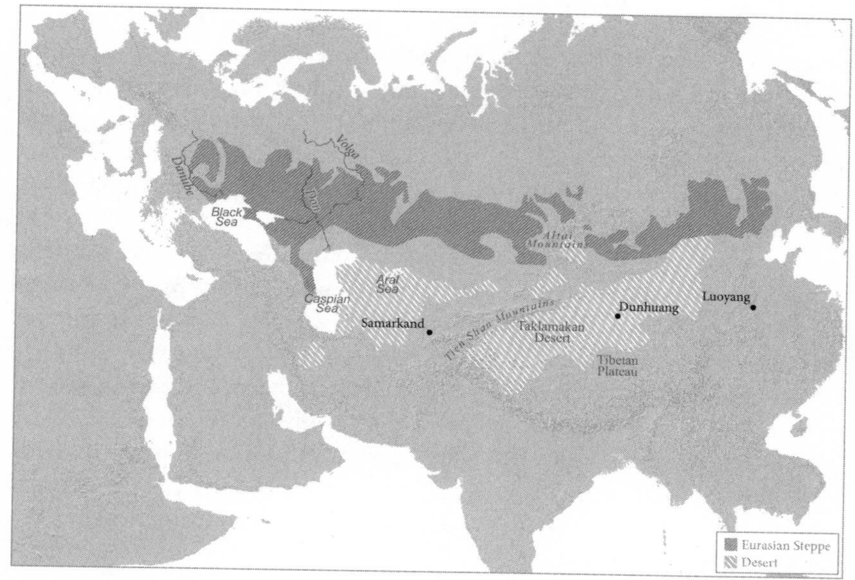

Map 16. The Eurasian Steppe

a desert. It unravels as a vast carpet of grass and scrubland. Its underbelly is striped with a series of desert regions unreached by the monsoons from the south. The deserts are dotted with oases, which have ever been positioned to serve as a relay system for the Silk Road. To the north of the steppe lies the cold belt of taiga; beyond that awaits an even colder belt of tundra. For the delivery of water, the steppe depends on the westerlies, the prevailing storm tracks of the mid-latitudes running along the globe's longest east-west landmass from the Atlantic to the Pacific. As an ecological region, the steppe dwarfs the actual land surface of the Mediterranean climate zone.[54]

To the inhabitants of the classical Mediterranean, the steppe was beyond time and history. Everything beyond the Danube was swallowed up by the "measureless wastes of Scythia," peopled by nomads incapable of experiencing the cycles of development and decline. Ethnographic commonplaces reaching back to the father of history, Herodotus, required little refreshing. In the fourth century, Ammianus Marcellinus described the steppe peoples in terms that scarcely admit their full humanity: they "have no huts and care nothing for using the plowshare, but they live upon flesh and an abundance of milk, and dwell in wagons, which they cover with rounded canopies of bark and drive over the boundless wastes. And when they come to a place rich in grass, they place their carts in a circle and feed like wild beasts."[55]

The steppe was ecologically resistant to the plough and destined to be the roaming grounds of pastoral nomads. The thin soil kept steep social hierarchies from taking root easily. Only in the later first millennium BC did mounted warriors build the earliest empires on the steppe. The first great steppe empire was built by the Xiongnu, from around 200 BC. The Xiongnu state arose in dialectical antagonism with the Han empire in China. Here, as in the classical Mediterranean, nomadism was the ideological mirror of civilization. The great Chinese historian Ssu-ma Ch'ien wrote a sympathetic and informed account of the Xiongnu in the first century BC. "People eat the meat of their animals, drink their milk, and wear their hides; the animals eat grass and drink water, therefore they move about in seasonal cycles. . . . Most of their domestic animals are horses, cows, sheep, and they also have rare animals such as camels, donkeys, mules, hinnies, and other equines. . . . As children they are able to ride sheep, and can shoot birds and mice with bow and arrow." "The Xiongnu clearly make warfare their [main] occupation." The life of war was their "inborn nature." The description might have been written by Herodotus.[56]

For centuries, in the east, the nomads were an existential threat. The Xiongnu were a multiethnic federation brought under the rule of a powerful central elite who could project overwhelming cavalry force against the Chinese state. Perpetual friction between the Han Chinese and the Xiongnu generated energy and fueled state formation on both sides. The Chinese bore the brunt of the steppe for centuries. Nomadic state formation leaned east, its pressures spreading along the frontiers between the fertile valleys of inner China and the rugged uplands of central Asia. But from the later second century AD, central Asia entered a period of obscure turmoil. Somewhere in the midst of these troubles, the steppe would turn its face to the west.[57]

We have a small but invaluable shaft of light on the chaos that would reverberate in east and west. In 1907 at Dunhuang, Sir Aurel Stein discovered, still sealed, a group of letters stashed in a former Han guard tower along the western edges of Chinese control. The letters were written by merchants from Sogdia, the small but vital state of central Asia centered around Samarkand, a crucial node in the Silk Road network. The letter in question was posted from China back to Samarkand. Written ca. AD 313, it describes an apocalyptic scene of famine, destruction, and desertion in the heartland of the eastern Han empire. Violence forced the emperor to abandon his capital at Luoyang, leaving it to the mercy of invading nomads. Crucially, the Sogdian merchant names the agents of this unbridled violence: the *Xwn*, that is, the Huns. The philological work of Étienne de la Vaissière has established the close affiliation of the Xiongnu who were the nemesis of the Han Chinese and the Huns who overran central Asia in the fourth century. To what extent the Huns of the fourth century were the direct genetic ancestors of the Xiongnu, or to what extent they assumed a fearsome name upon seizing control of the steppe, is not entirely clear. The Xiongnu, the *Xwn*, the Huns: the most advanced social formation of the steppe was about to swing its violence toward the west.[58]

In the fourth century, events in eastern and western Eurasia were drawn closer in irreversible fashion. Henceforth, events on the steppe were hugely consequential in the west. For Ammianus, the warlike nomads who had appeared on the edges of the Roman frontier were the chosen instrument of Fortune's caprice. It is a view that has become respectable again. After a period of doubt, many historians of the later Roman Empire have begun to take seriously the narrative that Ammianus presents, in which people movements on the steppe figure prominently in the geopolitical dynamics of the fourth century. The barbarians are back, and the Huns have a crucial but defined role in the story. Specifically, "the intrusion of Hunnic military power overturned a Goth-dominated political order which had been established north of the Black Sea for several political generations." Migration and invasion tilted the fortunes of the Roman imperial project along its northern frontier, interrupting the fragile resurgence of Roman power.[59]

The migration of the Huns is shrouded in the obscurity that inevitably surrounds the history of a letterless people. But the natural archives have a contribution to make, because the migration of the Huns deserves to be considered, among other things, as an environmental event. The monsoon rains drench the southern half of Asia, but the lands north of the Tibetan plateau are dry and continental. The climate of inner central Asia hinges on the westerlies, the mid-latitude storm tracks that are strongly influenced by Atlantic air masses.

When the North Atlantic Oscillation is positive, the westerly jet stream steers north and leaves central Asia arid. When the NAO is negative, storm tracks are pulled toward the equator and the rains rumble across the steppe. The Medieval Climate Anomaly (AD 1000–1350), a period dominated by a positive NAO, was cruelly dry in the interior of Asia. In the fourth century, the elements were in place for a prolonged drought in the steppe. One of the best high-resolution paleoclimate proxies is a series of Juniper tree rings from Dulan-Wulan on the Tibetan plateau. These trees lie far enough south that continental and monsoonal influences are mixed. But the fourth-century signal is arresting. Here, as Ed Cook has shown, was a time of megadrought. The two decades from ca. AD 350 to 370 were the worst multidecadal drought event of the last two millennia. The nomads who called central Asia home suddenly faced a crisis as dramatic as the Dust Bowl.[60]

The Huns were armed climate refugees on horseback. Their mode of life enabled them to search out new pastures with amazing speed. We wish we knew more about the inner logic of Hun social development in the fourth century. Clearly the climatic turbulence intersected a people, or conglomerate of peoples, in a period of consequential state formation. The climate did not act alone, simply displacing a menace from one side of the steppe to the other. It acted in concert with the rise or renewal of aggressive and complex confederations among the nomads. But precisely in the middle of the fourth century, the center of gravity on the steppe shifted from the Altai region (on the borders of what is today Kazakhstan and Mongolia) to the west. By AD 370, Huns had started to cross the Volga River. The advent of these people on the western steppe was momentous.[61]

In the words of Ammianus, "The seed and origin of all the ruin and various disasters that the wrath of Mars aroused, putting in turmoil all places with unwonted fires, we have found to be this. The people of the Huns, but little known from ancient records, dwelling beyond the Maeotic Sea [Sea of Azov] near the ice-bound ocean, exceed every degree of savagery. . . . Although they have the form of men, however ugly, they are so hardy in their mode of life that they have no need of fire nor of savory food, but eat the roots of wild plants and the half-raw flesh of any kind of animal whatever, which they put between their thighs and the backs of their horses, and thus warm it a little. They are never protected by any buildings, but they avoid these like tombs. . . . They are not at all adapted to battles on foot, but they are almost glued to their horses, which are hardy, it is true, but ugly. . . . No one in their country ever plows a field or touches a plow-handle. They are all without fixed abode, without hearth, or law, or settled mode of life, and keep roaming from place to place, like fugitives, accompanied by the wagons in which they live."[62]

The initial wave of Hunnic migration into Europe was not a coordinated assault. Far from it, only "a series of independent Hunnic warbands" came at first. But they brought new cavalry tactics that terrorized the inhabitants of the trans-Danubian plains. Their horses were ferociously effective. In the words of a Roman veterinary text, "For war, the horses of the Huns are by far the most useful, by reason of their endurance of hard work, cold and hunger."[63]

What made the Huns overwhelming was their basic weapon, the composite reflex bow. A modern analyst writes, "Very hard to manufacture, the composite reflex bow is also very hard to use with any accuracy, because its power makes it correspondingly resistant." The Hunnic bow may have had an effective range of up to 150 meters. "Shapely bows and arrows are their delight, sure and terrible are their hands; firm is their confidence that their missiles will bring death, and their frenzy is trained to do wrongful deeds with blows that never go wrong." The lightning maneuvers and deep range of these horse-mounted archers were unnerving, even to a man who had seen as much blood on the battlefield as Ammianus: "You would not hesitate to call them the most terrible of all warriors."[64]

The lands lying to the north of the Danube had been dominated by coalitions of Goths for over a century. By the later fourth century, they had "remained quiet for long ages." A kind of equilibrium had prevailed along the Danube, but the Huns threw it into disarray. In AD 376, in flight from the Huns, Goths appeared *en masse* seeking asylum inside Roman borders. Upwards of 100,000 Goths—men, women, and children—may have sought help. The Romans saw this desperate human tide as an opportunity, an unexpected influx of military recruits. The situation was handled indecisively. Some Goths were given passage, ferried across the Danube under Roman supervision. The refugees were venally exploited. Starving Goths were given dogs to eat, in exchange for selling their children. Rebellion simmered, and soon the Goths were in open revolt. They even managed to enlist mercenary Huns to join their side. The eastern emperor, Valens, hastened to the scene with his elite field army. On August 9 of AD 378, outside the city of Adrianople, he joined battle without waiting for western reserves and with faulty battlefield intelligence. The result was the worst military loss in Roman history. Valens himself was killed in the massacre.[65]

According to Ammianus, the Roman side lost two-thirds of its men, and a death toll of up to 20,000 for the Romans seems realistic. The short-term ramifications were severe. The elite core of the eastern army was annihilated. The sudden loss of so many of the empire's best troops and experienced commanders was eviscerating. The western court, in desperation, summoned from retirement the first non-Danubian emperor since the days of Gallienus—Theodosius I. The blow to the army's strength was long felt. Some regiments were never replaced. The more desperate tenor of recruitment efforts—such as the dragnet that caught up the villagers in Upper Egypt—is evident for a generation. And in negotiating from a position of compromised strength, the Romans began to experiment with a novel kind of policy: the settlement of entire people groups on Roman soil in exchange for military service under native commanders. For half a millennium, the Roman army had been one of the most effective means of assimilating foreigners into the empire. Now, the barbarization of the army would begin in earnest.[66]

The reign of Theodosius must be considered a success under the circumstances. But after his death in AD 395, no individual would ever again control both halves of the empire. Power was divided between his two young sons, and a period of court intrigue between Rome and Constantinople undermined the empire's response to the ongoing frontier emergency at the worst possible time. The "Gothic problem" flared up, and in AD 395 an able king named Alaric united the Goths who had been settled in AD 382. He harassed the empire for greater concessions, at just the moment when the courts of the east and west were jostling for the upper hand. The western court rallied behind the regency of the *generalissimo* (and son-in-law of Theodosius) Stilicho, who was in effective control from AD 395 until his murder in AD 408. For a brief moment it seemed as if he had calmed the surging waters. In AD 400, he triumphantly celebrated his consulship in Rome. His poet Claudian claimed he had restored the "equipoise of the world." But the calm was illusory. Suddenly the dam burst, and the ability to dominate the geopolitics of Europe suddenly slipped from the grasp of the western empire.[67]

Stilicho may have already been playing a chess board without enough pieces, and at the decisive moment the board itself was tilted by forces beyond his control. As a military phenomenon, the "fall" of the western empire should be dated to the years AD 405–410. The careful work of the historian Peter Heather has shown that we should think of the events in these years on two levels. At the surface level, the empire faced a series of simultaneous invasions that broke its ability to control the frontiers. In 405 a new line of Goths from beyond the Roman border crossed Noricum and ravaged Italy. Stilicho snuffed out the threat. But on December 31 of AD 406, another conglomerate force of barbarians—including Vandals, Alans, and Suevi—crossed the Rhine, looted Gaul, and advanced into Spain. They would never be expelled. Henceforth control of the territories beyond the Alps—especially Britain, Spain, and parts of northern Gaul—was shaky or non-existent.[68]

Just behind this visible surface, an even deeper force was pushing events forward. These invasions were not mere raids; they were migrations, movements of people, with women and children in train. And these movements were stirred by a geopolitical development that we can see only dimly in our sources: the movement of the Huns' center of gravity to the west. If the disturbing arrival of freelancing Hun warbands in the 370s had stirred the first Gothic crisis, the chaos of the years 405–8 was triggered by the relocation of Hun power to the west. Masses of peoples, not so thoroughly assimilated by life alongside the Romans as the Goths had been, now fled from the middle Danube and entered the empire. For the first time we hear of Huns in truly large numbers operating as far west as the Hungarian plain. We begin to see, in the figure of Uldin, a Hunnic king who is something more than a mere name to us. The Hunnic empire sought its fortunes in the west, and the peoples in front of it fell like dominos.[69]

The crisis applied more pressure than the load-bearing columns of the frontier system had ever been designed to withstand. In the fog of crisis, the Roman government continued to believe that the Goths under Alaric could be maintained as loyal servants of the empire. They were bound by legal arrangement to obey the emperors. But late in AD 408, seeking terms, Alaric led his forces across the Alps and surrounded Rome. He cut off its food supplies and tried to extort unimaginable payment. For three consecutive years, Alaric held the ancient capital hostage, and finally on August 24 of 410, his armies entered Rome. For the first time since Celtic tribes had taken the city in 390 BC, the eternal city fell into hostile hands. Even if the Christian Goths of Alaric spared the city from unbridled pillage, the symbolic repercussions were violent. "The brightest light of the whole world was extinguished, or rather the head of the Roman Empire was cut off, or to put it most truly, in one city the earth itself perished." The shock of the event called forth Augustine's masterpiece, *The City of God*; the only consolation was the reminder of the transience of all human things.[70]

The inability to stop the unthinkable reveals how suddenly the western empire had lost its prerogative to coordinate military force. The history of the fifth century would see the progressive fracturing of this power and the piecemeal loss of territorial hegemony in the west. Former provinces like Britain simply fade from view, thrown on their local resources, while others, like Africa, were hijacked in the full light of day. Some settlements—the Goths in Aquitaine, the Burgundians in Savoy, the Ostrogoths in Italy—were administered with a certain degree of legal nicety. But the empire was dealing from a position of despair. Decisions were made to the advantage of the center. Provincials were dismayed and their loyalties rearranged. In all cases, the native Roman population outnumbered the new arrivals, but the barbarians commandeered the superstructure of the state. Everywhere but narrow corridors of Italy and Gaul, the machinery of power in the west ceased to be Roman.[71]

East and West: Divergent Fortunes

The final and most famous act of the Huns was more of an encore than a decisive scene. As the Roman Empire was reeling, their most notorious king, Attila, scaled up the Hun war machine. For over a decade, he posed an existential threat to the eastern empire as well as the remnants of the Roman west. Throughout the AD 440s, he wasted the Balkans and engorged his royal circle on plundered wealth. In AD 447, after a massive earthquake felled the great walls of Constantinople (fifty-seven towers collapsed), the eastern capital of the Roman Empire lay helplessly exposed. Only the ramparts of the local disease pool repelled the advancing menace. "Against the stone of sickness they stumbled and the steeds fell. . . . He who was skillful in shooting with the bow, sickness of the bowels overthrew him—the riders of the steed slumbered and slept and the cruel army was silenced." As its last line of defense, the Roman Empire was protected by the invisible ring of germs that lurked in wait for unsuspecting invaders.[72]

Attila saved his two grandest campaigns for Gaul and Italy. At the head of a huge, mixed army of Huns and Germans he crossed the Rhine in AD 451; his force was met in open battle by a Roman general, Aetius, at the head of a mixed army of Romans and Germans. The stalemate blunted the advance of the Hun empire, now clearly beyond its steppe ecozone. But Attila was not finished. In AD 452 the clattering horde rode into Italy. His horsemen plundered the Po Valley. Milan fell without resistance, and Attila occupied the imperial palace. Enraged by a depiction of dead Huns sprawled beneath the emperor's throne, the king found an artist "to paint Attila upon a throne and the Roman emperors with sacks on their shoulders pouring out gold at his feet." Realizing that nothing could stop a Hun advance into central Italy, and unable to muster any military resistance worthy of the name, the Romans dispatched a desperate embassy headed by Pope Leo himself.[73]

It is one of the curiosities of history that the column of Huns receded back across the Alps into the Hungarian Plain. Attila was nothing if not a shrewd calculator. "Beneath his great ferocity he was a subtle man." What actually repulsed the invaders was seen, from one perspective, as "heaven-sent disasters: famine and some kind of disease." The retreat was in fact the predictable biological consequence of intruders colliding with the indigenous disease ecology. The heartland of empire was a gauntlet of germs. The unsung savior of Italy in this affair was perhaps even malaria. Pasturing their horses in the watery lowlands where mosquitoes breed and transmit the deadly protozoan, the Huns were easy prey for malaria. All in all, it may have been wise for the king of the Huns to turn his cavalry back toward the high steppe beyond the Danube, cold and dry, where the *Anopheles* mosquito could not follow.[74]

As the Huns ebbed back into the steppe, the Roman world they left in the dust of their retreat was almost unrecognizable as the one that had met them in the days before Adrianople. Uprooted from the central administration, the ancient structures of empire quickly withered in the west. In one poignant instance, we know of a brave Roman regiment that held its post for decades in the borderland province of Noricum. When their pay stopped arriving, they sent a detachment to Italy to fetch their stipends, "but unknown to anyone . . . they had been slaughtered by barbarians on the road." It was in these very years that "the western Roman army ceased to exist as a state institution." A few years later, in AD 476, there ceased to be a Roman emperor in the west.[75]

In most of the fifth-century west, the imperial renaissance was violently reversed. The Roman efflorescence wilted. The cities dwindled. After the villa boom that progressed across the fourth century, it is nearly impossible to find new villa construction in the fifth century. What buildings were still inhabited reveal a style of occupation that seems changed. The circuits of wealth had been cut. The money economy held on tenaciously, but people were forced to make desperate use of old coins, which were clipped, recirculated, and imitated in an economic world that was disarticulated. Elite trade and local networks never completely evaporated. But in all, it was a simpler world, with starker cleavages between the haves and have-nots. After the collapse of the great private fortunes, built on the fusion of markets and imperial service, the church unexpectedly found itself the wealthiest landowner in society—and commensurately powerful.[76]

The most unrestrained change was in the city of Rome. The population collapsed. It was obvious to observers in the early sixth century that Rome was a husk of its lost glory. "It is evident how great was the population of the city of Rome, seeing that it was fed by supplies furnished even from far off regions. . . . The vast extent of the walls bears witness to the throngs of citizens, as do the swollen capacity of the buildings of entertainment, the wonderful size of the baths, and that great number of water-mills which was clearly provided especially for the food supply." Depopulation radically altered the disease ecology of the city. Even the seasonal mortality profile shifted, as we can see from the now smaller set of Christian tombstones. The overall amplitude of seasonal variation became simply less extreme. The young remained the most vulnerable to the ravages of summer diseases, and as ever the frosts of winter swept away the frail. But for adults, a less pronounced and now bimodal spring-autumn pattern appears. Quite possibly, a Rome that had once been a city of immigrants, who met the city's native disease pool without immunities acquired in childhood, was now a more purely "local" population, vulnerable as ever to malaria but more able to resist the array of indigenous pathogens that reared up each summer. Rome was just another town.[77]

These patterns describe the course of change across most of the northwestern provinces of the empire. In Africa, the change was less sharp, and in the east, thanks in no small part to the blunt determinism of physical geography, the empire remained safe behind its natural barricades. We should not euphemize the events of the fifth century. But we should be careful all the same not to consign these western provinces in the immediate decades following the fall of Rome to a Dark Age. Certainly the eastern capital never gave up on the dream of imperial unity, even if its policies were self-interested and its attentions repeatedly distracted. The state of the western lands in the decades around AD 500 is hard for us to define, perhaps precisely because they were in disequilibrium. The logic of a post-Roman order was never allowed to settle, and its possibilities were not allowed to be played out, before the world was turned upside down again by the irredentist ambitions of an eastern Roman emperor and the turns of nature that roiled his plans. If the environment had receded for a little while and let human actions take center stage, nature was about to reassume the protagonist's role.

The Wine-Press of Wrath

A Ceremony at the Heart of Empire

In the fifth and early sixth centuries, the cord of empire was cut in the western provinces, and the forces of political entropy prevailed. In the east, the imperial administration continued to tighten its grip. The centripetal force put in motion by Diocletian and Constantine pursued its destiny, concentrating power in the capital, the bureaucracy, the court—and at the very center of it all, in the divinely chosen figure of the emperor himself. The power of the aristocracy, the administration, and the army all flowed from his sacred energy. For a long season, in the eastern empire, this model of autocratic power seemed charmed. Lifted on the buoyant prosperity of the eastern provinces, in the early sixth century, the horizons of the Roman Empire appeared endless. The imperial state centered on Constantinople was still fully late Roman, and it is only in light of later history that we see shades of the Byzantine future visibly streaking across the empire.

Diocletian had put an end to the constitutional charade that required the emperor to be first among equals, a fellow citizen with the virtue of *civilitas*. He wrapped the figure of the emperor in remote and awesome majesty. In consequence, or compensation, late Roman statecraft became a gratuitously ceremonial business. We can measure the extent of its ceremonial impulse, for instance, in the state's ability to make a pageant of something so bureaucratic as the audit of the imperial grain stores. On the day of this ritual, the emperor ascended into his chariot, where the Praetorian Prefect, the second most powerful man in the empire, kissed his feet. He paraded past the great horse track and the public baths, winding his way down to the bustling market district of Constantinople and the monumental public storehouses on the Golden Horn, where ships at anchor filled the ancient ports hugging the north shore of the old city. The Director of the Granaries met the emperor and presented the accounts. The emperor personally inspected the inventory, and if all was to his satisfaction, the director and his accountant were each rewarded with ten pounds of gold and a tunic "made completely of silk." The city's food supply safe and adequate, the emperor made a stately return to the palace.[1]

Ceremonies such as this were a vital medium of communication in the later Roman Empire. The inspection of the granaries was an exhibition of the emperor's power, in the staged performance of his most primordial obligation: to feed his people. In a city whose population had reached some 500,000 souls, food security was nothing to take for granted. The food system mobilized the resources of an entire empire. A vast bureaucracy, controlled by the palace officials, coordinated the delivery of taxes to the capital and the army. Since the days of Constantine, 80,000 subjects in the eastern capital had been entitled to the receipt of free bread, and the threat of urban riot meant that enough wheat to feed a half a million mouths had to arrive smoothly at the docks. As ever, Egypt remained the breadbasket of empire. In the early days of the emperor Justinian's reign (r. AD 527–565), we know that, every year, ships from Alexandria carried 8,000,000 *artabai*—310,000,000 liters—of wheat to the capital.

We do not know whether or not Justinian was the emperor who devised the ceremonial grain inspection, but it would entirely fit his style. "We deem even the smallest things worthy of our care," he trumpeted in one law. "Much less do we leave matters that are important or undergird our republic without attention."[2]

This ceremony brings to life the global networks centered on Constantinople. The grain supply connected the city to farms and fields stretching into the remotest reaches of upper Egypt. Constantinople in the sixth century was a vortex for the world's peoples and goods. "A throng of men of all conditions comes to the city from the whole world. Each of them is led to come by some errand of business or by some hope or by chance." Latin remained the official language of empire, but in the city streets you would hear Syriac and Aramaic, Coptic and Ethiopic, Gothic and Hunnic, Persian and Arabic, and of course Greek. It was no exaggeration to consider the capital a global hub. It drew inward the wares of the known world, such as the silk that had become the reward of the emperor's faithful servants. And where people and goods move, so too germs.[3]

The real ecological lesson of the emperor's ceremonial inspection of the grain inventory is hiding just out of sight, in the great storehouses hulking over the landscape. Granaries were everywhere in the later Roman world. The stockpiling of grain was deeply rooted in the Mediterranean psyche. In the empire, the vast network of cities, ships, and stores of grain created an

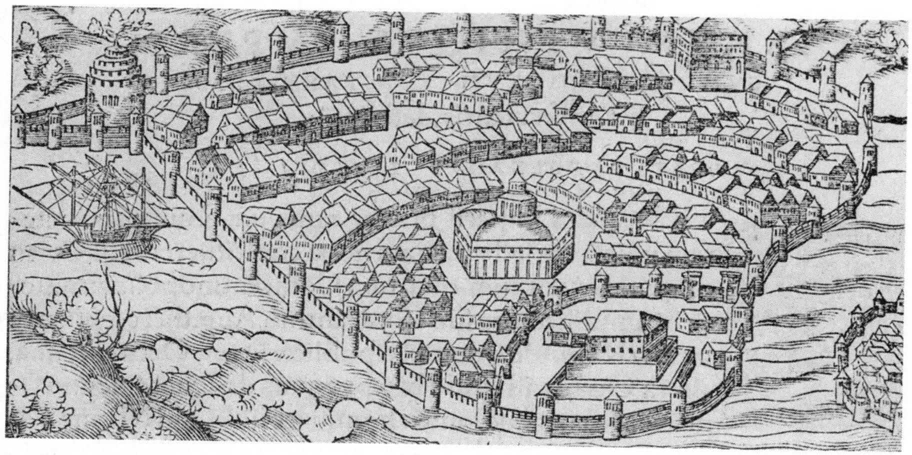

Figure 6.1 Constantinople, As Represented in the Notitia Dignitatum *(Sixteenth-Century Printing, University of Oklahoma History of Science Collections)*

ecosystem. This ecosystem served as an invitation for a species uncannily evolved to be commensal—literally, to "share a table"—with us: *Rattus rattus*, the black or ship rat.

We can be sure that as Justinian and his entourage neared the warehouses, thousands of rats scuttered into the darkness. "They steal along as quietly as spooks in the shadows close to the building line, or in the gutters, peering this way and that, sniffing, quivering, conscious every moment of what is going on around them." Those are the words of a New Yorker, writing in the mid-twentieth century, before pest control had quite gotten the upper hand in the modern city (to the extent that it has). In the ancient city, the struggle against infestation was futile. Black rats are prolific breeders. Food is the limiting factor on population size, and they love grain. Equipped with long tails, black rats are crafty climbers and willing travelers. They ride in ships by the hundreds. From a rat's perspective, the Roman Empire was an unimaginable blessing. The Roman world was crawling with rats.[4]

The fusion of global trade and rodent infestation was the ecological precondition for the greatest disease event human civilization had ever experienced: the first pandemic of plague. Norman Cantor wrote of the medieval Black Death, "It was as if a neutron bomb had been detonated." The *first*, late antique Black Death is less famous. Its relative obscurity is unmerited. In 541 plague appeared on the shores of Egypt. It diffused throughout the Roman world and beyond. For two centuries, it stayed, and then just as mysteriously it receded. The trauma of the fourteenth-century pandemic in many ways marks the threshold between the medieval and modern worlds, and the disintegrating force of the first plague pandemic deserves to be reckoned as the passage from antiquity to the middle ages. In wider perspective, the experience of humanity over the last millennium and a half has been imponderably shaped by the violence of the singular microbial agent that causes bubonic plague, a bacterium known as *Yersinia pestis*.[5]

The plague is an exceptional and promiscuous killer. Compared to smallpox, influenza, or a filovirus, *Y. pestis* is a huge microbe, lumbering along with an array of weapons. But, it is in constant need of a ride. Its diffusion in epidemic phases is dependent upon a delicate arrangement of hosts and vectors. A plague pandemic is an intricate concert, arduous in preparation, disturbingly unforgettable in its performance. Once it gains momentum, the plague is an overwhelming biological force. In the sixth century, the alignment of evolutionary history and human ecology precipitated a natural disaster that dwarfed, in both its intensity and duration, even the plagues of the second and third centuries. Of course, the plague pandemic was a natural disaster in much the same sense as the destruction of a hurricane that erases a settlement built precariously overhanging the sea. The pandemic was an unintended conspiracy between wild nature and the constructed ecology of the empire.

Hopefully, the detailed exploration of Roman disease history leading down to this point will cast into even greater relief what an epoch-making event the first bubonic plague pandemic was. *Y. pestis* is a truly extraordinary antagonist, almost impossibly evolved to become a global killer. Genetic study of this microbe is unlocking clues to its history and biology at an exhilarating pace. The biology of this single bacterium is one of the dominant facts of the history of the world in the last millennium and a half. Yet, even here, the course of its intercontinental rampage depended on the most intricate alignment of human networks, rodent populations, climate change, and pathogen evolution. From our vantage, we can feel a tingling drop of wonder at the sheer contingency that allowed this deadly microbe to trace a path of destruction from inner Asia to the edges of the Atlantic.

The arrival of the plague bacterium on Roman shores heralded a new age. Its persistence for two centuries created a prolonged epoch of demographic stagnation. In combination with the deterioration of the physical climate known as the Late Antique Little Ice Age—the subject of the next chapter—the pandemic washed out the last foundations of the ancient order.

Reconquista and Renaissance

Justinian reigned as emperor from AD 527 to 565. Less than a decade into his reign, he had already accomplished more than most who had ever held the title. The first part of his reign was a flurry of action virtually unparalleled in Roman history. Between his accession in AD 527 and the advent of plague in AD 541, Justinian made peace with Persia, reattached vast stretches of the western territories to Roman rule, codified the entire body of Roman law, overhauled the fiscal administration, and executed the grandest building spree in the annals of Roman history. He survived a perilous urban revolt and tried to forge orthodox unity in a fractious church, through his own theological labors. By AD 540, only his religious policy could be deemed unsuccessful.[6]

Justinian's uncle, Justin, who had seized the throne in AD 517, was not an altogether likely emperor. He was of the humblest stock. His detractors liked to claim that he was completely illiterate. He was seventy years old and childless on assuming the throne. But his nephew Petrus Sabbatius had been summoned to the capital and adopted, taking the name Justinian. He was groomed to rule and took sole command of the empire in AD 527.

Already in antiquity, he was loved and loathed. Indefatigable, Justinian worked day and night. He was ruthless and completely confident. In the independent and forceful Theodora, he found a worthy match. She was an actress and a demimondaine (as even her sympathizers admitted). Justinian unflinchingly repealed centuries of law prohibiting *mésalliance* with scandalous persons. The law survives. "We believe that we can thus imitate, as much as it is possible for us to do, the benevolence and great clemency of God to the human race, who condescends to pardon the daily sins of men, to receive our repentance and to lend us back to a better condition." It would be as though a sitting president married a Kardashian. No emperor generated so much literate hatred in his own day. In the *Secret History* of Procopius, a lurid critique of the Justinianic regime, the imperial pair are wantonly depraved, maybe even demons. Yet, in the orthodox tradition, Justinian and Theodora are saints.[7]

Opposition quickly congealed against Justinian's rule. To the establishment, his administrative reforms were odious. The moneyed elite and the central bureaucracy had an understanding. They were in cahoots. Taxes were collected, palms were greased. Justinian brought the zealous intolerance of an outsider to the fight against corruption. The architect of his clean-up operation was a figure known as John the Cappadocian. John sought efficiency, transparency, and direct lines of control. The sale of governorships was prohibited, provinces were reorganized, and the discretion of local elites was reduced in scope. John rankled the mandarins in Constantinople; he was painted as violent, greedy, and uncouth. Seething opposition boiled over in AD 532, when the famous Nika Revolt erupted in the capital. It was a *putsch* led by the disenfranchised aristocratic faction. Whole regions of the city were scorched, including the old Hagia Sophia. The regime survived, by gruesome measures: thousands were hacked to death. The regime of Justinian was uncowed.[8]

These machinations have a fundamental bearing on how we view the overall trajectory of Justinian's rule. The emperor produced alienation and bitterness among his cultured despisers. His reign is the rare case where history was written by the losers. They have left us a portrait of a regime careening out of control. The wars and indulgent building program were overreach, bought with the blood of the provinces, doomed to ultimate failure. The hangman John was the instrument of the emperor's hubris. It is not an altogether credible portrait. Justinian sought fiscal equilibrium for his ambitious designs. His reforms impressed A. H. M. Jones, the most accomplished scholar of the later Roman administration. And, reading between the lines even of the detractors, the prodigious talents of Justinian's agents are unmistakable. Justinian's greatest gift may have been his unerring eye for talent. John, his prefect, Tribonian, his lawyer, Anthemius, his architect, Belisarius, his general, and Theodora, his wife—all are transcendent figures, plucked by Justinian. Maybe not since the days of Augustus had there been such a sudden upwelling of sheer talent.[9]

The monuments of their achievement are plain to see. Chief among these is the *Corpus iuris civilis*, the landmark codification of Roman law. In the words of Gibbon, "The vain titles of the victories of Justinian are crumbled into dust; but the name of the legislator is inscribed on a fair and everlasting monument. Under his reign, and by his care, the civil jurisprudence was digested in the immortal works of the Code, the Pandects, and the Institutes." Justinian was not unaware of the dimensions of his achievement. "The task appeared to us most difficult, indeed impossible. Nevertheless, with hands stretched up to heaven, and imploring eternal aid, we stored up this task too in our mind, relying upon God, who in the magnitude of his goodness is able to sanction and to consummate achievements that are utterly beyond hope." Led by Tribonian, Justinian's team synthesized a thousand years of law and legal writing into a systematic and consistent whole. By AD 534, the edifice was triumphantly complete.[10]

Justinian's building program speaks for itself. The Hagia Sophia is a technical marvel. The largest dome built in the ancient world, "it soars to a height to match the sky." The dome of the Pantheon pales in comparison to the vaulting structure of the Hagia Sophia, which blends the principles of an axial basilica with the symmetry of a quadrangle, lifting the dome 182 feet from the floor. It is possible that Justinian was the greatest patron of the church in its history. He built thirty churches around Constantinople alone. The Nea Church in Jerusalem, dedicated to the Mother of God, was a marvel; if it still stood it would be a match for any of the greatest monuments of antiquity. He built hospitals and poor houses across the empire. Procopius registers some 600 military sites touched by Justinian in the Balkans, and the frontier with Persia was heavily fortified. Justinian's building program reveals a practical flair. The grain ships sailing from Alexandria often had to wait for favorable winds to cross the narrow channel of the Hellespont. Justinian built granaries on the island of Tenedos just to the south, large enough for the entire fleet to unload their cargo; barges carried it from there to the capital. Without having to wait for southerly winds, ships could make two or three runs during a single sailing season.[11]

Justinian was the last of the great Roman environmental engineers. The muscular state still bent nature to its will, on a scale that would have impressed Trajan. Flood control was a major preoccupation across Greece and Anatolia and into northern Mesopotamia. After a devastating flood at Edessa, Justinian reshaped the entire local landscape to create a new channel for the River Skirtus. Similarly he carved a new bed for the Cydnus around Tarsus. The remains of his Sangarius bridge in Bithynia are still imposing. The River Drakon flowing into the Sea of Marmara was flooding the valley around its mouth; Justinian cleared a forest and sculpted the plain to allow its flow to be contained. Justinian repaired dilapidated aqueducts and built new ones. In Constantinople, he constructed an enormous cistern to store fresh water for the dry summer season.[12]

The campaign to recover the western provinces was his boldest enterprise of all. Justinian was a native Latin speaker, from the old stock of the Danube march. Dreams of reconquering the western heartland fired his revanchist agenda. In AD 532, he signed the optimistically named Eternal Peace with his Persian adversary, Khusro I, and turned west. In 533, Belisarius led an expedition against the Vandals. A crack force comprised of 15,000 regulars sailed with a fleet of 500 transport ships. The victories were swift. By 534, Belisarius was back in Constantinople, leading the fallen Vandal king in a triumph. North Africa remained a secure Roman possession until it was wrested away in the Islamic conquests.[13]

The eviction of the Ostrogoths from Italy proved less decisive. In AD 536 Belisarius was dispatched west; he seized Sicily, Naples, and Rome with haste. By AD 540, Belisarius had withstood the counterattack and gained control of Ravenna. He took the royal treasure and the king, Vitiges, and again returned to Constantinople in glory. But Belisarius was recalled to respond to an emergency situation on the frontier with Persia, and real control of Italy was more elusive. Serious resistance dragged on to the middle of the 550s. Then a short, precarious peace was interrupted in AD 568 by the invasion of the Lombards. For centuries the Byzantines would control the outposts of Rome and Ravenna, and stretches of southern Italy. In the end, "Justinian's dream of restoring the western empire had brought Italy little more than misery." But that was far from the obvious destiny of events in AD 540.[14]

Renewed hostility with Persia divided the empire's strength. In the spring of AD 540, Khusro I blindsided the Romans and launched the most aggressive Persian invasion since Shapur I in the midst of the third-century crisis. He captured town after town, helplessly exposed to his march. Antioch was sacked—"a city that was ancient, of great importance, and the first of all the cities of the Romans in the East in terms of its wealth, size, population, beauty, and prosperity of every kind." Khusro bathed in the Mediterranean. But Belisarius was sent into the breach, and after one season, Khusro wheeled back toward Persia. In this indeterminate hour, the bomb went off.

In AD 541 the plague arrived at Pelusium, a town on the shores of the Mediterranean. By the spring of the next year the invisible enemy was in the capital. It was a point of immense fracture. The great plague inaugurated what has been called "the *other* age of Justinian." For the next twenty-three years, his reign trundled forward in the shadow of pestilence. The state struggled to field robust armies. Taxes rose to unseen heights. A new darkness hung over the emperor, himself a survivor of bubonic plague. It was an age of shocking reversals. "I cannot understand why it should be the will of God to exalt the fortunes of a man or place, and then to cast them down and destroy them for no cause that is apparent to us."[15]

The Making of a Murderer: A Natural History of *Yersinia pestis*

We have always had testimony from the sixth century insisting on the dramatic upheavals of nature in the reign of Justinian. Modern historians have struggled to know exactly what to make of reports that, inevitably, lack scientific precision and reflect the assumptions and prejudices of a very different age. Now, the agent of the Justinianic pandemic has been identified beyond all doubt as the bacterium *Yersinia pestis*. Labs observing the strictest of protocols have sequenced its genome from the archaeological remains of plague victims. This knowledge is an anchor in the storm. It tethers our speculation, even as it lets us plumb more deeply into the nature of the historic collision between the Roman Empire and the plague bacillus.

The bacterium known as *Yersinia pestis* has been the agent of three historic pandemics. The first erupted in the reign of Justinian. The medieval pandemic started with the Black Death in AD 1346–53 and lasted nearly half a millennium. A third pandemic erupted in AD 1894 in Yunnan China and spread globally. These three episodes are, in fact, colossal accidents. Humans are merely incidental victims caught in the crossfire of what is really a disease of rodents. From the bacterium's perspective, we are sorry hosts, since we are prone to die before the concentration of bacteria in our blood becomes sufficient for fleas to carry it to future victims. Most of the time, a human being infected with plague is a terminus, not a transmitter. Today *Y. pestis* is enzootic (permanently established in an animal population) around the world in rodent colonies. It is out there, lurking.[16]

Y. pestis evolved to be an extraordinarily lethal and promiscuous killer, with a strong preference for certain kinds of vehicles. To understand how humans could become collateral damage on a pandemic scale requires a sense of the biology of *Y. pestis*. Its genetic history and microbiology are perhaps more extensively studied than any other major pathogenic agent. *Y. pestis* is a re-emerging infectious disease and officially categorized as a bioterror threat. By sheer luck *Y. pestis* was present at the birth of paleomicrobiology; in 1998 a French lab sequenced *Y. pestis* genes from an eighteenth-century mass burial site, launching the study of ancient DNA. And, to add to the embarrassment of riches, the genus *Yersinia* is considered a "model" for pathogen evolution. Its microbiology has been the object of unusual scientific attention.[17]

The genus *Yersinia* belongs to the family Enterobacteriaceae, a group of gram-negative, rod-shaped bacteria encompassing common gut pathogens like *Salmonella*, *E. coli*, and *Shigella*. The genus *Yersinia* includes eighteen species. Fifteen of these are harmless to humans—they live in soil or water and lack the ability to cause sickness in mammals. Three species of *Yersinia* have evolved the ability to infect mammals: *Y. enterocolitica*, *Y. pseudotuberculosis*, and *Y. pestis*. These three species acquired genes that let them stand up to powerful immune systems. But these genes were acquired outside the chromosome, in what are called plasmids. Plasmids are floating wheels of

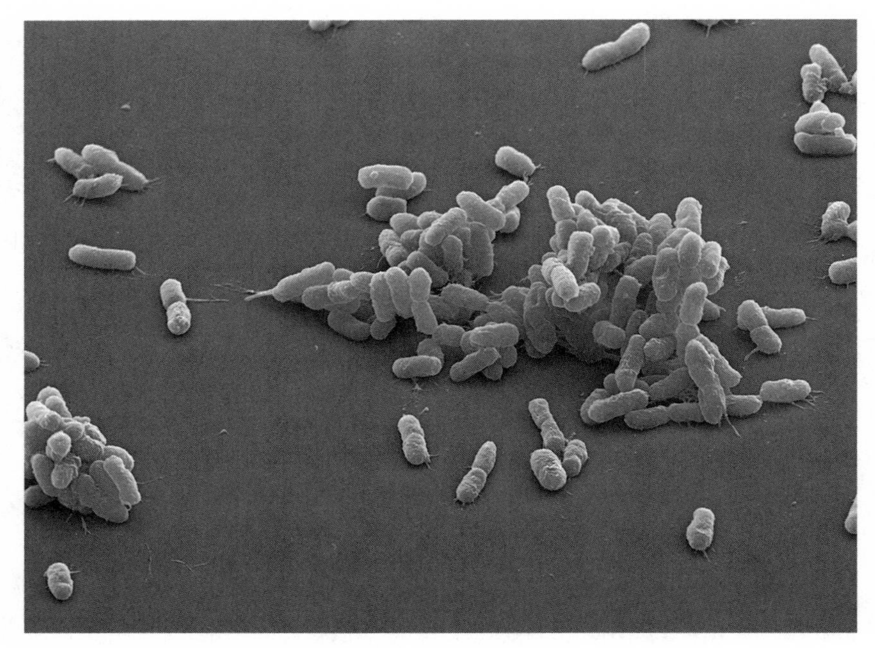

Figure 6.2. Yersinia pestis. *The most deadly bacterium ever. (Scanning Electron Microscope, Science Source)*

genetic material that encode a few specialized genes: it would not be misleading to think of them as genetic apps. The biography of *Y. pestis* could be summarized as a story of three plasmids. The first, known as yPV (plasmid of *Yersinia* virulence), is shared with *Y. enterocolitica* and *Y. pseudotuberculosis*. yPV builds a deadly weapon: a needle that injects specialized proteins into host cells on contact and is critical in disabling the host's innate immune system. The maneuver has been called the "*Yersinia* deadly kiss." The acquisition of this tool was the first step in the evolution of *Yersinia* toward a more deadly destiny.[18]

But with the acquisition of yPV, the genus *Yersinia* had not yet spawned its monster, *pestis*. Both *Y. enterocolitica* and *Y. pseudotuberculosis* still exist as pathogenic microbes. They cause self-limiting gastroenteritis in humans: they invade via the fecal-oral route, multiply in the intestines, cause diarrhea, and ultimately lose a battle with the immune system. *Y. pestis* evolved from *Y. pseudotuberculosis*. It diverged approximately 55,000 years ago through the addition and deletion of genes. *Y. pestis* actually lost about 10 percent of the genes of *Y. pseudotuberculosis*. The critical evolutionary step was the acquisition of a second deadly plasmid known as pPCP1. It transformed a heretofore mild enteric pathogen into a killer. pPCP1 builds an enzyme (known as pla, plasminogen activator) that renders *Y. pestis* a wildly destructive force, capable of deep tissue invasion.[19]

With the acquisition of the plasmid pPCP1, *Y. pestis* was able to infect humans via droplet and cause pneumonic plague. Pneumonic plague is a disease caused by *Y. pestis* and characterized by acute febrile illness. It overwhelms the body's defense mechanisms within two to three days, and fatality rates approach 100 percent. For ~55,000 years, *Y. pestis* has had the ability to cause this exceptionally lethal respiratory disease. It is also possible that early *Y. pestis* could be spread by the bite of ectoparasites, like fleas; but the bacterium did not yet have the genetic tools to survive in the gut of the flea, so any infection via this route probably relied on what is called "mechanical transmission," in essence the passage of germs on the infected proboscis used to suck blood, not totally unlike a dirty needle. In the case of *Y. pestis*, the efficiency of this kind of transmission is limited. But archaeological DNA recently recovered from Bronze Age skeletons scattered across northern Eurasia suggests that, one way or another, we have a long history with plague.[20]

Frankly, the epidemiology of ancestral *Y. pestis* is not yet well understood. To become the agent of pandemic plague, a third plasmid, known as pMT1, had to evolve a gene (*ymt*) that codes for a protein known as *Yersinia* murine toxin. It plays one indispensable role: it protects the bacillus in the mid-gut of the flea. Now, the bacterium could build a biofilm in the gut of the flea where it rapidly multiplied; with the digestive path blocked, starving fleas went desperately biting in search of blood, in the process regurgitating bacteria into new victims. This genetic adaptation allowed *Y. pestis* to ride arthropod vectors much more easily from host to host. It made the bacterium a stupendously efficient traveler. *Y. pestis* was now a flea-borne disease. It has long been known to be especially adapted to the oriental rat flea, *Xenopsylla cheopis*, although in recent years it has been recognized that the plague bacterium can infect and block a variety of fleas. By stowing itself away inside fleas, *Y. pestis* had the ability to become a runaway killer. The transmission by flea bite is also essential to the most characteristic pathology of the *bubonic* plague: the swollen lymph nodes known as buboes. Introduction into the dermis, rather than inhalation of infectious droplets, results in the invasion of the lymph nodes and the development of buboes.[21]

Table 6.1. The Evolution of a Monster

Ancestral Yersinia	*Y. pseudotuberculosis*	*Early Y. pestis*	*Modern Y. pestis*
		ca. 55,000 years ago	ca. 3,000 years ago
Non-pathogenic	Self-limiting enteritis	Pneumonic	Pneumonic/bubonic
	pPV (builds T3SS)	pPCP1 (builds pla)	pMT1 (builds ymt)
	Fights generic immune defenses	*Aggressively invades and destroys tissue*	*Survives in flea gut*

Modern *Y. pestis* evolved not long before ~951 BC, since a genome recovered from an archaeological victim of this date reveals the existence of all three plasmids, with the critical genes to cause an outbreak. As a disease of rodents and fleas that occasionally spills over into humans, *Y. pestis* is an evolutionary newborn. Certainly it has been an *enfant terrible*.

Modern *Y. pestis* genomes have been extensively studied, and the distribution of genetic variation within the species around the globe today provides critical clues to the history of the germ. The most basal and diverse strains of *Y. pestis* are found in central Asia, and it is virtually certain that the genetic events leading to the evolution of modern *Y. pestis* took place there. The Qinghai-Tibet Plateau in China looks like the ancestral home of the plague bacillus, at least based on the genetic data now available. For most of its history, *Y. pestis* lurked in what is called a maintenance phase, subsisting in nature through transmission among wild hosts. *Y. pestis* can probably infect any mammal, but rodents are its principal reservoir. *Y. pestis* thrives among social burrowing rodents, such as marmots and gerbils. Their lifestyle is conducive to flea-borne transmission. The Great Gerbil of central Asia and the Asian marmot seem to have partial resistance to the disease, helping sustain *Y. pestis* during lengthy enzootic periods. Given its versatility, *Y. pestis* need not have been overly reliant on any single host.[22]

For three thousand years, modern *Y. pestis* has been an enzootic disease of burrowing rodents in central Asia. It probably has a more eventful and tumultuous history among rodents than we will ever know. Its ability to travel by flea let it spill over from its maintenance hosts into inviting but unstable rodent worlds. During amplification events, the bacillus found new hosts, where it could briefly explode in epizootic flare-ups. The black rat or ship rat, *Rattus rattus*, seems uncannily designed to facilitate plague amplification. Its habits, its personality, and its massive populations make it both a helpless victim of the plague and an involuntary conscript in the spread of the bacterium. It is not an ideal permanent reservoir for plague, but it is especially important in facilitating human plague pandemics. The black rat is inseparable from the story of plague as we know it.[23]

Black rats are commensal, happy to live in close quarters with humans. They love the food and shelter we unintentionally provide. The rat is omnivorous but has a few strong preferences, like grain. Its long tail makes the black rat an adept climber. It often lives high above ground. And *Rattus rattus* loves to travel. It is known as the ship rat for a reason, for it insistently colonized sea-going vessels and grew fat on the stores of sailors. The black rat does not move far on its own, and it is territorial. A prolific creature, the black rat will breed year round, and adult females can produce 5 litters in a year; gestation is 3–4 weeks and newborns reach reproductive maturity within 3–5 months. Food is usually the limiting factor on population size for a small mammal with explosive demographic potential. Its predators—cats, owls, other small carnivores—are a modest control. Where food is abundant, black rats proliferate.[24]

The life of the black rat is terrorized by a small flea that lives in its fur and eats its blood, *Xenopsylla cheopis*, the oriental rat flea. During an amplification event, the flea is the primary vector for *Y. pestis*, which the flea ingests from infected rats and spreads to others. The formidable immune system of the black rat fights back, but this only allows the bacteria to concentrate in the blood before the rat succumbs. As the rat population dwindles, hungry fleas become desperate for blood and deign to feed on humans. A plague epidemic in humans is thus a two-stage event. First, *Y. pestis* must spill over from its wild enzootic reservoir into a runaway epizootic event. Then, it leaps from commensal rodents to humans. The human epidemic is the side effect of the epizootic event among rodents.[25]

That, in any case, is the classic model. For decades it has been challenged on various grounds. Most fundamentally, the identity of the pathogenic agent of the Black Death was hotly contested. In no small part, the doubts derived from the thought that the medieval pandemic was just too widespread and too explosive to be a disease dependent on rodents and fleas. The DNA evidence has now laid the dispute over the pathogen's identity to rest, but the epidemiological questions still linger. There is constructive debate about the possibility of other pathways traveled by the bacterium in the course of a pandemic. Some of these routes—such as transmission by other ectoparasites like human fleas or lice—would have bypassed the rats and seem increasingly plausible as an additional level of the plague's diffusion. The human flea, *Pulex irritans*, is looking increasingly guilty as an accomplice, and this mode of transmission may have overlaid the "classic" model in complementary fashion.[26]

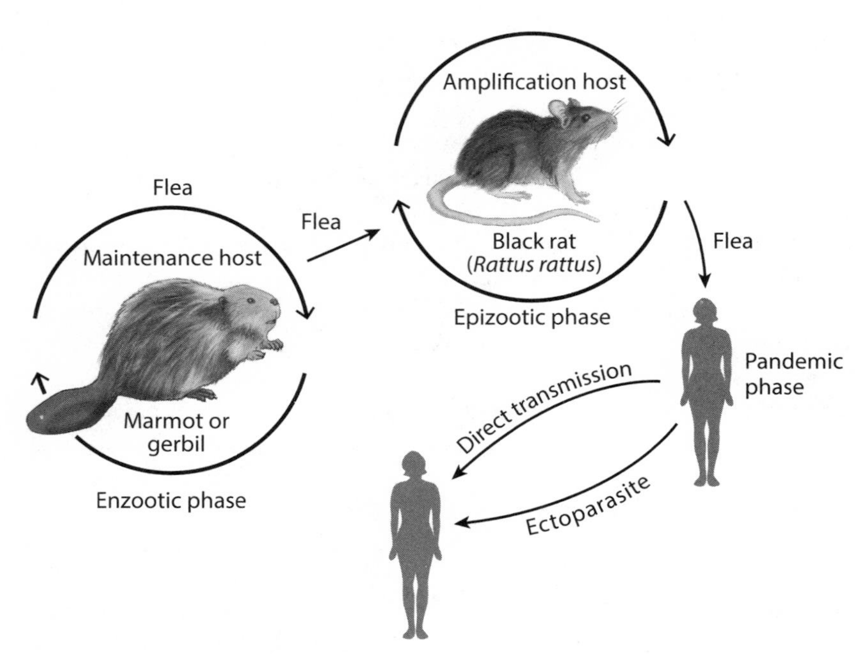

Figure 6.3. Classic Model of the Plague Cycle

Other routes of transmission—such as significant levels of direct transmission of pneumonic plague between humans—still seem on balance unlikely as a major part of the plague's powers of dispersion. But we should not underestimate the versatility of *Y. pestis*. The plague bacterium can ensnare a wide range of rodents and other mammals. It may turn out that the role of other small mammals, like lagomorphs, has been underappreciated. They might have been quiet links in the chain explosion of the plague pandemic. While we should emphasize the centrality of *Rattus rattus* and the oriental rat flea as the principal layer of conduction during the pandemics, *Y. pestis* may have taken advantage of its versatility, spreading through other mammals and human parasites during the great explosions. The plague pandemic was a germ supernova.[27]

Before the plague could become pandemic, then, an intricate ecological platform had to be in place. The colonization of the west by the black rat was a prerequisite. Rats have not lived in the territories ruled by Rome time out of mind. The black rat is a native of southeast Asia, and it drifted west in the recent past. It is an invader, and its final major thrust to the west was greatly accelerated by the Roman Empire. In the words of Michael

McCormick, "the diffusion of the rat across Europe looks increasingly like an integral part of the Roman conquest."

The very earliest remains of the black rat from the western Mediterranean belong to the age of the later Roman republic, the second century BC. It used to be doubted that the black rat had made enough progress by the time of Justinian to account for the first pandemic, but fifteen years ago, McCormick showed that the black rat had advanced to a degree not previously recognized; even though rat bones are easily missed by archaeologists, the intervening years have added still more evidence to the rat atlas of the Roman Empire. In England, for instance, the black rat followed the Roman conquest. It penetrated deep into the countryside. The dependence of the imperial system on the transport and storage of grain made the Roman Empire a heaven for the black rat. From the rat's point of view, the Roman Empire was a trophic bonanza.[28]

The Roman Empire prepared the ecological landscape for pandemic plague. We should note a small but curious detail. The Greeks and Romans were not totally unacquainted with the bubonic plague before it erupted under Justinian. It is absent in the early corpus of Hippocratic medical writings. But Rufus of Ephesus, writing at the end of the first century, knows of "pestilential buboes." He cites other authorities who had observed outbreaks of plague in Libya, Syria, and Egypt. Another contemporary, Aretaeus of Cappadocia, also made passing reference to pestilential buboes. But these must have been decidedly local or limited expressions of the disease. Galen, with his vast clinical knowledge and experience, betrays no familiarity with bubonic plague. Oribasius, a fourth-century doctor who produced a sprawling medical encyclopedia, excerpted Rufus on the bubonic plague. But when he produced a shorter manual of medical practice, plague did not make the cut. It was not practical knowledge.[29]

Plague may have knocked at the doors of empire before the reign of Justinian, but the hour of pandemic was not yet at hand. Down to the sixth century, the circumstances were not aligned for the great event. Some combination of genetic or ecological factors precluded the pandemic explosion. It is worth canvassing the possibility that a minor genetic transformation provided the final impetus. The DNA of *Y. pestis* from the most recent of the Bronze Age victims had all of the genetic tools it needed. But a crucial virulence factor, pla, built by the pPCP1 plasmid, still lacked one miniscule tweak that enhanced its deadly potential. Sometime before the Justinianic outbreak, there was a single mutation at amino acid 259 in the pla protein. In laboratory tests, this small substitution turns a dangerous bacterium into a savage one. This mutation, or another one like it, might

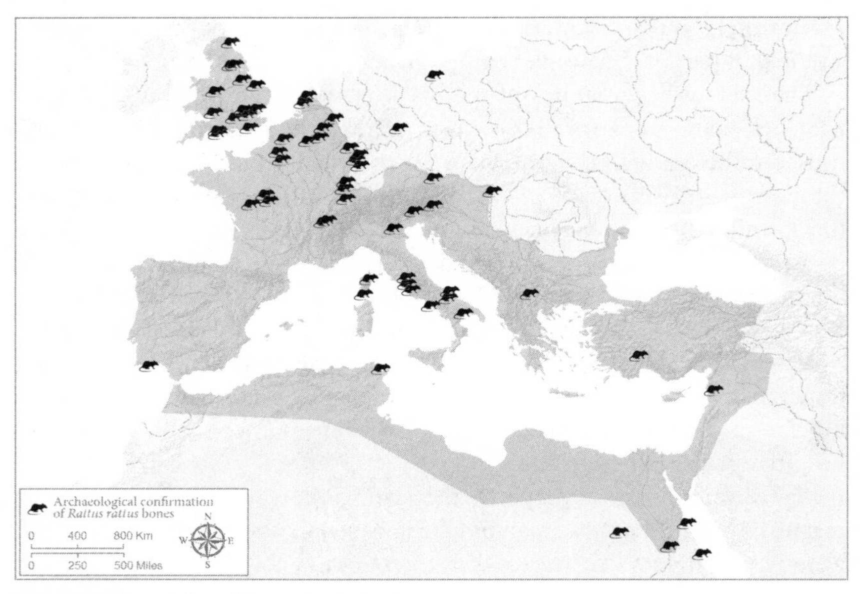

Map 17. A Rat Atlas of the Roman Empire

account for the new explosiveness of the bacterium. By the sixth century, *Y. pestis* was, in its genetic makeup, the virulent agent that would cause the great pandemics.[30]

In the sixth century, the genetic and ecological preconditions fatally aligned. The sparks would become a conflagration. The diffusion of the black rat, and the connectivity of the empire, laid the infrastructure for the spread of a lethal strain of *Y. pestis* on a pandemic scale. *Y. pestis* had only one more hurdle: it had to travel from the east. The strain of *Y. pestis* that caused the first pandemic diverged from an ancestral haunt in the uplands of western China. The closest known relatives of the *Y. pestis* lineage recovered from the sixth century have been found in present-day gray marmots and long-tailed ground squirrels in the Xinjiang region. *Y. pestis* was a scourge from the east. In the words of Monica Green, "All narratives of plague's history must be connected to that place of origin."[31]

The plague could have raced to the west along any number of routes. But the contemporary evidence leaves us an unambiguous clue about its itinerary. The disease first appeared on the empire's southern shores, at Pelusium, on the eastern lip of the Nile delta. Only by reconciling the molecular and human testimony can we retrace the voyage of an emerging pathogen that was ready to launch a lethal pandemic.

The Global Context: The World of Cosmas

In his *Christian Topography*, the sixth-century trader known as Cosmas Indicopleustes relayed the belief, held by the Brahman philosophers of India, that if you were to stretch a string from China to Rome, it would run through Persia and bisect the world. From the vantage of Cosmas, China, the "land of silk," was "beyond furthest India," at the opposite end of the earth. The shortest route might lay overland, through Persia. "That is why there is always found an abundance of silk in Persia." But for Cosmas, the more familiar passage to the far east was clearly across the waters. China was "toward the left part for those entering the Indian sea," past the Persian Gulf, beyond "Taprobane," our Sri Lanka. Cosmas knew the commerce in silk impelled men to journey to "the ends of the earth." In the sixth century, the ends of the earth were drawn together on threads of silk.[32]

"Cosmas Indicopleustes" means Cosmas the Voyager to India. This was not his real name. He may never even have been to what we call India. Medieval scribes bestowed the name on a writer who called himself simply "a Christian." More important, India in late antiquity was a rather broader concept than we are used to. It referred indiscriminately to the lands ringing the Indian Ocean, from Ethiopia to India proper. All that we know about Cosmas derives from his own writings. He was a merchant in Alexandria who worked the Red Sea trade; he had travelled far and wide. He claimed to have sailed three seas—the Mediterranean, Red, and Persian. He had certainly voyaged through Ethiopia, where he transcribed a historic inscription and saw a wild rhinoceros. Cosmas was scrupulously honest, and he nowhere claimed to have traveled the subcontinent. But his *Christian Topography* is a prime artifact of the interconnected Indian Ocean world in late antiquity.[33]

After a lull in the third century, Roman trade in the Red Sea and Indian Ocean rebounded in late antiquity. Berenike remained a bustling entrepôt. The twin ports at the northern end of the Red Sea, at Clysma (at Suez) and Aila, seem only to have grown in importance. The southern end of the Red Sea, on either side of the straits at the Bab-el-Mandab, was a hot zone of geopolitical tension. The powerful Axumite Kingdom in Ethiopia stared across at the rival kingdoms of southern Arabia. The contrast between Roman consumer demand, on the one hand, and the deficit of Roman power, on the other, was a fact of life for traders in the Indian Ocean. The Romans could barely control the Red Sea, their watery backyard. Power projection into the Ocean beyond was simply beyond the means of the empire. As

Cosmas candidly describes, the Red Sea trade connected the Romans to a wild and wooly world of adventurous traders and minor potentates. The Romans were players with a strong sense of their own civilizational superiority, but no on-the-spot advantages. Cosmas knew a maritime zone of exchange shared by Greeks, Ethiopians, Arabs, Persians, and Indians.[34]

The Christian Topography marshaled practical information about the movement of people, goods, and ideas. Pepper and silk were the prized articles of exchange. The spice trade remained big business in late antiquity. We have the haphazard fact that Constantine's donations to the Church of St. Peter's in Rome included 755 pounds of pepper—per year. The remarkable eleventh book of *The Christian Topography* even included passable sketches of the pepper tree.[35]

Alongside the trade in spices, silk had become big business by late antiquity. Silk was synonymous with China, where the secrets of the silkworm were closely guarded. The Romans imported silk over land and across the southern seas. The state was a promiscuous consumer, but aristocratic and ecclesiastical demand fueled the private market too. The importance of the silk trade is measured by its political dimensions. It is hard to think of another commodity that assumed real geopolitical importance in Roman history. In late antiquity, the silk trade had global clout. The Persians used it as leverage. Justinian actively sought to control or circumvent the trade. Late in his reign, Christian monks from India, who had "spent much time in a land situated far to the north of the nations of India which is called 'Serinde' [i.e., China]," offered to betray the secrets of silk production and smuggle un-hatched silkworm eggs from the east. They were dispatched to China and made the return, and "from that time onward silk has been produced in the lands of the Romans." Only chemical analysis of Byzantine silks will eventually reveal if this act of daring corporate espionage was really successful.[36]

Silk and pepper were complemented by an array of trade goods that filled cargoes moving across the ocean. Ivory and aromatics, aloe, cloves, clove-wood, sandalwood, gold, and slaves were all part of the trading system known to Cosmas. Slaves were not a negligible commodity. Chattel slaves have been mostly invisible in modern histories of the trade, but Cosmas casually assumed that "most slaves" imported into the Roman Empire came from Ethiopia. And, more ethereally, ideas moved across the waters. Christians (many originating from Persia) enjoyed missionary success throughout the east. Indian forms of philosophy and asceticism continued to fascinate and draw seekers. An India of the mind, peopled by otherworldly sages, was carried back to the west.[37]

Figure 6.4. Pepper Trees: Sketch in Text of Cosmas Indicopleustes (Florence, The Biblioteca Medicea Laurenziana, ms. Plut. 9.28, f. 269r. Reproduced with permission of MiBACT. Further reproduction by any means is prohibited).

The real scale of this commerce eludes us. The Romans fought to control an island in the Red Sea where the state levied a toll on imports from India. The revenue was described as "massive." The scattered finds of late Roman coins in India stretch from the fourth century down to the reign of Justinian. Maybe most revealing of all is the sudden importance of the Red Sea theater in the alternating phases of cold and hot war between the Romans and Persians. The Christian Axumite Kingdom of Ethiopia was ascendant in the early sixth century. The Himyarite Kingdom of south Arabia converted to Judaism—of an unusually militant stripe. Religious animus stoked ancient rivalry, and in AD 525 the Axumites invaded the Himyarite kingdom with Roman military aid. The conflict drew in the great powers. Over the next two decades, the Ethiopians and Himyarites were clients of the Romans and Persians. A generation later, Muhammad was born into this world, which has now been evocatively described as the "crucible of Islam." Religion, politics, and commerce intertwined to make this region strategically valuable. The Romans were keen to maintain a stable bridgehead into the waters beyond.[38]

Consumer demand for silk and spices drew together east and west. Ideas and animals, money and metals, moved across the seas. So, too, came germs. In AD 541, an unwelcome stowaway was smuggled into the empire from the world beyond. No one who has read the sources closely denies that

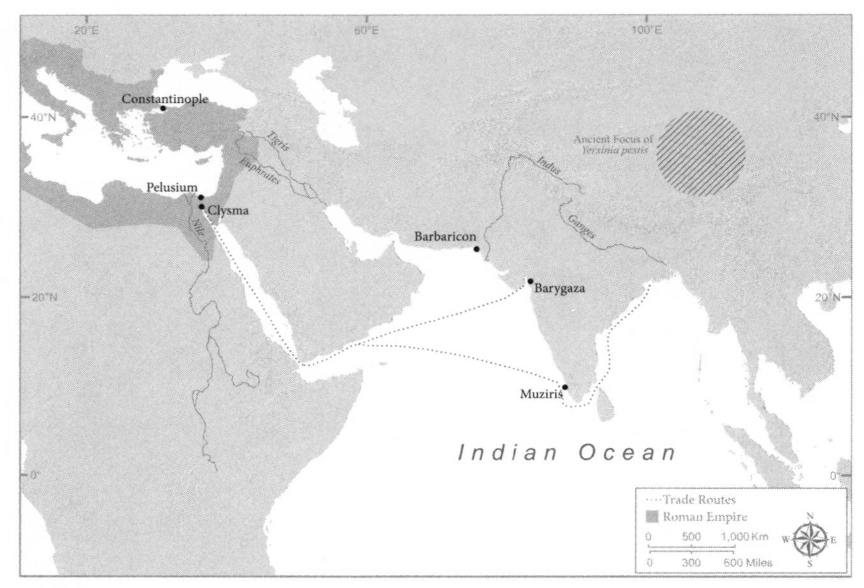

Map 18. The Itinerary of Y. pestis*: From China to Pelusium*

the Justinianic Plague first appeared in Egypt. Our star witness, Procopius, pinpointed the origin of the outbreak in Pelusium. John of Ephesus, who was on the scene in Alexandria when it appeared there, claimed that it came "from the regions to the southeast of India, of Kush, the Himyarites, and others." The dispersal from Pelusium, in combination with the genetic evidence of the plague's eastern origins, guarantees an Indian Ocean passage for the first plague pandemic. Pelusium lay due north of the port of Clysma, a primary terminus of Red Sea trade. Ships all the way from India were docked at its port. From Clysma, Pelusium was a short leap. It was only a few days' journey overland or a short sail down the old Canal of the Pharaohs, rebuilt by Trajan, connecting Clysma to the Nile just upriver from Pelusium. The first pandemic made its debut right at the hinge of the empire and the Indian Ocean world.[39]

It required one last twist of fate for the bacterium to make its grand entrance into the Roman world. The Asian uplands had prepared a monster in the germ *Y. pestis*. The ecology of the empire had built an infrastructure awaiting a pandemic. The silk trade was ready to ferry the deadly package. But the final conjunction, what finally let the spark jump, was abrupt climate change. The year AD 536 is known as a "Year without Summer." It was the terrifying first spasm in what is now known to be a cluster of volcanic explosions unmatched in the last three thousand years. Again in AD 540–41 there was a gripping volcanic winter. As we will see in the next chapter, the AD 530s and 540s were not just frosty. They were the coldest decades in the late Holocene. The reign of Justinian was beset by an epic, once-in-a-few-millennia cold snap, global in scale.[40]

The climate disturbance in the moments preceding the Plague of Justinian is a sudden, blinding flash that we instinctively know must be connected to the crash that immediately follows. We do not know just exactly how the one caused the other. A plague epidemic is a chain reaction involving at least five different species. It is a great biological domino event, encompassing the bacterium, the sylvatic host (e.g., marmots), the amplification host (the black rat), the arthropod vector (the oriental rat flea), and us. Minute changes in temperature and precipitation can affect the habitats, behavior, and physiology of each organism involved in the cycle. Still today, small vibrations in the climate trigger visible effects on plague cycles in rodent populations. Even within the relatively small bounds of year-to-year variability, the climate is a governor on the heat of enzootic plague.[41]

One thing is certain: the relation between climate and plague is not neat and linear. As with so many biological systems, it is marked by wild swings, narrow thresholds, and frenzied opportunism. Rainy years foster vegetation growth, which in turn sparks a trophic cascade in rodent populations. In excess, water can also flood the burrows of underground rodents and send them scurrying for new ground. Population explosions stir the emigration of rodents in search of new habitats. Today, there is a strong connection between El Niño and the outbreak of plague in China. It is entirely likely that these relationships held into the Holocene past. Given that there is a strong correlation between volcanism and El Niño, the volcanic eruptions of the AD 530s may have stirred the Chinese marmots or gerbils carrying *Y. pestis* out of their familiar subterranean colonies, triggering an epizootic that reached the rodents of the seaborne trade routes heading west. Altogether, the most likely scenario is that the climate patterns of the early sixth century—now dominated, as we will see in the next chapter, by a negative regime of the North Atlantic Oscillation – brought greater rain to the semiarid homelands of the reservoir species; vegetative growth sparked a demographic explosion of burrowing rodents, and *Y. pestis* spilled into new host populations.[42]

The climate also regulates the plague by acting on the fleas that carry the bacterium between hosts. The sensitivity of the flea to ambient temperatures creates the basic seasonal pattern of plague. The flea is picky about reproducing within a certain band of temperature. And the fatal blockage in its gut, that causes it to regurgitate infected blood, is preciously averse to temperatures that are too low or too high. The familiar result is a seasonally specific plague cycle. Epidemics gain momentum in the spring. But the high heat of summer can suddenly squelch the outbreak. In early twentieth-century India, the oppressive heat of late summer knocked down the incidence of bubonic plague to virtually nil. The sharp cooling of the 530s and 540s may have opened geographical possibilities that *Y. pestis* had never seen. Mild summers might have opened the gates across the balmy southern passage. The mean temperatures along the Spice Coast fall precisely along the threshold of what the plague cycle will tolerate.[43]

The precise sequence of events that caused the plague to spill out of its mountainous haunts and to explore new routes across the southern waters is likely to stay just beyond our grasp. Through the shadows we sense the enormous contingency of the fatal moment. The alignment of natural history and human history in the making of this moment baffles our distinctions between chance and structure. What we can say is that the deadly germ found its way, perhaps by the slimmest of margins, to the rats of the Roman Empire.

Near to Annihilating the Human Race

Procopius and John—each was quintessentially a man of the Justinianic age. Yet they were representatives of entirely different cultural worlds that had come to exist in uneasy proximity. Procopius of Caesarea was a traditionalist to the core. Trained in law, he entered the ranks of imperial service and became the legal advisor to the great general Belisarius, in whose gravitational field he trailed for the first half of Justinian's reign. Procopius wrote the most important history of the sixth century, a classicizing account of high politics. He is equally notorious for the lubricious *Secret History*, one of the greatest hit jobs in the annals of literature. Religion was not to his taste. The theological squabbling of his age wore his patience thin. "I consider it a sort of insane stupidity to investigate the nature of God, asking what sort it is. For man cannot, I think, accurately understand even human affairs, much less those pertaining to the nature of God." Procopius preferred to inhabit the sphere of classical Greek culture that deliberately stood a little to one side of the sweep of time.[44]

It is almost hard to believe that John of Ephesus was his contemporary. Precisely the bitter ecclesiastical conflicts that Procopius brushed aside were the consuming affair of John's life. Born in Amida, in the Syriac-speaking stretches of the eastern frontier, he was sent to a monastery in his childhood. He became a leader in the Miaphysite movement, caught up in the profound theological controversies about the nature of Christ that had riven the east since the doctrinal formulas of the Council of Chalcedon (AD 451). John arrived in Constantinople a religious exile. The churchman is best known as the author of an ecclesiastical history and a rich body of stories about eastern saints, preserved in his native Syriac. His world was framed by the shape of Biblical history. He had no doubts he was living in the stream of events foretold in the narratives of scripture.[45]

Procopius and John are an unlikely pair. They are bound forever by the happenstance that both of them witnessed the first visitation of the bubonic plague and lived to write vivid accounts of its devastation. We thus have two very different perspectives on the same event. For Procopius, this plague "that came close to wiping out the whole of mankind" was simply unaccountable. His report is, like that of Thucydides, dominated by a cool interest in the pathology of the disease and the immediate social trauma of mass mortality. For John, the plague was a chastisement. God's wrath fell upon the cities like a "wine-press and pitilessly trampled and squeezed all their inhabitants within them like fine grapes." The sins of the people, above all their greed, had called down the slaughter from heaven, "like a reaper upon standing wheat," who "mowed and laid down innumerable people of all ages, all sizes and all ranks, all together."[46]

We must approach our ancient accounts with a healthy balance of respect and caution. Our knowledge of the biology of *Y. pestis* is a towering advantage, and we are within our rights to use it. The biology of the *Y. pestis* strain that caused the first pandemic was closely related to the agent of the Black Death. That reality imposes certain expectations and limits. At the same time, the shape of a mortality event, especially on a pandemic scale, is influenced by the ecological and social circumstances that form the background of its spread. We should be alert to its particularities and open to the possibility that our eye-witnesses may record irreducibly unique insights about the behavior of the pathogen in a past context. The Justinianic Plague only happened once, and they were there.

It will help us to remember that *Y. pestis* is a versatile killer. Much depends on the means of infection. There are two principal routes: inoculation of the dermis via flea-bite and inhalation of aerosol droplets. The signature expression of the disease is bubonic plague, so called from the hard painful swellings of the lymph nodes, *boubones* in Greek. The bubonic form of the disease typically originates with the flea bite. The plague bacterium is injected into the dermis, where it multiplies and blackens the local tissue. The lymphatic system drains the bacteria into the nearest lymph node. There the bacteria dodge the immune response and replicate explosively. The lymph node swells. The location of the flea bite determines where the buboes form; the neck, armpits, and especially the groin are often the site of the swelling. After 3–5 days the victim is symptomatic. The course of the illness is another 3–5 days. Fever, chills, headache, malaise, and delirium move quickly. The buboes grow, like tumid oranges or grapefruits hanging off the body. *Y. pestis* overwhelms the victim's immune response, and sepsis ensues. In a world without public health infrastructure or antibiotics, case fatality rates were high, ~80 percent.[47]

There are variations on this theme of infection by flea bite. In some cases the bacteria skip the lymphatic route and dive directly into the bloodstream. The patient develops *primary* septicemic plague, and the immune system has little time even to begin mounting a response. It is a terrifying eventuality. The victim dies of overwhelming sepsis before external signs of the disease are visible. The end can come within hours of the initial infection. It is also possible for an infection that begins in the lymphatic system to leap to the circulatory system. When the plague enters the bloodstream from an infected lymph node, the victim develops *secondary* septicemic plague, so called because it is a consequence of the primary infection of the lymphatic system. In a case of secondary septicemic plague, the bacteria clot capillary vessels, causing small hemorrhages that appear as petechiae, pinpoint spots of discoloration. Bloody vomiting and diarrhea follow. In this course of the disease, also, sepsis is astonishingly swift and uniformly fatal: the splotches betoken death within a day.[48]

There is still another possible course of the disease that begins by flea bite. In a case of bubonic plague, the bacteria may find their way from the lymphatic system into the lungs. This pathology is known as *secondary* pneumonic plague. Pneumonic plague is a respiratory syndrome. The patient quickly develops a cough, with bloody sputum. The body's hyperinflammatory response floods the lungs with fluid, impairing pulmonary function. Pneumonic plague would have been invariably deadly in the ancient pandemic.[49]

Y. pestis can also travel in aerosol droplets. If the microbes become lodged in the upper respiratory tract, they can enter the lymphatic system and cause bubonic infection. If they are inhaled into the lungs, *primary* pneumonic plague develops. The incubation phase is short, 2–3 days, followed by a bronchopneumonia with fever, chest pain, and bloody cough. Case fatality rates approach 100 percent. Infected aerosol droplets can be discharged by patients with primary or secondary pneumonic plague. The importance of direct contagion via pneumonic plague in the historical pandemics is not entirely clear. It was not a very efficient means of transmission. On balance, primary pneumonic infection was probably more of a complementary dynamic than a fundamental one.[50]

The bacterium has still other ways of entering new victims. It can actually be ingested (just one reason not to eat rodents, especially in places where the disease is enzootic). But, it is the flea bite that enjoys pride of place as the principal route of infection in major plague epidemics.

Means of infection	*Pathway*	*Expression of disease*
Flea-bite	Lymph → Lymph node	Bubonic plague
	Bloodstream	Primary septicemic plague
	Lymph → Bloodstream	Secondary septicemic plague
	Lymph → Lungs	Secondary pneumonic plague
Aerosol droplet	Upper respiratory tract	Bubonic plague
	Lungs	Primary pneumonic plague

In AD 541, the rattle of war between the great powers suddenly seemed to hush before the roar of a strange new mortality. It started in the middle of summer, in Pelusium. Even before the affirmation of forensic DNA evidence, the fingerprints of *Y. pestis* were all over the pandemic. According to Procopius, its onset was marked by a mild but creeping fever. Then "a bubonic swelling appeared." The tumid growth extruded from the groin, principally, sometimes the armpit, ears, and thighs. Procopius observed that "in cases where the buboes grew very large and discharged pus, the patients overcame the disease and survived." It is an astute clinical note. In late stages the buboes can suppurate, and these patients may survive. Procopius also witnessed the permanent debilitation of survivors. The aftereffects of tissue necrosis can cause lifetime impairments. For John too the swelling in the groin was the strange signature of this plague. He observed that other animals—including wild animals—were struck with the disease. There were "even rats, with swollen tumours, struck down and dying."[51]

When victims of plague did not die immediately, "black blisters" the size of a lentil bloomed all over the body. Death followed the same day. John too observed black spots that appeared on the hands. "On whomsoever these appeared, the moment they did so the end would come within just one or two hours, or it might happen that the person had one day's delay." He considered this outcome a common course of the disease. Similarly, Procopius noted, some patients vomited blood, another sign of imminent death.[52]

The rapid, grim course of primary septicemic infection, when the bacterium enters the bloodstream directly, may account for contemporary reports that observed what seemed like virtually instantaneous death. "As they were looking at each other and talking, they began to totter and fell either in the streets or at home, in harbours, on ships, in churches and everywhere. It might happen that a person was sitting at work at his craft, holding his tools in his hands and working, and he would totter to the side and his soul would escape."[53]

Nothing in the surviving record suggests that pneumonic plague was prominent in the first pandemic. Respiratory symptoms might have been too pedestrian to deserve mention. But our ancient witnesses carefully chronicled other common symptoms like fever and malaise, so the absence is telling. And severe respiratory pathologies in the summer might not have been beneath notice. Other clues militate for the dominance of the flea vector in the first pandemic. Procopius observed that doctors and caregivers were at no special risk of contracting the disease. The poor were killed first. As we will soon see, the spatial and temporal patterns of dissemination are consistent with the predominance of the rat-flea mechanism in the overall diffusion of the first wave. In short, everything leads us to believe that the Justinianic Plague was dependent on the invisible epizootic catastrophe underlying the pandemic, the wave of animal death in which humans were incidentally swept away.[54]

From Pelusium, the contagion split into two branches. One headed west, to Alexandria. According to Procopius, it only then infected the rest of Egypt—an incisive observation excluding the Nile as the conduit of the plague into the empire. The plague also headed east, to Palestine. In a remarkable stroke of fortune, John was traveling across the east along an arc that passed from Alexandria through Palestine, Mesopotamia, and Asia Minor. On the edges of Egypt, one city "perished totally and completely with [only] seven men and one little boy ten years old remaining in it." Through "the whole of Palestine," both villages and cities "were left totally without inhabitants." The plague laid hold in Syria and Mesopotamia. As John trekked through the heartland of Asia Minor toward Constantinople, the disease pursued his convoy. "Day by day we too—like everybody—knocked at the gate of the tomb." "We saw desolate and groaning villages and corpses spread out on the earth, with no one to take up [and bury] them."[55]

The plague moved at two speeds: swiftly by sea and slowly by land. The mere sight of ships stirred terror. John recorded the grisly specter of "ships in the midst of the sea whose sailors were suddenly attacked by God's wrath and the ships became tombs for their captains and they continued adrift on the waves carrying the corpses of their owners." The waters were haunted. "Many people saw shapes of bronze boats and figures sitting in them resembling people with their heads cut off . . . black people without heads sitting in a glistening boat and travelling swiftly on the sea, so that this sight almost caused the souls of the people who saw it to expire." In a more clinical observation, Procopius noted that "the disease always spread out from the coasts and worked its way up into the interior."[56]

Once infected rats made landfall, the diffusion of the disease was accelerated by Roman transportation networks. Carts and wagons carried rodent stowaways along Roman roads. McCormick has shown the importance of rivers as effective conduits of plague in sixth-century Gaul. But, *Y. pestis* is insidiously diffusive because its transmission is also independent of humans. It could spread anywhere that rats could travel. Procopius noted the slow burn of the plague's advance in each area it reached. It was "always moving along and advancing at set intervals. For it seemed to move as if by prearranged plan: it would linger for a set time in each place, just enough to make sure that no person could brush it off as a slight matter, and from there it would disperse in different directions as far as the ends of the inhabited world, almost as if it feared that any hidden corner of the earth might escape it. It overlooked no island or cave or mountain peak where people happened to live." The disease spread deep into the recesses of the ancient countryside.[57]

The pace of metastasis was intricately timed to the progress of the underlying animal epizootic. Everywhere it spread, *Y. pestis* first diffused quietly through the rat colonies. As rat populations collapsed, fleas hunted desperately for blood. The historian of the Black Death, Ole Benedictow, estimates this cycle took on average two weeks. Then, the hungry fleas became less choosy and turned to humans. The human epidemic began. In an outbreak of the plague in Marseilles, the Gallic bishop Gregory of Tours described the arrival of a plague ship from Spain that immediately killed a household of eight people. Then there was a lull, which we can recognize as the epizootic ticking time bomb, after which the human plague erupted. "Like a wheat-field set alight, the town was suddenly ablaze with pestilence." After two months, the plague burned out, possibly as summer temperatures rose. Believing all was clear, people returned. But the plague ramped up again.[58]

Inevitably, the parlous living conditions of the poor put them in close contact with rodents. In the Black Death, the poor succumbed first, but eventually the affluent were swept up too. In the Plague of Justinian, the disease first "eagerly began to assault the class of the poor, who lay in the streets." The carnage was, ultimately, promiscuous. It fell "on houses large and small, beautiful and desirable, which suddenly became tombs for their inhabitants and in which servants and masters at the same time suddenly fell dead, mingling their rottenness together." "People differ from each other in the places that they live, the customs that govern their lifestyle, the manner of their personality, their professions, and many other ways, but none of these factors made the slightest difference when it came to this disease—and to this disease alone."[59]

From Alexandria, the dispersal of *Y. pestis* was ineluctable. If the grain trade was the bloodstream of the empire, Alexandria was its pumping heart. The pestilence in Alexandria stirred prophecies of doom across the sea. The plague was feared at Constantinople before it was felt. "The visitation came upon it by hearsay from all over the place for one or two years; only then did it reach the city." It seems probable that a state ship must have braved the stormy waters of winter to bear the news of emergency to the capital. The plague itself arrived in Constantinople in AD 542, by late February. The earliest surviving notice of the entire pandemic, suitably enough, is an edict promulgated by Justinian. The guild of bankers needed help securing debts in the mass mortality. "The danger of death has penetrated to every place, and it is unnecessary for anyone to hear what each one has experienced . . . when so many unexpected things have happened, as hardly any other time brought about." This was on March 1 of AD 542. Far worse was to come.[60]

The first visitation in Constantinople lasted four months. Both Procopius and John were on the ground. Their testimonies, from different mental worlds, are stunningly convergent. The first victims were the homeless. The toll started to rise. "At first only a few people died above the usual death rate but then the mortality rose higher until the toll in deaths reached five thousand a day, and after that it reached ten thousand, and then even more." John's daily counts are similar. The peak hit five thousand, then seven thousand, twelve thousand, and sixteen thousand dead per day. At first, there remained a semblance of public order. "Men were standing by the harbours, at the crossroads and at the gates counting the dead." According to John, the grisly tally continued until 230,000 had been numbered. "From then on the corpses were brought out without being counted." John reckoned that over 300,000 were laid low. A tally of ca. 250,000–300,000 dead within a population thought to number half a million on the eve of the calamity would fall squarely within the most carefully derived estimates for the death rates in places hit by the Black Death, at 50–60 percent.[61]

The social order wobbled and then collapsed. Work of all kinds stopped. The retail markets were shuttered, and a strange food shortage followed. "A true famine was careering about in a city that nevertheless abounded in all goods." "The entire city then came to a standstill as if it had perished, so that its food supply stopped. . . . Food vanished from the markets." Money could not be changed. Dread overshadowed the streets. "Nobody would go out of doors without a tag upon which his name was written and which hung on his neck or his arm." The palace succumbed. Its army of ministers was reduced to a few servants. Justinian himself contracted the plague. He was lucky to number among the one-fifth or so of those who survive infection. The state apparatus receded into invisibility. "The whole experience may be summed up by saying that it was altogether impossible to see anyone in [Constantinople] wearing the chlamys," the vivid costume of those who represented the face of imperial order.[62]

The city was soon saturated with corpses. At first the tenacious impulse to bury the dead was carried on by the families of the departed. Then it was like trying to stand in a mudslide. "Confusion began to reign everywhere and in all ways." Solemn rituals and even basic environmental control were pulled under. The emperor struggled just to clear the corpses from the streets. Both Procopius and John relate the detail that Justinian appointed his personal clerk, a man named Theodoros, to organize an emergency response. Pits were dug in the fields around the city. Then, they filled. The dead were dragged to the shore on tarps and ferried across the strait. According to Procopius, military towers in Sykai were filled with corpses "in a tangled heap." John was rather more graphic. The dead were cross-hatched in layers, like "hay in a stack." The victims were "trodden upon by feet and trampled like spoiled grapes. . . . The corpse which was trampled sank and was immersed in the pus of those below it." This was not lurid voyeurism. Quite literally, John thought he was watching "the wine-press of the fury of the wrath of God," that was a sign of the end times.[63]

The vivid, sensory record of the outbreak in Constantinople contrasts with the gaping silence that hangs over the rest of the empire. Our informants insisted that the pandemic engulfed "the whole world." It raged across the Roman Empire and beyond, including the Persians and "other barbarians." It swept across the entire east, including "Kush" and southern Arabia; the plague overran Palestine, Syria, Mesopotamia, and Asia Minor. Other chronicles assure us that the plague reached the Danubian provinces, Italy, North Africa, Gaul, Spain, and the British Isles. These reports have less nuance than a color-by-numbers picture. But we cannot ignore them.[64]

Our map of the first pandemic is full of shadows, from time to time broken by small shafts of illumination. We must read our patchy evidence with an alert eye for clues about the epidemiology of the first pandemic. We

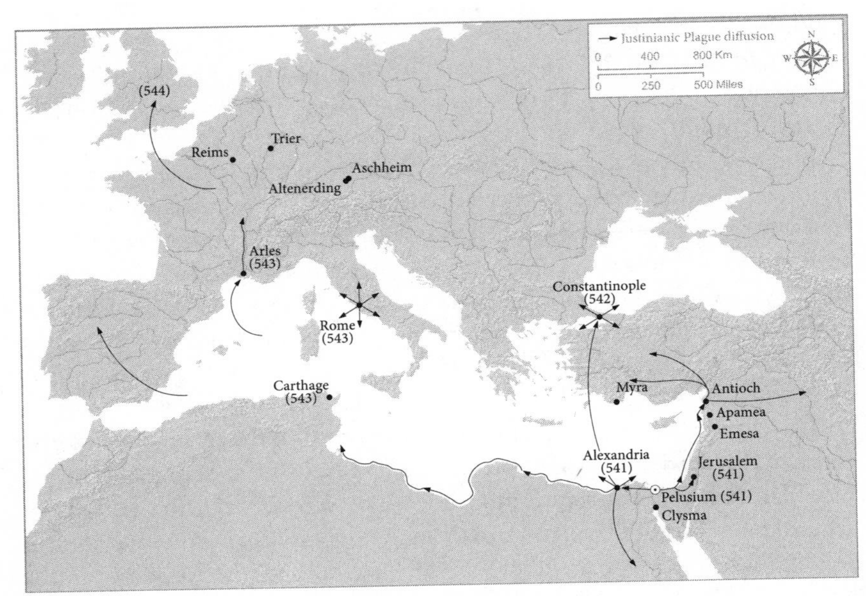

Map 19. The Itinerary of Y. pestis*: From Pelusium to Pandemic*

should ask two critical questions. First, *where* did the first pandemic spread, in terms of both physical and human geography? Second, *what happened* in places where the plague reached? The biology of *Y. pestis* was overwhelmingly the dominant factor, but not the only one. To a certain extent, the course of the disease was sensitive to human factors, the social and economic context of the pandemic. By asking the right questions, we can more clearly define the boundaries of what we know and at least constrain our speculation.

The cities of the eastern Mediterranean were hit hard. Alexandria was "ruined and deserted." Other casualties named in the haphazard record include Jerusalem, Emesa (70 km inland as the crow flies), Antioch, Apamea, Myra, and Aphrodisias. It is a relatively meager list. There are no startling patterns. Most cities in the east were probably struck, but strict caution requires the caveat that we do not know. The mortality itself has swallowed most of the testimony that ever existed.[65]

Before the Justinianic Plague, the limiting factor in the scale of ancient disease events was the mobility of the pathogen. Most people were protected by the viscosity of ancient travel and communications. Even in the interconnected Roman Empire, life moved to the dull pace of unmechanized transport. The demographic dominance of the countryside defused the impact of any mortality crisis; cities were the most vulnerable to pathogens transmitted directly between humans like the smallpox virus. In the modern literature on the Plague of Justinian, it is a casually entrenched assumption that the worst damage was in the cities. But nothing is more apt to lead us astray from the secret of the plague bacterium's violence.[66]

The plague is different. *Y. pestis* does not rely on travel via direct transmission between human victims. Nor is it spread by environmental contamination. Human population density is mostly beside the point, except insofar as it bears on the concentration of rats. Rodent vectors were abundant in the settled countryside and in the wild. The human networks of trade and communication were an accelerant, speeding the dispersal of the bacterium into far-flung rat colonies. But *Y. pestis* relentlessly diffused through the thick, ubiquitous networks of rodents. And given that the pandemic could also harness other small mammals and human parasites as vectors of transmission, its versatility was an additional force of propulsion.

In the first pandemic, *Y. pestis* spilled over without restraint across the countryside. Its implacable spread defied normal expectations. Crops lay unharvested in the field, and grapes rotted on the vine. In the eastern Mediterranean, the tentacles of the pandemic reached into the villages that dotted the countryside. The holy man Theodore of Sykeon contracted the bubonic plague when he was a boy of twelve. The pest had reached his village, which lay on the Roman highway cutting across central Anatolia, some 11 miles from the nearest town. A holy man who lived atop a pillar near Antioch watched the pestilence sweep "the whole country . . . every place in the country." It attacked the hinterland of Jerusalem. An inscription places the bubonic plague in Zoraua, a village in the Transjordan. And in Egypt, a holy man twenty-four miles upriver from Alexandria, at a cell "in the desert of Mendis," was carried off by the plague.[67]

In the west, the evidence is even patchier. The plague made it to North Africa, Spain, Italy, Gaul, Germany, and Britain. But the routes of its transmission and the depth of its infiltration are murky. The disease "took fire" in Africa. "The plague had begun to destroy men and women and the tottering world around them." In the Iberian peninsula, "almost all of Spain" was invaded by the pestilence in the first visitation. In Italy, an eerie silence hangs over the land. One lonely report affirms the outbreak in Italy. In Gaul, and only Gaul, we are better informed. The prolific bishop Gregory of Tours offers us glimpses of a world stricken by plague. What he tells us is invaluable. Infected rats reached Gallic shores at Arles in AD 543. The plague spread north, propelled by riverine transport networks. The first visitation did *not* reach Clermont, in the Auvergne, where Gregory was a young boy. It crept far to the north, to Trier and Rheims. The plague seems to have crossed the British Channel and reached the western limits of Europe by AD 544. An outbreak in Ireland in AD 576 is recorded in the annals, but the severity of the plague experience in the islands is ambiguous until a major episode in the AD 660s.[68]

Paradoxically, the degraded systems of connectivity in the west could have slowed the dispersal of the plague bacillus. But this line of argument cannot carry too much conviction. The fact is that plague is attested everywhere it might be expected. And the most stunning evidence of all now comes from a place we might have thought was beyond the reach of the pandemic. Two cemeteries outside Munich in southern Germany, at Aschheim and Altenerding, have yielded paleomolecular evidence for *Y. pestis*. The cemetery at Aschheim was in use across the sixth and seventh centuries. It humbly served a settlement of less than a hundred residents. The unusual frequency of multiple interments, datable to the middle decades of the sixth century, looked suspiciously like an episode of crisis mortality. DNA extracted from the dead has conclusively determined that these victims died of *Y. pestis*. The beast was here. It is hard to overstate the ramifications of finding the plague in a remote, rural outpost in the west. If the plague was *here*, it must have been in many other places which lie in the dark zones on our map.[69]

The marvels of molecular analysis may continue to relieve us of our ignorance. Other fragments of genetic material are out there. It has too often been repeated that the Justinianic plague did not leave archaeological traces in the form of mass graves. The prodigious labors of McCormick have now shown that the opposite is true. In a catalogue of some eighty-five archaeological features, he builds an overwhelming case that a sudden upsurge in mass graves is to be connected with the bubonic plague. Violence and other natural disasters surely account for some of the multiple burials from late antiquity. But the hard confirmation of genetic evidence from Bavaria seals the conclusion that *Y. pestis* reshaped something so intimate and conservative as the solemnity of burial, from the British Isles to the edges of Palestine. The Justinianic Plague had a vast orbit.[70]

For contemporaries of the first pandemic, it was newsworthy when any people was spared the plague's destruction. The Moors, the Turks, and the Arabs inhabiting the desert were reportedly exempt from the global catastrophe. A poetic account of the plague in Africa emphasized that the

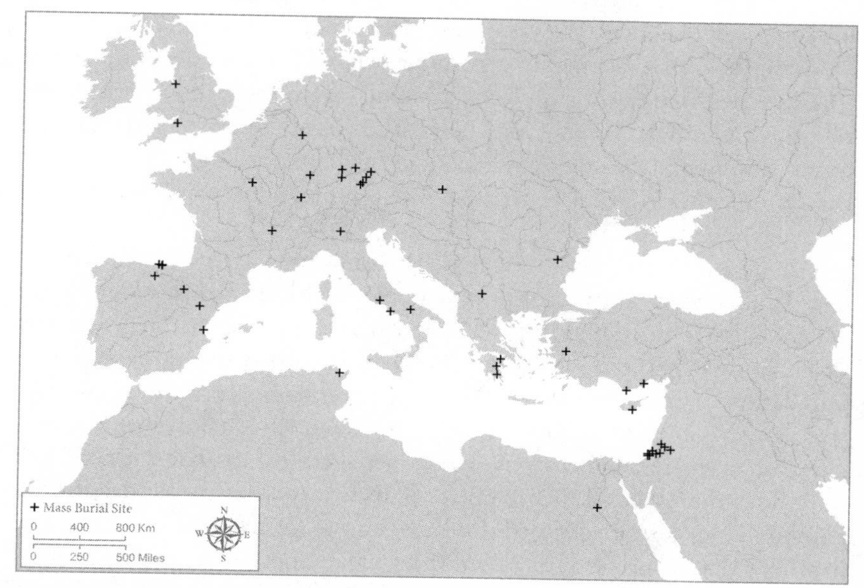

Map 20. The Geography of Mass Mortality (based on McCormick 2015 and 2016)

plague had annihilated the Romans but not "affected the rancorous tribes." The Turks themselves boasted "that from the very beginning of time they have never witnessed an epidemic of plague." And it has been the conventional wisdom that the plague passed over the Arabian heartland. "Neither Mecca nor Medina were affected by any of the plague epidemics which

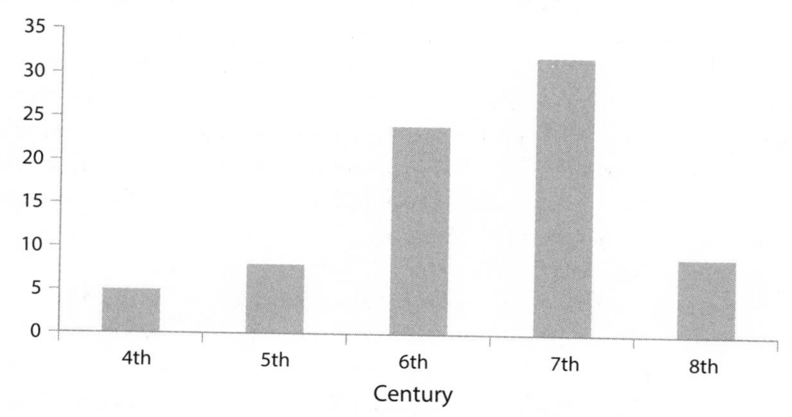

Figure 6.5. Mass Graves by Century (based on McCormick 2015 and 2016)

broke out elsewhere in the Near East." In the seventh century, Anastasius of Sinai, the abbot of the famous monastery of Saint Catherine's, noted that the "deserted and dry" places inhabited by unbelievers "never experience plague." The Moors, Turks, and inhabitants of central Arabia—all shared a nomadic lifestyle. The ecological explanation suggests itself: that the non-sedentary social formation was a protection against the deadly rat-flea-plague nexus.[71]

The plague was a thief in the night. In an instant, it reversed the collective, painful efforts of two centuries of demographic growth. Death tolls are elusive. John claimed that not 1 in 1000 survived the plague. That stretches all credibility. In the *Secret History*, Procopius insisted that about half the population died in the plague. "The plague broke out as well . . . and carried away half" the total population. It wiped out "the majority of farmers." "At least as many people survived it as perished in it, either because they were not infected at all or because they recovered after their infection." A Palestinian tombstone claimed that one-third of humanity was swept away, in a later sixth-century outbreak of the plague. These are our only explicit testimonies to the global mortality rates of the first pandemic.[72]

Ancient societies were always tilted toward the countryside. By now some 85–90 percent of the population lived outside of cities. What set the plague apart from earlier pandemics was its ability to infiltrate rural areas. It made the plague pandemic far more deadly than anything that had come before. Once the disease became epidemic, the overwhelming biology of the killer *Y. pestis* took over. The plague is basically indiscriminate, as our ancient authors were keen to emphasize. Young and old, men and women, rich and poor fell before its deadly march. But death preyed especially on the frail. Even against an enemy so formidable as *Y. pestis*, the underlying biological status of the population was not completely immaterial. The wrenching climatic anomalies of the years leading into the Justinianic Plague had thinned the food supply. The insalubrious disease environment in the Roman world left its inhabitants weakened and their immune systems depleted. All of these variables point to the frailty of the Roman population on the eve of the first pandemic. The first pandemic cut through a people hungry and weak.[73]

Under the strictest scrutiny, the scandalous mortality estimates attributed to the medieval Black Death have held up or even been revised upward. The late medieval documentation is considerably richer, and there are lessons to be learned from the death tolls reconstructed from a much denser record. Broadly, historians concur that "The Black Death killed an estimated 40

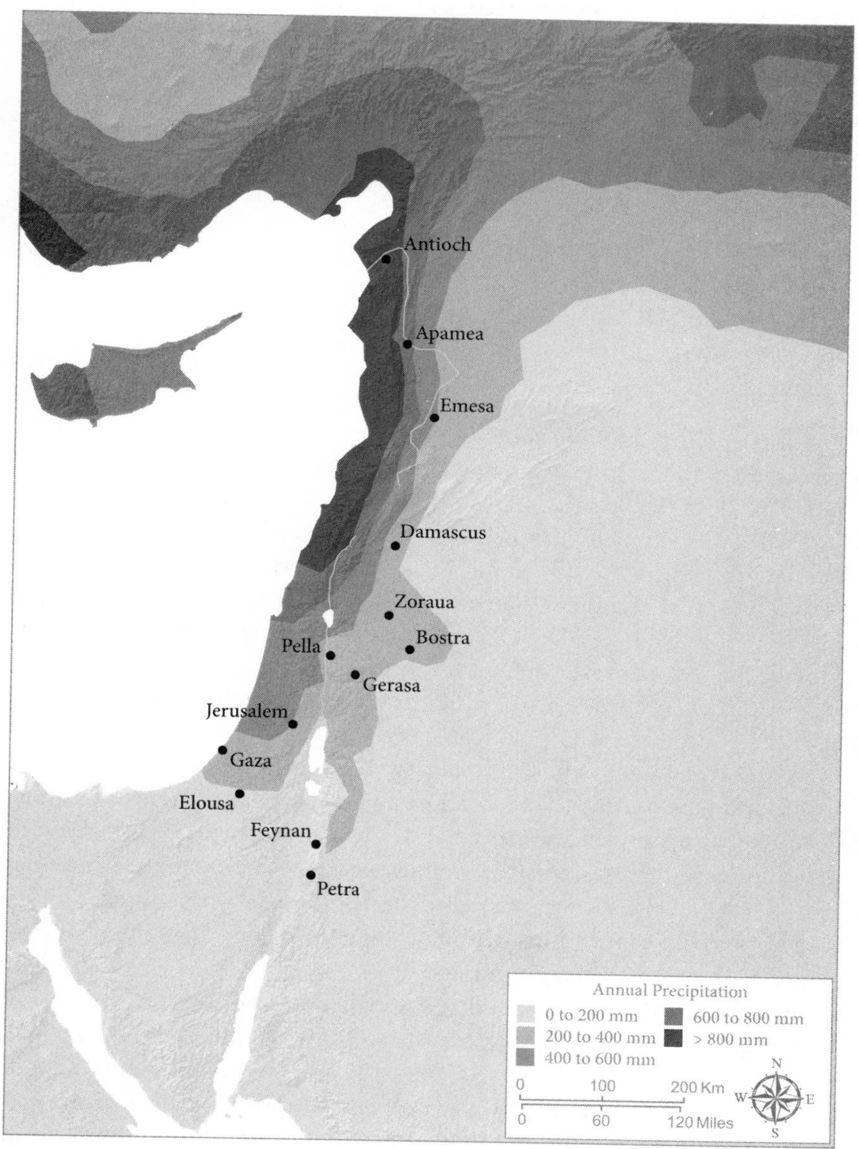

Map 21. Plague Ecology in the Near East

percent to 60 percent of all people in Europe, the Middle East, and North Africa when it first struck there in the mid-fourteenth century." Mortality tolls varied only modestly from nation to nation. The figures from Benedictow's careful synthesis are revealing.[74]

Region	*% Mortality*
England	62.5
France	60
Savoy	60
Languedoc/Forais	60
Provence	60
Italy	50–60
Piedmont	52.5
Tuscany	50–60
Spain	60

Fundamentally everything about our knowledge of the Justinianic Plague is consistent with the conclusion that the mortality also carried off an unfathomable half of the population.

The plague immediately convulsed the normal rhythms of life. The harvest rotted in the fields. Food was scarce. Then, with fewer mouths to feed, it was more abundant than usual. The price of wheat collapsed. Wages, by contrast, soared. In AD 544, Justinian decreed that "It has come to our knowledge that even after the correction meted out according to the Lord God's love of humanity [that is, the plague], men engaged in dealing and scheming and those who practice different crafts and those who work the land, and even sailors, who ought to have been made better, have turned to greed, and seek double or triple prices and wages against ancient custom." The inheritance system was thrown into disarray, and in an economy with extensive networks of credit, the banks were desperate to enforce debt obligations on successors. Except for churches, building activity ceased.[75]

The state was sent reeling. Justinian issued gold coins that fell below the sacred target of 1/72 of a pound. It was the first manipulation of the gold currency since Constantine, and it scandalized the mandarins. The army was already dangerously overextended, and now its ranks were decimated. The plague heralded the beginning of an unprecedented fiscal-military crisis. In the coming generations, the Roman state struggled to mobilize an army, or more often to pay for it. Justinian refused to forgive tax arrears in the years following the demographic catastrophe, until finally in AD 553 he relented. He declined to reduce the overall tax burden. The survivors were crushed with fiscal pressure. By the middle of his reign, the empire was probably charging the highest tax rate ever imposed in Roman history. Procopius' critique of the regime rests on the charge of fiscal rapacity. Meanwhile, the reform agenda creaked to a virtual halt. Peter Sarris counted 142 edicts and constitutions issued between AD 533 and 542 (14.2 per year). From AD 543 to 565, there was a total of 31 (1.3 per year). As we will see in the next chapter, there is a relatively uncomplicated line from demographic collapse to the failure of the eastern empire.[76]

But the shock of the first outbreak was only the beginning.

Two Centuries of Death: The Persistence of Plague

Once the first thrust of the disease was complete, *Y. pestis* deployed evasive maneuvers. Whereas a virus like smallpox leaves an imprint in the immune system that confers strong and lasting immunity on survivors, the plague bacterium probably conferred only partial and temporary immunity on any victims who survived. The question is not entirely settled, especially in the case of the historical pandemics. In the first pandemic, Evagrius Scholasticus reports that some who had been infected once or even twice fell in a later visitation of the plague. There are parallel examples from the Black Death. The human body's adaptive immune system should preserve a memory—specific B-cells and T-cells that recognize the bacterium from the previous fight—and in studies from modern China, it has been found that plague survivors indeed carry around the tools of acquired immunity to aid them in the event of re-infection. But these memory cells are not a guarantee of invincibility. Acquired immunity is more like an additional weapon in a grinding and multifront war, not an impenetrable seal against invasion.[77]

Plague had another, even more insidious stratagem in the long run. An obligate human parasite like smallpox lacked an animal reservoir where it could hide between outbreaks. Plague was more patient. As the wave of the first visitation pulled back from a ravaged landscape, small tidal pools were left behind. The plague lurked in any number of rodent species. These biological weapons of the plague—the fact that it does not confer strong immunity and that it has animal reservoirs—allowed the first pandemic to stretch across two centuries and cause repeated mass mortality events. To take its true measure, we should think of the first pandemic not as a big bang, but as a chain explosion that sounded for two centuries.

The medieval pandemic that followed the Black Death lasted for four centuries in Europe. We have lately come to see, a little more clearly, how it endured for so long. *Y. pestis* became enzootic in the west: plague was able to sustain itself among commensal or sylvatic species. The periodic recurrence of the plague did not rely on repeated reintroductions from the central Asian home of *Y. pestis*. Here the most traditional and novel forms of evidence are converging. Ann Carmichael has made a brilliant case for an enzootic plague focus in the foothills of the Alps and the Alpine marmot as a maintenance host. And new genomic evidence from plague victims has established that the bacterial agents of later outbreaks were direct lineal descendants of the Black Death. Once introduced into the west, this bacterial guest long overstayed its welcome, before mysteriously vanishing again.[78]

The first pandemic lasted from the arrival of *Y. pestis* in AD 541 down to its last violent gasp in AD 749. For two centuries, at irregular intervals, plague spilled out of its reservoirs in sudden, explosive outbursts. It is traditional to treat these outbursts as serial "waves" of plague. We will studiously avoid doing so here. The scholarship on the first pandemic has become a prisoner of this metaphor. The first visitation *was* wavelike, entering the empire from the outside and careering outward in a widening arc, churning through the superabundant rodent populations until it reached the western ocean. Thereafter the pattern was more complex and asymmetrical. If we hope to understand the ecology of persistence, we will have to retire the old metaphors.[79]

After the first visitation, the plague did not need to come from the outside. The initial amplification left the seeds of renewed disaster hidden behind its wake of destruction. For the next two centuries, we should imagine amplification events of variable magnitude arising from interior plague foci. Appendix B provides a catalogue of thirty-eight such events, some of which are probably interconnected. Some episodes of amplification seem local and transient, others far-reaching. We will never be able to draw a complete map of the sinuous path of plague across these two centuries. But we can look for the interplay of nature and society in these centuries of upheaval. The ecology of plague persistence—the hidden life of plague among animals—determined when and where a spillover could emerge. The shifting human configurations of empire and connectivity subtly and unwittingly channeled the explosive force of each new outbreak.

The first period of the plague's afterlife, down to ca. AD 620, was dominated by Constantinople. Major amplification events recurred with frequency. Seaborne connectivity amplified the force of outbreaks. The true location of the plague reservoir is obscure. The rat colonies of Constantinople may have incubated the seeds of an outbreak even during periods of quiescence. But it is more likely that plague was introduced from the provinces. Throughout the sixth century, Constantinople remained the nerve center of the entire eastern Mediterranean, with tentacles stretching far to the west. One of the real legacies of Justinian's reconquest was to ensure that the western Mediterranean remained linked into the disease system of the east. Amplification events could have originated almost anywhere and found their way to Constantinople; the capital was a relay station, gathering the germs of empire and acting as an engine of metastatic dispersal.[80]

Plague Outbreaks in Constantinople

542
558
573
586
599
619?
698
747

The first reappearance of plague came sixteen years after it first raged in the capital. It seemed as though it had never fully disappeared. "It had never really stopped, but had simply moved from one place to another, giving in this way something of a respite to those who had survived its ravages." Some victims "simply dropped dead while about their normal business at home or in the street or wherever they happened to be." The historian Agathias noted that men were more affected than women, perhaps because rodent populations had recovered in the commercial and industrial zones of the capital. Three years after this attack on the capital, the plague amplified in a zone stretching from eastern Anatolia, through Syria and Mesopotamia, into the Persian kingdom. Whether this was an extension of the outbreak in the capital, or an amplification arising from a plague reservoir in the east, is unclear. Again, about fifteen years later, in AD 573–4, an interregional amplification swept across the eastern Roman Empire. For a third time, the capital was staggered by the plague. The daily death toll reached 3000 in the city. In 586, a pestilence was said (hyperbolically) to have killed 400,000 people in the capital alone, but the outbreak is not attested beyond Constantinople, and so may have been a local phenomenon.[81]

Around 597, a plague outbreak swept through Thessalonica and the surrounding countryside. We can follow this amplification event in some detail. The mortality was so desolating that the Avars, enemy barbarians who had moved into eastern Europe, were inspired to take advantage. The next year, though, in occupying Thrace, the plague caught up to them; their leader was said to have lost seven sons in one day. By the following plague season, in AD 600, the outbreak reached the capital. The mortality was grotesque. A Syriac chronicle reports 380,000 dead in the capital. Once the plague reached the capital, it seemed to be everywhere. It moved overland into Bithynia and throughout Asia Minor and into Syria. It also moved west, striking up the Adriatic, across to North Africa, and up the west coast of Italy, with horrifying consequences at Rome. An amplification event that may have started in a sylvatic reservoir somewhere in the Balkan highlands found its way to the imperial city, whence it embarked for ports across the Mediterranean.

Yet, this outbreak was the last act of Constantinople as a great dispersal mechanism. Between AD 542 and AD 619, plague struck the capital on average every 15.4 years. Thereafter, it struck twice in 128 years, or once every 64 years. This abrupt shift follows the waning dominance of Constantinople in the eastern Mediterranean. From the middle of the seventh century, the city came to play a peripheral and passive role in the epidemiology of the first pandemic.[82]

In the west, the record is stubbornly meager. The incipient middle ages had dropped a curtain over this world. The shadows break only just enough for us to see the broadest outlines of the pandemic's career. We cannot exclude the possibility that our eyes have been tricked in the dim light. But, if the thin source record that we do have is credible, then for two generations the experience of the west was under the influence of Constantinople. Plague was repeatedly reintroduced by sea, from the east. Subsequently, in the first half of the seventh century, there was a lull. Finally, the last century of plague in the west may have been shaped by a plague focus in Iberia or the reintroduction of the plague from the Islamic world into Al-Andalus.

Western Plague Phases

Byzantine Phase	542–600
Seventh-Century Lull	600–660
Iberian Phase	660–749

The first wave of the Justinianic Plague swept all the way to the Atlantic edges of the world. Afterward, the plague fell silent in the west for more than two decades. Then repeated reintroduction of the plague by sea triggered a series of amplification events. The first renewal of plague started sometime between AD 565 and 571 in Liguria, a coastal strip under the

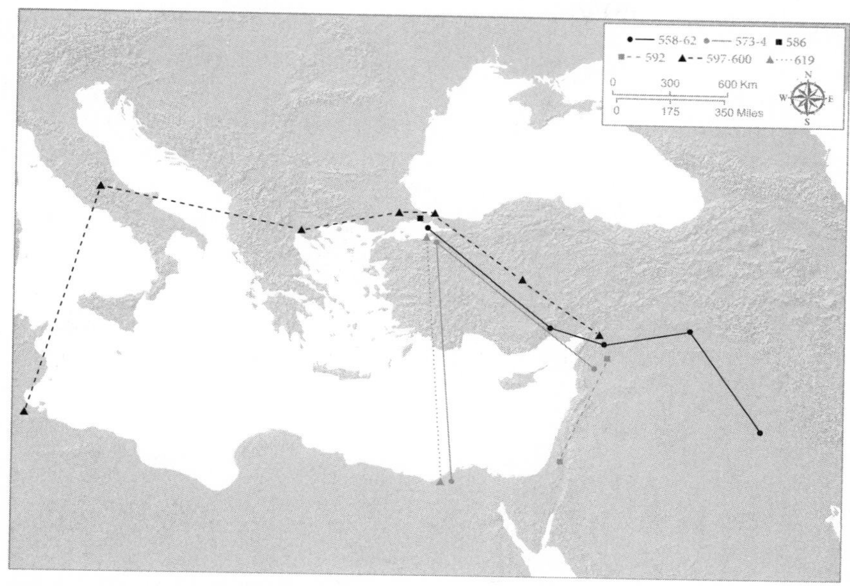

Map 22. Plague Amplifications in the East, AD 550–620

control of Byzantine forces. Thence it spread across Northern Italy, over the Alps to the old borderlands of the Roman Empire. The description of the historian Paul the Deacon, writing some two centuries after the fact, is vivid. "There began to appear in the groins of men and in other rather delicate places a swelling of the glands, after the manner of a nut or date, presently followed by an unbearable fever, so that upon the third day the man died." The effects were catastrophic. "You might see the world brought back to its ancient silence; no voice in the field; no whistling of shepherds. . . . The crops, outliving the time of harvest, awaited the reaper untouched. . . . Human habitations had become places of refuge for wild beasts."[83]

The amplification in Northern Italy was connected with a nearly simultaneous reappearance of the disease in Gaul. The outbreak in Gaul was severe, striking regions like the Auvergne which had been spared before. Lyon, Bourges, Chalon-sur-Saône, and Dijon were hit. This roll call traces the importance of riverine networks of communication in the dispersal. The recurrent plague events may not have penetrated much beyond the riverways. But the pandemic continued to rumble in Gaul. Minor plagues hit southwestern Gaul in AD 582–4. In AD 588, an outbreak started when a ship from Spain docked at Marseilles. The plague blazed like fire for two months. It also spread with lightning speed up the Rhone. But there is no suggestion that this amplification was anything beyond a confined event in the Rhone corridor.[84]

Twice in the last decade of the sixth century the plague amplified in the west. In 590–1, an outbreak famous for ushering in the pontificate of Gregory the Great struck in Rome. The Pope claimed that much of the population was carried off. It was not a local affair. The plague reached inland at least as far as Narni, and whether by land or by sea it reached the eastern shores of Italy. The plague travelled to Gaul, where again the transport networks along the Rhone River flung it inward, to Avignon and Viviers. But thereafter we hear of no outbreaks in Gaul. From the end of the sixth century, Gaul looked north rather than south. Its center of gravity shifted away from the Mediterranean and toward a more continental European future. In the short term, this isolation was a biological breakwater.[85]

The amplification in AD 599–600 was the last to emanate from Constantinople. It was a far-reaching disaster in the west, striking the Adriatic, North Africa, and the western shores of Italy, including Rome. Pope Gregory knew the plague had come from the east. But what he could not have imagined was that a time of peace lay at hand. The plague thereafter relented in the west, at least in our patchy sources. A Latin epitaph of AD 609 from Córdoba commemorates a victim of the disease in the early seventh century. It is a lonely reminder of how much we do not know. It also suggests the possibility of a plague focus in Iberia. So too does a seventh-century handbook of Christian sermons from Toledo. Four ready-made homilies grapple with the moral dilemmas of a bubonic plague epidemic. And when the plague did rear its head again in the west, from the end of the sixth century, Iberia was always involved. Spain is the only region in the west where there was recurrent plague and no obvious introduction from the outside by sea.[86]

If the plague found an enzootic reservoir in Iberia, it would have been a suitable staging point for the two subsequent episodes of plague in the British Isles, in AD 664–6 and AD 684–7. The first visitation in the time of Justinian had reached the Atlantic, but thereafter we lack solid indications of bubonic plague across the Channel. Archaeologists have observed that rat bones, after a peak in Roman times, virtually disappear from sixth- and seventh-century sites. This absence is meaningful. Standards of archaeological practice in Britain let us infer from the absence of evidence that the rat population crashed. The slow recovery of commensal rat colonies, here and elsewhere, was a factor impeding the spread of bubonic plague after the first visitation. When bubonic plague did return to Britain ca. 664, appearing first in Kent, it may well have been an Iberian import. An early medieval zone of Atlantic exchange, brought to light from both archaeological and textual evidence, connected England to the continent. Germs may have been an unintended consequence of this exchange at the western edge of the medieval world.[87]

Plague Amplification in the West

Italy	*Gaul*	*Iberia*	*Britain*
543	543	543	543
571	571		
	582–84		
	588		
590	590		
599			
		609	
			664–66
680			
			684–87
		693	
		707–9	
745			

By rights, the entire first pandemic might be considered an event in the history of Syria. Syria was a hot zone of plague activity across the entire two centuries of the pandemic, from the first to the last. There is the risk that, in looking for plague in Syria, we are the drunk looking for his keys under the lamppost—because that's where the light is. The chronicle tradition of Syria is a rich resource. But the annals of Constantinople are an equally unbroken record of major events in the capital. The prominence of the plague in the Levant is not a mirage. Syria gained a reputation as a reservoir of plague in these centuries. The epigraphic evidence provides independent confirmation. And the ecology is plausible. Northern Syria repeatedly appears as a source of amplification. Christian settlements dotted the plains and the sloping hillsides in the arc stretching from the Orontes valley to upper Mesopotamia. Plague has often found hosts among high-altitude rodents in semiarid regions. The illuminating study of the Black Death in the Ottoman world by Nükhet Varlık has shown how plague focalized in rodent populations in precisely these regions. The dry uplands of eastern Anatolia could well have been ground zero for plague maintenance in the first pandemic.[88]

Plague Amplification in the East

Egypt	*Palestine*	*Syria*	*Mesopotamia*
543	543	543	543
		561–62	561–62
573–74		573–74	
	592	592	
		599–600	
	626–28		626–28
	638–39	638–39	638–39
			670–71
672–73	672–73		672–73
		687–89	687–89
689–90			
		698–700	698–700
		704–6	704–6
		713	
714–15			
		718–19	718–19
		725–26	725–26
		729	
732–35	732–35	732–35	732–35
743–49	743–49	743–49	743–49

Plague may have searched out hiding places in the east immediately. The first recurrence struck Cilicia, Syria, Mesopotamia, and Persia in AD 561–62. It is unclear if this amplification was an extension of the outbreak in Constantinople in AD 558 or an independent event. A chronicle records great mortality in Cilicia. The outbreak might have spread outward from the Taurus Mountains. Certainly, in AD 592 there was an amplification in the east that was not coordinated with an outbreak in the capital. The great event of AD 599–600 was synchronized across regions. But thereafter, the Levantine plague zone and the Byzantine Empire become disarticulated. Plague outbreaks recurred in Syria, often reaching Palestine and Mesopotamia. Two amplifications—in AD 626–8 (the "Plague of Shirawayh") and AD 638–9 (the "Plague of ʿAmwas")—are remembered in early Islamic sources. The latter was effectively the first Muslim encounter with bubonic plague. After an intermission of about a generation, plague recurred at an even higher frequency in Syria and Mesopotamia, down to the very end of the pandemic.[89]

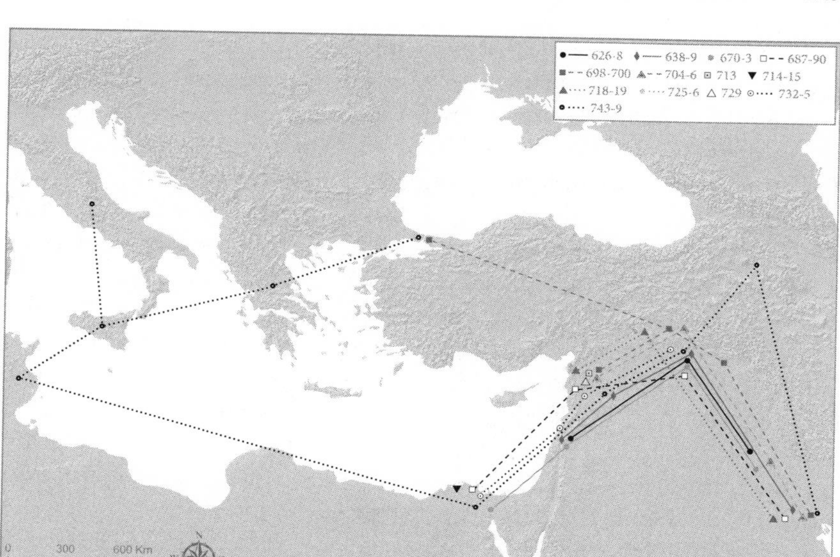

Map 23. Plague Amplification in the East, AD 620–750

Some recurrences were severe: the inscriptions of AD 592 refer to one-third of the universe dying. Others may have been more limited and local. Clearly, the connectivity of the Levant, and the fact that this was a zone of energy along political and cultural frontiers in late antiquity, escalated the effects of the plague's focalization close to the heartland of the Muslim world. The very frequency of the plague outbreaks may have dulled the severity of the later episodes. The recovery of rodent populations may have been partial and uneven. But the reality of plague was a lively background to the rise and fall of the Umayyad Caliphate.[90]

The first pandemic went out with a furious finale. The last amplification event, in the AD 740s, was wider in its geographic reach than any outbreak since the first visitation. It started in the caliphate and spread outward across its southern tentacles. But from Ifriqiya, it leapt north, possibly ferried by the slave ships running the sea lanes between Carthage, Sicily, and Italy where once the Roman grain fleet had sailed. The renewal of plague in Rome, for the first time in sixty-five years, was vicious. From there the plague raced back *eastward*, along the northern rim of the Mediterranean. It found Constantinople by AD 747. Death undid so many that again it was impossible to bury them all. The emperor had to repopulate the city by forced migration.[91]

This final outbreak of the first pandemic followed the contours of a new, medieval Mediterranean. The germ's itinerary is a measure of the distance traveled since the days of Justinian. By the middle of the eighth century, the medieval rebound had begun. A distinctly new order was dawning in the west, oriented around the Carolingian kings who would build a new empire—Christian in faith, Roman in name, but fully European in its genesis and dimensions. There was a strange and uneasy reconnection between the eastern and western Mediterranean. Biological history is not always tidy. But in this case there is a certain symbolism in the fact that the amplification of the AD 740s was the grand climax of the first pandemic. The plague was not destined to be a part of the new, medieval Mediterranean. It disappeared, for centuries, laying quiescent in the distant high country of central Asia.[92]

The offshoot of *Y. pestis* that immigrated to the Roman Empire in AD 541 and caused such wrenching devastation for two centuries was an evolutionary dead end. The pathogenic agent of the Justinianic Plague is an extinct branch of the species. Its disappearance is as mysterious as its arrival. Perhaps it is even more elusive. The hidden dynamics of rodent populations, and the overarching force of climate change, caused the plague to relent. It is probably important that the Late Antique Little Ice Age was giving way to the warmth of the high middle ages when the first pandemic ended, but we cannot say precisely how. The first era of plague ended as suddenly and unexpectedly as it had begun.[93]

Toward the End of the World

We struggle to comprehend biological events of this magnitude. The rise of *Y. pestis* was a landmark event in the history of the human species. Perhaps never before has humanity stared down an enemy so lethal and crafty. The two great plague pandemics that bookended the middle ages were, in relative terms, the most severe biological catastrophes in history. The violence of the initial wave reversed two centuries of demographic expansion in the blink of an eye. Then the persistence of plague for two centuries strangled hopes of recovery. If we imagine, for instance, a normal growth rate of 0.1 percent per annum leading into the first wave, 50 percent total mortality in an eastern Roman population of 30,000,000, and thereafter a combination of quick recovery rates (0.2 percent per annum) and smaller mortality events (10 percent mortality events every 15 years, which seems characteristic of the Constantinople phase of the pandemic), the power of the subsequent

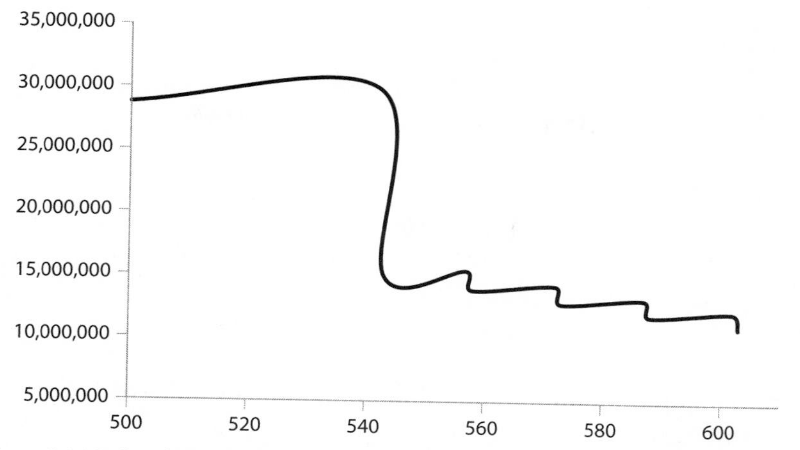

Figure 6.6. Notional Model of the Eastern Roman Population, ~AD 500–600

amplifications to maintain the population at low levels is apparent. It was as though the mass of the atmosphere had suddenly grown oppressively heavy, and human societies stooped beneath its invisible weight.[94]

But nature's caprice was not satisfied with the introduction of the deadliest germ it has ever conjured. If the shock of plague left Justinian's dream of reunifying the old empire stalled in quagmire and desolation, the final stages of the Roman Empire's dissolution did not represent the triumph of the bacterium alone. We cannot try to measure the plague's impact in isolation from the history of the climate. The fall of the Roman Empire was equally decided by the unwelcome arrival of a new climatic regime that is starting to be called the Late Antique Little Ice Age. The combination of plague and climate change sapped the strength of the empire. The unaccountable grief and fear left the survivors with the shuddering feeling that time itself was drawing to a close. "The end of the world is no longer just predicted, but is revealing itself."[95]

Judgment Day

The World of Gregory the Great

Pope Gregory the Great grew up in the world that Justinian had made. He was born in war-torn Rome, shortly after the city was first retaken by the army of Belisarius. The plague soon followed. Gregory's Rome was battered by generations of disease and war, but, at least in his formative years, the old capital was not bruised beyond recognition. The city was an imperial possession, and Gregory was an imperial man. He was one of the last faces of the old Roman nobility, the scion of a patrician line looking back to the ancient aristocracy. He still confidently moved across an imperial Mediterranean. His family owned estates scattered in the boom-lands of Sicily. His alliances with Africa were a reserve of strength. Gregory spent seven years in Constantinople, in the diplomatic service of the pope. His charge was to secure military aid from the emperor, Maurice. He failed. But his genteel piety impressed the ladies of the eastern capital, and he earned a hard sense of geopolitical reality that served him well in his own papacy. He also became the godfather of the emperor's son. Gregory was the last of a dying breed, but he was a remarkable specimen.[1]

Gregory has often been made to stand as a sentinel guarding the boundary between antiquity and the middle ages. In the course of his lifetime, the most recognizable features of the ancient landscape faded from view. Gregory saw the Roman senate, a corporate body that had proudly sat for more than a millennium, quietly dissolve. Already in his lifetime it was something of a phantom limb. We know this from Gregory's own letters, which reflect his efforts to sustain some semblance of public order by his own exertions. His career was not, however, a self-conscious attempt to fashion the medieval papacy. Gregory himself would have been incapable of such a conception. He operated within the mental framework of the Roman Empire, the "sacred republic." And, most importantly, Gregory lived in the full conviction that time itself was drawing to a close.[2]

Figure 7.1. Nurnberg Chronicle: Gregory the Great. (Fifteenth-Century Printing, University of Oklahoma History of Science Collections)

Gregory's eschatology is the thread that holds together the entire fabric of his thought and career. If we wish to understand his view of the world, we must appreciate his certainty that it was in its last hours. This sensibility was a direct response to his experience of the natural environment. Nature itself was writhing in anticipation of the end. Gregory's papacy had been born in a moment of dire natural emergency. Late in AD 589, torrential rains inundated Italy. The Adige flooded. The Tiber spilled its banks and crept higher than Rome's walls. Whole regions of the city were under water. Churches collapsed, and the papal grain stores were ruined. No one remembered a flood so overwhelming. Then followed the plague, in early AD 590. It came from the east and carried off Pope Pelagius II. The city turned to Gregory. Against the roiling backdrop of natural disaster, Gregory was installed on the throne of Peter.[3]

The incipient plague called forth an energetic liturgical response. Gregory instituted elaborate ritual processions—mournful prayer parades known as rogations—to stay the ravages of the pestilence. But if they seemed to work, the respite was brief. By AD 599, the west was swept again by pestilence from the east. "We suffer plagues without ceasing." The weary bishop could not stop the age from plunging toward its end. "I sigh longingly for the remedy of death. So much sickness of fevers has assaulted the clergy and people of this city that practically no free man, no slave remains who is good for any work or service. From the neighboring towns, the devastations of the mortality are announced to us every day. . . . People arriving from the East describe worse desolations still. By all these things, as the end of the world draws near, you know that the affliction is general."[4]

Gregory's eschatology was pressurized by the relentless and whimsical violence of the physical environment. Gregory felt he was in the presence of "novelties in the atmosphere, terrors in the sky, and storms out of their orderly seasons. . . ." We must be careful not to treat these as generic ravings. It is all too easy, from our lofty position, to brush past the naïve credulity of an ancient churchman and to dismiss his anxieties as so much stock in trade. It is sometimes suggested that, after all, plagues and earthquakes and storms were a constant presence in the ancient Mediterranean. But the natural archives urge us to pause and consider these fears with a little greater sympathy. The Plague of Justinian was the greatest mortality event up to that point in human history. The period was, in plain fact, rattled by an unusually violent spasm of earthquakes. And, the Late Antique Little Ice Age was a regime of climate as inhospitable to the political project of the Roman Empire as the climate optimum had once been favorable to the ventures of the great pope's distant ancestors. Gregory's life spanned an age of climatic deterioration on a par with anything in the late Holocene.[5]

The Late Antique Little Ice Age straddles the threshold between antiquity and the onset of the middle ages. It was an environmental event of the first order. Its origins lay far outside the sphere of human influence, but its human ramifications were immense, and they were inseparable from the consequences of the first pandemic. Together, climate change and disease exhausted the remnants of the Roman imperial order. The demographic consequences were primary. Gregory's Rome may have been home to as few as 10–20,000 souls huddled inside its walls; they would barely have filled a corner of the Colosseum. Across most of the old Roman world, ancient landscapes of settlement shriveled up. The state was deprived of metabolic energy, and painful atrophy set in.

Exactly one century lies between the capture of Rome by Belisarius and the retreat of the empire's armies behind the lightning advance of the Islamic conquests. Over that span of time, the Roman state exerted itself, with all its might, against the inexorable pull of the tides. It refused to go quietly into the deep. We do no disservice to human agency by trying to understand the currents that overpowered those who lived through the chaotic final events of antiquity. In fact, we can respect their experience all the more by trying to understand why they believed they lived on the very edges of time. For it was the eschatological mindset that, far from leaving these last generations passive against the stream of events, inspired their most surprising and enduring acts. The sense of impending doom was not a weight around the neck; it was more like a hidden map, a way of orienting motion in confused times. For the first time in history, an apocalyptic mood came to permeate a large, complex society. Gregory's sense of the approaching end was hardly his alone. The apocalyptic key transcended traditions, languages, and political boundaries in late antiquity. By listening carefully for it, we can draw the seemingly disparate parts of the late ancient world closer together, and at the same time restore a little warm blood to the final scenes of antiquity.[6]

Each of the great environmental convulsions in the Roman Empire provoked unpredictable spiritual reverberations. The Antonine Plague turned the imagination toward the possibilities of an archaic and increasingly universal Apollo cult. The Plague of Cyprian cracked the foundations of ancient civic polytheism and allowed Christianity to creep into the open. In the sixth and seventh centuries, the concatenation of plague and climate deterioration spawned an age of eschatology, within Christianity, Judaism, and that last offspring of late antiquity, Islam. The precise conjuncture of environmental damage, political disintegration, and religious ferment decided the final sequence of Rome's demise. In the seventh century, the most vital remnants of the empire were swallowed from the periphery, by a rising power that was neither fully within nor fully beyond the orbit of the classical Mediterranean. Materially and imaginatively, the ascent of Islam would have been inconceivable without the upheavals of nature.

This was the end of the world.

The Ice Age Cometh

In the intellectual life of the sixth century, two anciently contrasting views of nature faced one another with renewed and pointed intensity. One conception looked to nature as a model of order and regularity. Its undeviating perfection was a source of moral reason, and the best that humans could do was to live in tune with the harmonies of the cosmos. This benevolent view was given an elaborate metaphysics in Neo-Platonic philosophy, and among the mandarins who staffed the imperial bureaucracy it became a practical ideology. The empire they managed was a mirror of the orderly cosmos. The diametrically opposite view of nature held that the physical world was a source of flux, variety, and violence. This opinion had no firmer adherent than the emperor Justinian himself. In his eyes, nature was antagonistic, red in tooth and claw. This quarrel was no armchair disputation. It was a debate over how to rule the empire: through reason or will, tradition or reform. And it was a contrast of outlooks given special urgency by the blatantly unsettling signs of disturbance in the natural environment.[7]

In the age of Justinian, the fluctuations of nature created a period of Holocene climate history now being called the Late Antique Little Ice Age. It was the product of a dramatic convergence. In the Late Antique Little Ice Age, climate changes operative on multiple timescales conspired to spin up one of the most distinct phases of climate history in the last several millennia.[8]

The late Holocene was a time of cooling. From the warm peaks of the early Holocene until the recent onset of anthropogenic warming, the huge influence of orbital mechanics drove a gradual, millennial-scale cooling of the planet. But on its course to a colder earth, the climate has rocked and swayed, as the long slide was periodically stalled or reversed by warmer epochs such as the Roman Climate Optimum. The Holocene has also been punctuated by episodes of abrupt cooling, such as the famous Little Ice Age centered on the seventeenth century. The Late Antique Little Ice Age was one of these cooling episodes, when the forces impelling the Holocene toward its deeper destiny gathered momentum. If the RCO looked backward, to the mid-Holocene, the Late Antique Little Ice Age looked forward, to the next glaciation.

The RCO drew to a close around AD 150 and was followed by three centuries of instability and disorganization. From ca. AD 300 to 450, the most distinctive large-scale feature of the climate was the positive phase of the North Atlantic Oscillation. We have already observed signs of sharp drying in the lower mid-latitudes, a belt of aridity running from Spain to central Asia. From the AD 450s, this coherency broke, and the global climate regime showed signs of uncertain reorganization. Most notably, the phase of the North Atlantic Oscillation flipped. From the latter half of the fifth century, it was persistently negative, displacing winter storm tracks to the south. In Sicily, an age of much greater humidity began around AD 450. Across much of Anatolia an unmistakable switch from aridity to humidity progressed quickly. There were not yet signs of the epochal cooling that would soon impose itself, but it is important that the climate was in transition already before the great events on the horizon. We might consider the period ca. AD 450–530 the prelude to the Late Antique Little Ice Age.[9]

The subtle drift of the climate was then seized by planetary events. Strange stirrings in the sky have long been known from ancient accounts. In AD 536, contemporaries across the globe were awed by the "year without summer." Procopius, in Italy on campaign with Belisarius, described the "dread portent" of solar dimming. "During the whole year the sun gave forth its light without brightness, like the moon, and it seemed extremely like the sun in eclipse, for the beams it emitted were not clear like those it usually makes. From the time when this thing happened men were free neither from war nor pestilence nor any other thing that brings death." John of Ephesus offers parallel testimony from the east. "The sun darkened and stayed covered with darkness a year and a half, that is eighteen months. Although rays were visible around it for two or three hours (a day), they were as if diseased, with the result that fruits did not reach full ripeness. All the wine had the taste of reject grapes." Another, precise chronology associated the ominous disturbance with the visit of Pope Agapetus (an ancestor of Gregory the Great) in Constantinople—from March 24 of 536 to June 24 of 537.[10]

The disappearance of the sun was a disturbing omen under any circumstances; it also chanced to touch on some of the most sensitive ideological fault lines in contemporary Constantinople. For a man like the disgruntled career bureaucrat named John Lydus, it was more than a curious anomaly. It was a potential fissure in his worldview. In his treatise *On Portents*, he made a valiant effort at a naturalist explanation. He put down the solar anomaly to tractable physical causes in the atmosphere. "The sun becomes dim because the air is dense from rising moisture—as happened in the course of the recently passed fourteenth indiction [535–36] for nearly a whole year." It is a passable attempt to save appearances and rescue the regularity of nature.[11]

The most detailed report from the Year without Summer was by the hand of an Italian statesman named Cassiodorus. It survives in one of the last letters included in his collection of public documents known as the *Variae*. In 536, Cassiodorus was Praetorian Prefect in Italy under the Ostrogothic king. But, crucially, by the time he compiled his *Variae*, events had carried Cassiodorus to Constantinople. Thanks to an incisive study by Shane Bjornle, we have been taught that the *Variae* are anything but a neutral transcript of Cassiodorus' time in the civil service. They form, rather, a subtle polemical document, calculated to impress precisely such men as John Lydus or Procopius, the learned dissidents within the Justinianic regime. Among the bureaucratic corps in Constantinople, with their often subversive Neo-Platonist sympathies, the cosmos was the image of unchanging perfection and a source of moral order. Justinian was a monstrous religious fanatic, who had just survived an attempted coup by sheer butchery. Cassiodorus was finely attuned to these sensibilities, and his polished report of the solar dimming belongs to the strained political dialogue in the capital.[12]

"Nothing is done without a reason, nor is the world involved in fortuitous happenings," Cassiodorus wrote. Willful departures from tradition were painful enough. "Men are anxious [literally, tortured] when kings change their established customs, if they go forth in a guise that is other than what tradition has long implanted." We should imagine that Justinian is the real object of these barbed comments. "But who will not be disturbed, and filled with religious dread by such events, if something dark and contrary to custom seems to come from the stars? How strange it is, I ask you, to see the sun but not its usual brightness; to gaze on the moon, glory of the night, at its full, but shorn of its natural splendor? We are all still observing a sun as blue as the sea. We marvel at bodies that cast no shadow at mid-day and at the force of strongest heat reduced all the way to the impotence of extreme mildness. And this is not the brief absence of an eclipse but as one that has taken place for nothing short of almost the whole year. . . . We have had a winter without storms, a spring without mildness, a summer without heat."

Crop failure in Italy followed. But as Praetorian Prefect, Cassiodorus prudently ordered his deputy to relieve the shortage from the previous year's bountiful harvest. In the letter, he then returned to the philosophical problem of the sun's disappearance, and in a lengthy excursus offered a purely scientific explanation: a cold winter had created a lingering dense air, filling the vast space between the earth and the heavens, obscuring the sun. "What seems mysterious to the stupefied masses should be reasonable to you."[13]

Here was a virtuoso rhetorical performance, presenting a conservative image of wise and steady governance in the face of nature's predictable variability, and laced with subtle criticisms of Justinian. The polemical context only heightens the value of this testimony and assures us that the solar dimming had deeply rattled contemporaries. The year without summer reverberated worldwide. Irish annals testify to famines. Chinese chronicles report the disappearance of Canopus, the second-brightest star in the night sky, and snows in Shandong—on the same latitude as Sicily—in July. The event was unnerving, on a global scale.[14]

This impressive array of testimony was left hiding in plain sight until 1983. It took two NASA scientists to turn attention to the Year without

Summer, by drawing a connection between the literary sources and the physical evidence for volcanic activity in ice core records. Their intuition was pointed in the right direction. But the written evidence does not require an event of volcanic origin, and small but nagging inconsistencies in the ice core dates made definitive answers elusive. Ice cores don't come with time stamps, and it is an achievement to calibrate the age depth of ice layers. In the midst of uncertainty, other theories were canvassed, including asteroid impact. With the natural evidence inconclusive, the first detailed analysis of the written sources finally appeared, in 2005, and a rather minimalist hypothesis was put forward: maybe a local volcanic explosion. Questions lingered.[15]

The decisive breakthrough came from the dendrochronologist Michael Baillie, who insisted, on the basis of tree-ring evidence, that the ice core dates needed to be recalibrated. The accumulation of new cores and the ongoing refinement of the record has proven him right, and the paleoclimate community has achieved a remarkably satisfying alignment in the physical proxy record. There is now little doubt about the timing or magnitude of the events that so unsettled contemporaries: a cluster of volcanic eruptions that rivals anything in the Holocene. The AD 530s and 540s stand out against the entire late Holocene as a moment of unparalleled volcanic violence.[16]

Sometime in early AD 536, there was a massive volcanic eruption in the Northern Hemisphere, ejecting megatons of sulfate aerosols into the stratosphere. The precise identity of the volcano is so far unknown, but the effects were visible at Constantinople by late March. It remains not impossible that a meteorite impact at just this time contributed to the chaos, too. But the proxy evidence has clarified that there was a second, even more cataclysmic, explosion in AD 539 or 540. The second eruption was a tropical event that has left its traces at both poles. Twice in the space of four years, the earth belched historically massive clouds of sulfates into the stratosphere, blocking the intake of energy from the sun.[17]

If we only had the ice core evidence, we would observe a sequence of impressive volcanic eruptions. But the trees insist on the truly dramatic ramifications of these events. In series from across the Northern Hemisphere, AD 536 was the coldest year of the last two millennia. Average summer temperatures in Europe fell instantly by up to 2.5°, a truly staggering drop. In the aftermath of the eruption in AD 539–40, temperatures plunged worldwide. In Europe, average summer temperatures fell again by up to 2.7°. In proxies around the globe, the 530s and 540s stand out as cruelly frigid times. The decade 536–545 was the coldest decade of the last 2000 years. It was colder than the deepest trough of the Little Ice Age. In fact, the severity exceeds what might be expected of volcanic forcing alone. Somehow, the background conditions of the climate or the synergistic effects of the clustering rendered the impact of this volcanic outburst even more than the sum of its parts. The Late Antique Little Ice Age had arrived.[18]

The consequences were not immediately overwhelming. The harvests failed, but mercifully the previous year had been abundant and the inherent resilience of Mediterranean societies buffered them from instant famine. If there was an immediate effect of the sharp climate anomaly, it might be the hidden ecological trigger that led the plague bacterium to disperse in the years just following the spasm of volcanic activity. Whether the ice stirred contemporary human migrations in central Asia is unclear: drought events are more consequential than temperature anomalies. In sum, the cool years in the 530s and 540s did not elicit immediate social collapse or state failure in the Roman world. Rather, these harsh years quietly added stress to an imperial order already stretched by massive warfare and imminently to become the victim of *Y. pestis*.

The cooling in the 530s and 540s might have been sharp but transient. Instead the volcanic furor was overlaid by a longer and deeper decline in solar output. The sun's inconstant dynamo plummeted toward lower levels of energy output. Following a modest peak of solar activity around AD 500, a steep decline set in, reaching a low in the late seventh century. The beryllium isotope record measures solar energy output, *independent* of volcanic blocking. This tells us that at precisely the same moment when volcanoes layered the stratosphere with reflective aerosols, the sun began to eject less heat toward earth.[19]

The decline in solar output was deeper and more enduring than the volcanic forcing. A grand solar minimum, centered in the late seventh century, was the greatest plunge in energy received from the sun during the last 2,000 years. It was lower even than the famous Maunder minimum of the seventeenth century. One fitting measure of the profoundly colder times is found in the advance of Alpine glaciers. The glaciers swept down the mountain valleys. In the early seventh century, Alpine glaciers reached their first millennium maximum. The sun's diminishing output ensured that the cold spell was not a momentary shock, but an enduring background to the final scenes of the ancient world. The conjunction of natural variability, volcanic activity, and diminishing solar irradiance made the Late Antique Little Ice Age a distinct phase of Holocene climate.[20]

The coldest period stretched across a century and a half, from the middle of the 530s to the 680s. But even a global climate organization as sharply

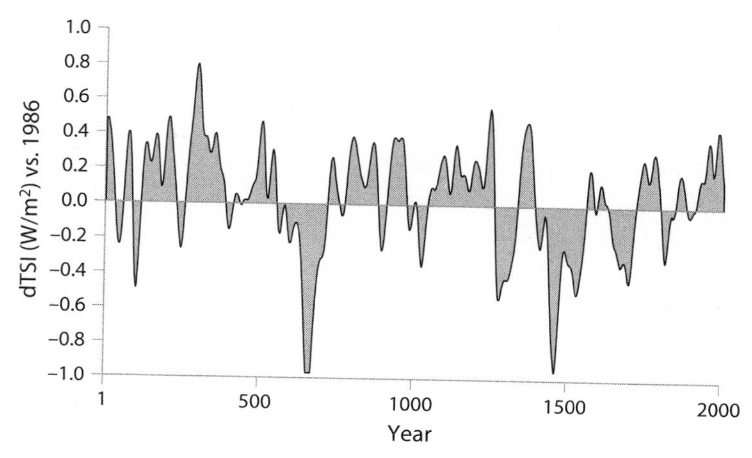

Figure 7.2. Change in Total Solar Irradiance v. 1986 (data from Steinhilber et al. 2009)

pronounced as the Late Antique Little Ice Age varied locally in its impacts. While temperature changes tend to be spatially coherent—it was colder nearly everywhere—moisture regimes are sensitive to regional and local climate mechanisms. The turn toward a more negative regime in the North Atlantic Oscillation index, which commenced before the triggers of volcanic forcing and lower insolation, was continued and perhaps even accentuated in the strongest period of the Late Antique Little Ice Age. The storm tracks pointed south, across southern Europe. In the Late Antique Little Ice Age, globally colder temperatures overlaid a phase of low pressure gradients in the North Atlantic, with intricate consequences across the northern hemisphere.[21]

Here the human and natural archives speak to one another. Gregory the Great's experience of the climate becomes less abstract. In Sicily, where the last remnants of the old land-holding order in the west clung to their habits of interregional property ownership, there was nothing short of an agricultural boom. The abundant rains brought renewed prosperity to the wheat economy for the last grandees of the Roman order. At the same time, the climate regime threatened to deliver waters in excess. The frequent flooding of sixth-century Italy is one sign. The destructive *winter* inundations across so much of Italy in AD 589 were an abrupt intensification of the climate regime that regularly steered precipitation over the Mediterranean.[22]

In Anatolia, the turns and subtleties of the Late Antique Little Ice Age are traced across the ecologically diverse subcontinent. In most regions the period of positive NAO ca. AD 300–450 had brought aridity. But in the course of the fifth century, the dry days of the past vanished. Winters were more intense, with heavier snows in the uplands. Flooding became a major preoccupation, across Anatolia and into northern Mesopotamia. Justinian redressed flood control from the western plains of Bithynia to the eastern foothills of the Taurus Mountains. Places like Edessa and Dara were ravaged by floods. Tarsus in Cilicia, the birthplace of Saint Paul, was overrun by snow melt and spring rains. The Cydnus river "wiped out completely all the suburbs . . . then it went roaring against the city itself, and tearing out the bridges, which were small, it covered all the market-places, flooded the streets, and wrought havoc by entering houses and rising even to their upper storeys." The wet cycle was a boon for wheat production in Anatolia, but the age of frost spelled trouble for the sensitive olive tree. Pollen records show that this quintessential Mediterranean plant was forced to retreat from everywhere except the lowlands and coasts, further back than at any other time since its arrival on these shores.[23]

In the south the story of the Late Antique Little Ice Age is more obscure and ambiguous. The march of aridification continued across North Africa, but its chronology is imprecise. The natural and human roles are hard to disentangle. To the south of the empire, groundwater tables dropped inexorably in the Saharan soils. The Garamantes in the Fezzan resorted to ever more desperate bids to retrieve water from the ground. The scaling-up of conflict between Romans and "Moors" from the later fifth century may represent the arrival of new peoples, fleeing from the dry south into the greener climes of North Africa.[24]

Inside Mediterranean Africa, changes in the water balance may have tilted the fate of societies. Archaeology testifies to a time of troubles in the later fifth and early sixth centuries, at a rhythm that does not easily align with the Vandal invasion or the Byzantine wars. Procopius reported the stark cumulative effects of climate change in North Africa. Ptolemais, a city in Cyrenaica, "in ancient times had been prosperous and populous, but as time went on it had come to be almost deserted owing to extreme scarcity of water." We may suspect his motives, in a passage whose purpose was to extol Justinian's water works. Further east, the great city of Lepcis Magna, home of Septimius Severus, "which in ancient times was large and populous," had been deserted and "largely buried in sand." Here, though, the most that could be said was that Justinian rebuilt a wall and several churches. Even in the most favorable light, the city does not cut an impressive figure. The dunes had irreversibly taken over the once proud outpost of civilization.[25]

In the Levant, the history of water is invested with all kinds of significance. The fractious history of the region has been made to lay a little heavily on the fundamentals of climate. The boundaries between humid rain-fed settlement and sparse dry desert are politically charged. And late antiquity occupies a special place in the climate history of the region, not least because of the enormous cultural realignments of the seventh century. Syria and Palestine were the heartland of the late antique east. They were an endlessly fecund source of religious energy and an economic engine. Settled agriculture was the source of tremendous wealth, and it crept outward further than ever before. But at some point the desert made a land-grab. The "dead villages" of Syria and the once-fertile wine country of Gaza were put beyond the reach of even irrigated agriculture. They stand as eloquent if haunting testimony to change. But the chronology and causes remain contested.[26]

We would do well to tread carefully and to pull apart the issue. It is worth underscoring the sheerness of the north-south gradient between the latitudes ~30°N and 40°N. These lines enclose the Near East, and nowhere else is every step from equator to pole so consequential. In late antiquity, we observe that the precipitation regimes in Anatolia and the Levant, far from moving in tandem, reveal an anticorrelation. While Anatolia was arid, Palestine was humid (i.e., ~AD 300–450/500). When Palestine started to turn more arid (~AD 500), Anatolia was soggy with rain. The divergences may have been driven by an upper atmospheric teleconnection that has been called the North Sea–Caspian Pattern. In winter, high-level pressure differences decide how air circulates over the eastern Mediterranean. When air is pushed from northeast to southwest, Israel is relatively wet; when air is pushed southwest to northeast, Israel is dry but Turkey wet. The dominant air flow may have flickered in late antiquity and shifted around AD 500. Regardless, it is worth remembering that the fates of all eastern Mediterranean societies are not bound together.[27]

If we hew to the evidence for climate, and defer momentarily the human responses to the environment, the natural proxies suggest that more arid conditions arrived in the Levant at some point between ~AD 500 and 600. We should not exclude the value of human testimony in offering refinement, even if human reports are subjective. In the early sixth century, a writer known as Procopius of Gaza described a scorching drought at Elousa in Palestine. The sands were scattered by the wind and the vines blown naked to their roots; the springs had become dry and salty, and Zeus no longer sent rain. The emperor Anastasius (r. AD 491–518) made significant and celebrated repairs to the aqueducts in Jerusalem. A bracing,

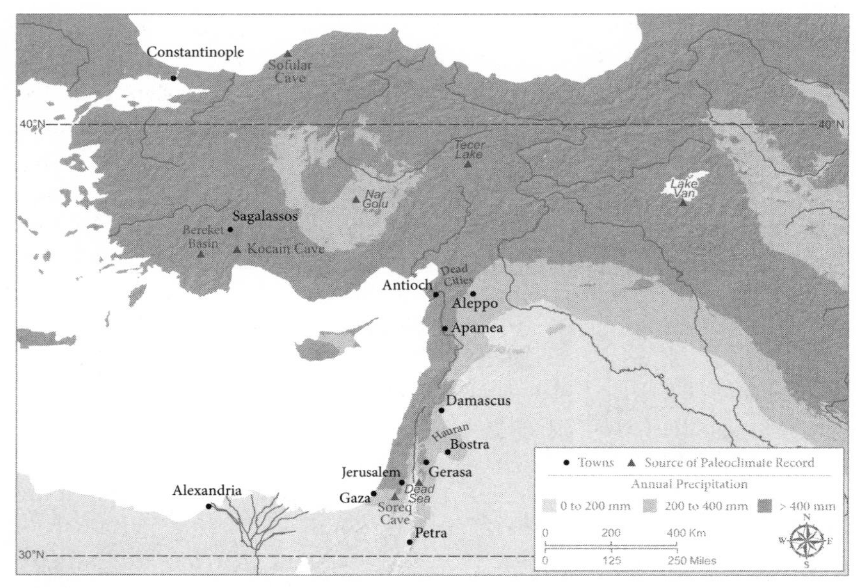

Map 24. The Late Roman Near East

four-year drought in Palestine commenced in AD 517. A Syriac chronicle, possibly referring to the same drought, measured it at fifteen years, and claimed that the Pool of Siloam in Jerusalem (where Jesus once sent a blind man to heal) dried up. Later in the century, a visiting saint found Jerusalem in the midst of a great drought that dried all the cisterns in the city. It is telling that overwhelming floods color the backdrop of literary scenes in sixth-century Anatolia, just when the stories of Palestine are full of baleful droughts. And yet, the slow desiccation of the region seems not to have withered the progress of civilization instantly; rather, the tension between human artifice and nature built up, to be released, suddenly, at a later time.[28]

The advent of the Late Antique Little Ice Age puts the construction works of Justinian in a somewhat different light. Justinian built cisterns and aqueducts, granaries and transport depots; he moved riverbeds and reclaimed floodplains. The outburst of environmental engineering was not an exercise of vain ambition. He applied the muscle of the state to the task of trying to control the flux of nature—at a moment of enormous flux. Justinian "joined forests and glens to each other" and "fastened the sea to the mountain." But even in his praise of Justinian's building program, the historian Procopius subtly compared the emperor to the ancient Persian monarch, Xerxes. It was not meant as a flattering parallel. The hubris of Xerxes led him to believe he could dominate nature like a docile subject. Justinian was to learn that nature was not easily subdued.[29]

Justinian's opponents erred in believing that the natural order is full of predictable harmony and regularity. Justinian's belief that nature is full of violent and ceaseless flux was closer to the truth. But the emperor's intellectual victory did little to steel his empire against the overwhelming power of the changing climate.

Final Trajectories: Zones of Decay, Zones of Energy

John the Almsgiver was born on the island of Cyprus, sometime toward the middle of the reign of Justinian. He was married and had a "bountiful crop" of children. They all died unseasonable deaths, "in the flower of their age," and John retreated to the religious life. He discovered a knack for church politics, and by AD 606 he was the patriarch in Alexandria. He would spend an eventful decade in that office. It is clear from his colorful biography that Alexandria remained, even at this late date, a center of robust commercial and cultural vitality. The trade networks of the eastern Mediterranean were vibrantly alive. The accouterments of the classical city continued to demarcate the urban landscape. Even in the early seventh century, Alexandria beamed against the dimming backdrop of the late classical world.[30]

Walking through the streets of John's Alexandria, we feel we have entered a time warp. That may be the calculated intent of his biographers. They lived through the next chapters of history. But they have not completely erased the subtle marks of change already in John's background. Reading the accounts of his life, we are rightly struck by the eager involvement of the church in the networks of sea-borne trade. When a famine closed in on the city, it was John who relieved it by dispatching "two of the Church's fast-sailing ships" to Sicily for grain. (Needing to import wheat to Egypt is almost the precise equivalent of Newcastle sending for coal.) Ship-captains and sailors—including those in the employ of the church—crowd the foreground of John's biography. The church owned a fleet of thirteen large ships, a detail we learn when they were forced to jettison their heavy cargo in a storm on the Adriatic—a cargo of grain, silver, and textiles. The famous almsgiving of John was underwritten by at least some measure of audacious church capitalism.[31]

John's world was already a receding circle of light in the gathering darkness. Alexandria and its fleet may have been the very last holdouts of the old order on the Roman seas. At the start of the seventh century, ceramic pottery still arrived from North Africa, Asia Minor, and Cyprus. The city was a hub of Mediterranean commerce. But by century's end, these last connections had been cut off, and the city relied on the Egyptian hinterland for its sharply truncated needs. John himself lived to see one of the decisive moments of collapse as the late Roman world folded in upon itself. As the Persians swept toward the city, in AD 616, he himself sailed back to his native island, where he died. In AD 618, the state-sponsored grain shipments to Constantinople ended forever. The spine of imperial connectivity was snapped.[32]

Historical change is neither sudden nor tidy. The twin catastrophes of plague and ice age did not collapse the Roman Empire in a clean blow. They did not even topple the regime of Justinian, who kept his grip on the levers of state to the bitter end of his life. But environmental degradation sapped the vitality of the empire. In the long run, the forces of dissolution prevailed. Sometime in the years spanned by John the Almsgiver's life, in the second half of the sixth century and the first years of the seventh, the empire crossed a tipping point. Different regions of the empire responded to the shocks of mortality and climate change at their own rhythm. Some wilted without delay, others withstood the winds of change for a time. Because the imperial system itself was a network, a connected system of vastly different ecological and economic territories, it could draw on the remaining zones of energy. Like a towering oak drawing its last nourishment from a decaying root system, the empire died from the inside, slowly. Only then was it felled by a swift blow, from without.

All too often, the historical change that counts is also silent. The pulses of demographic movement that determined the fate of empires were unheard beneath the din of battle. It is no wonder that students of antiquity have so often turned to archaeology to retrieve the past from the chill sentence of silence. The spadework of the archaeologist can trace the networks of trade that tied the Roman Mediterranean together. It can uncover the shifting landscapes of settlement and the biographies of cities whose rise and fall are the arc of civilization. The impact of environmental change must be sought in the intricate patterns traced by the archaeology of trade, settlement, and urbanism. The fall of the Roman Empire was, in most regions, a profound transformation in the basic circumstances of life. What must be sought in the archaeological record is raw evidence for people and prosperity, as well as more qualitative indices of complexity. People never disappeared from the old territories of the Roman Empire, but their ways of life were simplified and localized. Signs of that drama appear in the decline of towns and the recession of trade from one end of the old empire to the other.[33]

In the farthest west, the free fall was most undisguised. When Pope Gregory the Great, in light of the coming judgment, urgently dispatched missionaries to convert the pagans of the British Isles, his ministers found a land that the Romans would hardly have recognized. A fourth-century landscape dotted with Roman towns and prosperous farmsteads had been brutally erased. By the end of the fifth century, "there were no towns, no villas and no coins." Peasants of the Roman countryside had dined off industrially produced table ware; now even those of privileged station returned to the days of hand-thrown pottery. We should not underestimate a regression so basic; it would be as though we gave up refrigerators and returned to ice boxes. In many ways, the lifestyle of the early medieval elites compared poorly to middling persons of the late Roman Empire. The towns became shadows of their old selves. Britain was a backwater but never totally cut off: it is telling that Gregory's letters show him reacting to an emergent slave trade, carrying westerners to the markets of the wealthy east.[34]

In the Iberian peninsula, the Roman order did not give way easily, even in the face of Visigothic dominance. The settlement landscape of the fourth century was dominated by towns and villas, built by an aristocracy whose wealth derived from commercialized agriculture. The archaeology of the fifth and sixth centuries reveals, above all, fragmentation. It was the triumph of heterogeneity. New construction abated in town and countryside in the fifth century, but the towns and villas that existed remained operational. There are signs of demographic recession, especially in coastal Spain. Imported ceramics slowly disappear from the landscape. "The Mediterranean coast of Spain became increasingly marginalized from ca. 550 onwards, when politics focused on inland centers (Toledo, then Córdoba), and seaborne commerce steadily weakened." Towns did not disappear overnight, but from about AD 600, most of the major cities that still existed entered a terminal decline. In Spain, the disarticulation of the Roman order proceeded steadily across the later fifth and sixth centuries, and its later stages—around AD 550 and 600—may have been hastened by the sudden onslaught of the bubonic plague.[35]

In Gaul, the post-Roman world was bisected by a north-south divide falling along the line of the Loire River. In the north, the Roman order was rapidly transformed. The old fabric was torn apart. Coins nearly disappeared from the economy for a few generations in the later fifth and early sixth centuries. In the south, by contrast, life still revolved around the Mediterranean. The urban fabric held into the sixth century; villas remained inhabited, even if no new ones were being built; and eastern traders and eastern wares reached the shores of Gaul. Then, in the middle of the sixth century, the first wave of plague swept from the Mediterranean to the Atlantic. Some of the last bastions of Roman urbanism, such as Arles, vanished completely. Marseilles maintained a shadowy existence, the last outpost of connectivity. The repeated visitations of the plague may have affected the south of Gaul, even as the isolation of the north insulated it from later outbreaks. In the Frankish north, the seeds of a medieval order germinated. It was here that a new civilization started to grow, one not haunted by the incubus of plague.[36]

In Italy, the future was still indeterminate when the troops of Belisarius set sail for the campaign of reconquest. Already, urban markets shrank, and the villa economy slumped toward ruin. At different tempos, the cities became smaller, the church more prominent, the old monuments repurposed, and public space turned private; fortifications went up, often enclosing only parts of the old cities; towns were ruralized, as animals pastured in their streets. But ca. AD 500, the peninsula still presented a basically Roman face. The money economy prevailed. Ceramics from around the Mediterranean made their way not just to the old capital but to towns across the peninsula. The settlement hierarchy remained organized around the dispersed, lowland grid of villas and farmsteads. In the south, especially, life went on. The old order had not been overturned.[37]

In the decades of Ostrogothic dominion, a period of cautious optimism reigned in Italy. The dossier of the minister Cassiodorus reveals an intent to restore the prosperity of Italy, along ancient lines. "Our care is for the whole republic, in which, by the power of God, we are striving to bring back all things to their former state." Repairs to aqueducts, roads, and other elements of the public infrastructure were afoot. The Colosseum was spruced up, and games were still celebrated in the AD 520s. But in AD 536 came the forces of the eastern Romans, and in AD 543 came their germs. The combination of war, plague, and climate change proved overwhelming. The mid-sixth century represents a sharp turning point for most of Italy. A hesitant recovery was strangled in the cradle. The break is visible in both town and country. Most towns suffered a fate somewhere between hollowing out and utter annihilation. Rome is only the most famous and dramatic instance of the urban death spiral. Procopius claimed that by AD 547 there were only 500 people in the city: the number may not be entirely credible, but the point is made. The Colosseum fell silent. It was reclaimed for church use, already a bread dispensary in the days of Gregory the Great. By the end of the sixth century, the ancient practice of inscribing on stone comes to a whimpering end.[38]

The climate changes of the sixth century reversed centuries of human toil in Italy. The precocious cities and neat fields had been carved into nature, carefully harnessing its fickle powers. But depopulation and the withering power of the state undermined the control systems upon which the miracle of civilization depended. In the sixth century, a vicious circle exerted itself. Harsher environmental conditions—a colder and wetter climate—choked the demographic recovery, while the shortage of manpower put societies at a starker disadvantage against the natural environment. The floods we meet in the chronicles represent not so much the raw power of nature, as the untimely conjunction of environmental stress and social incapacitation. The terraces washed out. Ports silted up. Alluviation overran the valleys where the Romans had planted their farms and fields. The wild reasserted itself, as marshland and forest crept up on plough-lands that had been cultivated for centuries.[39]

Even if we imagine that the Justinianic Plague killed half the populace, there were still humans sprinkled in the landscape. But the truth is that in some parts of the empire they become uncannily hard to find. People eerily retreat from the material record in Italy. "The villages and farms which for a thousand years had underpinned a considerable level of civilization seem mostly to have gone." "In the seventh and eighth centuries, it is very difficult from field-survey and even from excavation to find any trace of settlement at all." "People are so much harder to recognize after c. AD 550." Making demographic measurements from the finds of survey archaeology is a notoriously risky enterprise, but one bold scholar hazards that the population of Italy was reduced to a half or quarter of its Roman levels.[40]

What happened in Italy was not mere decline; it was collapse and reorganization. Coins, once ubiquitous, vanished except from a handful of Byzantine outposts. The humble household products from overseas first receded, then disappeared. The vaulting hierarchies of the Roman social structure were involuted, leaving a drastically simplified binarism of haves and have-nots. The great wealth of the aristocracy evaporated, the middling element could no longer regenerate, and the Christian church found itself, unexpectedly, the richest inheritor in a less prosperous world. An entirely new logic of settlement imposed itself on the landscape, as the fertile lowlands—exposed to environmental stress and barbarian pillagers—were abandoned for the retreat of the hilltop village. As Brian Ward-Perkins has observed, Italy was sent reeling backwards, to levels of technology and material culture that had not been seen since before the Etruscans. The alliance of war, plague, and climate change conspired to reverse a millennium of material advance and turn Italy into an early medieval backwater, more important for the bones of its saints, than its economic or political prowess.[41]

North Africa sat between the sharp decline of the west and the continued vibrancy of the east. The Vandal conquest was not a great caesura. In many parts of Roman Africa, the fourth and fifth centuries witnessed a peak in settlement. In the eastern stretches, across Libya, this vibrancy was interrupted already in the fifth century. The tidewall of Roman civilization broke down, and new peoples encroached from the Sahara into the fringes of Roman settlement. But in the central axis of Tunisia, prosperity endured. African Red Slip Ware retained a huge market share around the Mediterranean. Carthage was a hub connecting the fertile hinterland to the wider world, and it prospered into the sixth century. But from the later sixth century, there was manifest recession all across the African heartland. The disarticulation of the seaborne commercial network is thought to have stalled the circulation of wealth into the African provinces, but plague, too, should be compassed as a possible culprit in an obviously wrenching population crisis. Again, here, the multifaceted and long-term dissolution of an old system was hastened by the blows of mortality crisis and climate change.[42]

We have learned to appreciate that the timing of change in the eastern Mediterranean was on a wholly different schedule. The territories connected to Constantinople thrived in late antiquity. Never before had the societies ringing the sea in a great arc from the northern Aegean to the shores of Egypt been drawn so close together or seen such broad prosperity. The only dead tissue in the empire Justinian first inherited was in fact the strip of Danubian provinces whence he originated. Battered repeatedly by invasion, the northern march struggled to recover its former economic vitality. Justinian made an aggressive push to protect the land of his fathers, unloading massive sums. But these expensive projects could not reverse the tide; the reconstructed towns ended up as grandiose *Fliehburgen*, little more than giant bunkers used by country people in times of emergency. The shock of the bubonic plague made these territories easy targets for infiltration by Slavs and Avars. Over the course of the second half of the sixth century, they slipped, little by little, from Roman control.[43]

To the south, in the core of Greece, we meet a world in the throes of roaring growth. Ancient cities flourished anew, with "unbroken continuity to 550 at least (with a Justinianic high point)." Spectacular churches went up in the fifth and sixth centuries. The countryside witnessed an explosion of settlement. Trade brought goods from far-off lands deep into the inland, mountainous folds. But in the middle of the sixth century, this efflorescence came to a screeching halt. The upswing was violently reversed. On the western edges of the Greek world, the city of Butrint, one of the most carefully excavated cities in the Mediterranean, shows steep decline after ~AD 550. Corinth was in decline before AD 600. Urban retreat moved in step with rural decay. In Macedonia, there was "a profound but 'silent' revolution during the latter part of Justinian's reign. A previously dynamic settlement-system characterised by a degree of monetisation, by professionally decorated church-building, and a degree of hierarchisation, lost all these features." In the south, the period after the mid-sixth century has been described as one of "utter desolation." In fact, so brutal and so complete was the collapse that it "has led to considerable hand-wringing on the part of scholars wondering where all the people have gone."[44]

The Greek case has great diagnostic value. The cities and valleys of Greece have been carefully combed. And here, attritional warfare and political turnover are not easy explanations. The corrosion reached remote corners of the Greek peninsula. The point of inflection in the middle of the sixth century is remarkably consistent from one site to another. It is possible to find some meager traces of habitation, and it is all but certain that humans, just less archaeologically visible than before, continued some level of occupation right through the early middle ages. There are even traces of overseas ceramics into the early seventh century. But this only shows that the demographic collapse preceded the breakdown of commercial circuits. The causes are here more cleanly isolated than elsewhere. Plague and climate change triggered a synchronous convulsion in the middle of the sixth century.[45]

The sixth century was also a turning point in Anatolia. The late Roman centuries had been a period of precocious development and population growth. In many regions, settlement structures peaked in the fifth or early sixth centuries. The great string of cities facing the Aegean formed one of the most heavily urbanized corridors in the entire late empire. This momentum was abruptly halted, precisely in the middle of the sixth century. The symbiotic town-hinterland systems went into simultaneous decline. At the site of Sagalassos, where the city and its countryside have been carefully surveyed, the rupture was dramatic. "Most probably as a result of the recurrent plague, Sagalassos seems to have become a completely altered city." Here, a coherent fabric came suddenly unwound.[46]

In some parts of Anatolia, there were in fact two critical pulses of change, one ca. AD 550 and another ca. AD 620. In the first, growth stalled, but the settlement system was not overturned. Survivors fought to maintain the ancient patterns of life, even in the face of repeated visitations of the plague. Wetter and colder conditions reduced the cultivated land, and monotonous arable farming occupied a more dominant place in the landscape. Several generations plodded forward in a clearly attenuated condition, until the onslaught of Persian invasions delivered the *coup de grâce* to a staggering society. By the middle of the seventh century, most features of the human landscape had been effaced truly beyond recognition. One of the heartlands of classical civilization was hurled back into a primitive, fragmented state unlike anything seen for over a millennium.[47]

The fate of Egypt in the age of crisis is something of a mystery. The unique ecology of the Nile valley always framed events in Egypt. The dynamics of change are captured in miniature in an episode described by Procopius. Just a few years after the first visitation of the plague, the Nile flooded to a height of eighteen cubits, in ordinary times a godsend of water and enriching silt. Upstream all seemed normal. Downriver, events took an unexpected turn. "As for the country below, after the water had first covered the surface, it did not recede but remained that way throughout the time of sowing, a thing that had never before happened in all of time." The excessive inundation must be put down to the conjuncture of natural and human causes. The Nile valley was the most heavily engineered ecological district in the ancient world. Every year, at the inundation, its divine waters were diverted through an immense network of canals to irrigate the land. The intricate machinery of dikes, canals, pumps, and wheels was a huge symphony of human ingenuity and hard labor. The sudden disappearance of manpower in lands upriver threw the network of water control into disrepair. The controlled flow of water in the valley had been interrupted, and the downstream inhabitants in the fertile delta were overwhelmed. Remarkably, these events were replayed almost exactly in the aftermath of the medieval Black Death.[48]

The Egyptian economy hinged on a vast machinery of water management. The dynamics of technology, and its ownership, may have played a quietly decisive role in the aftermath of the Justinianic Plague in the Nile valley. In the words of Procopius, "The Nile's swamping of the land became the cause of great misfortune *in the present time*." To make matters worse, Egypt depended on a commodity economy. The Egyptians were overspecialized in wheat. In the later sixth century, with fewer mouths to feed at home and abroad, the bottom fell out of the wheat market. The supply of wheat was more than enough to glut the market. Rents stagnated. Wage growth was modest, at best. As in Mamluk Egypt after the Black Death, the plague in Egypt was no gift to the peasantry. Lower levels of market

Figure 7.3. Mosaic Depiction of Nilometer from Sepphoris (Photograph by Zev Radovan)

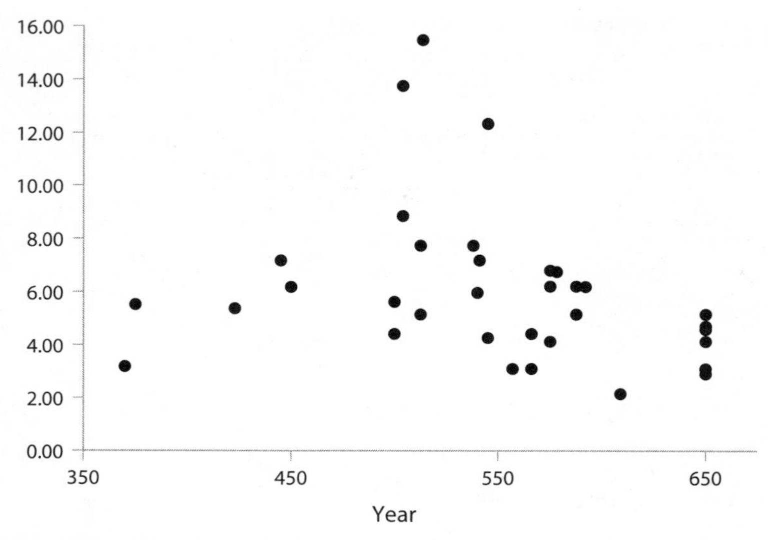

Figure 7.4. Wheat Prices in Gold (carats/hl)

integration reversed gains from trade, hurting everyone, and damage to technology reduced the productivity of labor. Moreover, by means fair and foul, rich landowners kept the labor force under their thumb.[49]

The single best-known aristocratic property in all of ancient Egypt, the Apion estate, was in its heyday in the fifty years after the first wave of plague. The estate grew dizzyingly in the generation after the pandemic. The sheer size of the Apion estate, already massive, seems to have doubled in the midst of crisis. This growth has not been fully explained. We could posit, behind this aggressive acquisition, the throes of disorienting population collapse that allowed the concentration of land-holding. Yet in the aftermath of plague the Apion estate squeezed out profits that seem astonishingly unhealthy. We may be watching, in slow motion, the erosion of the economic foundations of the aristocratic class. The estate managers have been described as "obsessed" by the problem of labor scarcity, and they tried to bind workers to the estate whenever possible. Judging by this one property, the elites of Egypt, using their control of capital, technology, and the fiscal system, acquired vast tracts of land, yet struggled to realize even modest levels of return. On the other hand, the very continuity of this family estate into the seventh century suggests some level of stability. The elites may not have been prospering, but they were able to hold on.[50]

Alexandria fared better than any other ancient metropolis into the seventh century. John the Almsgiver's brazen leadership as patriarch required a vibrant background. The vitality of Alexandria owed much to the sea. It pointed the metropolis toward the prosperity of the Levantine shore, behind which lay *the* great zone of energy in late antiquity. Syria and Palestine were the spiritual and economic heartland of the fifth and sixth centuries. The arc stretching from southern Palestine to the Taurus foothills was delirious with growth. From ca. 350 to 550, population persistently pushed outward. Cities boomed, led by Antioch and Jerusalem and followed by dozens of secondary towns, facing the networks of the Mediterranean Sea. Levantine traders dominated commerce from the Red Sea to the western Mediterranean. The wine of Gaza was an international success, the *grand cru* of late antiquity. (In one episode, the patriarch John the Almsgiver was made suspicious by the fine quality of the wine served at the Eucharist, and he was furious to learn it was imported from Gaza!)[51]

The cities of the Levant maintained their classical order. Baths, games, and theaters bristled with life. The new faith was seamlessly integrated into the urban fabric. The "Holy Land" was thoroughly Christianized, and no region enjoyed such an exuberant boom of church and monastery building in late antiquity. This prosperity derived from, and flowed back into, the countryside. Hardy villages filled the coastal plains, the hilly interior, and the dry ribbon of semidesert running from northern Mesopotamia to the Negev. Many of these villages lay beyond the easy grasp of the major towns. The ghostly stone-built villages of the limestone massif in Syria belonged not to coastal elites, but to substantial peasants.[52]

The demographic wave crested in the middle of the sixth century. Thereafter, new building slowed or became intermittent. In the north, the crisis was acute. The combination of plague, Persian destruction, and a series of earthquakes proved insuperable. Earthquakes were a bane of civilization in the seismically active Mediterranean, and right through the early sixth century, the response was to rebuild. But from the late sixth century on, societies struggled to rebound from natural disasters. Antioch faded from greatness in the latter half of the sixth century. The villages in its orbit suffered. In the Dead Cities, contraction and simplification—but not total demise—set in from the second half of the sixth century.[53]

In the southern Levant, the middle of the sixth century was more of a bump than a reversal. The face of cities began to change, in some cases drastically, losing the top-down political rationality of the classical city for a more frenetic, organic style of life. It may be that the coastal regions suffered worse than the interior predesert. The crisis slowed the building boom but did not end it. We should be cautious, though, in treating building construction as a measure of economic prosperity. This was no clean proxy for GDP. From ca.

AD 550, church building became virtually the only form of public construction. It was now weighted toward the villages. As we will see, not a few of these churches, many of them quite elegant, outfitted with magnificent mosaics, were built from the proceeds of pious fear. These were little communication portals, where humans could seek the help of powerful protectors in a chaotic world. They are a barometer of the apocalyptic atmosphere, as much as an index of economic vitality. But in all, the southern Levant proved the most resilient corner of the entire ancient Mediterranean world.[54]

The role of climate change in stalling the economic momentum of the Levant remains elusive. The stone remains of Syria or the lonely wine-presses in the unyielding desert behind Gaza seem to present a striking time-lapse photo of late Holocene climate change. The desert, to the east and the south, always loomed over the thin, semiarid strips of settled civilization along the coast. But human settlement and exploitation of the landscape triumphed in a semiarid world, always poised on the razor's edge of drought. The influx of capital and the integration of markets provided the means to colonize riskier environments. Scrupulous soil conservation and the mass-scale deployment of irrigation technology enabled the expansion of agriculture, straight into the teeth of ecologically forbidding circumstances. What was achieved in the Negev is nothing less than "one of the most successful landscape transformations in the Mediterranean in any period." But human and climate factors did not move in step. In fact, we might see a tension building between them in the sixth century. Agriculture intensified, even as aridification gradually set in, pushing in opposite directions. Farmers contrived ingenious ways to hold back the desert and forestall its inexorable advance.[55]

The plague roared through this world. But repeated mortality events did not evacuate the landscape of people, nor overturn its logic. What the focalization of plague in the eastern Mediterranean may have accomplished was the displacement of the greatest energy away from the coasts and deeper into the interior. The stark lands of the predesert, east of the Jordan River, enjoyed a lively existence deep into the age of crisis. Along the innermost ring stretching from Petra to Damascus, an entire world of Arab Christian societies flourished. They were deeply connected to the Roman Empire, if always along the edges of it. Here irrigated farming, oasis agriculture, nomadism, and caravan trade mingled cheek by jowl. In the late sixth century, these societies looked west, to the Roman Empire. But soon their empire would fail them. They would pass "quietly and almost willingly without even the slightest whimper into a new and momentous age, the significance of which was neither recognised nor appreciated at the time."[56]

From one end of the Mediterranean to the other, the diligent labors of archaeology have recovered these quiet histories of expansion and decline.

Each landscape surveyed, each town excavated, tells a slightly different story, colored by local facts. But there were deeper, shared patterns of change beneath this exquisite complexity. In recent years, magisterial syntheses have traced the kaleidoscopic patterns of interdependence between the imperial state, networks of exchange, regional aristocracies, and agrarian life. But the physical environment cannot be an inert backdrop to the story, and the earthy, biological foundation of production and reproduction must play more than a bit part. Without the deep movements of demography, models of the state and social order become weightless abstractions. The natural environment and human demography were acted upon by the state, the economy, and the social order, but they also acted and reacted upon them in turn, with a motive power of their own that had consequences at the highest levels of political organization—as events were soon to tell.

The harder climate and the malevolent germ had worked deep change across the territories of the empire, and the reckoning was at hand. When John the Almsgiver left Alexandria for his native Cyprus, he could watch the scaffolding of the ancient Roman order in the Mediterranean start to crumble around him. He lived through the grueling violence of total war between Rome and Persia in the early seventh century; it was devastating, for both sides. The end of the subsidized grain supply in Constantinople marked the conclusion of an era. But the armies of Persia were only a prelude to something of even greater moment, whose long-range consequences for Mediterranean and indeed global history were inestimably more profound. Just four years after the death of John, an apocalyptic prophet led his followers from Mecca to Yathrib, the *hegira*. Soon they would be on the borders of Roman Arabia. John's friend and biographer, Sophronius, patriarch of Jerusalem, would live to see the Roman world's last great zone of energy simply fall from the grasp of an exhausted empire. Perhaps the center of gravity had already subtly shifted to the dry, rugged interior. The events on the horizon would, for the first time in a thousand years, definitively turn the face of the Levant toward the east.

The Failure of Empire

In AD 559, by then in the thirty-third year of his rule, Justinian summoned his general Belisarius out of forced retirement. In the spring, the Danube had turned to ice, "as usual . . . to a considerable depth" (an unexpected comment on life in the Late Antique Little Ice Age, since the Danube freezes about once a generation today). Thousands of Kotrigurs, nomadic cavalrymen from beyond the Black Sea, had crossed the frozen river and set their sights on a lightning attack against Constantinople. Belisarius accepted his commission. The great commander "once more put on his breastplate and helmet and donned the familiar uniform of his younger days." With the main armies engaged on distant frontiers, Belisarius could muster only three hundred soldiers and a band of peasants ill prepared for combat. But with a mixture of discipline and deception, Belisarius turned back the column of invaders and spared the imperial capital the shame of defeat. Belisarius was the champion of his country again.[57]

This is an elaborate set piece by Agathias, the historian who continued the narrative of Procopius. It is a tale bearing a sharp point. The pitiful specter of a once mighty empire, cowering before a small band of horsemen, was meant to sum up the state of affairs. "The fortunes of the Roman state had sunk so low that on the very outskirts of the Imperial City such atrocities were being committed by a handful of barbarians." In the year before the rescue of the city by Belisarius, the second visitation of the bubonic plague had shocked an empire that remained in a stubborn position of unsustainable overreach. For Agathias, the military-fiscal death spiral was the central problem of Justinian's reign. He treated the reader to figures that still fascinate and befuddle scholars by their confident precision. An army that once fielded 645,000 men now measured only 150,000. The former number is implausibly high, the latter suspiciously though not inconceivably low. The point is altogether the same. "The Roman armies had not in fact remained at the desired level attained by the earlier Emperors but had dwindled to a fraction of what they had been and were no longer adequate to the requirements of a vast empire."[58]

In human demography, more is not always better. Population pressure can crowd the finite countryside and crunch resources. But demographic abundance is almost always a boon to the state. The state feeds on disposable bodies. The eastern Roman Empire was the great beneficiary of the long-term population growth leading into the first pandemic. In the early sixth century, the Roman army was able to replenish its ranks, once again with a light touch. Hereditary service and voluntary enlistment provided sufficient manpower. "There was a large reservoir of unemployed or underemployed men, particularly landless peasants, on which to draw." But the demographic bleeding in the age of plague marked a new era in Roman statecraft. From the time of the plague, the Roman Empire faced an ultimately irresolvable conundrum. It could not field the army its imperial geography required, and it could not pay for such an army as it was able to muster. The exact sequence of events through which this drama played itself out, in the desperate years between the reign of Justinian and the final calamities under Heraclius, were shaped by contingency. But the structural mechanics were ultimately determinative.[59]

The imperial crusades of Justinian stretched the fiscal-military capacity of the empire. The African campaign, launched in the heady days before the plague, stirred grave anxieties in his financial bureau. The reopening of conflict on the Persian front was costly, but Justinian was able to repair the situation on both his eastern and western flanks, though at great cost. Then the shock of AD 542 shifted the ground underneath him. The war in Italy stalled, and Belisarius was sent back to the west in AD 544. Troops could not be spared from the thinned ranks in the east, so Belisarius went on an enlistment campaign in Thrace and gathered some 4,000 men. The deeper problem was how to pay the troops. Belisarius implored the emperor for troops and money. "Even the few soldiers he had were unwilling to fight, claiming that the state owed them much money." They were "without men, horses, arms, or money, and no man, I think, would ever be able to carry on a war without a plentiful supply of these things." It was only the ominous beginning of a new crisis of statecraft.[60]

The power of the Roman Empire had always been constrained in a way that silently checked all polities until the seventeenth century: the state lacked the capacity to borrow money on any large scale. The absence of debt finance was constricting. In the days of silver money, the emperors could debase the currency. But by the sixth century, the soldiers were paid in kind and in gold. The desperate gambit of debasement was not available. In financial straits, the empire had two options. It could not pay its soldiers, or it could squeeze its taxpayers. The AD 540s initiate an era in which the Roman Empire often did both. We are told that Justinian "was always late in paying his soldiers and, generally, treated them in a heavy-handed way." He "began openly cheating the soldiers out of part of their pay and not paying the rest until it was long overdue." Justinian was said to have cancelled the gold bonus that soldiers received every five years, the basis of reciprocal loyalty since the very first barracks emperors. And he may have stripped the border units settled all along the frontiers of their commission altogether. There is utterly nothing like this in the long annals of Roman history. Justinian was the first deadbeat emperor.[61]

The troops felt the strain. So did taxpayers. At first Justinian refused to forgive tax arrears; sporadic jubilees were expected from emperors, but Justinian was pitiless. Finally, in AD 553, he grudgingly remitted back charges down to the year the first plague outbreak ended. Even his public attitude was not gracious. "Although now, if ever, many expenditures are necessary for the republic, which has been greatly increased through the kindness of God, and which carries on wars with the surrounding barbarians in proportion to such increase, nevertheless we ... remit to our subjects all delinquent taxes." It was a meager concession.[62]

Tax assessments were levied by district, and even though the number of laborers was greatly reduced, the aggregate charges were not adjusted, so that the real rate on survivors soared. "When the plague broke out ... and wiped out the majority of the farmers, this caused many estates to be deserted, as you can imagine. Yet he showed no leniency toward their owners. He never once waived the annual tax, demanding not only the sum that was assessed on each of them but also, from them too, that which was due from their deceased neighbors." In the village of Aphrodito in upper Egypt, which has proven the richest source of papyri in this period, we can catch tantalizing glimpses of the rising tax rates on the ground. The tax hike amounted to a staggering 66 percent. Tax rates were consistently higher in the later sixth century than at any time in all of Roman history.[63]

It is surprising that Justinian was not overthrown. But he had already survived a political coup early in his reign and had dimmed the enthusiasm for a new revolt. His ruthless treatment of Belisarius—the successful and faithful general—has seemed shocking. But Justinian would not risk allowing the dissatisfaction of his subjects to focus on its most natural candidate, and the general was as loyal as a dog. The emperor's prodigious talents let him keep a vice grip on power to the bitter end. The opposition failed to find a champion. A reign that had begun with such high hopes—the reform of Roman law, the overhaul of the administration, the building program, and above all the restoration of a Mediterranean empire—ended with the empire lying mortally wounded. When Justinian died at last, the state was exhausted. His successor, Justin II, inherited a treasury with a morass of uncollectable debts. He immediately cancelled arrears. He was handed control of an army that he admitted, in his public voice, "had gone to ruin through want of necessary things, so that the republic was injured by innumerable invasions and incursions of barbarians."[64]

The emperors after Justinian might plug holes in the dike, but they could do nothing to push back the gathering tidewaters. Justin II (r. AD 565–74) stopped diplomatic payments to barbarians, but the act only redoubled the diffuse violence along the frontiers. Each round of plague strangled the life from the state. In the immediate aftermath of the outbreak in AD 573, Tiberius II (r. AD 574–582) carried out desperate enlistment campaigns in east and west. Control in the Balkans faltered, and the possessions in Italy were whittled down. Maurice (r. AD 582–602), as capable as any who ever wore the purple, continued aggressive conscriptions. Even in the throes of this desperate period, the empire could field respectable field armies, and the military manual written by Maurice assumes the ability to put 15,000 soldiers into formation. But the military system was fiscally unsustainable. Maurice took the fateful step of directly cutting the pay scale. In former times, Roman emperors might have achieved as much by debasing the coinage, but that was at least a cut in disguise. No emperor had ever dared a straight pay cut. The utterly predictable happened at last. Maurice was overthrown, and his usurper was soon usurped in turn. The old scourge of civil strife was now inflicted on the empire again. It was to prove too much. The emperor Heraclius (r. AD 610–41) would preside over the failure of empire.[65]

To those who lived through it, it seemed like the final hour of the world was at hand.

The Hour: Muhammad's World

The monk and writer John Moschus was born in the middle of the reign of Justinian. He was probably born in Cilicia but heard the call of the Judean desert as a young man. Moschus was an exact contemporary of John the Almsgiver; along with his friend and fellow traveler Sophronius, he wrote a biography of the Alexandrian patriarch. These three belonged to the last generation that moved easily throughout a Mediterranean world held together by the glue of empire. This easiness of movement is the vital backdrop to the collection of edifying stories for which Moschus is best known, the *Spiritual Meadow*. John's enduring contribution to the monastic literature of late antiquity was this string of short, earthy vignettes that transport us to the Roman Empire in its last days, to a landscape dappled by the light of the waning sun.[66]

In one of the tales, we meet a lawyer from Palestine named Procopius. The lawyer happened to be in Jerusalem when an outbreak of plague erupted in the coastal town of Caesarea. He was terrified his children would die. "Should I send and bring them home? No man can flee the wrath of God. Should I leave them there? They might die without me seeing them." At a loss, the lawyer sought the counsel of a renowned holy man, Abba Zachaios. Procopius found him in the Church of Saint Mary the Bearer of God, praying. Abba Zachaios turned to the east and "continued reaching up towards heaven for about two hours without saying a word." Then the holy man turned to Procopius, assuring him that his children would live and that the plague would abate in two days, both of which came to pass.[67]

This was an affecting story and, at the same time, a parable of good behavior. It hoped to point the reader, gently, toward certain reassuring landmarks. The lawyer found Abba Zachaios praying in the Church of Saint Mary the Bearer of God. Contemporaries called it simply the Nea Ekklesia, the New Church. It had been built by Justinian, completed only a year after the first visitation of the plague. It was Justinian's definitive contribution to the architecture of Jerusalem. He had relandscaped the entire urban core, to align his church with Constantine's Holy Sepulchre. Justinian's church was, consciously, twice the size of the Temple of Solomon. Its huge stone masonry and imposing fire-red columns were an aggrandizing statement of the empire's power. It was the most visible contrivance of human art on the Jerusalem skyline, and it remained, into the seventh century, a monumental statement of the empire's presence in the Holy City. The lawyer sought counsel in what amounted to an imperially authorized holy place.[68]

There the lawyer found Abba Zachaios, arms lifted in prayer. We suspect his devotions were addressed to Mary, the Bearer of God. Here the empire's influence is more subtly present. Palestine was the cradle of Marian devotion. But in the fifth century, the cult of Mary was taken up by the central empire, and by the sixth century devotion to Mary emanated from Constantinople. The decades of plague transformed Constantinople into the city of Mary. The empire was under her protection. To understand the Mary who attained such spiritual prominence in late antiquity, we will have to dispel later, medieval images of Mary from our mind. The figure of Mary that dominated in late antiquity was not the tender *mater dolorosa*, whose sufferings were the awesome shared point of common humanity. Rather, the Mary who captured the empire's imagination was the Queen of Heaven. She was a formidable presence, busy in the grand sweep of events. On the day of judgment she would intercede on behalf of humanity before an angry God. The lawyer desperate to see his children was not granted a personal miracle or a private favor of compassion. Rather, through the medium of Abba Zachaios, he was granted a brief but calming glimpse of the cosmic events unfolding around him.[69]

The layman lawyer who sought help in the Nea Church confessed that "no man can flee the wrath of God." This was not the resigned fatalism of one pious individual, so much as the shared sensibility of an entire era. Inhabitants of the later sixth and seventh centuries felt they lived along the fast-crumbling cliff's edge of the present age. In such an environment, the inescapability of the plague was an existential fact. A Christian in Antioch claimed that anyone who fled a city under the sentence of plague would be hunted down by its implacable force. A monastic father in Sinai wrote a thoughtful reflection on the question of whether one could flee the plague. In Islam, a massive didactic tradition grew up around the inescapability of the plague. Except that they are in Arabic, some of the arguments seem virtually pulled from contemporary Latin, Greek, and Syriac texts. The similarities are not superficial. Behind them lays an ocean of shared eschatological sentiment.[70]

The human response to the spiraling environmental crisis of the sixth and seventh centuries activated the full apocalyptic potential of the surrounding religious atmosphere. Christianity is an eschatological faith. Apocalyptic notes run like a constant background music across the history of the church. But they have not always had the same level of intensity. After the fervor of the first Christian generations, expectations of imminent judgment were subdued. The conversion of the empire to Christianity further dulled anxieties about the end times. Events like the coming of the year AD 500 could fan transient millenarian speculation, but after its uneventful passing, triumphal tones again drowned out the pessimistic notes for a time.[71]

Then nature intervened. The natural catastrophes of the sixth century induced one of the greatest mood swings in human history. The occlusion of the sun, the rattling of the earth, and the advent of worldwide plague stoked the fires of eschatological expectation, across the Christian world and beyond. Signs of profound collective distress have been detected in places so far removed as Norse myth and Chinese Buddhism. We can only follow the billowing sense of imminent doom in any detail inside the Roman Empire. Already as the plague first approached, dark rumors flew. On the eve of the initial outbreak in Constantinople, a woman "went into ecstasy" and was taken into a church, saying "that in three days' time the sea would rise and take everybody." The mortality stirred feelings of ineffable dread, often older than Christianity. "According to the ancient oracles of the Egyptians and to the leading astrologers of present-day Persia there occurs in the course of endless time a succession of lucky and unlucky cycles. These luminaries would have us believe that we are at present passing through one of the most disastrous and inauspicious of such cycles: hence the universal prevalence of war and internal dissension and of frequent and persistent epidemics of plague."[72]

The mainline Christian reaction to the age of plague was sketched already by John of Ephesus, who tried to come to grips with the horror of the first visitation. The only possible conclusion, in the face of such unaccountable violence, was that the end times were drawing near. The plague was the sign of God's fury. John ransacked the prophetic and apocalyptic traditions to understand the plague. It was the wine-press of God's wrath promised in the Biblical *Apocalypse*. God's ravenous justice ensured that "the people should be astonished and remain in amazement about His righteous judgments which cannot be understood, nor comprehended, by human beings, as it is written, 'Thy judgments are like the great deep.'" The suffering inflicted by the plague was meant to be "a chastisement." Here is a word with peculiar depth in a society familiar with the dark extremities of the master-slave relationship; chastisement was the last, desperate, and most brutally corporal effort to reorient the interior will of a recalcitrant slave. Justinian publicly called the plague a sign of God's philanthropy, his "love for mankind." The mass mortality was a wake-up call to survivors, sent as a courtesy warning in advance of the great judgment to come.[73]

The fears of the sixth century generated an organized ecclesiastical response in the form of liturgical rogations, great communal rituals intended to ward off the pestilence. These rituals were pioneered in the fifth century, before the pandemic. They were first improvised as an all-purpose liturgy that could expiate a community's sins. They were a liturgy of last resort, and in the time of Justinian they still had the sheen of something new. In AD 543, the bishop of Clermont (the uncle of the chronicler Gregory of Tours) fended off the plague by leading his congregation on a lengthy prayer march in the middle of Lent to a remote rural shrine, singing the psalms. They were spared. These liturgies spread as easily, and anonymously, as a computer virus. The Church of the East in Syria, at the opposite end of the Christian world from Clermont, enacted nearly identical rituals of supplication.[74]

Most of the these desperate responses escaped the historical record. But we are vividly informed about the elaborate spiritual exercises conducted by Gregory the Great. He organized processions tracing the geography of piety that had come to overwrite the old civic coordinates of Rome. For three days, the city rang with prayers and chanting, as choirs sang the psalms and the *Kyrie eleison*. On a Wednesday, the people assembled at seven churches across the city. They processed in prayer lines crisscrossing the city until the great litanies converged at . . . the Great Church of Holy Mary, the famous Santa Maria Maggiore. "There we may at great length make our supplication to the Lord with tears and groans." One deacon witnessed eighty people fall dead during their prayers. "The Pope never once stopped preaching to the people, nor did the people pause in their prayers."[75]

These rogations are but one visible element in a vast religious *koine* that reacted to the plague with communal acts of intercessory ritual, tinged with apocalyptic fear. The imminent judgment was a call to repentance.

The plague was a last chance to turn from sin. And no sin weighed more heavily on the late antique heart than greed. As Peter Brown has shown, anxieties about wealth generated a perpetual moral crisis in late ancient Christianity. Earthly possessions were a trial of faith. Here the plague struck a tender nerve. The most memorable vignettes in John of Ephesus' history of the plague linger over individuals singled out for punishment because of their greed. From one angle, the plague was God's final, ghastly effort to pry loose our tight-gripped hold on material things.[76]

In some cases it worked. In a distant village in upper Egypt, we happen to see that the plague triggered an instantaneous effusion of pious giving. Elsewhere the thanksgiving of survivors was monumental in scale. Marvelous new buildings went up in fulfillment of promises muttered in fear. It is no accident that church building remained the most active form of public construction. The natural crises are in the near background of this wave of building. On the wall of a sixth-century church at Petra we find painted the *91st Psalm*. "His fidelity is an encircling shield. You need not fear the terror by night, or the arrow that flies by day, the plague that stalks in the darkness, or the scourge that ravages at noon." Many of the new constructions were dedicated to Mary or Michael. At Nessana, a town in the Negev, for instance, a new church (known as the South Church) was dedicated to Holy Mary the Bearer of God, in the immediate aftermath of a plague outbreak. Its dedicatory inscription is entirely typical: it implores her, "Help and have pity." The pattern held in the west, too. In AD 545 in Ravenna, a small church was built by two men and dedicated to the Archangel Michael, in thanks for the "benefits" he procured unto them, namely mercy amidst the ravages of plague. The mosaics in the apse of the church depicted Christ flanked by Michael and Gabriel. Other angels blared the trumpets of the apocalypse. It was an extravagant statement of thanks, from a wealthy survivor left standing after the first sounds of the judgment.[77]

The expression of gratitude to the Archangel Michael was not idiosyncratic. An anonymous Coptic sermon claimed that a copy of the New Testament, given in the name of the Archangel Michael, had promising talismanic powers for churches or households: "neither sickness, nor pestilence, nor ill luck shall enter the house wherein it is for ever." Eschatological fervor pushed Michael to the forefront of religious devotion. In the midst of the plague, the "Angel of God, with his hair white as snow," was seen to be among mankind, doling out judgments. The place of the archangel, ascendant already before the plague, now became fixed in the cultural landscape as never before. He was the instrument of God's last judgment. His business was at hand.[78]

The only greater beneficiary of the crisis was the Mother of God. She enjoyed a new prominence in the religious life of the later sixth century, especially in Constantinople. "The Virgin came to assume a dominant—perhaps *the* dominant—place in the religious life of the city." For the first time, in the midst of the plague, the Feast of the Hypapante was instituted in Constantinople. The eastern equivalent of Candlemas, it commemorates the purification of the Virgin at the temple. The Hypapante is celebrated on February 2, right on the cusp of plague season, and the day of purification may have touched primal religious feelings. Justinian commanded its celebration across the empire. Marian devotion became more widespread, across society. Images of Mary became more common on domestic artifacts, often with an apotropaic purpose. A dazzling pectoral of the later sixth century invoked the Virgin's succor. "Protect her who wears this." An armband implored, "Mother of God, help Anna." The liturgical prominence of Mary and the explosive proliferation of her images testify that the religious ideas we find in the literary texts are reflective of a broader cultural sensibility, apocalyptic in tone. In the final stanza of the great *Akathistos* hymn to Mary, one of the centerpieces of early Byzantine piety, she is supplicated, "Deliver from every evil and from the punishment to come all those who cry to you: Alleluia!"[79]

It was also during the age of plague and climate crisis that the veneration of icons assumed an intimate place in the religious practice of the church. Recently, Mischa Meier has built on the suggestion of Averil Cameron that the bewildering agony of the pandemic encouraged the spread of icon veneration. It is a convincing link. Maybe the most poignant spiritual artifact of the age is the great Byzantine icon of Mary known as the *Salus populi Romani*—the Deliverance (or Health) of the Roman People—hanging in Santa Maria Maggiore in Rome. It is probably a sixth-century original. It is a symbol of Marian devotion and the connections that bound east and west in the sixth century. For what it is worth, the late medieval *Golden Legend* portrays Pope Gregory the Great carrying an icon of the Virgin during his rogations. The Archangel Michael appeared atop the Castel Sant'Angelo, sheathing his sword and bringing an end to the plague. This may be the stuff of legend, with layers of medieval accretion. But the spirit is entirely in tune with the late sixth century.[80]

We should never forget that Gregory passed many years in the eastern capital. He was in Constantinople for at least one of the major recurrences of plague. There he would have witnessed the grand public litanies in times of distress. Gregory the Great's eschatological sensibility was colored by his experience of eastern Christianity. For Gregory, as for a figure like John of Ephesus, the woes of plague and war were a clarion call to repentance. "Those scourges of God which we fear when they are still far off must terrify

Figure 7.5. Salus populi romani*: Possible Sixth-Century Icon, "The Deliverance of the Roman People," Santa Maria Maggiore, Rome (Alinari / Art Resource, NY)*

us all the more when they are come among us and we have already had our taste of them. Our present trial must open the way to our conversion. . . . I see my entire flock being struck down by the sword of the wrath of God, as one after another they are visited by sudden destruction." The looming judgment was an impetus to action. It inspired Gregory to missionize the pagans in Britain, to bring them salvation as the time ran short. The miracles of the saints meant that this age was not yet "entirely deserted." But the natural disasters were a sure sign that the edifice of this age was quickly crumbling.[81]

The beliefs of Christian authority figures like John of Ephesus or Gregory the Great were framed by the narratives of scripture. The biblical canon generously provided the apocalyptic mind with a kit of authorized images and symbols. This tradition was by its nature kaleidoscopic. Its fragmentary and frankly weird symbols could be endlessly bent into new configurations. This tradition was also an invisible cordon around what could be said and thought. "Although patristic theology left no room for home-made prophecy, it allowed ample scope for creative interpretation of the relevant biblical texts." It is notable that commentaries on *Revelation* begin to appear in the sixth century. The book had always been a little to one side of the mainstream Christian tradition, but in the age of plague, it was combed with a new urgency. The boundaries of apocalyptic thought were being tested.[82]

The days of prophetic utterance, strictly speaking, had long been closed in Judaism and Christianity. But ecstatic experiences and religious visions had always hovered on the edges of orthodoxy. A holy man like Nicholas of Sion was personally visited by the Archangel Michael, to receive warning of the plague. Abba Zachaios communicated with the divine, in the authorized confines of the Nea Church. But the gifts of clairvoyance were not always safely corralled. In the later sixth and seventh centuries, the loose energy of apocalyptic expectation began to overflow the banks of the old textual traditions.[83]

This pattern is as evident in Judaism as in Christianity. In the midst of crisis, a fecund new era of Jewish apocalyptic writing opened up. Ceaseless natural disaster, combined with the epic confrontations of Rome and Persia, stirred a fresh sense of mysticism and expectation among Jews across the Mediterranean and Near East. "The Holy One, blessed be He, will introduce heat into the world from the heat of the sun, along with consumption and fever, many terrible diseases, plague, and pestilence. Every day there will die among the gentile nations one million people, and all the wicked ones among Israel will perish." The growing antagonism between the Roman state and its Jewish subjects, which reached a fever pitch in the forced baptisms ca. AD 630, fired wild messianic fervor. Under pressure, Jews looked for the "footprints of the Messiah." The Jewish forms of expectation were their own, but they clearly breathed the same apocalyptic air as those around them.[84]

In the early seventh century, the momentum of political events gave a new, unstable charge to apocalyptic ideas. The endless war between Rome and Persia was fuel to the fire. The clash of great empires known as "the two eyes of the world" seemed like the ultimate confrontation. The conflict took on overtones of holy war. Already under Maurice, the Roman army was using the "Virgin Mother of God" as a watchword. Between AD 602–628, the violence broke through its customary theaters. It became a total war. Persian armies cut into the deep tissue of the empire. The Holy Land fell. Syria was taken in AD 610 and Palestine by AD 614. The fall of Jerusalem was a moral shock and attended by wholesale slaughter. The relic of the True Cross was taken into Persian possession. The "psychological impact" of Jerusalem's fall was "perhaps only comparable to the trauma the Romans experienced when Rome was sacked in 410." Apocalyptic time sped up. Next Egypt fell and then Anatolia. In some places, such as Asia Minor, there was never recovery.[85]

The destruction was vast, and the worst was in view. By AD 626, the Persians were at the walls of Constantinople. An army of Avars had advanced upon the capital simultaneously. In the darkest hour, the people turned to the Virgin. Her icon was paraded in the streets and upon the great walls. The salvation of the city seemed supernatural. The emperor Heraclius, meanwhile, launched an aggressive counterattack. With icons of Christ and the Virgin in the van of his armies (and considerable aid from the allied Turks), he reclaimed the smoldering remnants of the eastern provinces by AD 628. Ever so briefly, the old political equilibrium was restored. The true cross was triumphantly returned to its home in Jerusalem. Political events were creatively enfolded in apocalyptic meaning, in a way they simply had not been since the prophecies recorded in the book of *Daniel*. And now, the entire world looked to political events with the bated breath of eschatological hope.[86]

The emperor Heraclius was hailed as a figure of cosmic significance. But his restoration was to prove short-lived. The speed of the next act has always astonished. While Rome and Persia were locked in bloody confrontation, something stirred to the south. In the space of just a few years, Arab invaders simply detached the prize territories of the Roman east from the nerve center in Constantinople. Turning the desert fringe that enveloped the Levant into a zone of conquest and control, the army of believers from Arabia dismembered the Roman Empire. The conquests were swift and pitiless, but great destruction was simply not necessary to effect one of the greatest geopolitical heists in history. After the defeat at Yarmouk (AD 636), the emperor Heraclius ordered his armies to draw back. It is one sign that the concatenation of plague, climate change, and endless war had drained the vitality of the Roman Empire. Syria, Palestine, and Egypt were taken in the space of a decade. New frontiers were drawn faster than contemporaries could make sense of the dramatic changes.[87]

In later times, propagandists at the Abbasid court would attribute the great conquest to the pure and stout sons of Arabia. It is an ingeniously seductive tale. But the Arabs were no strangers. The scholarly labors of Glen Bowersock have now given us a remarkable 360° view of the Arabian heartland at the dawn of Islam, and it is enveloped on all sides by the wider world. The Red Sea networks that surrounded the Arabs had been integral to the geopolitics of the great powers for centuries. Arabs had served as federate soldiers for the Romans and Persians, and they knew the commercial networks of the Near East intimately. There were Christian Arabs throughout the Roman desert; Christian missionaries fanned out across Arabia. For a time, there had been a Jewish kingdom in South Arabia. Even the Hijaz was not an exotic desert braved only by nomads. It was altogether less forbidding. Bedouins, traders, and settled agriculturalists alike called it home. In the seventh century, the Arab world was drawn into the epic confrontation between the great powers. It has even been suggested that Mohammad's *hegira* was engineered, through back channels and regional clients, by the court of Constantinople.[88]

The spark that lit the fire in Arabia was the rise of a new monotheistic religious ideology, one that would bind together a community of believers across ancient tribal divisions. The religious mission of Muhammad was not simply precipitated from the atmosphere of apocalyptic feeling across the near eastern world. Neither was it alien to the religious *koine* of late antiquity. It was a distinctive outgrowth of the apocalyptic fervor that set in with the arrival of pandemic plague and the ice age. The seeds of eschatological fear had floated on the wind, beyond the borders of Rome, and taken root in strange soil. What set the new religion apart was not so much its native Arabian elements, as its greater range of motion. Where Jewish and Christian eschatology was confined by the closed tradition of revelation, in Arabia a new prophet claimed to have a final revelation of God through the angel Gabriel. The message itself would not have seemed totally unfamiliar to John of Ephesus or Gregory the Great. The revelation was urgent: worship the one God, for the Hour is at hand.[89]

Critical study of Islam, peeling back the layers of subsequent centuries, has emphasized that monotheism and eschatological warning were central to the prophet Muhammad's religious message. "The coming judgment is in fact the second most common theme of the Qurʾan, preceded only by the call to monotheism." The Qurʾan proclaims itself to be "a warning like those warnings of old: that Last Hour which is so near draws ever nearer." "God's is the knowledge of the hidden reality of the heavens

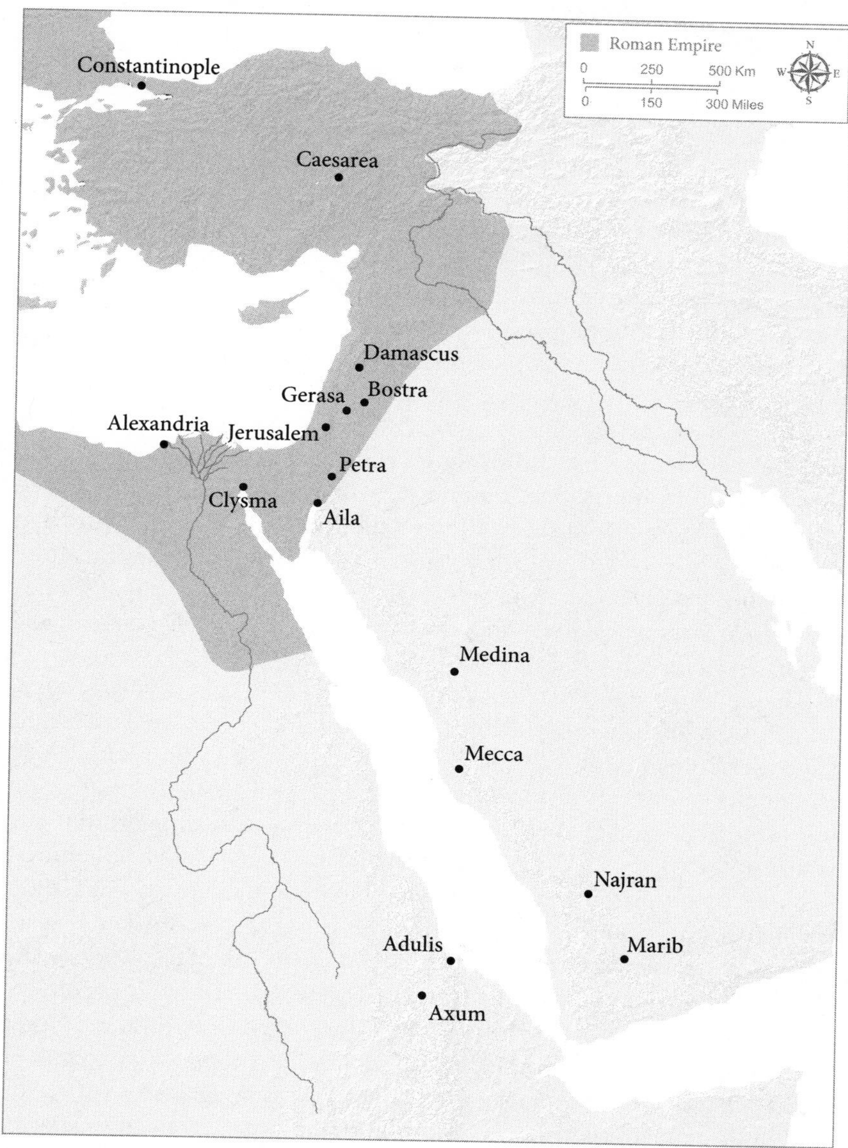

Map 25. The World of Early Islam

and the earth. And so, the advent of the Last Hour will but manifest itself like the twinkling of an eye, or closer still." The origins of Islam lie in an urgent eschatological movement, willing to spread its revelation by the sword, proclaiming the Hour to be at hand. Here, the eschatological energy of the seventh century found its most unrestrained development. It was electrifying. The message was the last element in the perfect storm. The southeastern frontier of the empire was erased almost overnight. Political lines of a thousand years were instantaneously and permanently redrawn.[90]

The Nea Church in Jerusalem, where we met the lawyer and Abba Zachaios, had pointed the political geography of the Holy Land toward the Roman Empire. The church last appears in history on Christmas day in AD 634, as the setting for a sermon by the patriarch Sophronius, friend of John Moschus and biographer of John the Almsgiver. Sophronius had outlived his friends and survived to witness the fall of Jerusalem to Islam. To him, the Arabs were the "abomination of desolation clearly foretold to us by the prophets." They were a chastisement, and only by God's will did they "add victory to victory." Still he was not without hope. "If we repent of our sins we will laugh at the demise of our enemies the Saracens and in a short time we will see their destruction and complete ruin. For their bloody swords will pierce their own hearts, their bows will be splintered, their arrows will be left sticking in them and they will open the way to Bethlehem for us." But Sophronius was the voice of a lost cause. The Nea Church disappears from history. It is possible that the giant stone masonry, once a symbol of Rome's power, was stripped out and incorporated in the Dome of the Rock, old stone to new building.[91]

The conquest of the eastern provinces in the AD 630s and 640s by a prophetic eschatological movement might be considered the final act in the fall of the Roman Empire. With the detachment of the eastern possessions, the empire's last great zone of energy was lost. The Mediterranean world was cut apart. The Roman Empire was reduced to a Byzantine rump state whose straggled possessions were meager and impoverished. The Islamic caliphate now claimed what were and would remain the most vibrant heartlands of cultural, spiritual, and scientific endeavor, as the Fertile Crescent, once again, reasserted its title as the core and crossroads of civilization. The fragmented territories of the Latin west became the backwater of Eurasia. They were fated to pass a long cycle in the outer orbit of civilization. Never again would there be a pan-Mediterranean empire, linking the energies of the Old World continents into a unified power. A new age had arrived.

Rome's empire was always poised uncertainly between fragility and resilience, and in the end the forces of dissolution prevailed. But the supreme sway of climate and disease in this story relieves a little of the temptation to find the hidden flaws or fatal choices that spelled Rome's demise. The fall of Rome's empire was not the inexorable consequence of some intrinsic fault that only worked itself out in the fullness of time. Nor was it the unnecessary outcome of some false path that wiser steps might have circumvented.

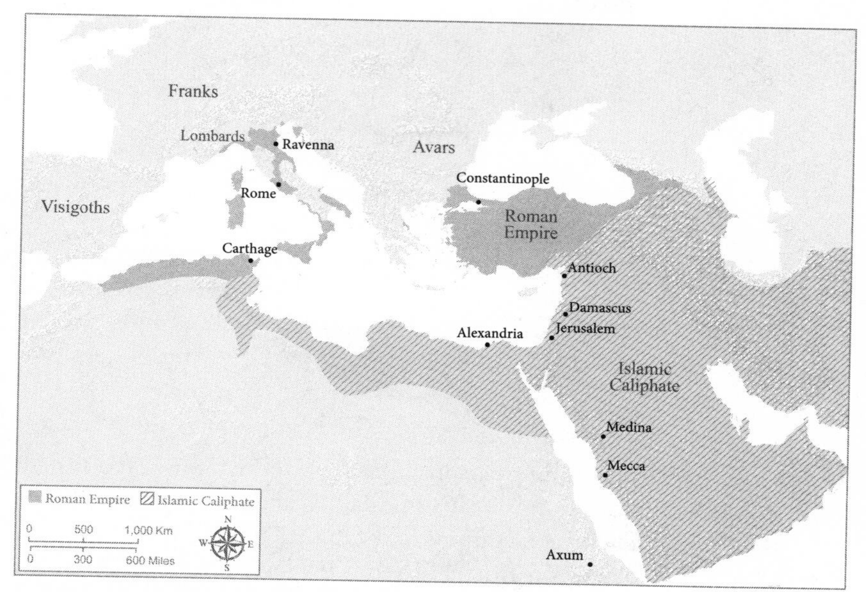

Map 26. The Early Medieval Mediterranean

Long reflection on the fate of Rome led Edward Gibbon to marvel not that the empire had fallen, but rather that it "had subsisted so long." All that we have learned about Rome in the intervening time, not least the exhilarating discoveries of recent years, only serves to confirm and even expand such a humane sentiment. In the face of relentless adversity, the empire held firm. Amid unaccountable sorrows, its people endured. Until, at last, the mortal frame of the empire could bear no more, and proud new civilizations arose from the rich soil left in its ashes.

Epilogue

HUMANITY'S TRIUMPH?

In 1798, an Anglican country parson published, anonymously, the first of what would be many editions of his scandalous and brilliant *Essay on the Principle of Population*. In later editions of the *Essay*, Thomas Robert Malthus added a lengthy chapter on Rome, offering his own contribution to the debate between David Hume and Robert Wallace on the "Populousness of Ancient Nations." That seemingly arcane dispute had marked a quiet watershed. Hume's negative assessment knocked the classical civilizations off their pedestal and, in a way, helped to bolster modernity's self-awareness and sense of superiority. In his *Essay*, Malthus simply places Rome among the broad, indistinct class of civilizations where "the population seems to have been seldom measured accurately according to the average and permanent means of subsistence, but generally to have vibrated between the two extremes." Malthus cannot be accused of especially original or profound insights on Roman history. Yet, the *Essay* has proven so influential and enduringly adaptable because of the essential rightness of its central doctrine: that human societies are dependent upon their ecological foundations. It is, still today, an inspired way to think about the human condition—and about our relationship to a civilization so distant as the Romans.[1]

Around the time Malthus published his first edition, somewhere on earth a child was born with a very special distinction. For the first time in the history of the species, there were now a billion humans alive. It had been a long trek. The expansion of human numbers started with the great dispersal out of Africa and the uncanny ability of our species to colonize nearly any environment on the planet. Still, there were only some five million humans in total, thinly strewn across the habitable continents, when our ingenious Stone Age ancestors discovered the possibilities of domestication. The rise of agriculture was an energy revolution, a way of converting solar radiation into consumable calories with an efficiency that changed everything. The explosive potential of the revolution was realized in the vertiginous increase of human numbers.[2]

Those first farming civilizations were not so different, in their energy basis, from the world that Malthus knew in 1800. In the England where Malthus was born, *per capita* wages were a little higher than they had been at the dawn of agriculture, but not radically so. In fact, average incomes in eighteenth-century England were vastly closer to Roman levels than to those we enjoy in the developed world today. It was far from clear, as Malthus wrote, that humanity had escaped the energy trap of preindustrial economies. And certainly not all societies had. On the cusp of the Industrial Revolution, for instance, wages and human welfare in the heartlands of Chinese civilization were roughly comparable to most European societies. But in the course of the eighteenth and nineteenth centuries, the Chinese population multiplied and outstripped its ecological capacities, inducing just the kind of vicious sequence of famine and social catastrophe that elementary Malthusian doctrine would have predicted.[3]

Ironically, Malthus, as a prophet, was most glaringly wrong in the case of his own country. With the English in the vanguard, humanity engineered another, even more sweeping energy revolution. The solar energy congealed underground in fossil form was tapped and harnessed to machines; scientific enterprise was mobilized behind the useful arts. The combination of more energy, more food, sanitary reform, and (late in the day) germ theory and antibiotic pharmaceuticals contributed to a population upswell unlike anything in the history of life on the planet. In just the last two centuries, humanity has added another six billion to its living ranks. Even though this revolution was stirring right under his nose, the Reverend Malthus did not appreciate the ways that technical innovation would liberate human societies from the dread implications of the energy trap. Most of the seven billion humans alive today enjoy levels of material well-being and life expectancy beyond anything the Romans could have comprehended.

So do we inhabitants of the modern world stand on the far side of a chasm, separated from the ancients by the seemingly endless horizons of our energy regime? In certain ways, yes. Our graver perils lie in the exhaust fumes of abundance rather than the razor's edge of scarcity. But this book has suggested some unexpected ways in which we are bound to the past, right across the great rift of modernity. And here, too, we might find inspiration in Malthus' essential lesson, even as we recognize that our position affords us a more capacious view. It has been central to the argument of this book that the rise of the Roman Empire catalyzed—and in turn, depended intimately upon—an economic efflorescence. Gibbon's "happiest" age was one of those phases of history when trade and technology outran the vengeful force of diminishing returns. For a long cycle, the Romans enjoyed real, intensive growth. The more general implication is that preindustrial economies were springy, and the "vibrations" of Malthusian theory could work themselves out over very long stretches of time. Modernity has been built on a singular energy breakthrough, but there were premonitions, and Rome was one.

We have also seen that nature, which creates the "means of subsistence" upon which premodern societies feed themselves, is anything but a static backdrop. On its own terms and tempo, nature alters the conditions within which human societies have sought to scratch out their livelihoods. Even in the relatively calm Holocene, the sun acted like a whimsical dimmer switch, modulating the amount of energy received; volcanoes and the earth's own erratic internal systems have further scrambled the prospects of human societies. This pulsing irregularity shakes and sways an already complex arrangement. Polities and societies are built on economic and demographic foundations, and these in turn grow and contract under the external influence of nature's capricious will.

The energy limits of preindustrial societies were malleable and in flux. These amendments expand rather than overturn the Malthusian laws. But this book has suggested another, deeper logic at work, just beyond anything Malthus imagined. The Malthusian regime describes the ecological constraints of plant energy. (Meat is just plant energy inefficiently but delectably converted to food.) Whenever this energy ran scarce, human populations were cut down to size by a deadly but generic and interchangeable array of terrible devices, including epidemic diseases. In reality, mortality has been a much wilder, more independent and unpredictable force than the strict laws of energy limits would predict. One reason is that epidemic disease depends so thoroughly on the biology of the pathogens whose business it has been to regulate human populations. Food scarcity *can* call forth and impel some infectious agents, but others are ravingly indifferent to the nutritional status of the societies they stalk. A cursory glance at the trajectory of human population growth, from the invention of agriculture down to the first billion people, reveals how imponderably decisive just a few microbial enemies have been in the destiny of human societies.[4]

Seen in this frame, the Malthusian laws are at last too narrow to endure. They fix all our attention on humans and plants. But the microbes are not simply an unruly inconvenience, a mild disturbance in the pattern. They belong to the deeper pattern, a fuller ecology of the earth in which our species competes and cooperates with others, including invisible ones. Bacteria,

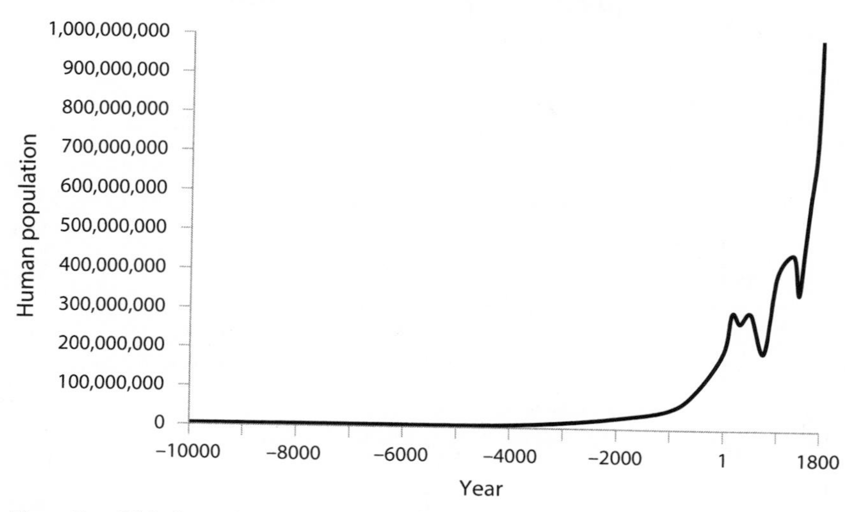

Figure E.1. Global Population Growth, Estimated

viruses, and other parasites are not an inert part of the machinery; they are, rather, agents operating in their own interest, seizing such opportunities as they happen to be presented. This perspective casts the triumphs of humanity in a more humbling and, perhaps, uncertain light.

The Anthropocene is the name persistently gathering acceptance for the current epoch of earth history, in recognition of the indelible effects of human civilization on the planet's physical and biological systems. In addition to accelerating climate change and leaving permanent signatures of our existence in the radioactive traces of our nuclear technologies, we have re-ordered the circumstances of competition and cooperation among nearly all species on earth. In the words of John McNeill, "for all species, on land and sea, the Anthropocene has revised the rules of evolution. Biological fitness—defined as success in the business of survival and reproduction—has increasingly hinged on compatibility with human enterprise. Those species that fit neatly into a humanized planet, such as pigeons, squirrels, rats, cattle, goats, crabgrass, rice, and maize prosper." But here a more sinister paradox is left unstated. The growth of human numbers has also rewritten the rules of the game for the microbial co-residents of planet earth.[5]

There are maybe a trillion microbial species in total; the average human lumbers around bearing some 40 trillion bacterial cells alone. They have been here for some three and a half billion years. It's a microbe's world—we're just living in it. Most of this wondrously diverse panoply is indifferent to us. There are only some 1400 microbes known to be pathogenic to humans. These have evolved the molecular tools—virulence factors—to menace us despite the defensive armory of our remarkable immune systems. The rise of a planet full of pathogens is very much the consequence of microbial evolution, which in turn has been profoundly determined by the explosion of human numbers and our species' pitiless transformation of landscapes across the globe. Evolution is propelled by the blind force of random mutation, but we have created the context in which evolution tinkers and experiments.[6]

Here, we are still only at the beginning of a new understanding, struggling to make order from the disorienting new data that arrive with gathering speed. The extreme youth of history's great pathogens is a fact that is still emerging from labs around the world. Future advances in microbial genomics are likely to underscore the drama of evolution in the last few millennia—and into the present day. Our urgent awareness of "emerging infectious diseases" is a recognition that the creative destruction of evolution continues—and perhaps even accelerates. But so far, most catalogues of emerging infectious diseases only go back a century or so. This time depth is arbitrary and misleading. The last few thousand years have been the platform for a new age of roiling evolutionary ferment among pathogenic microbes. The Roman Empire was caught in the turbulence of this great acceleration.

The ancients revered the frightful sway of the goddess Fortuna, aware, in their own way, that the presiding powers of history seem to be a volatile mix of structure and chance, laws of nature and sheer luck. The Romans lived at a fateful juncture in the human story, and the civilization they built was, in ways the Romans could not have imagined, the victim both of its own success and the caprice of the environment. The enduring power of the Romans to enchant us derives, at least in part, from the poignancy of our knowledge that they stood on the invisible edge of unsuspected change. The long, intertwined story of humanity and nature is full of paradox, surprise, and blind chance. That is why the particularity of history matters. Nature, like humanity, is cunning, but constrained by the circumstances of the past. Our story, and the story of the planet, are inseparable.

It might speak to us, in many ways, that the environment played such a part in the making and the undoing of one of history's most conspicuous civilizations. Rome is almost inevitably a mirror and a measure. But we should not see the case of Rome as the object lesson of a dead civilization. Rather, the Roman experience is important as part of an ongoing story. Far from marking the final scene of an irretrievably lost ancient world, the Roman encounter with nature may represent the opening act of a new drama, one that is still unfolding around us. A precociously global world, where the revenge of nature begins to make itself felt, despite persistent illusions of control . . . this might feel not so unfamiliar. The primacy of the natural environment in the fate of this civilization draws us closer to the Romans, huddled together to cheer the ancient spectacles and unsuspecting of the next chapter, in ways we might not have imagined.

ACKNOWLEDGMENTS

In the course of this project, I have incurred more debts than I can possibly acknowledge, and the gratitude I feel toward the colleagues, institutions, friends, and family whose support allowed me to write this book far outstrips my ability to express it. The project was generously supported early on by the Guggenheim Foundation. Over the last several years, I have benefitted from the opportunity to share different versions of the argument with thoughtful audiences at Berkeley, Columbia, Yale, Princeton, Indiana, the University of Oklahoma History of Science Colloquium and President's Associates, Stanford (where I was fortunate to spend a brief spell as a Visiting Scholar), and Harvard (on multiple occasions).

I am indebted to several colleagues who have shared their data with me, and I must single out Rebecca Gowland and Kristina Killgrove not only as exceptional scholars but also as models of generosity and openness. I am also grateful to the many colleagues who have shared their own work in progress or kindly offered thoughts on my own, including Cam Grey, Colin Elliott, Gilles Bransbourg, Laetitia Ciccolini, Clifford Ando, Peter Temin, Joseph Bryant, Adam Izdebski, Brent Shaw, Marcel Keller, Henry Gruber, and John Mulhall. I thank Maja Kiminko for assistance in tracking down an image; Jack Tannous for sharing wonderful, obscure references to the plague; Joseph Hinnebusch for answering questions about fleas; Hendrick Poinar and Ana Duggan for helpful conversation about smallpox.

The University of Oklahoma, under the leadership of President David Boren, is an amazing place. It has given me extraordinary opportunities for so much of my life, and I am deeply grateful for the constant support of my friends, teachers, and colleagues. I have been lucky to have capable research assistants like Skyler Anderson and Steven Thorn to help with this project. Todd Fagin is a talented cartographer, responsible for the maps throughout the book. The dedicated team in University Libraries deserves special notice. Kerry Magruder and JoAnn Palmeri provided expert assistance. The circulation and interlibrary loan teams have been helpful and endlessly patient. The Department of Classics and Letters has been a wonderful and enriching home for so many years, and I am also in the debt of the remarkable team that I have the honor to work with in the Provost's Office. Colleagues in numerous departments, including Meteorology, Anthropology, Biology, and History, have kindly endured my pestering inquiries and taught me much. To all my colleagues and friends and students, I am grateful. To Bill, David, Luis, Scott, and Andrew, thanks for everything. In short, from start to finish this is an OU book: Boomer Sooner.

It has been a privilege to work with Princeton University Press. Jay Boggis proved an ace copyeditor. Matt Rohal and Karen Carter have been constantly kind and helpful. And my editor Rob Tempio has, in so many ways, from start to finish, guided this book to publication; his good judgment is to credit for many improvements, small and enormous, that have made this a much better book.

I was fortunate to have in Walter Scheidel, John McNeill, and William Harris three generous expert reviewers, whose extensive and candid advice has saved me from numerous mistakes and strengthened the argument throughout. I am grateful to Ann Carmichael for reading parts of the manuscript and offering helpful conversations about the history of disease. Similarly, Michelle Ziegler kindly read my chapter on plague and offered invaluable suggestions. Daniel Sargent read the text and provided some of the most helpful advice I received. I am grateful to Chris May for reading the entire manuscript with extraordinary care and insight; his medical experience sharpened my thinking and expression throughout. Scott Johnson, too, is an amazing friend and generous colleague, and he commented on almost every page of the text. Thank you all.

It has been my great fortune to have had remarkable teachers throughout my life, from Edmond Public Schools to OU and Harvard, and I hope this book is a small tribute to their influence and inspiration. The late J. Rufus Fears introduced me to the "fall of the Roman Empire" in an undergraduate capstone at OU, and I have thought about him countless times while writing this book. My graduate mentor, Christopher Jones, taught me that there is still much to learn about the Romans; his scholarly example and constant friendship have been such an encouragement over the years. And, this particular book bears the unmistakable imprint of Michael McCormick. Thanks to his enterprise and creativity, I was surrounded as a graduate student by the exhilarating possibility of using the natural sciences to illuminate the human past. Mike's support for me over the years exceeds all accounting. The Initiative for the Science of the Human Past is the model of cutting-edge research at the intersection of the sciences and the humanities, and I am grateful that Mike has given me so many ways to be involved. This book would be unthinkable without him.

Finally, to my mom and my whole family, thank you for your love, sacrifice, and support. Michelle, you are my true partner and beloved companion in everything I do, and this book is our book. Sylvie, August, and Blaise—you're my best friends, and this one's for you.

APPENDIX A: FEMUR LENGTH DATA FROM HISTORICAL ITALIAN POPULATIONS

Site	*Ref.*	*Sample size (m)*	*Sample size (f)*	*Date Range*	*Male Femur mm*	*Female Femur mm*	*Raw/ Rec*
Spina	Marcozzi and Cesare 1969	6	–	1000–600 BC	448.8	–	REC
Atestino (Padova) = Este	Corrain 1971	5	1	9–6C BC	469.1	391.0	REC
Osteria dell'Osa	Becker 1992	47		900–650 BC	449.1		REC
Campovalano Abruzzo	Coppa et al. 1987	6	6	10C–4C BC	456.5	424.3	RAW
Monte Casasia (Sicily)	Facchini and Brasili Gualandi 1980	19	11	7–6C BC	443.1	414.5	RAW
Castiglione	Facchini and Brasili Gualandi 1977–9a	7	8	7–6C BC	434.4	409.0	RAW
Salapia	Corrain, Capitanio, and Erspamer 1972	9	8	9–3C BC	436.8	412.1	RAW
Sirolo (Numana, Marche)	Corrain and Capitanio 1969	7	1	8–4C BC	450.1	413.0	RAW
Camerano I	Corrain, Capitanio, and Erspamer 1977	27	7	6–5C BC	454.3	417.5	RAW
Selvaccia	Pardini and Manucci 1981	9	5	6–5C BC	455.9	408.3	REC
S. Martino in Gattara Ravenna	Facchini 1968	2	1	5C BC	465.0		RAW
Pontecagnano	Pardini et al. 1982	145	84	5–4C BC	452.0	416.6	REC

Site	*Ref.*	*Sample size (m)*	*Sample size (f)*	*Date Range*	*Male Femur mm*	*Female Femur mm*	*Raw/ Rec*
Certoso di Bologna	Facchini and Evangelisti 1975	4	4	5–4C BC	431.0	403.0	REC
Pantanello / Metaponto	Carter 1998	20	40	515–275 BC	427.3	410.5	RAW
Rutigliano (Bari)	Scattarella and De Lucia 1982	16	13	6–4C BC	438.0	416.0	RAW
Satricum (S Lazio)	Becker 1999	6	4	5–3C BC	474.0	411.0	RAW
Tarquinia	Mallegni, Fornaciari, and Tarabella 1979	5	5	6C–2C BC	455.5	417.3	REC
Camerano II	Corrain, Capitanio, and Erspamer 1977	30	14	4–3C BC	450.3	410.3	RAW
Tarquinia	Becker 1993	13	11	4–3C BC	455.9	420.0	REC
Dos dell'Arca (Valcamonica)	Corrain and Capitanio 1967	3	4	5–2C BC	453.7	431.4	REC
Monte Bibele (Bologna)	Gruppioni 1980, Brasili Gualandi 1989	10	4	4–2C BC	445.3	417.5	REC
Castellaccio Europarco (republican)	Killgrove 2010a	6	4	4–1C BC	431.0	393.0	RAW
Valeggio (sul Mincio, Verona)	Capitanio 1986–7	12	6	1C BC–1C AD	422.0	415.0	REC
Collelongo (Aquila)	Borgognini Tarli and La Gioia 1977	14	10	1C BC–1C AD	422.0	407.0	REC
Pompeii	Lazer 2009	148	?	79	440.0	407.5	RAW
Pompeii	Henneberg and Henneberg 2002	?	?	79	444.7	408.0	REC
Pompeii	Gowland and Garnsey 2010	?	?	79	433.2	407.5	REC
Herculaneum	Capasso 2001	?	?	79	423.6	395.1	REC

Site	*Ref.*	*Sample size (m)*	*Sample size (f)*	*Date Range*	*Male Femur mm*	*Female Femur mm*	*Raw/ Rec*
Via Collatina	Buccellato et al. 2008	?	?	70–200	452.1	412.6	REC
Le Palazzette (Ravenna)	Facchini and Brasili Gualandi 1977–9b	12	11	1–3C AD	448.7	410.6	REC
Potenzia	Capitanio 1974	9	6	1–3C AD	443.2	425.0	RAW
Via Basiliano	Buccellato et al. 2003	?	?	70–240 AD	452.1	416.2	REC
Urbino	Corrain, Capitanio, and Erspamer 1982	29	12	1–3C AD	450.2	396.0	RAW
Casal Bertone	Killgrove 2010a	20	7	1–3C AD	439.0	410.6	RAW
Castellaccio Europarco (imperial)	Killgrove 2010a	19	6	1–3C AD	443.5	383.3	RAW
Tomba Barberini	Catalano et al. 2001a, 2001b	12	7		445.3	405.7	REC
Quadraro	Catalano et al. 2001a, 2001b	9	7		448.3	413	REC
Serenissima	Catalano et al. 2001a, 2001b	9	7		445.3	403.2	REC
Vallerano	Catalano et al. 2001a, 2001b; Cucina et al. 2006	8	3		452.5	421.5	REC
Casal Ferranti/ Osteria Curato	Catalano 2001a, 2001b	7	2		447	417.4	REC
Fano	Corrain, Capitanio, and Erspamer 1982	7	5	2–3C AD	451.7	401.7	RAW
Bagnacavallo (Ravenna)	Facchini and Stella Guerra 1969	6	3	2–3C AD	434.0	401.0	REC
S. Vittorino	Catalano 2001a, 2001b	4	3		456.5	414.2	REC

Site	*Ref.*	*Sample size (m)*	*Sample size (f)*	*Date Range*	*Male Femur mm*	*Female Femur mm*	*Raw/ Rec*
Velia	Gowland and Garnsey 2010				443.5	407.2	RAW
Isola Sacra	Gowland and Garnsey 2010			1–3C AD	437.4	409.0	RAW
Basiliano	Gowland and Garnsey 2010				449.1	404.2	RAW
Serenissima	Gowland and Garnsey 2010				437.7	395.2	RAW
Lucrezia Romana	Gowland and Garnsey 2010				451.0	410.0	RAW
Potenzia	Corrain, Capitanio, and Erspamer 1982	13	8	2–4C AD	441.4	418.6	RAW
La Marabina (Classe, Ravenna)	Martuzzi Veronesi and Malacarne 1968	4		2–4C AD	422.5		RAW
Mont-Blanc Aosta fase 2 (VAO)	Corrain, Capitanio, and Erspamer 1986; Corrain and Capitanio 1988	46		2–4C AD	438.0		REC
Castellecchio di Reno (BO)	Belcastro and Giusberti 1997	21	11	2–4C AD	457.0	419.3	REC
Civitanova Marche (MAR)	Corrain, Capitanio, and Erspamer 1982; Erspamer 1985	23	23	4C AD	451.2	406.3	RAW
Vadena (Laimburg) Bozen	Capitanio 1981	6		350–410 AD	–	439.0	REC
Mont-Blanc Aosta fase 2 (VAO)	Corrain, Capitanio, and Erspamer 1986; Corrain and Capitanio 1988	39		4–5C AD	438.8		REC
Agrigento	Carra 1995	7	7	mostly 350–450	444.1	400.6	REC

Site	*Ref.*	*Sample size (m)*	*Sample size (f)*	*Date Range*	*Male Femur mm*	*Female Femur mm*	*Raw/ Rec*
Chieri (PIE)	Mallegni et al. 1998	15	8	5–6C AD	428.1	414.2	REC
Dossello di Offanengo (Cremona)	Capitanio 1985	4		5–8C AD	474.0		RAW
Centallo (PIE)	Mallegni et al. 1998	36	13	6–7C AD	414.7	400.0	REC
Mola di Monte Gelato	Conheeney 1997	3	8	early medieval	447.3	418.2	RAW
Mont-Blanc Aosta fase 2 (VAO)	Corrain, Capitanio, and Erspamer 1986; Corrain and Capitanio 1988	27		6–7C AD	441.5		REC
Rivoli (PIE)	Mallegni et al. 1998	7	2	6–8C AD	421.8	391.1	REC
Mont-Blanc Aosta fase 2 (VAO)	Corrain, Capitanio, and Erspamer 1986; Corrain and Capitanio 1988	47		7–8C AD	442.5		REC
Acqui (PIE)	Mallegni et al. 1998	15	8	7–11C AD	418.4	386.2	REC
Atesino	Corrain 1971	5	1	1000–300 BC	469.1	391.0	RAW
Fermo	Corrain and Capitanio 1972	4	5	9–6C BC	455.2	426.4	RAW
Monte Saraceno (Mattinata, Gargano)	Corrain and Nalin 1965	5	3	7–6C BC	434.6	402.7	RAW

APPENDIX B: AMPLIFICATION EVENTS IN THE FIRST PANDEMIC (AD 558–749)

In this catalogue of amplification events, I have noted at the head of each entry when I believe there is a plausible connection to a previous entry.

1.
   Date: 558
   Regions affected: Constantinople
   Notes: Agathias provides a good description of the symptoms of bubonic and septicemic plague. According to Agapios, it also affected surrounding countries.
   Sources:
      Agathias, *Hist.* 5.10
      John Malalas, *Chron.* 18.127 (489)
      Theophanes, *Chron.* AM 6050
      Agapios, *Kitab al-ʾUnwan*
      See also: Stathakopoulos no. 134
2. *Possibly an extension of Event 1*
   Date: 561–62
   Regions affected: Cilicia, Syria, Mesopotamia, Persia
   Notes: In 561, according to Theophanes, there was a great mortality (not specified as bubonic plague) in Cilicia and Anazarbos (contra Stathakoupolos and Conrad, I do not believe he includes Antioch in AD 561). But Stathakoupolos provides compelling reasons to believe that the outbreak in Antioch described in the *Vita* of Simeon the Younger at 126–29 belongs around AD 561. A Syriac chronicle written by a Mesopotamian priest named Thomas described a plague starting in April of AD 562, presumably in western Syria. This outbreak is the best candidate for the bubonic plague events in Syria and the Sassanian kingdom during the tenure of Joseph as *katholikos*, whose memory was blackened by its association with this visitation. This visitation should also be the second (of four) mentioned by Evagrius, but with Stathakopoulos, I agree that

Evagrius provides no grounds to date it to AD 558. Thus, the evidence supports a second amplification of bubonic plague starting in Cilicia and spreading east in AD 561–62. Possibly, this amplification was connected with the resumption of plague in Constantinople three years prior, or it could have originated from a reservoir in eastern Anatolia.

Sources:

Theophanes, *Chron*. AM 6053

Vita Symeon Stylites Junior, 126–29

Chron. ad a. 640 (tr. Palmer, *The Seventh Century in the West-Syrian Chronicles*, p. 15)

Barhadbsabba, PO 4, p. 388–89

Chron. Seert, PO 7, pp. 185–86

Amr ibn Matta, ed. Gismondi p. 42–43

See also Stathakopoulos no. 136

3.

Date: between 565 and 571

Regions affected: Liguria, Northern Italy

Notes: In perhaps the most evocative of all western reports of plague, Paul the Deacon describes an outbreak beginning in Liguria and sweeping with devastating effect toward the north, stopping at the boundaries of the Bavarians and Alamanii and affecting only the Romans. The chronological clues are that it occurred toward the end of Narses' activity in Italy and around the first years of Justin II's reign. Thus, with Stathakoupolos, it is tempting to place this amplification of the plague around 570–71 and associate it with the next event.

Sources:

Paul the Deacon, *Hist. Langobardorum* 2.4

See also: Stathakopoulos no. 139

4. *Possibly an extension of Event 3*

Date: 571

Regions affected: Italy, Gaul

Notes: Marius chronicles a plague killing many across Italy and Gaul. Unlike the first visitation, this time the plague reached Gregory's Clermont in the Auvergne. It also hit Lyon, Bourges, Chalon-sur-Saône, and Dijon. It is thus tempting to see this amplification, connected with #3 above, as a broader event, with the plague arriving along the Italian Riviera and penetrating inward, and arriving in southern Gaul and moving along the Rhone River.

Sources:

Marius of Avenches, an. 571
Gregory of Tours, *Lib. hist.* 4.31–32
See also: Stathakopoulos no. 144

5.

Date: 573–74

Regions affected: Constantinople, Egypt, the East

Notes: Again bubonic plague ravaged Constantinople, as John of Biclaro's eyewitness report emphasizes; Michael the Syrian has 3000 dying per day in the capital. All sources agree that the plague was severe in the capital. John of Nikiu, from Egypt, has it affecting "all places." Agapios and Michael the Syrian also claim it was general. This is presumably the third of four visitations that Evagrius counted.

Sources:

John of Biclaro, an. 573 (MGH AA 11, p. 213)
Agapios, *Kitab al-'Unwan*
John of Nikiu, 94.18
Chron. ad an. 846
Michael the Syrian, 10.8 (346)
See also Stathakopoulos no. 145

6.

Date: 582–84

Regions affected: southwestern Gaul

Notes: Gregory had heard that in AD 582 bubonic plague was raging in Narbonne. In AD 584, he reported again a pestilence in various places but especially Narbonne, whose inhabitants returned in the third year after its original appearance; falsely believing they were safe, they died. The city of Albi also suffered. Narbonne is a coastal city, suggesting again introduction of the plague by sea and inland penetration, though if Gregory's report is complete, the amplification was limited and patchy.

Sources: Gregory of Tours, *Lib. hist.* 6.14 & 6.33

7.

Date: 586

Regions affected: Constantinople

Notes: There is no specific information that this plague was bubonic, but Agapios reports a death toll of 400,000 in the capital in the fourth year of the reign of Maurice. While this number should mean no more than "a very great many people," the previous belief about "waves" of plague has probably underestimated the possibility that this was an amplification of bubonic plague in Constantinople, which remains uncertain.

Sources: Agapios, *Kitab al-ʾUnwan*

8.

Date: 588

Regions affected: Gaul

Notes: Gregory provides a remarkably vivid and epidemiologically plausible account of an amplification that began when a ship from Spain docked in Marseilles. One family fell almost instantly, then a delay, then the whole town was ablaze for two months; the plague ceased, then started again, which could be related to the heat of summer abating. Moreover, the plague at Marseilles quickly moved up the Rhone to a village outside Lyon.

Sources:

Gregory of Tours, *Lib. hist.* 9.21–22

9.

Date: 590–91

Regions affected: Rome, Narni, Rhone Valley.

Notes: Following extreme flooding, a severe outbreak of plague struck Rome. Pelagius II died, and Gregory the Great became pope. Gregory also reports an outbreak of bubonic plague in Avignon and Viviers, again underscoring the importance of riverine transport networks in introducing the bubonic plague into Gaul. In *Ep*. 2.2, Gregory refers to an epidemic in 591 in Narni, suggesting inland penetration in Italy.

Sources:

Gregory of Tours, *Lib. hist.* 10.1 & 10.23
Gregory the Great, *Dial.* 4.18, 4.26, 4.37; *Ep*. 2.2
Paul the Deacon, *Hist. Langobardorum* 3.24
Liber pontificalis 65
See also: Stathakopoulos no. 151

10. *Possibly related to #9.*

Date: 591

Regions affected: Ravenna, Grado, Istria

Notes: Paul the Deacon reports a visitation of the plague in three places on the Adriatic.

Sources:

Paul the Deacon, *Hist. Langobardorum* 4.4
See also: Stathakopoulos no. 154

11.

Date: 592

Regions affected: Syria, Palestine

Notes: For the fourth time, bubonic plague hit Antioch, in this case killing Evagrius' daughter and grandson. The funerary inscription from Feinan refers to one-third of the universe dying. And this amplification is probably the one described in the poetry of Hassan ibn Thabit, although that is not certain.

Sources:

Evagrius, *Hist. eccl.* 4.29
Inscriptions from Palaestina Tertia Ib, nos. 68–70
Hassan ibn Thabit (Conrad 1984)
See also Stathakopoulos no. 155

12.

Date: 597

Regions affected: Thessalonica and countryside

Notes: The author of the *Miracles of Demetrius* claims that God sent the bubonic plague not just in the city but in the whole countryside and caused mass mortality. The Avars heard of the depopulation and attacked the city. Stathakopoulos provides very convincing reasons to date this outbreak to 597.

Sources:

Mir. Demetr. 3 & 14
See also: Stathakopoulos no. 156

13. *Possibly related to #12.*

Date: 598

Regions affected: Thrace

Notes: The invading Avars were struck by the bubonic plague, and the Chagan supposedly lost seven sons in one day.

Sources:

Theophylact Simocatta, 7.15.2
See also Stathakopoulos no. 159

14. *Possibly related to #12 and #13.*

Date: 599–600

Regions affected: Constantinople, Asia Minor, Syria, North Africa, Italy

Notes: Michael's chronicle reports incredible mortality figures from Constantinople (3,180,000) and claims the outbreak swept Bithynia and all of "Asia." The *Chronicle of 1234* reports 380,000 victims in Constantinople. In *Ep.* 9.232, Gregory describes a devastating mortality sweeping Rome, other cities in the region, Africa, and the east. He is explicit that the disease started in the east, where even worse desolation was reported. Without great chronological specificity, Paul places another outbreak in Ravenna and then Verona. Elias and Thomas confirm the outbreak in Syria, too.

Sources:

Michael the Syrian, 10.23 (387)
Chronicon ad an. 1234
Gregory the Great, *Ep.* 9.232, 10.20
Paul the Deacon, *Hist. Langobardorum* 4.14
Elias of Nisibis, an. 911
Thomas of Marga, *Book of Governors* 11
See also Stathakopoulos no. 160

15.

Date: 609
Regions affected: Spain
Notes: A Latin epitaph from Córdoba describes a victim who died of bubonic plague in an otherwise unknown amplification.
Sources:
CIL II 7.677

16.

Date: 610
Regions affected: China
Sources: see Twitchett 1979

17.

Date: 610–41
Regions affected: Constantinople
Notes: A deadly plague occurred in the capital in the reign of Heraclius. No other outbreaks are known. Stathakopoulos connects this outbreak with a pestilence observed by John the Almsgiver in Alexandria.
Sources:
Mirac. sanct. Artemii 34
See also Stathakopoulos no. 173

18.

Date: 626–28
Regions affected: Palestine, Mesopotamia
Notes: Michael records a severe plague in Palestine. Eutychius places it also in the kingdom of the Persians, as do al-Tabari and numerous other Arabic sources. al-Tabari claimed that most of the Persians perished.
Sources:
Michael the Syrian, 11.3 (409)

Eutychius, *Annales*
al-Tabari 1061
Arabic sources in Conrad, p. 159ff.
See also: Stathakopoulos nos. 177, 178

19.

Date: 627–28
Regions affected: Hami (Xinjiang)
Notes: Whether this plague recorded in Chinese sources among the Turks was bubonic is uncertain.
Sources:
Julien 1864, p. 231

20.

Date: 638–39
Regions affected: Palestine, Syria, Mesopotamia
Notes: A plague struck Palestine, Syria, and Mesopotamia. Known as the Plague of 'Amwas, it is widely remembered in the Arabic tradition.
Sources:
Michael the Syrian, 11.8 (423)
Elias of Nisibis, (AH 18)
Chronicle of 1234, 76 (AH 18)
Arabic sources in Conrad, pp. 167ff.
See also Stathakopoulos no. 180

21.

Date: 664–66
Regions affected: England and Ireland
Notes: Bede describes a plague that started in southeast England and spread over the island as well as Ireland. Adamnan describes this as the first of two outbreaks of pestilence that were global in scope. The description of buboes, and the widespread nature of the epidemic, speak for the likelihood of bubonic plague.
Sources:
Adamnan, *Vita Columbae* 47
Bede, *Hist. eccl.* 3.23, 27, 30; 4.1, 7, 8
Bede, *Vit. Cuthb.* 8 (*Two Lives of Saint Cuthbert*, 180–85)
See also Maddicott 2007

22.

Date: 670–71
Regions affected: Kufa (Mesopotamia)
Notes: An outbreak of bubonic plague occurred in Kufa but is not attested elsewhere.

Sources:

Arabic sources in Conrad, pp. 250–53
See also Stathakopoulos no. 185

23.

Date: 672–73
Regions affected: Egypt, Palestine, Mesopotamia
Notes: Theophanes records a mortality (not specified as bubonic in his laconic report) in Egypt. Agapios claimed that bubonic plague struck Egypt and Palestine. In Mesopotamia, it is specifically attested at Kufa and al-Najaf.
Sources:

Theophanes, *Chron.* AM 6164
Agapios, *Kitab al-'Unwan*
Arabic sources in Conrad, pp. 253ff.
See also: Stathakopoulos no.186

24.

Date: 680
Regions affected: Rome, Pavia
Notes: Paul describes a severe epidemic lasting three months (July–September) in Rome and Pavia. Stathakopoulos plausibly argues this was bubonic plague.
Sources:

Paul the Deacon, *Hist. Langobardorum* 6.5
Liber pontificalis 81
See also Stathakopoulos no. 192

25.

Date: 684–87
Regions affected: England and Ireland
Notes: Bede describes a plague that ravaged "many provinces." Adamnan describes this as the second of two outbreaks of pestilence that were global in scope.
Sources:

Adamnan, *Vita Columbae* 47
Bede, *Hist. eccl.* 4.14
See also Maddicott 2007

26.

Date: 687–89
Regions affected: Syria, Mesopotamia
Notes: In a highly apocalyptic vein, John bar Penkaye describes a devastating outbreak of bubonic plague. A contemporary famine is widely described. Arabic sources detail the heavy toll of this amplification,

known as the "Plague of the Torrent." As Conrad argues, it is somewhat unclear whether we should envision a single amplification or a series of outbreaks in a short span of time in the 680s.

Sources:

John bar Penkaye, *Rish melle*
Arabic sources in Conrad, pp. 263ff.
See also: Stathakopoulos nos. 194, 195

27. *Possibly connected with #26.*

Date: 689–90
Regions affected: Egypt
Notes: An outbreak of bubonic plague is recorded in Egypt. Conrad p. 272 does not believe the grounds are strong for a connection with #26.

Sources:

Arabic sources in Conrad, pp. 271ff.
See also: Stathakopoulos no. 196

28.

Date: 693
Regions affected: Spain, southwestern Gaul
Notes: The *Mozarabic Chronicle of 754* reports an outbreak of bubonic plague in the days of King Egica, which is likely to be associated with a mortality caused by bubonic plague recorded in a law affirming the acts of the sixteenth council of Toledo; in the royal acts of the seventeenth council, Narbonensis is described as depopulated.

Sources:

Mozarabic Chronicle of 754, 41
See also: Kulikowski 2007, p. 153–54

29.

Date: 698–700
Regions affected: Constantinople, Syria, Mesopotamia
Notes: Plague appeared in the same year in Constantinople and Syria. The emperor Leontius dredged the Neorion harbor, suggesting a belief in the port and its waters in the etiology of plague. The *Chronicle ad an. 819* places it in "all regions of Syria." From there it spread east. As Stathakopoulos notes, our information is not detailed enough to know whether the plague traveled from Syria to Constantinople or vice versa, although an amplification beginning in Syria and spreading outward in both directions, as Conrad hypothesized, is attractive.

Sources:

Elias of Nisibis (AH 79 and 80)

Chron. ad an. 819, AG 1011

Arabic sources in Conrad, pp. 274ff.

Theophanes AM 6190 & 6192

Nikephoros, *Brev.* 41

Leo Grammaticus, *Chron*, ed. Bekker p. 167

See also: Stathakopoulos nos. 198 and 199

30.

Date: 704–6

Regions affected: Syria and Mesopotamia

Notes: Michael's chronicle claims a grave pestilence killed a third of the population, possibly in Syria. From 706, the plague reached Iraq, striking Basra and Kufa. It was known as the Plague of the Maidens.

Sources:

Michael the Syrian, 11.17 (449)

Chron. Zuqnin (AG 1016)

Arabic sources in Conrad, pp. 278ff.

See also: Stathakopoulos nos. 201, 203

31.

Date: 707–9

Regions affected: Spain

Notes: In 707, 708, and 709, an outbreak of plague killed half the population of al-Andalus, preparing the way for its conquest.

Sources:

Akhbar majmu'a, 7.BkS, tr. James 2012

See also Kulikowski 2007

32.

Date: 713

Regions affected: Syria

Notes: Among a series of other disasters, God was said to have sent the bubonic plague, hitting Antioch.

Sources:

Chronicle of Disasters (AG 1024)

Michael the Syrian, 11.17 (452)

Chron. ad an. 819 & *ad an. 846* (AG 1024)

See also Stathakopoulos no. 205

33.

Date: 714–15

Regions affected: Egypt

Notes: According to Severos, the plague recurred in successive years, causing massive mortality, in the time of the patriarch Alexander II. The identification of the pathogenic agent is circumstantial, but both Stathakopoulos and Conrad associate the outbreak with bubonic plague.

Sources:

Severos, *History of the Patriarchs*, 17

See also Stathakopoulos no. 207

34.

Date: 718–19

Regions affected: Syria, Mesopotamia

Notes: While this outbreak of plague could be related to a pestilence that affected Arab troops besieging Constantinople, army epidemics are so common that it is dangerous to infer a relationship between those events and the certainly attested outbreaks of bubonic plague in Syria and Mesopotamia. What can be securely stated is that a plague in Syria again spread to Iraq.

Sources:

Arabic sources in Conrad, pp. 286ff.

See also: Stathakopoulos no. 209

35.

Date: 725–26

Regions affected: Syria, Mesopotamia

Notes: A number of sources place a severe outbreak of bubonic plague in Syria, including the western pilgrim Willibald, traveling in the Holy Land. Michael the Syrian indicates that Mesopotamia was also struck. An epizootic also occurred.

Sources:

Theophanes, *Chron.*, AM 6218

Vita Willibaldi, 4

Michael the Syrian, 11.19 (436)

Agapios, *Kitab al-ʾUnwan*

Elias of Nisibis (AD 107)

Chron. ad an. 819 (AD 1036)

See also Stathakopoulos no. 213

36.

Date: 729

Regions affected: Syria

Notes: Michael records an outbreak of bubonic plague in Syria.

Sources:

Michael the Syrian, 11.21 (463)

37.

Date: 732–35

Regions affected: Egypt, Palestine, Syria, Mesopotamia

Notes: An outbreak stretched across Egypt and Palestine (Agapios) to Syria (Theophanes) and Mesopotamia (Arabic sources)

Sources:

Theophanes, *Chron.* AM 6225

Agapios, *Kitab al-'Unwan*

Arabic sources in Conrad, pp. 291ff.

See also Stathakopoulos no. 214

38.

Date: 743–49

Regions affected: Egypt, North Africa, Syria, Mesopotamia, Sicily, Italy, Greece, Constantinople, Armenia

Notes: The last amplification during the first pandemic was one of the most geographically widespread since the first phase of the plague. The Arabic sources place the origins of the outbreak in northern Mesopotamia, although it was raging in Egypt just as early, and it raged in Egypt yearly for several years. As it spread westward, it hopped from North Africa to Sicily and from there infected mainland Italy, including probably Rome, and spread rapidly back east to Constantinople, where it caused a devastating mortality over the course of several years.

Sources:

Severos, *History of the Patriarchs* 18

Michael the Syrian, 11.22 (465–66)

Chron. Zuqnin, an. 1055–56, an. 1061–62

Chron. ad an 1234

Theophanes, *Chron.* AM 6238

Nikephoros, *Brev.* 67

Nikephoros, *Antirhetikos* 3

Theodore Studites, *Laud. Platonis* (PG 99: col. 805)

Glycas, *Annales*, p. 527

John Zonaras, *Epit. hist.* 15.6

John of Naples, *Gesta episcoporum neapolitanorum* 42 (with McCormick 2007, p. 292)

Arabic sources in Conrad, pp. 293ff.

See also: Stathakopoulos no. 218–22

NOTES

Prologue: Nature's Triumph

1. On the ceremony surrounding this visit and others like it, McCormick 1986, 123–24. On the population of late antique Rome, see Van Dam 2010; the essays in Harris 1999b; Sirks 1991; Durliat 1990. The public inventory is preserved in two related documents known as the *Curiosum* and *Notitia*, edited in Nordh 1949. The reader should be aware they are closer to a chamber of commerce brochure than a rigorous census. See Arce 1999; Reynolds 1996, 209–50; Hermansen 1978.

2. "A city greater": Claudian, *Stil.* 3.130–34, adapted from the translation of Platnauer 1922.

3. "Equipoise": Claudian, *Stil.* 3.10, tr. Platnauer.

4. On Claudian generally, see Ware 2012; Cameron 1970; on the poet's origin, see Mulligan 2007. "Sprung from humble beginnings": Claudian, *Stil.* 3.136–54, tr. Platnauer.

5. "Inscribed in the annals": Claudian, *Stil.* 2.475–76, tr. Platnauer. "Glory of the woods": Claudian, *Stil.* 3.317, tr. Platnauer. "Marvels of the south": Claudian, *Stil.* 3.345–46, tr. Platnauer. On the animals imported for Roman games, see Toner 2014; Van Dam 2010, 23–24; Guasti 2007; MacKinnon 2006; Jennison 1937.

6. On Claudian's statue: CIL 6.1710; Ware 2012, 1. "In one city": Jerome, *Comm. In Ezech.* pr. On the sack of Rome, see Chapter 5.

7. Gibbon 1781, vol. 3, chapter 38, "General Observations on the Fall of the Roman Empire in the West." On this passage, Bowersock 2009, 28.

Chapter 1: Environment and Empire

1. On the archaeology of early Rome, Holloway 1994. See Carandini 2011 for a provocative reading of the archaeological evidence. For the lead-up to Rome's rise, in the very long perspective, see the brilliant survey of Broodbank 2013.

2. On the rise of Greece, see recently Ober 2015. Carthage: Ameling 1993.

3. On Rome's republican constitution, see Mouritsen 2013, with older literature, and Lintott 1999. Harris 1985 remains the best treatment of Roman militarism. See also Eckstein 2006 on the wider geopolitical context. In general, Beard 2015, here 257.

4. Braudel 1972–73, for a wonderful panorama of the premodern Mediterranean.

5. "Of all the contiguous empires": Scheidel 2014, 7.

6. On the Mediterranean generally, Broodbank 2013; Abulafia 2011; Grove and Rackham 2001; Horden and Purcell 2000; Sallares 1991.

7. On the concept of a "grand bargain," see Scheidel 2015a and 2015b. For the relations between Rome and the provinces, Noreña 2011; Mattingly 2006; Ando 2000; Woolf 1998. For the case of Judea, Isaac 1992. On the spread of citizenship, see now Lavan 2016.

8. See Chapter 2.

9. Malthus 1798 Chapter 7. On wages, Harper 2016a. See further Chapter 2.

10. Goldstone 2002.

11. For the conceptual framework of growth, see Temin 2013. The concept of an "organic economy": Malanima 2013; originally framed by Wrigley 1988. Trade and technology are treated more fully in Chapter 2.

12. Gibbon 1776, Ch. 3. On Gibbon's intellectual world, see Matthews 2010; Bowersock 2009.

13. Brown 1971 gave shape to the idea of "late antiquity" as a distinct period worthy in its own right. His chronological bounds—from Marcus Aurelius to Muhammad—are adopted in this book. For the social development index, see Morris 2010 and 2013.

14. The catalogue of hypotheses in Demandt 1984, with a second edition in 2014. For a thoughtful overview of the dynamics of premodern empires, see the essay of Goldstone and Haldon 2009. For an up-to-date and compelling synthesis of Rome's sequential imperial crises, see now esp. Kulikowski 2016.

15. For these new approaches, Izdebski et al. 2015 (focused on climate); McCormick 2011. See also Scheidel forthcoming; Harper 2016b; Harris 2013a. A valuable overview of Mediterranean paleoclimate studies is found in Lionello 2012. Bioarchaeology: Larsen 2015; Killgrove 2014; MacKinnon 2007. Archaeogenetics: Krause and Pääbo 2016.

16. It should be evident how much I owe to the trailblazing scholarship of environmental historians, especially those working in the Mediterranean, who for many decades now have urged greater attention to environmental issues. In addition to the works cited in footnote 6, we might briefly note, e.g., Meiggs 1982; Hughes 1994; Shaw 1995. The structure of the argument owes a great deal to the recent work of historians who emphasize the role of environmental change in human history, esp. Campbell 2016; Knapp and Manning 2016; Brooke 2014; Cline 2014; Broodbank 2013; Parker 2013; White 2011; Lieberman 2003. Finally, it is worth noting that isolated voices, often outsiders, have sometimes claimed dramatic environmental change as a factor in Rome's decline, e.g., Huntington 1917, Hyams 1952. See Demandt 2014, 347–68 for a full treatment.

17. Pleistocene: Brooke 2014; Ruddiman 2001. Solar variability: Beer et al. 2006.

18. Holocene variability: Mayewski et al. 2004; Bond et al. 2001. See Ruddiman 2005, for the view that humans started impacting the climate system with the spread of agriculture, rather than the Industrial Revolution.

19. These periods are described in detail throughout the book. For the sake of clarity, the Roman Climate Optimum is in common parlance but lacks consensus around its boundaries. I propose 200 BC–AD 150. The Roman Transitional Period is little studied. The term is my own. As becomes apparent in later chapters, I believe it divides somewhat into an early, dry period (AD 150–300) and a later period dominated by a positive NAO (AD 300–450). The Late Antique Little Ice Age is a term just coming into common use. I date its beginnings somewhat earlier than others (ca. AD 450) but agree its starkest phase occurs ca. AD 530–680. For overviews, Harper and McCormick forthcoming; McCormick et al. 2012; Manning 2013; Luterbacher et al. 2013.

20. In Harper forthcoming, I outline more fully some of the different ways that climate and disease history overlap.

21. McNeill 1976. Wolfe, Dunavan, and Diamond 2007; Diamond 1997; Crosby 1986; Le Roy Ladurie 1973. See Carmichael 2006 for a broad overview. It is a testament to McNeill's genius that his structure, built in an age without paleomolecular evidence, has endured so well. I have drawn inspiration from Landers 1993 on the historical ecology of infectious disease, and Hatcher 2003, on the powerful role of mortality.

22. See esp. Shah 2016; Harkins and Stone 2015; Barrett and Armelagos 2013; Harper and Armelagos 2013; Quammen 2012; Jones et al. 2008; Garrett 1994.

23. See esp. Sallares 2002 and further in Chapter 3.

24. On the disease ecology of the Roman Empire and its demographic implications, see Scheidel 2001a and 2001b.

25. The essays in Green 2014b provide an exciting overview of the possibilities opened by this perspective, in the context of the medieval world, and see now also Green 2017. For the importance of the Roman period in globalization, in broader perspective, Belich, Darwin, and Wickham 2016, 9. Also Pitts and Versluys 2015, on "globalization" and the Romans.

26. Wilson 1998, for the idea of consilience. McCormick 2011, for its application to premodern history.

27. Butzer 2012; Scheffer 2009; Folke 2006. The essays in McAnany and Yoffee 2010 show how the paradigm can be applied in historical and archaeological contexts. Carmichael 2006, esp. 10. Recently applied with great richness to the environmental history of medieval Europe by Campbell 2016, esp. 23.

28. Cf. in a different context Cronon 1983, 13: "Environment may initially shape the range of choices available to a people at a given moment, but then culture reshapes the environment in responding to those choices. The reshaped environment presents a new set of possibilities for cultural reproduction, thus setting up a new cycle of mutual determination."

Chapter 2: The Happiest Age

1. Students of Galen are well served by two recent overviews: Mattern 2013; Schlange-Schöningen 2003. Nutton 1973 is fundamental. Bowersock 1969, 59–75, put Galen in his cultural context. "All Asia": Philostratus, *Vita Apoll.* 4.34. Hadrian and Pergamum: Birley 1997, 166–68; Halfmann 1986, 191. On the medical aspects of the cult in general, Steger 2016.

2. Jones 2012a on Galen's travels. Alexandria: Galen, *Anat. Admin.* 1.2, tr. Singer.

3. Rome the epitome and "because of the large number": Galen, *Hipp. Artic* 1.22, quoting the rhetor Polemo. For Rome as the epitome, see also Athen. 1.20b. "Daily ten thousand": Galen, *Purg. Med. Fac.* 2. On these passages, Mattern 2013, 126.

4. "Despite being scoffed at": Galen, *Praecog.* 3.19, tr. Nutton. Boethus: *Praecog.* 2.25, tr. Nutton. Pig: Galen, *Anat. Admin.* 8.3, tr. Singer. Mime: Galen, *Anat. Admin.* 7.13. See Mattern 2013, 183–86.

5. "Great was the name": Galen, *Praecog.* 5.4, tr. Nutton.

6. For the bibliography of the Antonine Plague, see Chapter 3.

7. Galen's lengthiest clinical description is at *Meth. Med.* 5.12, tr. Johnston and Horsley. There he discusses milk from highland cattle. Armenian dirt: *Simp. Med.* 9.1.4 (12.191K). Urine: *Simp. Med.* 10.1.15 (12.285K). On Apollo's role, see Chapter 3.

8. ILS 2288.

9. "Boundaries of the nations": Appian, *Hist. Rom.* pr.1. See in general Luttwak 2016; Millar 2004, esp. 188; Mattern 1999; Whittaker 1994; Isaac 1992. Quadi: Cassius Dio, *Hist. Rom.* 71.20.2.

10. On the friction of space and the challenges of coordination, Scheidel 2014 is especially thought-provoking; for the later empire, Kelly 1998, 157. For the number 160 senators: Eck 2000a, 227.

11. "Marginal costs": Whittaker 1994, 86. "Holding power": Appian, *Hist. Rom.* pr. 26. "What the Celts": Pausanias 1.9.6, tr. Levi.

12. "Not only": Hassall 2000, 321, and generally on the scale and make-up of the Roman army: Le Bohec 1994. On the origins of soldiers, see now Roselaar 2016; Ivleva 2016.

13. On pay rates, see most recently Speidel 2014, differing from Alston 1994 in some details. Campbell 2005a, 20–21. Percent of GDP: Scheidel 2015b; Hopkins 1980, 124–25.

14. "During the days": Herodian, 2.11.4–5, tr. Whittaker. Low levels of mobilization: Bang 2013, 421–23; Hopkins 2009a, 196; Mattern 1999, 82–84.

15. Aelius Aristides, *Or.* 26.58, 63, 67, 99, tr. Behr. We share the opinion of Birley 1987, 86, against more skeptical appraisals: "The tribute remains remarkable, when all allowances have been made."

16. Aelius Aristides, *Or.* 26.76, tr. Behr.

17. Aelius Aristides, *Or.* 26.6, tr. Behr.

18. On Hume, Mossner 1980, 266–68.

19. Specialist attempts to reconstruct the size of the Roman population begin with Beloch 1886 (34–36 on Hume). Brunt 1987; Lo Cascio 1994 and 2009; Frier 2000; for the state of the debate up to 2001, see especially the contribution of Scheidel 2001b. More recent contributions include Launaro 2011; De Ligt 2012; Hin 2013. Any global figure for the imperial

population is a composite of regional estimates, and over the last century scholars have built upon the solid foundations laid by Beloch to develop more refined and plausible numbers for various parts of the Roman world. While precision is specious, the maximum of 75 million derives from the distribution in Table 1.1, each tending toward the high end of what is reasonable and likely (largely following Scheidel 2001b, 48). The best evidence for aggregate regional totals comes from Italy and Egypt. In Italy, we are largely dependent on a series of census figures reported down to the reign of Augustus. Hin 2013 produced a breakthrough, with a convincing reading that proposes a figure of 11–12 million. Premodern populations could grow at average rates of 0.1–0.15% per annum over long periods (Scheidel 2007, 42), in the absence of mortality shocks, so that the best estimate for Italy in AD 166 is ca. 14 million. This figure has the added satisfaction of being slightly above the high medieval maximum and just below the early eighteenth-century extent of Italy's populace. For Egypt, we have an enticing report from the Jewish historian Josephus that the population, exclusive of the great metropolis of Alexandria, was 7.5 million. It has been shown, from a range of evidence, that this is altogether unlikely, and that ca. 5 million was more probably the apex in the second century. See Bowman 2011 with previous literature and esp. Rathbone 1990. In general, the estimates for other regions have even wider margins of error, but thanks to much patient labor and careful reasoning from a number of scholars in recent years the range of 70–75 million inhabitants commands cautious confidence. See Bonsall 2013, 17–18; Mattingly 2006 on Britain.

20. On fertility in the Roman world generally, see Hin 2013; the essays in Holleran and Pudsey 2011; Scheidel 2001b, esp. 35–39; and Bagnall and Frier 1994. Age at marriage: Scheidel 2007; Saller 1994; Shaw 1987. Widows: Krause 1994. Roman marriage: Treggiari 1991. "Women are usually married": Soranus, *Gyn.* 1.9.34, tr. Owsei.

21. Harper 2013a, on the importance of status in shaping reproductive culture. Contraception: Frier 1994, though see Caldwell 2004.

22. Maddison 2001, 28, estimates global population at the beginning of the first millennium at ~230,000,000. See also Livi-Bacci 2012, 25; McEvedy and Jones 1978. Ptolemy, *Geogr.* 1.11, 1.17, 6.16. Chinese population: Marks 2012, 106; Deng 2004; Sadao 1986. "In the provinces": quoted in Lewis 2007, 256–57.

23. The case that there were "too many" Romans: Scheidel 2012; Lo Cascio 2009; Frier 2001. Food crisis: Garnsey 1988.

24. As tempting as the proxy evidence is (see De Callataÿ 2005 for one application and Wilson 2009 for a vigorous defense), the methodological objections of Scheidel 2009 seem incisive and insuperable to me. However, the evidence of animal bones (e.g., Jongman 2007) does seem to sit especially poorly with any theory of Roman "overpopulation." The assemblages of pig, cow, and sheep bones run into the tens of thousands. The main limit is that animal consumption is subject to tastes: to become Roman entailed acquiring a taste for pork. Nevertheless, the appeal of these data is irresistible, for meat consumption is a direct rebuttal to claims of Malthusian immiseration. And, even giving some allowance for the Romanization of taste and the imprecisions of dating (especially the tendency to use the "second century" rather than before/after the plague as a chronological boundary), the expansion of meat consumption on a mass scale points away from a prolonged descent into subsistence crisis.

25. The essays in Bowman and Wilson 2009 are stimulating. The wage evidence is discussed in Harper 2016a and the Data-file is available at darmc.harvard.edu under "Data Availability." It seems to me to be the best way to demonstrate intensive growth available to us at this time.

26. At last there is a comprehensive study of Roman urbanism, in the remarkable new synthesis of Hanson 2016.

27. On the size of Roman elite properties, Scheidel 2017; Harper 2015b; Duncan-Jones 1990, 121–42; Duncan-Jones 1982. On the distribution of wealth, Scheidel and Friesen 2009.

28. "The wealth of the Roman Empire": Frier 2001, 158. Kinds of growth: Temin 2013, 195–219. For the concept of efflorescence: Goldstone 2002.

29. "There was never": Greene 2000, 754.

30. "The large number": Wilson 2002, 11. In general, Schneider 2007.

31. "The size": Greene 2000, 756. Lateen: Whitewright 2009.

32. "So many merchant ships": Aelius Aristides, *Or.* 26.11, tr. Behr. "The merchants of the earth": *Apoc.* 18:11–13 (KJV). For a general overview, see Harris 2000.

33. The role of institutions in facilitating exchange: Frier and Kehoe 2007. See also Kehoe 2007. Credit in the Roman economy: Rathbone and Temin 2008; Harris 2008; Harris 2006; Temin 2004; Andreau 1999; Bogaert 1994.

34. Harris 2000, 717 (grain) and 720 (wine). The wine trade: Morley 2007; Tchernia 1986; Purcell 1985. For the later period, Pieri 2005. U.S. wine production statistics are available at www.wineinstitute.org.

35. See esp. Morris 2010 and 2013 for the inadequacy of the "hockey stick" view of social development. Goldstone 2002 is also insightful. For an environmentally inflected view similar to the one presented here, see Campbell 2016.

36. See Lehoux 2007, 119–20, for sensible reflections on Ptolemy's observations; Sallares 2007a, 24–25.

37. The "strange parallels" in medieval Asia: Lieberman 2003. See also Lee, Fok, and Zhang 2008, for Chinese population growth and climate. In the European middle ages, Campbell 2016 and 2010 have overwhelmingly demonstrated the power of climate in the rhythms of growth and recession. The Roman Climate Optimum is explored in Hin 2013; Manning 2013; McCormick et al. 2012.

38. See esp. Burroughs 2005; Taylor et al. 1993 for the image of the flickering switch.

39. Burroughs 2005; Ruddiman 2001. Human impact: Brooke 2014.

40. "How it might have been": Broodbank 2013, 202–48.

41. Division of the Holocene: see Walker et al. 2012. Mid-to-late Holocene: Finné et al. 2011; Wanner 2008. Seasonality: see esp. Magny et al. 2012a.

42. Holocene climate change: Mayewski et al. 2004. "In the galactic scheme of things": http://science.nasa.gov/science-news/science-at-nasa/2013/08jan_sunclimate/. Hallstatt: Usoskin 2016.

43. The southern reaches of Roman power: Alston 1995, 34–35; Strabo *Geogr.* 17.1.12. See Scheidel 2014.

44. See esp. Horden and Purcell 2000 and Sallares 1991 on the dynamics of the Mediterranean and its impacts on human society. On the Mediterranean climate more generally, Lionello 2012; Xoplaki 2002.

45. For overviews, Harper and McCormick forthcoming; Manning 2013; Lionello 2012; McCormick et al. 2012. Lamb 1982 had already reconstructed many of the outlines, with far less evidence.

46. Proposed start dates for the RCO include: 550 BC, 450 BC, 400 BC, 200 BC, and 1 BC; end markers have included 50 BC, 50 AD, 250 AD, 300 AD, and 350 AD. If not an inducement to despair, these divergences suggest the complexity of the problem and the need for deeper efforts at synthesis that go beyond individual proxies sensitive to local conditions.

47. See Usoskin 2016; Steinhilber et al. 2012; Gray et al. 2010; Beer et al. 2006; Usoskin and Kromer 2005; Shindell et al. 2003; Shindell 2001; Bond et al. 2001; Beer, Mende, and Stellmacher 2000.

48. Data source for Fig. 3: ftp://ftp.ncdc.noaa.gov/pub/data/paleo/climate_forcing/solar_variability/steinhilber2009tsi.txt.

49. For the timing and scale of volcanic activity, Sigl et al. 2015 is now fundamental.

50. The written and botanical evidence was already brilliantly exploited by Lamb 1982. Glaciers: Le Roy et al. 2015; Six and Vincent 2014; Holzhauser et al. 2005; Hoelzle et al. 2003; Haeberli et al. 1999.

51. In general, Manning 2013. The Alpine record: Büntgen et al. 2011.

52. Speleothems generally: McDermott et al. 2011; Göktürk 2011; McDermott 2004. Spannagel: Vollweiler et al. 2006; Mangini, Spötl, and Verdes 2005. Iberian: Martín-Chivelet et al. 2011. Kocain: Göktürk 2011. Uzunturla: Göktürk 2011. Grotto Savi: Frisia et al. 2005.

53. Iberian-Roman Humid Period: Pérez-Sanz et al. 2013; Currás et al. 2012; Martín-Puertas et al. 2009. Klapferloch cave in far southern Austria shows a moist phase from 300 BC to 400 AD: Boch and Spötl, 2011. A series of Italian lakes show a high stand in the late republic: Magny et al. 2007; Dragoni 1998. "Fabulous artificial landscape": Aldrete 2006, 4. In general, see the valuable study of Aldrete 2006 for Tiber flooding; Wilson 2013, 269–71; Camuffo and Enzi 1996. Pliny, *Ep.* 8.17.

54. Deforestation: Harris 2013b and 2011; Hughes 2011; Sallares 2007a, 22–23. See further below. Flood data for Figs. 4 and 5 from Aldrete 2006.

55. The sample is small (n = 11), and a fluke cannot be excluded. However, most of the testimony we do have from the Roman Empire is specific and credible. Ovid, *Fasti* 3.519–20: the races were moved from the Campus Martius to the Caelian Hill. The hydrological evidence from Rome also puts Ptolemy's weather report from coastal Alexandria in a different perspective. Like the evidence of the summer Tiber floods, the casual assumption of regular rain in the summers is impossible in today's climate regime. Both point to deep changes in the late Holocene hydrological system in the Mediterranean. Columella: see Hin 2013, 80. In general, also Heide 1997, esp. 117.

56. Irrigation: Leone 2012. Institutions: e.g., Kehoe 1988, 81–88.

57. Elephants: Pliny the Elder, *Hist. Nat.* 8.1. See esp. Leveau 2014. Wilson 2013, 263: the "long-running debate" is still not settled, but the accumulated evidence is now quite convincing. Jaouadi et al. 2016; Essefi et al. 2013; Detriche et al. 2008; Marquer et al. 2008; Bkhairi and Karray 2008; Faust et al. 2004; Slim et al. 2004; Ballais 2004; Stevenson et al. 1993; Brun 1992. Gilbertson 1996, for a good discussion of the older literature.

58. Mattingly 2003–13.

59. "It is even possible": Wilson 2012. Cremaschi and Zerboni 2010; Cremaschi et al. 2006; Burroughs 2005, 231: "The onset of the present superarid conditions did not occur until around 1.5 kya."

60. Talmud: Bavli Ta'anit 19b, tr. Sperber 273. Dead Sea: Bookman 2004; cf. Migowski et al. 2006. See Hadas 1993 for archaeological evidence. McCormick et al. 2012 for discussion. Hirschfeld 2006, focused on the later period, but with valuable insights. Soreq: Orland 2009.

61. See esp. Magny et al. 2012a for the earlier seasonal patterns, positing weaker Hadley cell formation in the earlier period.

62. "Day by day": Lucretius, *De Re. Nat.* 5 lines 1370–71. Lucan, *Pharsalia*, 9. Hadrian: Harris 2013b, 182–83. Ando forthcoming, on these cultural models of deforestation.

63. Climate models of the effects of changes in Mediterranean ground cover: Gates and Ließ 2001; Gaertner et al. 2001; Reale and Shukla 2000; Reale and Dirmeyer 2000. *Contra*, Dermody 2011. Deforestation: Ando forthcoming; Harris 2013b; Hughes 2011; Harris 2011; Sallares 2007; Chabal 2001. Harris 2013 is especially nuanced, arguing for temporal and regional variations, e.g., using archaeological and palynological evidence to show that long timber, especially in areas easily connected to sea-borne transport networks, was more heavily depleted in the Roman Empire. Britain: Dark 2000, 115–19.

64. Hin 2013, 85ff. is the best attempt to link expansion and climate so far. In the mountains: Pliny, *Nat. Hist.* 18.12.63. 5M hectares: Lo Cascio and Malanima 2005, 219. See also Spurr 1986, 17: "The advice against 'steep places' is surely a recognition of the fact that cereals *were* grown in some areas under such conditions."

65. Response to temperature: Dermody et al. 2014; Spurr 1986, 21. Response to precipitation: Touchan et al. 2016. Yields: Spurr 1986, 82–88.

66. Viability: see Garnsey 1988, 10–12. Cf. Leveau 2014; Mattingly 1994, 9–11: "erratic variation is the norm."

67. "Relentless violence": Columella, *De Re Rust.* 1.1.4–5. Remote olive presses: e.g., Foxhall 1990, 109. See also Waelkens et al. 1999.

68. For the notion of an "enabling background," see Campbell 2016. See also Campbell 2010; Galloway 1986.

69. Hadrian's itinerary: Halfmann 1986, 192. Inscription: ILS 2487. "Virtually no emperor": *Hist. Aug.*, *Vita Hadr.* 13.5.

70. "When he came": *Hist. Aug.*, *Vita Hadr.* 22.14. Inscriptions: CIL 8.2609–10. Wheat prices: see Harper 2016a. Carthaginian water supply: Leveau 2014; Di Stefano 2009; Wilson 1998.

71. Fronto, *Ep.* 3.8.1. For this image, see Jones 1972, 143–44.

72. Butzer 2012; McAnany and Yoffee 2010; Scheffer 2009; Folke 2006.

73. In general, Horden and Purcell 2000; Sallares 1991; Garnsey 1988. "Barley": Galen, *De Subt. Diaeta* 6. "After taking": Galen, *Alim. Fac.*, [page 93], tr. Grant. Emergency foods: Galen, *Alim. Fac.*, [page 95], tr. Grant. Storage: Garnsey 1988, 52–54.

74. Euboean oration: Dio Chrysostom, *Or.* 7. "Twin notions": Garnsey 1988, 57.

75. For Pliny as a patron, see esp. Saller 1982.

76. "Cities on the sea coast": Gregory of Nazianzus, *Fun. Or. in Laud. Bas.* 34.3, tr. Jones 1940.

77. Roman intervention: e.g., at Antioch in Pisidia in AD 92–93: AE 1925, tr. Levick 2000b, 120. Private interventions: see Garnsey 1988, 14. E.g., SEG 2.366 (Austin 113); Syll-3 495 (Austin 97); I. Priene 108; I. Erythrai-Klazomenai 28; IGR 3.796; IGR 4.785; IG 4.944; 5.2.515. Macedonia: SEG 17.315 = Freis 1994 no. 91.

78. Trajan: Pliny, *Pan.* 32. Hadrian: Cassius Dio, *Hist. Rom.* 69.5.3.

79. Septimius Severus: *Hist. Aug.*, *Vita Sev.* 23.6. "If, as we pray": I. Ephesos 2.211, tr. Garnsey 1988, 255. See also Boatwright 2000, 93–94. Signal boats: Seneca, *Ep.* 77. On the grain supply generally, Erdkamp 2005; Garnsey 1988, esp. 218–70; Rickman 1980.

80. Exposure: Harper 2011, 81–83; Corbier 2001; Bagnall 1997; Harris 1994; Boswell 1988. Migration: Hin 2013, 210–59. See Chapter 3.

81. Senate: Eck 2000a and 2000b. "Came from families": Hopkins 2009a, 188–89.

82. "Load-bearing": Shaw 2000, 362. Grand bargain: Scheidel 2015a and 2015b. Thin layer: Hopkins 2009a, 184.

83. Luttwak 2016, on the transition to a territorial empire.

84. The point about Cassius Dio is made by Saller 2000, 818.

85. This discussion is much indebted to Hopkins 2009a. See also the important contributions of Scheidel 2015a and 2015b.

86. Vespasian: Suetonius, *Vesp.* 23 and Cassius Dio, *Hist. Rom.* 65.14.5. Domitian: Cassius Dio, *Hist. Rom.* 67.4. Griffin 2000, 79–80. Hadrian and Marcus: Birley 2000, 182.

87. Luttwak 2016. See also Mattern 1999; Whittaker 1994; Le Bohec 1994, 147–78; Ferrill 1986. Trouble under Antoninus: *Hist. Aug.*, *Vita Anton.* 5.4–5; De Ste. Croix 1981, 475; *Hist. Aug.*, *Vita Marc.* 5.4; CIL 3.1412 = ILS 7155. The speech: Aelius Aristides, *Or.* 35.14, accepting the arguments of Jones 1972. See now Jones 2013. Debasement: Butcher and Ponting 2012, 74.

88. On Galen's first arrival in Rome, see Schlange-Schöningen 2003, 140–42; Lucius' itinerary: Halfmann 1986, 210–11. On the campaign: Ritterling 1904.

89. On the course and leadership of the campaign, see Birley 2000, 161–65; Birley 1987. Senatorial military command in general: Goldsworthy 2003, 60–63.

90. On Seleucia, see Hopkins 1972. "Polluted everything": Ammianus Marcellinus, *Res gest.* 23.6.24, tr. Rolfe.

91. "Runaway slave" and "caught the first boat": Galen, *Praecog.* 9.3 (14.649K), tr. Nutton.

Chapter 3: Apollo's Revenge

1. Nile: Aelius Aristides, *Or.* 36. Sickness: 48.62–63, tr. Behr. On Aristides in general: Downie 2013; Israelowich 2012; the essays in Harris and Holmes 2008; Bowersock 1969; Behr 1968.

2. On his ailments, see esp. Israelowich 2012. Galen: Jones 2008, 253; Bowersock 1969, 62. "The doctors": Aelius Aristides, *Or.* 48.63, tr. Behr. Hypochondriac: e.g., Marcone 2002, 806, "un sofista ipocondriaco." Beard 2015, 500: "hypochondriac."

3. Smyrna: Aelius Aristides, *Or.* 19; Philostratus, *Vita. Soph.* 2.9. The "normalcy" of Aristides is brought out by Israelowich 2012 and Jones 2008.

4. The plague episode is described at Aelius Aristides, *Or.* 48.38, tr. Behr.

5. The idea that the *Sacred Tales* were intimately related to the experience of plague: Israelowich 2012. The Antonine Plague was once considered a fundamental event in Roman history (by, e.g., Barthold Niebuhr, one of the founders of modern historiography) and remained prominent through the theory of manpower shortage advocated by Boak 1955. It lost its prominence, for a variety of reasons, including the distaste of Moses Finley for numbers and for demography. Gilliam 1961 among others offered an influential minimizing view. But Duncan-Jones 1996 turned attention to the plague once again, and a lively conversation has continued over the past 20 years, represented by the essays in Lo Cascio 2012; Bruun 2007; Jones 2006; Jones 2005; Gourevitch 2005; Bruun 2003; Greenberg 2003; Zelener 2003; Marcone 2002; Bagnall 2002; Scheidel 2002; van Minnen 2001; Duncan-Jones 1996; Littman and Littman 1973.

6. Valuable introductions to infectious diseases in human history include Barrett and Armelagos 2013; Oldstone 2010; Crawford 2007; Goudsmit 2004; Hays 1998; Karlen 1995; McKeown 1988; McNeill 1976.

7. "Great Tree": Darwin 1859, 130.

8. McNeill 1976; already Le Roy Ladurie 1973. See Armelagos et al. 2005.

9. History of diseases from genetics: Harkins and Stone 2015; Trueba 2014; Harper and Armelagos 2013; Pearce-Duvet 2006; Brosch et al. 2002. Measles: see Newfield 2015, esp. for the attractive suggestion that a near ancestor of measles was active in Europe in late antiquity. TB: see below.

10. Picornavirus: Lewis-Rogers and Crandall 2010. Sleeping sickness, yaws: Harkins and Stone 2015.

11. Healthier Paleolithic: Brooke 2014, 213–20.

12. Latitudinal species gradient: Jablonski et al. 2017; Fine 2015; Davies et al. 2011. For pathogens: Stephens et al. 2016; Hanson et al. 2012; Dunn et al. 2010; Martiny et al. 2006; Guernier et al. 2004.

13. Neolithic turning point: Brooke 2014, 220–42. Harkins and Stone 2015 is one of the best syntheses to draw on the new, genomic evidence.

14. Plague: Rasmussen 2015; Valtuena forthcoming; and see Chapter 6. TB: below.

15. "The great city": Talmud Bavli, *Pesahim* 118b, from Hopkins 2009a, 192. "The power of Rome": Josephus, *Bell. Jud.* 2.16.4 (362) tr. Whiston. Imperial Rome is vividly sketched by Purcell 2000.

16. Mortality rates in Roman societies have been extensively studied without definitive resolution. The most famous of Roman lawyers, Ulpian, developed an annuities schedule based on a life table that paints a bleak picture of Roman mortality (see Frier 1982). But its relationship to reality has been inconclusively debated. The ages at death recorded on countless ancient tombstones are hopelessly skewed by the selective habits of Roman commemoration. These have been disappointing dead ends. I am in agreement with Walter Scheidel about the role of epidemic mortality and its influence on the value of model life tables: Scheidel 2001c, as well as the urban graveyard effect, Scheidel 2003. See Hin 2013, 101–71; Bagnall and Frier 1994, 75–110; Frier 1983; Frier 1982. Life expectancy: Scheidel 2001b, 39. Roman weaning: Prowse et al. 2008. Population of Rome: Morley 1996, 33–39.

17. Bagnall and Frier 1994; Scheidel 2001c.

18. Emperors: Scheidel 1999. Faustina and Marcus: Levick 2014, 62–63; Birley 1987.

19. For an introduction, Larsen 2015. For a holistic application to samples from Rome, Killgrove 2010a. For the state of bioarchaeology in Roman studies, Killgrove 2014.

20. For example, porotic hyperostosis is often seen as an index of biological stress in populations, caused by pathogenic infection, malnutrition, or congenital anemia, but the data from the Roman Empire at present lack sufficiently strong standardization, and drawing conclusions would be unwarranted.

21. For an overview of stature as a proxy of well-being, Steckel 2013; Floud et al. 2011.

22. Steckel 2013, 407.

23. Some methodologies reconstruct stature from the entire skeleton, but most anthropologists have used mathematical formulas to predict stature from the measurement of long bones, the humerus, radius, tibia, or particularly the femur. Long-bone dimensions correlate with overall stature: tall people have longer femurs. But the formulas used to convert bone lengths into height are derived from different modern populations (in particular an overly influential set of whites and African-Americans from the mid-twentieth century United States), introducing uncertainties, especially since distal elements like the tibia and radius may be more elastic in populations under stress. Worst of all, anthropologists over the decades have used different formulas, which produce different results. See Klein Goldewijk and Jacobs 2013.

24. Britain: Gowland and Walther forthcoming. Bonsall 2013, 228–29. Roberts and Cox 2003 is still a valuable meta-study.

25. This meta-analysis is my own, an attempt to replicate and update Kron 2005. See Appendix A for the underlying data. I hasten to add that I believe the value of this analysis is severely compromised by the limitations of the original studies, which derive predominantly from older Italian bio-anthropological traditions. The original studies sometimes report raw femur lengths and sometimes only average statures, based on various regression formulae. When raw lengths are reported, I use those; when only stature is reported, I extrapolate femur length by solving the regression equation as though it were based on femur length, which was not always the case; therefore I distinguish between raw and reconstructed data in the Figure. The original studies almost never include standard deviations, and I have not weighted the analysis based on the number of observations: Figure 3.2 makes no claims to statistical validity. The time variable is the mid-point of the reported range. In short, I place more trust in the single large study of Giannecchini and Moggi-Cecchi 2008, carefully controlled and without the risk of interobserver differences, than other metastudies, including my own.

26. The most important study by far is Giannecchini and Moggi-Cecchi 2008. Its authors gained access to a variety of Italian collections and then analyzed and reported actual long-bone lengths. Older studies include Koepke and Baten 2005 and Kron 2005.

27. See esp. the work of Garnsey 1999, 1998, and 1988.

28. Roman diet: Killgrove 2010; Cummings 2009; Rutgers et al. 2009; Craig et al. 2009; Prowse et al. 2004. Archaeology of animal bones: Jongman 2007; King 1999. Britain: Bonsall 2013, 28: "it is now thought that meat comprised a greater proportion of an average individual's diet than once believed." Cummings 2009; Muldner and Richards 2007.

29. Cucina et al. 2006; Bonfiglioli et al. 2003; Manzi 1999.

30. Dorset: Redfern et al. 2015; Redfern and DeWitte 2011a; Redfern and DeWitte 2011b. York: Peck 2009.

31. Dutch: Maat 2005. Antebellum paradox: Treme and Craig 2013; Sharpe 2012; Zehetmayer 2011; Komlos 2012 (presenting a different view, arguing for the importance of declining wages); Alter 2004; Haines, Craig, and Weiss 2003. "There certainly seems": Malthus 1826, 408.

32. Urbanization rates: Hanson 2017; Morley 2011; Wilson 2011; Lo Cascio 2009; Scheidel 2001b, 74–85; Morley 1996, 182–83. Regional disease ecologies: Scheidel 2001a and 1996. Galen's Pergamum: Galen, *Anim. Affect. Dign.* 9. On the urban graveyard effect, see recently Tacoma 2016, 144–52; the essays in de Ligt and Tacoma 2016, especially Lo Cascio 2016; Hin 2013. The extent of the graveyard effect in Rome is still debatable. But it seems to me that the evidence for extremely high levels of urban mortality are supported by the best evidence: (1) the growing bioarchaeological evidence for high levels of in-migration (see Prowse 2016;

Bruun 2016; Killgrove 2010a; Killgrove 2010b; Prowse et al. 2007) in the early empire; (2) the converging evidence for endemic malaria; (3) the stature evidence; (4) the archaeological evidence that sanitation systems barely blunted the poor hygienic conditions of the city (Mitchell 2017; Koloski-Ostrow 2015). Nonetheless, I would accept the arguments of, e.g., Lo Cascio 2016 that the population of Rome may have grown between Augustus and Marcus Aurelius. I would also argue that, in many places, rural mortality rates were high too, so the urban-rural differences may not have been extreme. As Hin 2013, 227, notes, urban and rural mortality rates fell along a spectrum.

33. For arguments against the "urban graveyard effect" in Rome, see Lo Cascio 2016; Kron 2012; Lo Cascio 2006. On Roman toilets and sewers, see esp. now Koloski-Ostrow 2015, "hallmark" at 3; van Tilburg 2015; Hobson 2009.

34. See Koloski-Ostrow 2015, 88–89 on chamber pots. "The hygienic implications": Scobie 1986, 411 and waste volume at 413. Mitchell 2017, 48, presents an important archaeological synthesis concluding that "public sanitation measures were insufficient to protect the population from parasites spread by fecal contamination."

35. All seasonal mortality data presented is based on my own dataset, compiled from the corpora of Christian inscriptions from Rome. Harper 2015c; Scheidel 2001a and 1996; Shaw 1996. The graphic representation is expressed as an index normalized for the varying number of days in a month (if there were equal mortality across the year, a straight line at 100 would appear). On the causes of seasonal variation, see Grassly and Fraser 2006.

36. Harper 2015c.

37. "This irregularity": Galen, *Temp*. 1.4.528 tr. Singer.

38. On acute diarrheas, see DuPont 1993, 676–80. In general, Scheidel 2001a. The seasonal indices of cause-specific mortality in Italian towns in 1881–82 is drawn from Ferrari and Livi Bacci 1985, 281, and reports the data for "malattie respiratorie" and "enterite e diarrea."

39. "Awesome power": Sallares 2002, 2. For the role of malaria past and present, Shah 2010. On the global genetic diversity of *Plasmodium*, Faust and Dobson 2015. The seasonal mortality data for malaria in Rome, 1874–76, come from Rey and Sormani 1878, using their category "pernicious intermittent fevers."

40. Age and genetic history of malaria: see Loy et al. 2017; Pearce-Duvet 2006, 376–77; Sallares 2004. In Italy, in later times, Percoco 2013. DNA: Marciniak 2016.

41. Sallares 2002 is fundamental. The awareness of the first-century doctor Celsus (e.g., *De medicinia* 3.3.2) is important testimony. "We no longer": Galen, *Morb. Temp.* 7.435K. "Most of all": Galen, *Hipp. Epid.*, 2.25, 17.A.121-2K. See Sallares 2002, 222.

42. Seasonal patterns: Shaw 1996, 127.

43. See O'Sullivan et al. 2008; above all Sallares 2002, 95. "If anyone": Pliny, *Hist. Nat.* 36.24.123.

44. "Why do men?": Ps.-Aristotle, *Prob*. 14.7.909, tr. Sallares 2002, 282. Monica: Sallares 2002, 86. Synergistic malaria: Scheidel 2003.

45. Agronomist: Palladius, *Op. Ag.* 1.7.4. Damp spring: Ps.-Aristotle, *Prob.* 1.19.861. For an analysis of climate controls on malaria in eighteenth century France, Roucaute et al. 2014.

46. "When the entire": Galen, *Temp*. 1.4.531, tr. Singer. Epidemic every 5–8 years: Sallares 2002, 229. "If one person": Seneca, *De Clem.* 1.25.4.

47. If we went back further in time, the historian Livy lists epidemics, mostly in Rome or among the army. It is worth mentioning, too, the Plague of Athens. A number of epidemics attested throughout the 420s–360s BC could be related, but their periodicity and epidemiology do not require any connection between them.

48. Pliny the Elder, *Nat. Hist*. 7.51 (170). We cannot completely exclude the possibility of invasive pathogens rumbling through the empire. Arthropod-borne viruses, like Dengue Fever and Yellow Fever, both reached the Mediterranean in later centuries; they could have done so in antiquity, and they would have largely blended in with the other malarial and gastroenteric epidemics. Influenza epidemics are a better possibility in the abstract, although positive

evidence is curiously lacking. A tantalizing passage in the writings of Rufus of Ephesus, a doctor writing in the reign of Trajan, shows definite familiarity with "pestilential buboes," which can be nothing other than some early form of *Yersinia pestis*, true bubonic plague. That is a problem we will return to in due time. Bubonic plague clearly never became pandemic in his day, and the corpus of Galen betrays no awareness of the disease whatsoever. For Rufus, a pestilence was a concatenation of "every terrible thing": diarrheas, fevers, vomiting, delirium, pain, spasms, etc., but not buboes. Epidemics in the empire, by all appearances, were explosions from within.

49. Spread of intestinal parasites and Roman conquest: Mitchell 2017. TB genome: Achtman 2016; Bos et al. 2014; Comas et al. 2013; Stone et al. 2009. Historical significance of TB: Roberts 2015; Müller et al. 2014; Holloway et al. 2011; Stone et al. 2009; Roberts and Buikstra 2003. Britain: Taylor, Young, and Mays 2005. "Watershed": Eddy 2015.

50. The state of the question is nicely summarized in Green 2017, 502–5. Genetic history of leprosy: Singh et al. 2015. History of leprosy: Donoghue et al. 2015; Monot et al. 2005; Mark 2002 (esp. for the spread from India to Egypt, with older theories discussed); the essays in Roberts, Lewis, and Manchester 2002, esp. Lechat 2002, 158. Pliny the Elder, *Nat. Hist.* 26.5; Plutarch, *Mor.* 731b–34c. Rufus apud Oribasius, *Coll. Med.* 4.63. Roman cases: Inskip et al. 2015 (sub-Roman); Stone et al. 2009; Mariotti et al. 2005; Roberts 2002. Child skeleton: Rubini et al. 2014. Phylogeny: Schuenemann et al. 2013.

51. Plutarch, *Moralia* 731b–734c, "seeds" at 731d.

52. For the phylogeny, see Duggan et al. 2016; Babkin and Babkina 2015; Babkin and Babkina 2012. Updating Shchelkunov 2009; Li et al. 2007. The biogeography of the gerbil and the camel strongly suggest an African origin of human *Variola*. Independently, the fact that African strains of *Variola* reveal the greatest genetic diversity also demonstrates the evolutionary beginnings of smallpox in Africa. On the introduction of camels into Africa, see Farah et al. 2004.

53. In general, see the invaluable collection of sources in FHN volume 3. Augustus: Strabo, *Geogr*. 16.4.22–27. Purcell 2016; Seland 2014; Tomber 2012; Cherian 2011; Tomber 2008; Cappers 2006; De Romanis and Tchernia 1997; Casson 1989; Raschke 1978. Farasan: Phillips, Villeneuve, and Facey 2004, with Nappo 2015, 75–78; Speidel 2007. For the role of the Roman state, see esp. Wilson 2015. A rather skeptical view of the extent of Indian Ocean trade, embodied especially in Raschke 1978, has long existed, but the obsolescence of the dogmatic Weberian primitivism on which it was based, the discovery of the Muziris papyrus, and the accumulation of archaeological evidence (in the Roman Red Sea ports and more generally the Indian Ocean rim, as carefully synthesized in the work of Tomber), seem to me to have laid to rest the minimalizing view.

54. Growth of Myos Hormos: Strabo, *Geog*. 2.5.12. Berenike: Sidebotham 2011. Hard to find information: Strabo, *Geog.* 15.1.4. Pliny the Elder, *Hist. Nat.* 6.101. Seneca the Younger dedicated a book, now lost, to the subject of India, and he shows a very un-Roman awareness of the possibility that equally grand empires existed in the far corners of the globe: Pliny the Elder, *Hist. Nat.* 6.60. See Parker 2008, 70. Poet: Statius, *Silv.* 5.1.603. "Those who are accustomed": Ptolemy, *Geogr*. 1.9, tr. Stevenson. "So many merchant ships": Aelius Aristides, *Or.* 26.11–12, tr. Behr. Nile: Aelius Aristides, *Or.* 36.1, see FHN 3.198ff.

55. "Commercial exchange": Frankopan 2015, 16. Periplus: Casson 1989, 10. Pliny the Elder, *Hist. Nat.* 12.84. Muziris papyrus: De Romanis 2015; Rathbone 2000 on economic significance; Casson 1990.

56. Alexandrian tariff: Dig. 39.4.16.7. On spices in the cookbook of Apicius, Parker 2008, 151–52. Spice quarter: Parker 2008, 153. Pepper prices: Pliny the Elder, *Hist. Nat.* 12.28. Hadrian's Wall: Vindolanda Tablet #184. Tomber 2008, is an invaluable overview of the archaeological evidence for trade in the Red Sea and Indian Ocean.

57. On the routes connecting the Indian Ocean and China, Marks 2012, 83.

58. For Roman coins, see now esp. Darley 2013. Tamil poetry: Power 2012, 56; Parker 2008, 173; Seland 2007, on the later date of these poems. Colony: Casson 1989, 19ff. China:

McLaughlin 2010, 133–34. The visitors to China and Chinese awareness of Rome generally: *Hou Hanshu* 23, tr. Hill.

59. Adulis: *Periplus Mar. Eryth.* 4, tr. Casson. Rhinoceros: Buttrey 2007. Zoskales: *Periplus Mar. Eryth.* 5, tr. Casson.

60. "All the gates": Aristides, *Or.* 26.102, tr. Behr. Dio, *Or.* 32.36 and 39, tr. FHN III, 925. Socotra: Strauch 2012. Indian Ocean in the long-term perspective, see Banaji 2016.

61. Jenkins et al. 2013. It should be noted that very new work is showing the surprising geographical range of diseases of zoonotic origin: Han et al. 2016.

62. Rossignol 2012; Marino 2012. See esp. Bowersock 2001, on the blackening of Lucius' reputation. Alexikakos: Ritti, Şimşek, Yıldız 2000, 7–8; MAMA IV.275a. Aelius Aristides, *Or.* 48.38, tr. Behr. For the literary evidence, see Marino 2012; Marcone 2002. Plague of Athens: Pausanias, 1.3.4.

63. Pestilence in Arabia: *Hist. Aug.*, *Vita Ant.* 9.4. Sabaic inscriptions: Robin 1992. Connected with the Antonine Plague by Rossignol 2012; Robin 1992, 234, southern Arabia "le foyer initial de la contagion."

64. On Galen's behavior at the outbreak of plague, see most recently Mattern 2013, 187–89.

65. "Like some beast": Pseudo-Galen, *Ther. Pis.* 16 (14.280-1K), tr. Mattern 204. Gaul, Germany: Ammianus Marcellinus, *Res Gest.* 23.6.24. Athens: Philostratus, *Vit. Soph.* 2.561, tr. Wright. Jones 2012b, 82–83; Jones 1971, 179. SEG 29.127, 60–63; SEG 31.131. Ostia: OGIS 595. Beyond Danube: ILS 7215a.

66. Temple of Apollo: Ammianus Marcellinus, *Res Gest.* 23.6.24; HA, *Vit. Luc.* 8. Democratization: Brown 2016. Satire: Lucian, *Alex.* 36.

67. London: Tomlin 2014. Inscriptions: Jones 2006 and 2005. Their geographical locations are indicated in Map 10. Kissing: Jones 2016.

68. The inscriptions are from Callipolis (I. Sestos, IGSK 19 no. 11); Pergamum (IGRR 4.360); Didyma (I. Didyma 217); Caesarea Troketta (Merkelbach and Stauber I, Klaros no. 8); Odessos (Merkelbach and Stauber I, Klaros no. 18); Sardis/Ephesus (Graf 1992 = SEG 41, 481); Hierapolis (Merkelbach and Stauber I, Klaros no. 12); Pisidia (Anat. St. 2003 151–55). While none of these are precisely dated, there are converging reasons to believe they all belong to the outbreak of the pestilence under Marcus Aurelius. They are treated collectively in Oesterheld 2008, 43–231; Faraone 1992, 61–64; Parke 1985. "Woe! Woe!": Callipolis, tr. Parke 1985, 150–51. Fumigations: Pinault 1992, 54–55. "You are not alone": Hierapolis, tr. Parke 1985, 153. "Which destroys": Hierapolis, tr. Parke 1985, 154. There is provincial coinage from Asia Minor, from the reign of Commodus forward, depicting APOLLO PROPULAEUS: Weinreich 1913. See also an inscription from Antioch, Perdrizet 1903.

69. Bubonic plague is beyond consideration on clinical and epidemiological grounds. So is exanthemous typhus, occasionally mentioned, but a poor fit clinically (e.g., its extreme fever, lack of pustular lesions), epidemiologically (it is a "camp" fever spread by lice), and historically (not attested until centuries later). Measles is not totally unreasonable, for it is extraordinarily contagious—more so than smallpox. But the dried pustules that fell off like scales from victims are a smallpox rash, and the most common complications of measles are respiratory, absent in our accounts. On measles, see Perry and Halsey 2004. Recent molecular clock analysis suggests that measles probably emerged a few centuries after the Antonine Plague, but see Wertheim and Pond 2011 for a measure of healthy skepticism about the precision of such dates. For Galen and the plague generally, see Marino 2012; Gourevitch 2005; Boudon 2001. Countless: Galen, *Praes. Puls.* 3.4 (9.357K). Black bile: Galen, *Atra Bile* 4 (5.115K). Fever: Galen, *Hipp. Epid.* 3.57 (17a.709K) and *Simp. Med.* 9.1.4 (12.191K). Rash and ulcerations: Galen, *Meth. Med.* 5.12 (10.367K) and *Atra Bile* 4 (5.115K). Stool: Galen, K17a.741 and *Hipp. Epid.* 3.57 (17a.709K). Dry: *Meth. Med.* 5.12 (10.367K). Great and longest-lasting, e.g., at Galen, *Praes. Puls.* 3.3 (9.341–42 and 357-8K); 17a.741K; 17a.885K; 17a.709K; 17a.710K; 17b.683K; 12.191K; 19.15, 17-8K.

70. Galen, *Meth. Med.* 5.12 (10.367K), tr. Johnston and Horsley.

71. "There was no need": Galen, *Meth. Med.* 5.12 (10.367K), tr. Johnston and Horsley. Galen, *Atra Bile* 4 (5.115K).

72. See esp. Fenner 1988. Fenn 2001 provides an accessible overview.

73. Hemorrhagic: Fenner 1988, 32, 63.

74. Conjunctival irritation: Galen, *De Substantia Facultatum Naturalium* 5 (4.788K). Fever: Galen, *Hipp. Epid.* 3.57 (17a.709K) and *Simp. Med.* 9.1.4 (12.191K). Confluent: Galen, *Atra Bile* 4 (5.115K).

75. Genomics and age of smallpox: Duggan et al. 2016; Babkin and Babkina 2015; Babkin and Babkina 2012.

76. I have provided a reasonably complete catalogue of the historical evidence for smallpox or smallpox-like viruses down to AD 1000 or so online at http://www.kyleharper.net/uncategorized/smallpox-resources-and-thoughts/. I would note the important evidence from India in the *Aṣṭāṅgahṛdayasaṃhitā* of Vagbhata in the seventh century and particularly the *Madhava nidanam* written by Madhava-kara in the early eighth century. The earliest possible Chinese evidence is Ge Hong (aka Ko Hung), *Chou hou pei chi fang*, "Handy Therapies for Emergencies." See Needham et al. 2000, 125–27. The *Chronicle* of Pseudo-Joshua the Stylite, sections 26 and 28, describes what may well be epidemic smallpox in late fifth-century Edessa: see Harper forthcoming. Finally, a series of medical writers from seventh-century Alexandria to tenth-century Iraq, culminating in the remarkable physician Al Rhazes, show deep familiarity with smallpox (as well as chickenpox and measles). For the later sources, see Carmichael and Silverstein 1987.

77. On ancient concepts of *miasma*, see the essays in Bazin-Tacchela et al. 2001.

78. Entire world: *Hist. Aug.*, *Vit. Ver.* 7.3. "Entire army": Eutropius, *Brev.* 8.12.2. Spread: Jerome, *Chron.*, an. 172. "Polluted everything": Ammianus Marcellinus, *Res Gest.* 23.6.24 tr. Rolfe. "Through many provinces": Orosius, *Hist. Adv. Pag.* 7.15.5–6. Death toll estimates: Zelener 2012 (20–25%); Paine and Storey 2012 (over 30%); Jongman 2012 (25–33%); Harris 2012 (22%); Scheidel 2002 (25%); Rathbone 1990 (20–30%); Littman and Littman 1973 (7–10%); Gilliam 1961 (1–2%).

79. The most sophisticated (frankly, the only) epidemiological study of the Antonine Plague is Zelener 2003, whose conclusions are accessible in Zelener 2012. Riley 2010 and Livi Bacci 2006 are especially helpful.

80. Riley 2010, 455. Cf. Brooks 1993, 12–13: "despite all the language of it whip-sawing up and down the continent, spreading like wildfire, or going on the rampage, smallpox in the real world tends to infect only people who are in the same house or hospital."

81. "The pathogen load": Livi Bacci 2006, 225.

82. Galen, *Hipp. Epid.* 3.57 (17a.710K). On the crisis in the delta, see Elliott 2016 and Blouin 2014.

83. See esp. Zelener 2003. Noricum: AE 1994, 1334. Egypt: see below. Rome: Cassius Dio, *Hist. Rom.* 73.14.

84. P. Thmouis 1. See Elliott 2016; Blouin 2014, 255; Marcone 2002, 811; Rathbone 1990. Banditry: Alston 2009.

85. SB XVI.12816. Hobson 1984. See esp. Keenan 2003; van Minnen 1995; Rathbone 1990.

86. Greek recruitment: Jones 2012b. Army: see esp. Eck 2012.

87. Specialists will observe that I have side-stepped some terms of the debate as it has emerged in the aftermath of the fundamental article of Duncan-Jones 1996 and the important contribution in Scheidel 2002. I believe the skeptical work of, e.g., Bruun 2012, 2007, and 2003 has been constructive and has helped define the limits of some of the evidence, but in the current terms the conversation has reached an impasse. Duncan-Jones 1996 builds a strong circumstantial case that many of the interruptions (e.g., of building inscriptions) during this period point to a severe medical crisis. Some of the evidence holds up, but this kind of analysis can be no more than suggestive, since it does not pinpoint the cause of crisis. Nonetheless, it has served its purpose in reviving the debate. The case presented here starts from what has been

missing: a clearer sense of the background and scale of "epidemics" and the epidemiological possibilities of what was clearly a novel pathogen. I think this case adds credibility to the literary evidence, and when the epigraphic evidence is added, it becomes increasingly difficult to minimize the impact of the pestilence. Moreover, my study of the papyri largely vindicates Scheidel 2002 (certainly in the case of rents and land prices, though wages are more complex). My interpretation agrees with Elliott 2016 that climate change was a part of the crisis, but without diminishing the importance of the disease factor.

88. Mines: Wilson 2007. Egyptian coinage: Howgego, Butcher, and Ponting 2010. Civic mints: Gitler 1990–91 and Butcher 2004. Prices: Harper 2016a (wheat); Rathbone and von Reden 2015; Rathbone 1997 and 1996.

89. Harper 2016a. See also Scheidel 2002 and Bagnall 2002.

90. 22–24%: Zelener 2012.

91. Incubation phase, Fenner 1988, 5, and further below. Smallpox generally, Hopkins 2002.

92. "The long-term impact": Livi Bacci 2006, 205. See also Cameron, Kelton, and Swedlund 2015; Jones 2003.

93. Marcus: *Hist. Aug.*, *Vita Marc.* 17.5. Galen, *Bon. Mal. Succ.* 1 (6.749K). Cf. Orosius, *Hist. Adv. Pag.* 7.15.5–6. The "continuous famines" that Galen describes in *Bon. Mal. Succ.* 1, most likely belong to the period *after* the Plague. A severe subsistence crisis is documented ca. 165–171: Kirbihler 2006, esp. 621; Ieraci Bio 1981, 115. De Ste. Croix 1981, 13–14.

94. 1804 petitions: P. Yale 61.

95. Birley 1987.

96. Lo Cascio 1991.

97. "He did not meet": Cassius Dio, 72.36.3, tr. Cary. "As soon": Marcus Aurelius, *Med.* 4.48.

Chapter 4: The Old Age of the World

1. In general on these games, Körner 2002, 248–59. The animals: *Hist. Aug.*, *Tres Gord.* 33.1-3. On the secular games in general, Ando 2012, 119; Pighi 1967. Gibbon 1776, vol. 1, Chapter 7. On the abandonment of the secular games, see Chapter 5 herein. The reader may immediately note some problems with the math, since Philip's reign does not fall on a hundred-year anniversary of Augustus' reign. Starting with the emperor Claudius, the Romans had a convenient disagreement about how to calculate the secular cycle, and in consequence celebrated the century games on two different loops.

2. On the *umbilicus*, see Swain 2007, 17, and below. The prayer: Lane Fox 1987, 464, from an earlier celebration. Coins: *RIC* Philip, 12–25.

3. On Philip, Ando 2012, 115–21; Körner 2002.

4. This chapter draws on Harper 2016b, 2016c, and 2015a. The "crisis" of the third century has generated a massive bibliography, but some major guideposts are Ando 2012; Drinkwater 2005; Potter 2004; the essays in Hekster, de Kleijn and Slootjes 2007; the essays in Swain and Edwards 2004, esp. Duncan-Jones 2004; Carrié and Rousselle 1999; Witschel 1999; Strobel 1993; Bleckmann 1992; MacMullen 1976; Alföldy 1974.

5. "New Empire": e.g., Harries 2012; Barnes 1982; a formula that follows the lead of Edward Gibbon. "First fall": Scheidel 2013.

6. Marcus and Faustina: Levick 2014, 62–63.

7. For the background and rise of Septimius, Campbell 2005a, 1–4; Birley 1988.

8. "Iron and rust": Cassius Dio, *Hist. Rom.* 72.36. The Severans "more continuators of the Antonines than precursors of Diocletian": Carrié 2005, 270.

9. Septimius "was not a 'military emperor'": Campbell 2005a, 10. "Get along": Cassius Dio, *Hist. Rom.* 77.15.2.

10. "Stormed the heights": Birley 1988, 24.

11. Proposal to Julia Domna: Birley 1988, 75–76.

12. Antonine Constitution: P. Giss. 40; Dig. 1.5.17; Buraselis 1989, 189–98. The hostile report of Cassius Dio imputes pecuniary motives to the act: it expanded the ranks of those paying certain taxes. The actual motives may have been religious, as the new citizens widened the circle of civic worship. Diffusion of Roman law: Garnsey 2004; Modrzejewski 1970. Macedonia: ISMDA no. 63. Levant: Cotton 1993. "Since the laws": Menander Rhetor, *Epid*. 1.364.10.

13. Ibbetson 2005. For the career of the great Severan jurist Ulpian, Honoré 2002.

14. On the Severan administration, Lo Cascio 2005b, "deficit" at 132. Campbell 2005a, 12–13. It is true that Septimius placed equestrians in offices that were the hallowed reserve of senators, most prominently as legates of his three new legions, the *I*, *II*, and *III Parthica*. But in each case there were extenuating circumstances, and he artfully avoided the dissonance of placing senatorial legates under equestrians.

15. Domitian: Griffin 2000, 71–72. Army raise: Herodian, 3.8.4; Campbell 2005a, 9.

16. Soldiers marry: Birley 1988, 128.

17. Severan high culture was an extension of the Antonine era, but also its own expression. The career of Galen is exemplary: he lived until AD 216/7 and spent virtually as much of his public career under Septimius as Marcus. Philosophy produced new stars like Alexander of Aphrodisias, one of the greatest Aristotelians after the master himself. Undoubtedly the Severan age witnessed the ascent of Platonism at the expense of the once dominant Stoic and Epicurean schools. Greek prose fiction was a vividly energetic field. An entire continent of Syriac culture, once hidden, suddenly comes into view for the first time. The period is full of surprises, like the wonderfully named poet Septimius Nestor, whose lipogrammatic *Iliad* (where every book of the poem was recast, without the use of the letter that stood for the book's number) was a virtuosity that won him fame in east and west. Philostratus of Athens flourished, canonizing the literary movement he called the "second sophistic" and composing his biography of the wonderworking sage, Apollonius of Tyana. See in general the essays in Swain, Harrison, and Elsner 2007. On Severan Rome, Lusnia 2014; Wilson 2007; Reynolds 1996. Watermills and grain: Lo Cascio 2005c, 163.

18. Monumental building draws to a close with last Severan: Wilson 2007, 291. Tertullian, *De anima* 30.

19. See Bowman 2011, 328; Keenan 2003; Alston 2002 and 2001; van Minnen 1995; Rathbone 1990.

20. Pay: Southern 2006, 108–9. Septimius seems to have pulled off a major debasement in AD 194 (to ~45% silver) without major repercussions. Butcher and Ponting 2012; Corbier 2005a and 2005b; Lo Cascio 1986. Fiduciary/price levels: Haklai-Rotenberg 2011; Rathbone 1997 and 1996.

21. On Maximinus, Syme 1971, 179–93. More recently, Campbell 2005a, 26–27. His overthrow: Drinkwater 2005, 31–33; Kolb 1977, in detail.

22. For the narrative of events in these years, see Drinkwater 2005, 33–38. "Calmly": Peachin 1991.

23. "So extreme": Duncan-Jones 2004, 21.

24. Cyprian, *Ad Demetr*. 3.

25. See esp. Zocca 1995 and Fredouille 2003, 21–38, on the metaphor. "Sex fuel": Ach. Tat. 2.3.3. "Dry nature": Galen, *Temp*. 2.580–81, tr. Singer. "Since death": Galen, *Temp*. 2.582, tr. Singer. "The falling rays": Cyprian, *Ad Demetr.* 3.

26. In light of the perspective provided by new studies, I would perhaps put less emphasis on volcanic forcing at the end of the RCO than some other valuable studies of Roman climate (e.g., Elliott 2016 and Rossignol and Durost 2007). A volcanic eruption in AD 169 may have provoked some cooling, and an even larger one in AD 266 could have done the same. See now Sigl 2015 for a reliable study of the timing and magnitude of volcanic events. Insolation: Steinhilber et al. 2012. Glaciers: Le Roy et al. 2015; Holzhauser et al. 2005. Spanish temperature

record: Martín-Chivelet et al. 2011. Austrian speleothem: Vollweiler et al. 2006; Mangini, Spötl, and Verdes 2005. Thracian speleothem: Göktürk 2011.

27. Cyprian, *De Mort.* 2. Cyprian, *Ad Demetr.* 7. Rich: Cyprian, *Ad Demetr.* 10. "If the vine fails": Cyprian, *Ad Demetr.* 20.

28. This draws from Sperber 1974. On Ḥanina bar Ḥama, see Miller 1992. Climate evidence: see Issar and Zohar 2004, 210, esp. the report of a villa at 'Ein Fashkha near Qumran.

29. Marriner et al. 2013; Marriner et al. 2012; Abtew et al. 2009; Jiang et al. 2002; Krom 2002; Eltahir 1996.

30. The ancients and the Nile: Bonneau 1971; Bonneau 1964. Nilometer: Popper 1951. Variability in the Nile: Macklin et al. 2015; Hassan 2007. On the hydrology of the Nile generally, Said 1993.

31. Raw data: McCormick, Harper, More, and Gibson 2012. Applications: Izdebski et al. 2016; McCormick 2013b.

32. Marriner et al. 2013; Marriner et al. 2012; Abtew et al. 2009; Hassan 2007; De Putter et al. 1998; Eltahir 1996. ENSO, moreover, is one of the most powerful global climate mechanisms, and its influence extends to the Mediterranean and Near East. It has complex effects in the region stretching from North Africa to the Levant, often moving in opposition to the effects in the Nile valley: ENSO years can bring rain to the semiarid southern Mediterranean. However, in the exit phase of the ENSO pulse there is a correlation with sharp aridity. In any case, if there was a connection between droughts in North Africa, Palestine, and Egypt, it would require a global mechanism like ENSO, and we can hypothesize that one of the deep shifts in the Roman Transitional Period was a greater frequency of El Niño events. The breadbasket of the Roman Empire depended on large-scale climate patterns of the Southern Hemisphere: Alpert et al. 2006; Nicholson and Kim 1997. Here El Niño frequency is reconstructed from a sedimentation record in Ecuador by Moy et al. 2002, with data at ftp://ftp.ncdc.noaa.gov/pub/data/paleo/paleolimnology/ecuador/pallcacocha_red_intensity.txt.

33. P. Erl. 18 (BL III 52); P. Oxy. XLII 3048; P. Oxy. 38.2854. See Rathbone and von Reden 2015, 184: "This is the worst grain shortage attested in Roman Egypt." Tacoma 2006, 265; Casanova 1984. Bishop: Eusebius, *Hist. Eccl.* 7.21. The reverberations of this crisis in Alexandria echo in the thirteenth Sibylline oracle: *Orac. Sibyll.* 13.50–51.

34. Yields: Rathbone 2007, 703; Rowlandson 1996, 247–52; Rathbone 1991, 185, 242–44. Gross production: Rathbone 2007, 243–44. 4–8 million *artabai*: Minimum from Scheidel and Friesen 2009. Maximum from Justinian's *Edict* 13. Value calculation based on 1 *artaba* = 12 *drachmai* = 3 *denarii* = 3/25 *aureus*. Famines: Borsch 2005; Hassan 2007.

35. Brent 2010; Sage 1975. Christian numbers: see below.

36. McNeill 1976, 136–37. See, e.g., Brooke 2014, 343. "One which affected": Corbier 2005b, 398.

37. For a comprehensive treatment, Harper 2015a and 2016c. We might add now a twenty-fourth testimony, if an oblique one, in the *ex eventu* prophecy preserved in the *Martyrdom of Marian and James*, 12. I thank Joseph Bryant for calling this reference to my attention.

38. Thebes: Tiradritti 2014. Chronology: Harper 2015a.

39. Fifteen years: this report goes back to a third-century historian of Athens named Philostratus, on whom, see Jones 2011. Evagrius Scholasticus, *Hist. Eccl.* 4.29; *Excerpta Salmasiana* II (ed. Roberto = FHG 4.151, 598); Symeon the Logothete (Wahlgren 2006, 77); George Kedrenos, *Chron. Brev.* vol. 2, 465–66; John Zonaras, *Epit. Hist.* 12.21.

40. "There was almost": Orosius, *Hist. Adv. Pag.* 7.21.5–6. "Blighted": Jordanes, *Get.* 19.104. "Affected cities and villages": Zosimus, *Hist. Nov.* 1.26.2, tr. Ridley.

41. "The pain": Cyprian, *De Mortalitate* 8. "These are adduced": Cyprian, *De Mortalitate* 14. See also Grout-Gerletti 1995, 235–36.

42. "Carrying off": Pontius, *Vit. Cypr.* 9, tr. Wallis. "The affliction fell": Gregory of Nyssa, *Vit. Greg. Thaum.* 956–57, tr. Slusser. On this text, Van Dam 1982.

43. See Harper 2016c. Ps.-Cyprian, *De Laud. Mart.* 8.1.

44. "Unusually relentless": Orosius, *Hist. Adv. Pag.* 7.22.1.

45. "Air": Orosius, *Hist. Adv. Pag.* 7.27.10. "Disease was transmitted": *Excerpta Salmasiana* II (ed. Roberto = FHG 4.151, 598). This source goes back to Philostratus of Athens. Ideas of ocular extromission: Bartsch 2006.

46. "This immense city": Eusebius, *Hist. Eccl.* 7.21, tr. Williamson 1965. 62%: Parkin 1992, 63–64. 5000/day: *Hist. Aug.*, *Vit. Gall.* 5.5. "The human race": Cyprian, *Ad Demtr.* 2, tr. Wallis.

47. Eusebius, *Hist. Eccl.* 9.8. Exanthemous typhus: Grout-Gerletti 1995, 236.

48. Barry 2004.

49. Barry 2004, 224–37. I thank my colleague Tassie Hirschfeld for encouraging me to think more carefully about influenza.

50. On VHFs in general, see Marty et al. 2006. For Yellow Fever, see McNeill 2010, on its impact in the New World; Cooper and Kiple 1993.

51. Harper 2015a lays out the possible case for an arenavirus, and while still possible, the transmission dynamics now seem to me to require direct human transmission.

52. Genetics of filoviruses: Aiewsakun and Katzourakis 2015; Taylor et al. 2010; Belyi et al. 2010. Ebola generally: Quammen 2014; Feldmann and Geisbert 2011. Case fatality: see the very helpful collection of data at http://epidemic.bio.ed.ac.uk/ebolavirus_fatality_rate.

53. Drinkwater 2005, 38–39.

54. "An empire requires": Cassius Dio, *Hist. Rom.* 52.28–29.

55. "The Alemanni": Eutropius, *Brev.* 9.8. For the sequence of events, Drinkwater 2005, 28–66; Wilkes 2005a; Potter 2004, 310–14. Todd 2005, esp. 442.

56. *Orac. Sibyll.* 13.106–8, 147–48, with Potter 1990. Zosimus depends heavily on the good contemporary source Dexippus of Athens for these sections of his *New History*, and the direct relationship between plague and insecurity was prominently drawn in his picture of the age. Shapur: *Excerpta Salmasiana* II (ed. Roberto = FHG 4.151, 598).

57. The Goths: see Todd 2005. The Persians: Frye 2005.

58. "Technology convergence": Todd 2005, 451.

59. "People of Rome": Zosimus, *Hist. Nov.* 1.37.2. Altar of Victory: see Ando 2012, 161. Gallic Empire: Drinkwater 1987. In general, Drinkwater 2005, 44–48.

60. In general, Corbier 2005a and 2005b. Silver: Estiot 1996; Walker 1976. Prices: Harper 2016a. Bankers: P. Oxy. 12.1411. Hoards: De Greef 2002; Duncan-Jones 2004, 45–46; Bland 1997. Fiduciary: Haklai-Rotenberg 2011. The silver in Fig. 4.3 data follow the suggestions of Pannekeet 2008 [available in English on academia.edu], with Gitler and Ponting 2003; Walker 1976.

61. Villages: Keenan 2003; Alston 2001; van Minnen 1995. Census records: Bagnall and Frier 1994, 9. Endowments: Corbier 2005b, 413. Epigraphy: MacMullen 1982. Temples: Bagnall 1988 and below. Ateliers: e.g., Corbier 2005b, 419.

62. Gallienus' background: Syme 1983, 197; Drinkwater 2005, 41. "Feared that the imperial power": Aurelius Victor, *Caes.* 33.33–34, tr. Bird. On the reforms of Gallienus, see Piso 2014; Cosme 2007; Lo Cascio 2005c, 159–60; Christol 1986; Christol 1982; Pflaum 1976; de Blois 1976.

63. Scheidel 2013. On the region, Wilkes 2005b; Wilkes 1996. Lack of senators: Syme 1971, 180.

64. Scheidel 2013. "Zones of energy": Syme 1984, 897.

65. Legal program: Johnston 2005; Corcoran 2000. For administrative reforms, see Chapter 5. Which is not to say that there was not patronage of the region (e.g., Diocletian's palace).

66. Turning point: Bastien 1988. See also Callu 1969. Distributions in person: Lee 2007, 57–58. Arras, aka the Beaurains hoard: Bastien and Metzger 1977.

67. Coins of Apollo Salutaris: RIC IV.3: Trebonianus Gallus, nos. 5, 19, 32, 103 and 104a–b; RIC IV.3 Volusianus, nos. 188, 247, 248a–b; RIC IV.3 Aemilianus, no. 27; RIC V.1, Valerianus, no. 76. Manders 2012, 132. "The peace of the gods": *Hist. Aug.*, *Vit. Gall.* 5.5. On religious response to plagues, generally, see Reff 2005.

68. On the question of "persecution," see Ando 2012, 134–41; Manders 2011; Luijendijk 2008; Bleckmann 2006; Selinger 2002; Rives 1999.

69. Porphyry: in Eusebius, *Praep. Ev.* 5.1.9.

70. Rate of Christian spread: see Schor 2009; Harris 2005; Hopkins 1998; Stark 1996; Lane Fox 1987; MacMullen 1984; Barnes 1982. Onomastics: Frankfurter 2014; Depauw and Clarysse 2013; Bagnall 1987b; Wipszycka 1988 and 1986; Bagnall 1982. It should be obvious that I basically accept the thrust of Depauw and Clarysse, following in turn Bagnall.

71. "World full of gods": Hopkins 2009b. New *ethnos*: Buell 2005. Networks: Schor 2011; Brown 2012. See Stark 1996, which has been too breezily dismissed.

72. "How does one describe": Eusebius, *Hist. Eccl.* 8.1.5. The first known church building in Oxyrhynchus dates to AD 304: Luijendijk 2008, 19. First identifiable Christian: P. Oxy. 42.3035. Sotas: Luijendijk 2008, 94ff.

73. In general, Rebillard 2009; Bodel 2008; Spera 2003; Pergola 1998. Early burials: Fiocchi Nicolai and Guyon 2006; Ferrua 1978; Catacombs of Priscilla, *ICUR* IX 24828ff. Callixtus: *ICUR* IV 10558.

74. Olympian Zeus: Pausanias 1.18.6; Levick 2000a, 623. "Eyes": Libanius, *Or.* 30.9; see Fowden 2005, 538. Bagnall 1988, which remains the most stimulating treatment of the sputtering of civic paganism in the later third century, at 286: "Après cela, le silence tombe sur les temples d'Égypte." Bagnall 1993, 261–68. Against total stagnation, see Lane Fox 1987, 572–85, building on the work of Louis Robert.

75. On late paganism, see esp. now Jones 2014 for a balanced approach. Ephesus: Rogers 1991.

76. "An altar": Apuleius, *Flor.* 1, tr. Fowden 2005, 540. See the vivid evocation in Watts 2015, 17–36. On Constantine's conversion, see Chapter 5.

77. Aurelian: Drinkwater 2005, 51–53.

78. Nadir: Duncan-Jones 2004, is a useful summary.

Chapter 5: Fortune's Rapid Wheel

1. Claudian, *Carm.* 20. For the antecedents and parallels of the poem, see Røstvig 1972, 71.

2. Date: Cameron 1970, 391.

3. "The frame": Prosper of Aquitaine, *Carm. Ad Uxorem* 7–8, see Santelia 2009; Roberts 1992, 99–100.

4. "Few things": Harris 2016, 220.

5. Officials: Kelly 1998, 163.

6. "Very little use": Jones 1964, 48. "Chopped": Lactantius, *De Mort. Pers.* 7.4, tr. Johns. For general treatments, Lo Cascio 2005a; Corcoran 2000; Barnes 1982; Jones 1964, 37–76, esp. 42–52.

7. We have an intriguingly precise and eminently plausible figure preserved by a sixth-century bureaucrat putting the strength of Diocletian's army at 435,266: John Lydus, *Mens.* 1.27. Campbell 2005b, 123–24; Whitby 2004, 159–60; Lee 1998, 219–20; Treadgold 1995, 43–64; Ferrill 1986, 42; Jones 1964, 679–86, esp. 679–80. Fiscal policy: Bransbourg 2015; Carrié 1994; Cerati 1975; Jones 1957. Generally, Bowman 2005.

8. *Edictum De Pretiis Rerum Venalium*, pr., ed. Lauffer. On the Edict, Corcoran 2000, 205–33. On the economic background, see the essays in Camilli and Sorda 1993 and Bagnall 1985.

9. For assessments, see esp. now Lenski 2016, with a tour de force overview of the various Constantines of modern historians. Major recent treatments, in just the last decade, include Potter 2013; Barnes 2011; Van Dam 2007; the essays in Lenski 2006; Cameron 2005. Constantine as Augustus: Harper 2013b; Matthews 2010, 41–56; Van Dam 2007.

10. Administration/senate: Harper 2013b; Kelly 2006; Heather 1998b; Jones 1964, 525–28. Constantinople: Dagron 1984. The equestrian order all but disappeared, with only one grade

left intact, as those who had once risen through talent and service could now hope for the highest honors. Constantine regularized the ad hoc body of the emperor's court companions, the *comites*, into an official system of honors with three grades, each with special privileges. He audaciously revived the archaic title of *patricius*, patrician, now reserved not for those with the most blue-blooded heritage but bestowed purely by grant from the emperor. And under Constantine, the emperor's *own* family name, Flavius, was widely conferred as something of a cross between a name and a title for certain government officials. Now the corps of the imperial servants shared even the family name of the emperor. The use of Flavius would endure across late antiquity, a reflection of Constantine's success as the founder of a new order: Keenan 1973 and 1974.

11. McGinn 1999 is especially stimulating, as is Evans Grubbs 1995. "So that the multitude": CT 11.16.3, tr. Pharr. Slavery: Harper 2013b, drawing from Harper 2011, Part 3. Illegitimate children: Harper 2011, 424–62. Divorce: Harper 2012; Memmer 2000; Arjava 1988; Bagnall 1987a.

12. Barnes 2011, for one view of the emperor's religious policy.

13. Herodian, 1.6.5. See esp. now the marvelous collection of essays in Grig and Kelly 2012; Mango 1986; Dagron 1984.

14. Profligate: Zosimus, *Historia nova* 2.38.2–3. "Repeatedly and ceaselessly . . .": Eusebius, *Vit. Const.* 4.1–2. "The lot": Eusebius, *Vit. Const.* 4.71.

15. Volcanism: Sigl et al. 2015. Insolation: Steinhilber, Beer, and Fröhlich 2009. Glaciers: Le Roy et al. 2015.

16. NAO: Burroughs 2005, 170–75; Hurrell et al. 2003; Marshall et al. 2001; Visbeck et al. 2001.

17. Compare esp. the long-term effects of a persistently positive NAO in the medieval climate anomaly (aka "medieval warm period"): Trouet et al. 2009. Manning 2013, 107–8.

18. Baker et al. 2015; Another paleoclimate record of the NAO has been derived from a high-resolution lake sediment record from Greenland: Olsen et al. 2012. Spain: Martín-Puertas et al. 2009; Currás et al. 2012. Oaks: Büntgen et al. 2011.

19. It is worth noting that the frequency of Tiber flooding declined precipitously in the fourth and fifth centuries, but only relative to the exceptionally high levels of the RCO. "An emptied landscape": Brown 2012, 100. Greater moisture in the Kapsia cave record from Greece: Finné et al. 2014. And in the Shkodra lake record from eastern Albania: Zanchetta et al. 2012.

20. Overviews of eastern Mediterranean climate mechanisms: Finné et al. 2011; Xoplaki 2002. Arid Anatolia: Haldon et al. 2014. Sofular: Göktürk 2011; Fleitmann et al. 2009. Bereket Basin: Kaniewski et al. 2007. Nar Gölü: Dean et al. 2013; Woodbridge and Roberts 2011. Tecer Lake: Kuzucuoğlu et al. 2011. Israel: Migowski et al. 2006; Bookman et al. 2004. There is some uncertainty regarding the timing of the onset and ending of a more humid period. See McCormick et al. 2012 for a discussion. See also Chapter 7.

21. Stathakopoulos 2004 is most comprehensive; Telelēs 2004, focused on climate events; Patlagean 1977 still has much of value. Brown 2002 and Holman 2001 help us appreciate the new kinds of perspectives we have in late antiquity.

22. See Holman 2001; Garnsey 1988, 22–23. Basil of Caesarea, *Dest. Horr.* 4. Harper 2011, 410–11.

23. "The hopes of all the provinces": Symmachus, *Rel.* 3.15, tr. Barrow.

24. Stathakopoulos 2004, no. 29, 207. "Provinces come to the relief": Symmachus, *Rel.* 3.17, tr. Barrow. "Vicissitudes" and "kept alive": Symmachus, *Rel.* 3.16, tr. Barrow. Ambrose, *Ep.* 73.19, tr. Liebeschuetz. The famine is also mentioned by the contemporary writer known as Ambrosiaster and alluded to by Prudentius.

25. Problems in Egypt: P. Lond. 3.982. See Rea 1997; Zuckerman 1995, 187. The following year saw massive hunger in Antioch. Stathakopoulos 2004, no. 30, 209. The wrenching climate events of AD 451 similarly reverberated across the Mediterranean. See below, for the intersection with the invasion of Attila.

26. See Chapter 3.

27. Stathakopoulos 2004. Eusebius, *Hist. Eccl.* 9.8. Ambrose of Milan claimed that the warfare among the barbarians ca. AD 378 had spread horror—and famine and pestilence—over the whole world. In AD 442, the chronicler Hydatius records the appearance of a comet followed by a pestilence spreading over the world. Neither testimony finds any contemporary confirmation, and both add to the rhetorical effects of doom-saying. They are safely minimized.

28. For this outbreak of malaria, see below.

29. The episode is vividly recorded by Ps.-Joshua Stylites 38–46. Stathakopoulos 2004, no. 80, 250–55. Garnsey 1988, 1–7, 20–36. See also Harper forthcoming.

30. For a week of travel, see orbis.stanford.edu. For Palladius' trip, *Hist. Laus.* 35.

31. "Personal oracle": Zuckerman 1995, 193. Nile: *Hist. Monach. in Egypt.* 11.

32. See most recently Sheridan 2015 and van der Vliet 2015, 165–67, for John. The identification of Apa John in these papyri with John of Lycopolis is made by Zuckerman 1995. Not all are convinced: see Choat 2007. It would not materially alter the argument here. Conscription and self-mutilation: CT 7.13.4 (AD 367); CT 7.13.5 (AD 368); CT 7.13.10 (AD 381), tr. Pharr, adapted here. The provinces would only receive half-credit for mutilated recruits: two counted as one in their levy.

33. Lo Cascio 1993; the state carried out forced requisitions of precious metals, reimbursed in overvalued billon coinage, which began under Diocletian and has been called "le trait le plus original de la fiscalité de Dioclétian," Carrié 2003; Carrié 2007, 156. The Price Edict seems to have "undervalued" gold, and the state would continue to undervalue gold as a matter of policy for the next quarter century. On the inflation, see esp. Bagnall 1985. The wheat prices are from Harper 2016a.

34. Constantine's "liberalization" of gold: Lo Cascio 1998 and 1995. Taxes: the *collatio lustralis* or *chyrsargyron* was a tax on commerce levied every five years in gold, obviously to underwrite the costs of quinquennial donatives to the soldiers. He also instituted the *collatio glebalis*, an annual tax on senatorial property paid in gold. Both of these taxes would have been progressive, falling disproportionately on the wealthier classes; and though both were wildly unpopular, they endured. The new gold economy: Carrié 2007; Banaji 2007; Corbier 2005a, 346; Brenot and Loriot 1992; Callu and Loriot 1990; Morrisson et al. 1985, 92–95. Stray finds of *solidi* sharply increase after Constantine's reign: Bland 1997, 32–33.

35. "Particular late Roman fusion": Banaji 2007, 55. Heliodorus: Libanius, *Or.* 62.46–8.

36. Disappearance of banks: Andreau 1998; Andreau 1986. CJ 5.37.22 (AD 329) is particularly revealing. The revival of banking has not been treated in full, but see Barnish 1985; Petrucci 1998; Bogaert 1973. John Chyrsostom, *In Pr. Act.* 4.2 (PG 51: 99). See Bogaert 1973, 244, 257–58, who calls it the most explicit definition of a bank in all of Greek literature. "The merchant": John Chrysostom, *Hom. In Io.* 1.3 (PG 59: 28). See also his *Hom. In 1 Cor.* 14.3 (PG 61: 117); *De Laz.* 1.3 (PG 48: 966). "Sailing and trading": Augustine, *Enarr. In Ps.* 136.3, tr. McCormick 2012, 57.

37. The bibliography on ARS is massive. See Fentress et al. 2004. "Like the merchant": Ps.-Macarius, *Serm.* 29.2.1, tr. McCormick 2012, 57. *Expositio* as "practical guide": McCormick 2001, 85.

38. Harvester of Mactar: ILS 7457, tr. Parkin and Pomeroy 2007, 39. See Shaw 2015; Brown 2012, 4–6. Coppersmiths, sausage-makers: Libanius, *Or.* 42 with Petit 1957. Augustine: *Conf.* 6.13 and 6.15. See Shanzer 2002, 170.

39. In general, Harper 2011, with further literature. Melania: Palladius, *Laus. Hist.* 61; Gerontius, *Vit. Mel.* 10–12. With Harper 2011, 192; Clark 1984.

40. Harper 2011, 46–49. "This man": Libanius, *Or.* 47.28. A law of AD 383 permitted the town councils of Thrace to recruit from the common people "abounding in the wealth of slaves" who had avoided curial service through "the obscurity of a low name": *CT* 12.1.96.

41. "Even the household": John Chrysostom, *In Ephes.* 22.2 (PG 62: 158). Priests, etc.: Harper 2011, 49–56. Linenworker: *CIL* 15.7184; *CIL* 15.7175. With Thurmond 1994, 468–69. Assistant Professors: Libanius, *Or.* 31.11. "The ownership of a small number": Bagnall 1993, 125.

42. "The western senatorial elite": Wickham 2005, 156. See also Jones 1964, 778–84. Olympiodorus: *Frag*. 41. Harper 2015b. Matthews 1975. Melania: see note 34 above.

43. See esp. Bagnall 1992 and Bowman 1985, on the land registers from Hermopolis. Placed in context in Harper 2015b.

44. For the institutional framework, see esp. Kehoe 2007. Constantine: CT 11.16.3, tr. Pharr. Synesius, *Ep*. 148, tr. Fitzgerald.

45. On poverty in late antiquity, the essays in Holman 2008; the essays in Atkins and Osborne 2006; Brown 2002; Holman 2001; Neri 1998; Patlagean 1977. Antioch: Libanius, *Or*. 27; Stathakopoulos 2004, no. 30, 209. Martin: Sulpicius Severus, *Dial*. 2.10, tr. Hoare. "Their roof": Gregory of Nyssa, *De Benef*. 453, tr. Holman 2001, 194.

46. "Shall we?" Gregory of Nazianzus, *De pauperum amore*, 15, tr. Vinson. "You see": Gregory of Nyssa, *In illud: Quatenus uni ex his fecistis mihi fecistis*, ed. van Heck, 114. "The sick man": Gregory of Nyssa, *De Benef*. 453, tr. Holman 2001, 195.

47. For Rome in late antiquity, in general, Grig and Kelly 2012; Van Dam 2010; Harris 1999b. "So little concern": Aurelius Victor, *Caes*. 28.2. Also Zosimus, *Hist. Nov*. 2.7. "On every side": Ammianus Marcellinus, *Res Gest*. 16.10.13, tr. Rolfe. On the entitlements, Sirks 1991 (bread at 308; oil at 389–90; wine at 392–93; pork at 361ff.).

48. Population: Van Dam 2010, 55; Zuckerman 2004. "Dry land": Van Dam 2010, 55. On Constantinople in general: Dagron 1984.

49. Alexandria: Fraser 1951 and generally Haas 1997. Late antique urbanism is a vast and controversial topic. See in general Liebescheutz 2001, for an authoritative and detailed treatment, which is in general followed in the presentation here and in Chapter 7.

50. Wickham 2005 is a magisterial survey of the archaeology and its economic implications. See also Brogiolo and Chavarría Arnau 2005; Chavarría and Lewit 2004; Lewit 2004; Bowden, Lavan, and Machado 2003; Brogiolo, Gauthier, and Christie 2000. Christie 2011, 20, on frontier landscapes as now "scarred and thus uninviting." For eastern dynamism, see esp. Decker 2009. See also Chapter 7.

51. Heritable: Jones 1964, 615. "The huge army": Jones 1964, 933. Half the size: Van Dam 2010, 27. Recruitment generally: Campbell 2005b, 126–27; Lee 1998, 221–22.

52. See esp. Ferrill 1986.

53. In general, see Christie 2011, 70–73. Ferrill 1986, 78–82; Jones 1964. "When an army": Thompson 1958, 18.

54. Di Cosmo 2002, 13–43.

55. "Measureless wastes": Ammianus Marcellinus, *Res Gest*. 31.2.13, tr. Rolfe. "They have no": Ammianus Marcellinus, *Res Gest*. 31.2.17, tr. Rolfe.

56. See in particular Di Cosmo 2002, esp. 269–77.

57. Ying-Shih 1986, 383–405. On the unwinding of the Han dynasty, Mansvelt Beck 1986, 357–76.

58. De la Vaissière 2015; 2005a, 2005b, and 2003. Translated in Juliano and Lerner 2001, no. 8, 47–49. The best treatment of the Huns broadly, though now dated, remains Maenchen-Helfen 1973 (who doubted the identifications accepted here). See also Thompson 1996.

59. Ammianus Marcellinus, *Res Gest*. 31.1.1, tr. Rolfe. "The intrusion": Heather 2015, 212. This was an enduring turning point in the history of the steppe. Maas 2015, 9: "The doors to populations from further east would remain open, most immediately for Avars and Turks . . ."

60. Dulan-Wulan: Cook 2013, who develops the climate framework in detail. I would emphasize the role of the NAO more strongly than ENSO, which he puts in the foreground. Arid central Asia: see Campbell 2016, 48–49; Oberhänsli et al. 2011; Chen 2010; Oberhänsli et al. 2007; Sorrel et al. 2007.

61. Heather 1995. Frankopan 2015, 46: "Between about 350 and 360 there was a huge wave of migration as tribes were shunted off their lands and driven westwards. This was most likely caused by climate change, which made life on the steppe exceptionally harsh and triggered intense competition for resources."

62. Ammianus Marcellinus, *Res Gest.* 31.2.1, 3, 6, and 10, tr. Rolfe.

63. "Series of independent": Heather 2015, 214; also Heather 1998a, 502. Hunnic horses: Vegetius, *Mul.* 3.6.2 and 5, tr. Mezzabotta 2000. Their appearance made an indelible impression. "The Hunnish horses have a large, hooked head, protuberant eyes, narrow nostrils, broad cheeks, a strong, inflexible neck, manes hanging down below the knees, larger-than-usual ribs, a bent spine, a bushy tail. . . ."

64. "Very hard to manufacture": Luttwak 2009, 25. Though cf. Elton 2015, 127, against its total novelty. Maenchen-Helfen 1973, 221–28. "Shapely bows and arrows": Sidonius Apollinaris, *Carm.* 2.266–69, tr. Anderson. "You would not hesitate": Ammianus Marcellinus, *Res Gest.* 31.2.9, tr. Rolfe.

65. "Remained quiet": Ammianus Marcellinus, *Res Gest.* 31.5.17, tr. Rolfe. Numbers: Heather 2015, 213. Maenchen-Helfen 1973, 26–30. The battle: Ferrill 1986, 56–63.

66. Adrianople: Hoffman 1969–70, 440–58. Barbarization: Whitby 2004, 164–70, for a very cautious interpretation, with previous literature; Curran 1998, 101–3; Lee 1998, 222–24; Elton 1996, 136–52; Ferrill 1986, 68–70, 83–85.

67. "Equipoise": Claudian, *Stil.* 3.10, tr. Platnauer.

68. Heather's reconstruction: Heather 2015; 2010; 2006; 1995.

69. Whereas the Huns in AD 395 had launched a major assault through the Caucasus, east of the Black Sea, in AD 408/9 they crossed the Danube, west of the Black Sea. In just these years, the great Theodosian walls of Constantinople went up, the mighty double-wall system that would protect the town for a thousand years. Uldin: Maenchen-Helfen 1973, 59–72.

70. "In one city": Jerome, *Comm. In Ezech.* pr.

71. Fifth century: Kulikowski 2012 is a helpful and up-to-date overview of the questions.

72. "Against the stone of sickness": Isaac of Antioch, *Homily on the Royal City*, tr. Moss, 61, 69. On Attila's operations in the 440s, see Kelly 2015, 200–1; Maenchen-Helfen 1973, 108–25.

73. "To paint Attila": Priscus of Panium, tr. Kelly 2008, 260.

74. "Beneath his great savagery": Jordanes, *Get.* 186, tr. Mierow. "Heaven-sent disasters": Hydatius, *Chron.* 29, tr. Burgess 103. Maenchen-Helfen 1973, 129–42. One of the most chilling finds in the annals of Roman archaeology, the infant cemetery at Lugnano: Soren and Soren 1999, 461–649. Some 60 miles north of Rome, on the site of a rural villa, at least 47 fetuses and infants were buried, right in the middle of the fifth century. It is evident that they were buried in a short interval, of some weeks or months, reflecting the ravages of epidemic mortality. The excavations uncovered the debris of dark magical rites that still ruled in the deep countryside. And two, independent scientific methods—DNA sequencing and the recovery of a signature chemical byproduct called hemozoin—have proven that malaria was killer. Even though it lies south of the furthest advance of the Hunnic invasion, archaeologists have not implausibly linked this cemetery with the precise circumstances that motivated Attila's retreat. Malaria outbreaks depend on cycles of mosquito reproduction, which in turn are sensitive to the background pulse of the climate and can cover large regions: Roucaute et al. 2014. Hemozoin: Shelton 2015. DNA: Sallares et al. 2003; Abbott 2001. Generally: Bianucci et al. 2015.

75. Last paycheck: Eugippius, *Vita Severin.* 20. For the archaeology of the region, see Christie 2011, 218. "The western Roman army": Whitby 2000b, 288. cf. Ferrill 1986, 22.

76. Coins: McCormick 2013a. Church: Brown 2012.

77. "It is evident": Cassiodorus, *Var.* 11.39.1 and-2, tr. Barnish. Seasonality: Harper 2015c.

Chapter 6: The Wine-Press of Wrath

1. *De cerem.* 2.51, tr. Moffatt and Tall. This opening invokes McCormick 1998, where study of the ecology of the first pandemic really begins.

2. Wheat levels and "We deem": Justinian *Edictum* 13, tr. Blume. Jones 1964, 698.

3. "A throng": Procopius, *De aedific.* 1.11.24, tr. Dewing. Languages: Croke 2005, 74–76, for a vivid evocation of the "crowded and boisterous" city. Feissel 1995, on immigrants attested by epitaphs.

4. "They steal": Mitchell 1992, 491 [orig. 1944].

5. "Neutron bomb": Cantor 2001, 25. The fine collection of essays in Little 2007a represents the state of the field. Major treatments include Meier 2016; Mitchell 2015, 409–13, 479–91; Horden 2005; Meier 2005; Meier 2003; Stathakopoulos 2004; Sarris 2002; Stathakopoulos 2000; Conrad 1981; Durliat 1989; Allen 1979; Biraben 1975, 22–48; Biraben and Le Goff 1969. More popular studies, but with worthwhile insights, include Rosen 2007 and Keys 2000.

6. Maas 2005 provides a good overview of the reign of Justinian. For a highly negative assessment, see O'Donnell 2008. Meier 2003 convincingly foregrounds the role of natural disasters in wrecking the projects of Justinian and coloring the entire age.

7. CJ 5.4.23, tr. Blume. See Daube 1966–67.

8. Opposition: see Bjornlie 2013; Kaldellis 2004; Maas 1992; Cameron 1985, 23–24. Haldon 2005 is a helpful sketch of the administrative structure of Justinian's empire. An overview of the Nika rebellion: Cameron 2000a, 71–72. Liebeschuetz 2000, 208 and 220: the "notables" were in control of the tax system.

9. Jones 1964, 278–85. See also Stein 1968, II.419–83. Theodora: see recently the insightful biography of Potter 2015, from an ample literature. Tribonian: Honoré 1978.

10. Gibbon 1788, Vol. IV, Ch. 44. In the very first days of his reign, Justinian commissioned a task force, led by John the Cappadocian, to gather and harmonize the imperial laws from the time of Hadrian to the present. A first edition was promulgated in AD 529. But the exercise, already ambitious, exposed how disparate and complex the body of Roman law remained. The project of codification was expanded in scope to encompass all of Roman legislation and jurisprudence. See Honoré 2010, 28; Humfress 2005. "The task": Justinian, *Deo Auctore* 2, tr. Watson. Three million lines: Cameron 2000a, 67.

11. "It soars": Procopius, *De aedific.* 1.1.27, tr. Dewing. For an overview of Justinian's patronage, Alchermes 2005, 355–66; Cameron 1985, 86–87. Tenedos: Procopius, *De aedific.* 5.1.7–17.

12. Edessa: Procopius, *De aedific.* 2.7.4. See also 2.8.18 on the Euphrates and 2.10.6 on the Orontes. Tarsus: Procopius, *De aedific.* 5.5.15–20. Sangarius: Procopius, *De aedific.* 5.3.6. See Whitby 1985. Drakon: Procopius, *De aedific.* 5.2.6–13. Aqueducts: Procopius, *De aedific.* 3.7.1 (Trabezond); 4.9.14 (Perinthus); 4.11.11–13 (Anastasiopolis); 5.2.4 (Helenopolis); 5.3.1 (Nicaea). Cistern: Procopius, *De aedific.* 1.11.10. Crow 2012, 127–29.

13. Cameron 2000a, 73–74.

14. Humphries 2000, 533–35, at 535.

15. "A city": Procopius, *Bell.* 2.8.23, tr. Kaldellis. Tens of thousands of captives were dragged to Persia and settled in a new city: "Khusro's Better Than Antioch." The *other* age: Meier 2003. "I cannot understand": Procopius, *Bell.* 2.10.4, tr. Kaldellis.

16. Slack 2012; Eisen and Gage 2009; Gage and Kosoy 2005.

17. First to sequence: Raoult et al. 2000; Drancourt et al. 1998. Model: McNally et al. 2016.

18. McNally et al. 2016; Hinnebusch, Chouikha, and Sun 2016; Pechous 2016; Gage and Kosoy 2005; Cornelis and Wolf-Watz 1997.

19. Zimbler et al. 2015; Chain et al. 2004.

20. Transmission: Hinnebusch 2017. Bronze Age: Rasmussen et al. 2015.

21. Ymt: Hinnebusch et al. 2002. Variety of fleas: Eisen, Dennis, and Gage 2015. Miarinjara et al. 2016.

22. Cui et al. 2013. See Varlık 2015, 19–20.

23. In general: McCormick 2003.

24. Varlık 2015, 20–28.

25. Varlık 2015, 28–38.

26. *Pulex irritans*: e.g., Ratovonjato et al. 2014. It has recently been confirmed that *Y. pestis* can block other fleas besides *X. cheopis* (Hinnebusch 2017). An important current question is whether the blocking mechanism, which makes transmission much more efficient, is possible in *P. irritans*. I thank Dr. Hinnebusch for generous conversation on this point.

27. Human fleas/ectoparasites: Campbell 2016, esp. 232–33; Eisen, Dennis, and Gage 2015; Eisen and Gage 2012; Audoin-Rouzeau 2003, 115–56, for a comprehensive treatment of the earlier literature. Other pathways: Varlık 2015, 19–20; Green 2014a, 32–33; Carmichael 2014, 159; Anisimov, Lindler, and Pier 2004.

28. McCormick 2003, 1. The data in Map 17 build on McCormick's database, available on darmc.harvard.edu, with updates (making no claims to comprehensiveness) discovered in the interim, which I have been able to locate. Previously, Audoin-Rouzeau 2003, 161–68.

29. Rufus apud Oribasius, *Coll. Med.* 44.41 and esp. 44.14. Rufus cites as authorities Dionysius the Hunchback and Poseidonius and Dioscorides. Despite modern speculation, these figures are totally unidentifiable. Aretaeus, *De Causis et Signis Acutorum Morborum* 2.3.2. I thank John Mulhall for very helpful conversation; his forthcoming work promises to clarify the history of bubonic plague in the ancient medical literature. Shorter manual: Oribasius, *Syn. ad Eust. Fil.* See also Sallares 2007, 251.

30. Rasmussen et al. 2015; Zimbler et al. 2015 on the importance of this virulence factor. Other minor evolutionary events are also possible. See Feldman et al. 2016 for other small features of the first-pandemic genome associated with virulence genes.

31. Green 2014a, 37. It is significant that the new high-coverage reconstruction from Altenerding (Feldman et al. 2016) confirms the "placement of the branch leading to the Justinianic strain on Branch 0 between two modern strains isolated from Chinese rodents (0.ANT1 and 0.ANT2)."

32. Cosmas Indicopleustes, *Top. Christ.* 2.46.

33. On Cosmas, see Darley 2013; Kominko 2013; Bowersock 2013, 22–43; Wolska-Conus 1968; Wolska-Conus 1962. On the Red Sea in late antiquity, Power 2012.

34. See, e.g., Cosmas Indicopleustes, *Top. Christ.* 11.15. Banaji 2016, 131, well draws out this dynamic.

35. Cosmas Indicopleustes, *Top. Christ.* 11.10, see Wolska-Conus 1973, 335. In the Greek *Martyrdom of St. Arethas*, an account of the martyrs of Najran, India is where "aromatics, pepper, silk, and precious pearls" come from: *Mart. Areth.* 2. See also *Mart. Areth.* 29 for a hint of the most active ports in the sixth century. Ports: Wilson 2015, 29–30; Power 2012, 28–41, esp. 41 on Berenike. Constantine: Seland 2012.

36. "Spent much time": Procopius, *Bell.* 8.17.1–6, tr. Kaldellis. On Rome and China generally, Ferguson and Keynes 1978.

37. Slaves: Harper 2011, 89–90. "Most slaves": Cosmas Indicopleustes, *Top. Christ.* 2.64. Ivory: Cutler 1985, 22–24. India of the mind: see the story of the Egyptian lawyer who traveled as far east as the Bay of Bengal and back, bringing stories of Brahmins recorded in Ps. Palladius, *De Gent. Ind. et de Brag.*, tr. Desantis. For concepts of India in late antiquity, see Johnson 2016, 133–37; Mayerson 1993.

38. "Massive": Choricius of Gaza, *Or.* 3.67 (Foerster p. 65). See Mayerson 1993, 173. Coins: see esp. the cautious work of Darley 2015 and 2013; Walburg 2008; Krishnamurthy 2007; Turner 1989. Geopolitics: Bowersock 2013, esp. 106–19; Bowersock 2012; Power 2012, 68–75; Greatrex 2005, 501.

39. Procopius, *Bell.* 2.22.6. John of Ephesus, in Michael the Syrian, *Chron.* 9.28.305, p. 235. The Syriac sources put the distant origin of the plague in Kush, the biblical term for the exotic lands to the south, which, as the *Christian Topography* makes clear, could mean the Himyarite territory of southern Arabia. Clysma: Tsiamis et al. 2009. Green 2014a, 47; McCormick 2007, 303 also convincingly suggests a southern entry after passage across the Indian Ocean. On the archaeology of Pelusium, Jaritz and Carrez-Maratray 1996: detectible activity draws to a close in the middle of the sixth century.

40. See the extended discussion in Chapter 7. Compare Varlık 2015, 50–53, for valuable thoughts on the Black Death. Also McMichael 2010.

41. On the relation between extreme climate events and infectious diseases generally, McMichael 2015; Altizer 2006.

42. See Ari et al. 2011; Kausrud et al. 2010, Gage et al. 2008, on the complex web of relations between plague and climate. Food availability regulates population sizes of rodents, so that climate can trigger wild swings, e.g., White 2008, 230. Rodent population dynamics in general, and emigration in particular: Krebs 2013. El Niño: Zhang et al. 2007; Xu et al. 2015; Xu et al. 2014; Enscore et al. 2002. For the Black Death, see now Campbell 2016.

43. Ari et al. 2011, 2; Audoin-Rouzeau 2003, 67–70; Cavanaugh and Marshall 1972; Cavanaugh 1971; Verjbitski, Bannerman, and Kápadiâ 1908.

44. On Procopius, see Kaldellis 2004; Cameron 1985, esp. 42–43 on the plague as a digression in the *Wars*. "I consider": Procopius, *Bell.* 5.3.6–7, tr. Kaldellis.

45. On John, see Morony 2007; Kaldellis 2007; Ginkel 1995; Harvey 1990.

46. The account of Procopius is in *Bell.* 2.22–23. The account of John is extant in later chronicles, most extensively in the work known as the *Chronicle of Zuqnin*. It is conveniently available in English translation in Witakowski 1996. See in general Kaldellis 2007 (esp. at p. 14), on the sophistication (rather than slavishness) of Procopius' use of his model, Thucydides.

47. Benedictow 2004, 26; Audoin-Rouzeau 2003, 50–55.

48. Sebbane et al. 2006; Benedictow 2004.

49. Benedictow 2004; Pechous 2016.

50. Ingestion: Butler et al. 1982.

51. Procopius: fever: *Bell.* 2.22.15–16; swellings: *Bell.* 2.22.17; "in cases": *Bell.* 2.22.37; debilitation: *Bell.* 2.22.38–39. John of Ephesus, in *Chronicle of Zuqnin*, tr. Witakowski 1996, p. 87.

52. Black blisters: Procopius, *Bell.* 2.22.30. "On whomsoever": John of Ephesus, in *Chronicle of Zuqnin*, tr. Witakowski 1996, p. 88. Vomiting blood: Procopius, *Bell.* 2.22.31.

53. John of Ephesus, in *Chronicle of Zuqnin*, tr. Witakowski 1996, p. 88. Also Agathias, *Hist.* 5.10.4. Sallares 2007b, 235.

54. In a later outbreak, Evagrius, *Hist. Eccl.* 4.29 (178) describes affliction in the throat. Doctors at no special risk (in direct contradiction to his model, Thucydides, it might be noted): Procopius, *Bell.* 2.22.23. Mainly bubonic symptoms: so also Allen 1979, 8. Sallares 2007b, 244, points out that pneumonic plague could have been more important than our (nonmedical) sources suggest.

55. Two branches: Procopius, *Bell.* 2.22.6. "Perished totally": John of Ephesus, in *Chronicle of Zuqnin*, tr. Witakowski 1996, p. 77. "Whole of Palestine": John of Ephesus, in *Chronicle of Zuqnin*, tr. Witakowski 1996, p. 77. "Day by day": John of Ephesus, in *Chronicle of Zuqnin*, tr. Witakowski 1996, p. 80.

56. "Ships in the midst": John of Ephesus, in *Chronicle of Zuqnin*, tr. Witakowski 1996, p. 75. "Many people saw": John of Ephesus, in *Chronicle of Zuqnin*, tr. Witakowski 1996, p. 77. "The disease always": Procopius, *Bell.* 2.22.9, tr. Kaldellis.

57. Rivers: McCormick 1998, esp. 59–61. "Always moving along": Procopius, *Bell.* 2.22.6–8, tr. Kaldellis. cf. also John of Ephesus, in *Chronicle of Zuqnin*, tr. Witakowski 1996, p. 85–86: "News was sent to every place in advance, and then the scourge arrived there, coming to a city or a village and falling upon it as a reaper, eagerly and swiftly, as well as upon other settlements in its vicinity, up to one, two, or three miles from it."

58. See esp. Benedictow 2004. "Like a wheat-field": Gregory of Tours, *Hist. Franc.* 9.22, tr. Thorpe (Americanizing "wheat" for "corn").

59. Poor in the Black Death: Benedictow 2004. "Eagerly began to assault": John of Ephesus, in Michael the Syrian, *Chronicle*, 235–36. "It fell": John of Ephesus, in *Chronicle of Zuqnin*, tr. Witakowski 1996, p. 74. "People differ": Procopius, *Bell.* 2.22.4, tr. Kaldellis.

60. Prophecies: John Malalas, *Chron.* 18.90, tr. Jeffreys; see Chapter 7. "The visitation came": John of Ephesus, in *Chronicle of Zuqnin*, tr. Witakowski 1996, p. 86. "The danger": Justinian,

Edictum 9.3, tr. Blume. See also Edict 7 and Novel 117. For the chronology, see Stathakopoulos 2004; McCormick 1998, 52–53. I do not find the dating in Meier 2003, 92–93, convincing, as it relies on later sources, discounts Procopius, and does not take into account McCormick 1998. That the Hypapante was moved in February 542 (see next chapter) could well have been an apotropaic measure in advance of the plague's arrival.

61. "At first": Procopius, *Bell.* 2.23.2, tr. Kaldellis. "Men were standing": John of Ephesus, in *Chronicle of Zuqnin*, tr. Witakowski 1996, p. 86–87. Black Death estimates: see below.

62. "A true famine": Procopius, *Bell.* 2.23.19, tr. Kaldellis. "The entire city": John of Ephesus, in *Chronicle of Zuqnin*, tr. Witakowski 1996, p. 88. "Nobody would go": John of Ephesus, in *Chronicle of Zuqnin*, tr. Witakowski 1996, p. 93. "The whole experience": Procopius, *Bell.* 2.23.20, tr. Kaldellis.

63. "Confusion began to reign": Procopius, *Bell.* 2.23.3, tr. Kaldellis. "In a heap": Procopius, *Bell.* 2.23.10, tr. Kaldellis. "Trodden upon": John of Ephesus, in *Chronicle of Zuqnin*, tr. Witakowski 1996, p. 91. "The wine-press": John of Ephesus, in *Chronicle of Zuqnin*, tr. Witakowski 1996, p. 96. Cf. *Apoc.* 14:19.

64. "Whole world": Procopius, *Bell.* 2.22.1 and John of Ephesus, in *Chronicle of Zuqnin*, tr. Witakowski 1996, p. 102. Persians: Procopius, *Bell.* 2.23.21. Other barbarians: Procopius, *Bell.* 2.23.21 and 2.24.5. Kush, South Arabia, and the west: Michael the Syrian, *Chron.* 9.28, p. 235 and 240. On the west, see Little 2007b.

65. Alexandria: John of Ephesus, in *Chronicle of Zuqnin*, tr. Witakowski 1996, p. 93. Michael the Syrian, *Chron.* 9.28, p. 236. Jerusalem: Michael the Syrian, *Chron.* 9.28, p. 238; Cyril of Scythopolis, *Vit. Kyr.* 10 (229). Emesa: Leontios of Neapolis, *Vit. Sym.* 151. Antioch: Evagrius, *Hist. Eccl.* 4.29 (177). Apamea: Evagrius, *Hist. Eccl.* 4.29 (177). Myra: *Vita Nich. Sion.* 52. Aphrodisias: Roueché and Reynolds 1989, no. 86 (a far from certain reference to the great pestilence).

66. Rightly, Sallares 2007b, 271.

67. Crops in the field: John of Ephesus, in Michael the Syrian, *Chron.* 240. Theodore: *Vita Theod. Syk.* 8. Antioch: *Vita Sym. Styl. Iun.* 69. Jerusalem: Cyril of Scythopolis, *Vita Kyriak.* 10 (229). Zoraua (marked on Map 21): Benovitz 2014, 491; Feissel 2006, 267; Koder 1995. Egypt: John of Ephesus, *Lives of the Eastern Saints*, 13, vol. 1, p. 212. See Harvey 1990, 79, on the location.

68. Africa: Corippus, *Ioh.* 3.343–89. Victor of Tunnuna, *Chron.* an. 542. Spain: *Consularia Caesaraugustana*, see Kulikowski 2007. Italy: Marcellinus Comes, *Chron.* an. 542. Gaul: Gregory of Tours, *Hist. Franc.* 4.5; Gregory of Tours, *Glor. Mart.* 50; Gregory of Tours, *Glor. Conf.* 78 (not certainly the first wave); Gregory of Tours, *Vit. Patr.* 6.6 and 17.4 (probably the first outbreak). British isles: Maddicott 2007, 174.

69. Aschheim: Wagner et al. 2014; Harbeck et al. 2013; Wiechmann and Grupe 2005. On the cemetery: Gutsmiedl-Schümann 2010; Staskiewicz 2007; Gutsmiedl 2005. Altenerding: Feldman et al. 2016. The genome of the medieval pandemic has also been sequenced: Bos et al. 2011; Schuenemann et al. 2011; Haensch et al. 2010.

70. McCormick 2016 and 2015.

71. Moors: Corippus, *Ioh.* 3., tr. Shea. Turks: Theophylact Simocatta 7.8.11, tr. Whitby and Whitby. "Neither Mecca nor Medina": Conrad 1981, 151. So also Little 2007b, 8. Anastasius of Sinai, *Ques. Resp.* 28.9 and 66, tr. Munitiz.

72. Not 1 in 1000: John of Ephesus, in Michael the Syrian, *Chron.* 9.28, p. 240. "The plague broke out": Procopius, *Anek.* 18.44, tr. Kaldellis. Majority of farmers: Procopius, *Anek.* 23.20, tr. Kaldellis. "At least as many": Procopius, *Anek.* 6.22, tr. Kaldellis. Tombstone: I. Palaestina Tertia, Ib, no. 68. See also nos. 69–70 and Benovitz 2014, 491–92.

73. DeWitte and Hughes-Morey 2012.

74. "The Black Death": Green 2014a, 9. Benedictow 2014, 383, Table 38. Campbell 2016, 14: by the 1380s, net decline in the European population of 50%; also at p. 310, with figures of 40–45% for the first wave in England. Borsch 2014, on Egypt in the Black Death. Toubert 2016, 27; DeWitte 2014, 101.

75. "It has come to our attention": Justinian, *Novella* 122 (AD 544). Prices/wages: Harper 2016a. Building activity: Di Segni 1999, but see esp. the important observations on the continuity of ecclesiastical building. See further Chapter 7. Most scholars who have worked on the first pandemic conclude that it was an event of epoch-making significance. Durliat 1989 is the only one to have argued at any length for the relative *insignificance* of the pandemic, although Horden 2005 is cautious and Stathakoupolos 2004 is measured. Scholars who work on the period more generally will sometimes brush off the plague or minimize it (e.g., Wickham 2016, 43–44; Wickham 2005). Durliat considered the paucity of evidence in inscriptions and papyri, among other things, but this is hardly probative. It might be useful to outline in brief why a minimalist case has become untenable. (1) The biological identity of the agent of the plague, *Y. pestis*, is now confirmed. Unless medically significant genetic variations are found between the Justinianic strain and the strain that caused the Black Death—and the opposite seems to be true—this virtually closes the case. (2) An epidemiological approach suggests the widespread impact of the pandemic. In particular, the pandemic's ability to penetrate rural areas is critical. The discovery of the plague bacterium in two isolated cemeteries outside Munich is the most important evidence to come to light since the discovery of John of Ephesus' text. The molecular evidence vindicates the literary sources. And if the plague was *here*, it was everywhere. (3) Ecological plausibility underlies a maximalist interpretation: the plausibility of the case for a pandemic has been carefully made. (4) A minimalist case must *entirely* discount the literary evidence. It requires a bold leap of faith to say that authors with entirely different worldviews, in radically different parts of the empire, simultaneously decided to exaggerate a mortality event. The sheer coherence of the literary evidence is remarkable. The quality of the observations is remarkably convincing. Hopefully, in the bigger picture, this book has helped to clarify that ancient sources did not report massive pandemics willy-nilly. There are really three well-documented pandemics in Roman antiquity: the Antonine, Cyprianic, and Justinianic Plagues. Each of these stands out against a background of ordinary, local, epidemic disease. Moreover, the consilience of the literary evidence with the physical evidence is powerful. Our ancient authorities have become more believable by the day. (5) The continuing accumulation of evidence—e.g., for plague in the skeletons of victims far from the center of the empire, inscriptions suggesting large-scale mortality, the large number of mass burials now discovered—undermines the minimalist case. (6) The next chapter traces the longer-range impacts of sudden demographic contraction. I believe this analysis provides a plausible model for how a sudden demographic catastrophe played out over 2–3 generations and issued in state failure. (7) The next chapter traces the cultural response to the pandemic. As Meier 2016 convincingly argues, the dramatic cultural shift suggests a deep underlying crisis, independent of all of the other evidence.

76. Procopius thought the failure of the gold coinage was a "thing that had never occurred in the past": *Anek.* 22.38, tr. Kaldellis. Morrison and Sodini 2002, 218. Hahn 2000. See Chapter 7, for the fiscal-military crisis. Laws: Sarris 2006, 219; Sarris 2002, 174–75.

77. Evagrius, *Hist. Eccl.* 4.29 (178). Black Death: Benedictow 1992, 126–45. China: Li et al. 2012. In general: Bi 2016.

78. Carmichael 2014; Varlık 2014. Genetic: Bos et al. 2016; Seifert et al. 2016.

79. For the chronology of its end, see McCormick 2007, 292.

80. Compare Varlık 2015, for the role of Istanbul, esp. 24 on the ecology of persistence.

81. "It had never really stopped": Agathias, *Hist.* 5.10.1–2, tr. Frendo. "Simply dropped dead": Agathias, *Hist.* 5.10.4, tr. Frendo. Men more than women: Agathias, *Hist.* 5.10.4. See Appendix B, Event #1; Stathakopoulos no. 134, 304–6. Anatolia, Syria, Mesopotamia: see Appendix B, Event #2; Stathakopoulos no. 136, 307–9. Amplification of AD 573–74: Appendix B, Event #3; Stathakopoulos no. 145, 315–16. Amplification of AD 586: Appendix B, Event #7; Stathakopoulos no. 150, 319–20.

82. Events of 597–600: Appendix B, Events #12-4; Stathakopoulos nos. 156, 159–64, 324–34.

83. Appendix B, Event #3; Stathakopoulos no. 139, 310–11. "There began to appear": Paul the Deacon, *Hist. Langob.* 2.4, tr. Foulke.

84. Event in Gaul in AD 571: Appendix B, Event #4. Gregory of Tours, *Hist. Franc.* 4.31–32. Marius of Avenches, an. 571. Event of AD 582–84: Appendix B, Event #6. Gregory of Tours, *Hist. Franc.* 6.14 and 6.33. Event in AD 588: Appendix B, Event #8. Gregory of Tours, *Hist. Franc.* 9.21–22.

85. Appendix B, Events #9–10.

86. Appendix B, Event #14. Inscription from Córdoba: CIL II 7.677. Homilary: Kulikowski 2007.

87. Appendix B, Events #21 and 25. See Maddicott 2007. Rats: Reilly 2010. Atlantic zone: Loveluck 2013, esp. 202–4.

88. See above all the invaluable dissertation, Conrad 1981. For upland focalization, Green 2014a, 18; Varlık 2014, esp. 208; Panzac 1985. For settlement in this region, see Eger 2015, 202–6.

89. Event of AD 561–62 in the east: Appendix B, Event #2; Stathakopoulos no. 136, 307–9. Event of AD 592: Appendix B, Event #11; Stathakopoulos no. 155, 323–24. Conrad 1994. See above all Conrad 1981 for these events. On the focalization of plague in these regions in the second pandemic, see Panzac 1985, 105–8.

90. One-third: I. Palaestina Tertia, Ib, no. 68. See also nos. 69–70 and Benovitz 2014, 491–92. See also the important Arabic sources brought to bear by Conrad 1994. Michael the Syrian claimed that one-third of the world died in the amplification of AD 704–5: *Chron.* 11.17 (449).

91. Appendix B, Event #38.

92. This medieval rebound is of course the subject of McCormick 2001.

93. Extinct branch: Wagner et al. 2014.

94. For a sample of mortality rates in later outbreaks of plague, Alfani 2013 has a wealth of information. Haldon 2016, 232, on estimates of eastern Mediterranean populations, with similar ranges as reported in Fig. 2.

95. Gregory the Great, *Dial.* 3.38.3.

Chapter 7: Judgment Day

1. Markus 1997.

2. Senate: Gregory the Great, *Hom. Ezech.* 2.6.22, with the cautions of Humphries 2007, 23–24. Gregory and empire: Dal Santo 2013. On Gregory's thought in general, Demacopoulos 2015, at 87–88 on identification with the empire; Straw 1988.

3. On Gregory's eschatology: Demacopoulos 2015, esp. 92–93; Kisić 2011; Markus 1997, 51–67; Dagens 1970. The epidemic of 589–90: Appendix B, Event #9. Gregory of Tours, *Hist Franc.* 10.1, 10.23; Gregory the Great, *Dial.* 4.18, 4.26, 4.37; *Reg.* 2.2; Paul the Deacon, *Hist. Langob.* 3.24; *Liber pontificalis* 65. See further below, on the physical evidence.

4. "We suffer": Gregory the Great, *Hom. In Ev.* 1.1.1. "I sigh longingly": Gregory the Great, *Reg.* 9.232, tr. McCormick 2001, 27, with mortality for "epidemic." On the "emptiness" of the city: Demacopoulos 2015, 92–93.

5. "Inversion": Gregory the Great, *Reg.* 9.232. Seismic clustering: Stiros 2001. Spiritual effects of the earthquakes: Magdalino 1993, 6; Croke 1981.

6. For general treatments of the truly striking amount of apocalyptic literature in the late sixth and seventh centuries, see Meier 2016; Meier 2003; Reeves 2005; Reinink 2002; Cook 2002; Hoyland 1997, 257–335; Magdalino 1993; Alexander 1985. Arguably, the subtle role of apocalyptic thinking in the context of infectious disease is with us still: see Carmichael 2006.

7. Bjornlie 2013, 254–82.

8. Older terms for it include the (misleading) "Vandal Minimum" or "Dark Age Cold Period."

9. NAO: Baker et al. 2015; Olsen et al. 2012. See also Brooke 2016 and 2014, 341–42, 352–53. Sicily: Sadori et al. 2016. Anatolia: see Izdebski et al. 2016; Haldon et al. 2014; Izdebski 2013.

10. Procopius, *Bell.* 4.14.5–6, tr. Kaldellis. John of Ephesus, in *Chronicle of Zuqnin*, tr. Witakowski 1996, 65. Four hours a day: Michael the Syrian, *Chron.* 9.26 (296) and the *Chronicle of 1234*. An account preserved in the chronicle of Agapios measured the duration of the event at fourteen months: Agapios, *Kitab al-'Unvan*, fol. 72v. Pope Agapetus: Pseudo-Zacharias of Mitylene, *Chron.* 9.19, tr. Greatrex. On this source, Brock 1979–80, 4–5. On the written evidence for the AD 536 event, see most fully Arjava 2005.

11. John Lydus, *De portentis* 9c, tr. Arjava 2005. It portended evil in Europe but not in the dry lands to the south and east, like India and Persia, since only in Europe was "the moisture in question evaporated and gathered into clouds dimming the light of the sun so that it did not come into our sight or pierce this dense substance."

12. Bjornlie 2013.

13. Cassiodorus, *Var.* 12.25 (MGH AA 12, 381–82). I have adapted the translation of Barnish from *Translated Texts for Historians* and benefited greatly from an unpublished translation by Michael McCormick.

14. For a global context, see the essays in Gunn 2000.

15. Newfield 2016, is a very helpful overview. NASA: Stothers and Rampino 1993. Impact: Keys 1999; Baillie 1999. Written evidence: Arjava 2005.

16. Sigl et al. 2015; Baillie and McAneney 2015; Baillie 2008.

17. Abbott et al. 2014 for the case that a cometary source also contributed to the contemporary dimming.

18. Toohey et al. 2016; Kostick and Ludlow 2016; Büntgen et al. 2016; Sigl et al. 2015.

19. Usoskin et al. 2016; Steinhilber et al. 2012.

20. Data source for Fig. 7.2: ftp://ftp.ncdc.noaa.gov/pub/data/paleo/climate_forcing/solar_variability/steinhilber2009tsi.txt. Glaciers: Le Roy et al. 2015, esp. Fig. 7 and p. 14; Holzhauser et al. 2005, Fig. 6. A high-resolution marine sediment record from the Gulf of Taranto shows a cooler period AD 500–750: Grauel et al. 2013.

21. Coldest period: Büntgen 2016.

22. Sicily: Sadori et al. 2016. For the floods, see Squatriti 2010 for an exhaustive treatment of the literary evidence. While we share the view that the memory of these floods largely survives *because* of their association with Gregory, and while the significance of a single flood episode of 589 can certainly be overstated, the fact remains that physical evidence for greater humidity and for flooding in mainland Italy from the late fifth to the mid-seventh century is convincing. Cremonini, Labate, and Curina 2013; Christie 2006, 487; Squatriti 1998, 68; Cremaschi, Marchetti, and Ravazzi 1994. Moreover, and just as importantly, the pattern evident in Italian landscapes also fits well with our reconstruction of the period ca. AD 450–650 as dominated by a negative regime of the NAO. There is no mystery about the strong connection between negative NAO and flooding in Italy: see Benito et al. 2015a; Benito et al. 2015b; Zanchettin, Traverso, and Tomasino 2008; Brunetti 2002. While flooding is always local, it is not irreducibly a microclimatic phenomenon. Large-scale atmospheric patterns, principally the NAO, have a huge determining influence.

23. Haldon et al. 2014, 137; Izdebski 2013, 133–43. Bereket Basin: Kaniewski et al. 2007. Nar Gölü: Dean et al. 2013; Woodbridge and Roberts 2011. Tecer Lake: Kuzucuoğlu et al. 2011. Justinian: Procopius, *De aedific.* 5.5.15–20, tr. Dewing. In far SE Turkey, patterns of soil erosion suggest the importance of extreme precipitation in the late Roman period: Casana 2008. Syntheses of Anatolia: Izdebski et al. 2016; Haldon 2016; Haldon et al. 2014; Izdebski 2013.

24. Fentress and Wilson 2016.

25. Ptolemais: Procopius, *De aedific.* 7.2.9, tr. Dewing. Lepcis: Procopius, *De aedific.* 6.4.1, tr. Dewing. Cf. Mattingly 1994, 2. Late antique droughts and climate change: Fareh 2007.

26. In general, see Avni 2014.

27. On the North Sea–Caspian Pattern, see Kutiel and Türkeş 2005; esp. Kutiel and Benaroch 2002. More generally, Black 2012; Manning 2013, 111–12; Roberts et al. 2012.

28. Natural proxies: Haldon et al. 2014, 123; Rambeau and Black 2011; Neumann et al. 2010; Leroy 2010 (climate shift around AD 550); Migowski et al. 2006; Bookman et al. 2004. Issar and Zohar 2004, esp. 211, esp. the dendrochronological evidence for drought sometime in 5C–early 6C. Gaza: Choricius, *Ep*. 81. Drought in Palestine: Stathakopoulos 2004, no. 85, 259–61. Syriac: Pseudo-Zacharias, *Chron*. 8.4. Visiting saint: *Vita Theod. Syk.* 50. Aqueducts: Jones 2007. Justinian repaired the aqueducts of Bostra (IGLS 13.9134). Generally, Decker 2009, 8–11.

29. Procopius, *De aedific*. 4.2.12, tr. Dewing. For other comparisons of Justinian to Xerxes, see Kaldellis 2004, 35.

30. "Bountiful crop": *Vit. Ioh. Eleem*. I.3, tr. Dawes and Baynes. Baths: *Vit. Ioh. Eleem.* II.1. Alexandria: Holum 2005, 99. John's high political connections: Booth 2013, 51.

31. "Two of the Church's": *Vit. Ioh. Eleem*. II.13, tr. Dawes and Baynes (Americanizing "wheat" for "corn"). Storm on Adriatic: *Vit. Ioh. Eleem.* II.28, tr. Dawes and Baynes.

32. Ceramics: Haas 1997, 343–44. End of annona: McCormick 2001, 110–11.

33. Major syntheses include Christie 2011; Ward-Perkins 2005a and 2005b; Wickham 2005; Morrison and Sodini 2002. On the fate of towns, Krause and Witschel 2006; Holum 2005; Lavan and Bowden 2001; Liebeschuetz 2001; Rich 1992. Mitchell 2015, 479–91, is the most convincing and up-to-date attempt to put the archaeology in the context of demographic contraction caused by the plague.

34. "There were no towns": Ward-Perkins 2000a, 350. Compared poorly: Ward-Perkins 2000b, 324. Slave markets: Gregory the Great, *Reg*. 3.16; 6.10; 6.29; 9.105; 9.124. See Harper 2011, 498.

35. In general, Kulikowski 2004 and 2006. "The Mediterranean coast": Wickham 2005, 491.

36. See Wickham 2005, esp. 666.

37. Arnold 2014 has now sketched this moment in late Roman Italy with particular vividness. Also O'Donnell 2008. For the archaeology, see Christie 2011.

38. "Our care": Cassiodorus, *Var*. 3.31, tr. Barnish. Colosseum: Christie 2006, 147. Turning point: Christie 2006, 459–60: "For many noted rural sites the Byzantine-Gothic Wars period indeed marks an apparent cut-off point in their recognizable settlement sequence." At pp. 185 and 250: "The picture generally registered for towns after c. AD 550 is one with large to small religious complexes, damaged and demolished ancient public edifices, scatterings and concentrations of houses, plus a variety of 'no-go' and open spaces." 500 people: see the comments of Christie 2006, 61. Decline in inscriptions: see Fig. 5.2 in Chapter 5. Population: Morrison and Sodini 2002.

39. Loss of public control over environment, etc.: Christie 2006, 200, 487–89.

40. "The villages and farms": Barker, Hodges, and Clark 1995, 253. "In the seventh and eighth centuries": Ward-Perkins 2000a, 355, cf. 325. "People are so much harder": Christie 2006, 560. Half to quarter: Ward-Perkins 2005, 138. For the methodological hazards generally, see Witcher 2011.

41. Coins: Ward-Perkins 2005, 113. Ceramics: Ward-Perkins 2005, 106. Uplands: see Gregory the Great, *Reg*. 2.17, 6.27, 10.13. Christie 2006, 461. Pre-Etruscan: Ward-Perkins 2005, 88, 120.

42. Urban peak in 4C: Lepelley 2006. Fentress and Wilson 2016, 17. Kasserine survey: Hitchner 1988; Hitchner 1989; Hitchner 1990. Cameron 2000b, 558.

43. Decker 2016, 9–11, 17; Whitby 2000, 97–98a.

44. "Unbroken continuity": Wickham 2005, 627. Butrint: Hansen et al. 2013; Bowden, Hodges, and Cerova 2011; Decker 2009, 93. Corinth: Scranton 1957; Brandes 1999. Macedonia: Dunn 2004, 579. "Considerable hand-wringing": Decker 2009, 131.

45. Decker 2016, 130–34; Pettegrew 2007; Mee and Forbes 1997.

46. Overviews in Haldon 2016 and Izdebski 2013. "Most probably": Waelkens et al. 2006, 231.

47. I am convinced by the careful arguments of Liebeschuetz 2001, esp. at 43, 48–53, 408. Waelkens 2006; Wickham 2005, 627. Fragmentation: Haldon 2016. Continuing rural occupation at a lower level: Vanhaverbeke et al. 2009 (a 'decapitated' landscape); Vionis et al. 2009.

48. Nile flood: Procopius, *Bell.* 7.29.6–7, tr. Kaldellis. Canals, etc.: Bagnall 1993, 17–18. Black Death in Egypt: Borsch 2014; Borsch 2005, 46–47: "The smooth functioning of this network depended not only on the exactness of its timing, but also on a substantial input of labor and raw materials for its annual maintenance."

49. "Nile's swamping": Procopius, *Bell.* 7.29.19, tr. Kaldellis. It is also worth noting that the great Ma'rib Dam in southern Arabia flooded three times in the sixth century and finally collapsed, suggesting that intense monsoonal activity may have characterized the mid-6C. See Morony 2007, 63. Wheat and rents: Harper 2016a. We might have expected laborers, now more scarce, to leverage lower rents. But it may be that real rents were customary and therefore sticky, or that landowners exerted extra-market power to hold down the power of laborers, or that on average higher-quality land was now farmed and laborers had reduced the rent rates on these lands. It is an unsolvable conundrum. Cash rents did drop, though not immediately, which might represent a cleaner market signal of falling commodity prices and rising labor costs.

50. See Hickey 2012; Sarris 2006; Mazza 2001. "Obsessed": Hickey 2012, 88.

51. In general, Decker 2009. Fine wine: *Vit. Ioh. Eleem.* I.10.

52. Cities: Liebeschuetz 2001.

53. Witakowski 2010; Kennedy 2007a; Foss 1997; Tate 1992; Sodini et al. 1980; Tchalenko 1953–58. See esp. Casana 2014, 214, for convincing archaeological evidence that continuity was less pronounced in the northern Levant than in the south.

54. Izdebski 2016; Hirschfeld 2006; Kennedy 2000. Church building as preparation for the Last Judgment: Magdalino 1993, 12.

55. Hirschfeld 2006, 2004, for the impact of climate change. See now, however, Avni 2014. "One of the most": Decker 2009, 196. Irrigation: Kamash 2012. Kouki 2013, emphasizes that human settlement and climate change in south Jordan did not move in step, and sees aridity progressing throughout the sixth and seventh centuries. Generally, see Mikhail 2013, on the importance of a dialectical understanding of humans and the environment in the region. Izdebski 2016, esp. 202, on the fact that some settlements are simply so far east in what is today desert that precipitation had to have been higher. Rubin 1989, for an older skeptical case.

56. Liebeschuetz 2001, 57; Walmsey 2007, 41. Negev: Avni et al. 2006: human activity "is super-imposed on the natural long-term trend leading toward desertification." Wadi Faynan: Barker, Gilbertson, and Mattingly 2007. "Quietly": Walmsey 2007, 47. Liebeshuetz 2001, 303.

57. Agathias, *Hist.* 5.15.7, tr. Frendo.

58. "The fortunes": Agathias, *Hist.* 5.13.5, tr. Frendo. "The Roman armies": Agathias, *Hist.* 5.13.7, tr. Frendo. On Agathias and the plague, Kaldellis 2007, 15–16. Lee 2007, 117–18.

59. "There was a large reservoir": Jones 1964, 670. But for traces of conscription, see Whitby 2000b, 302–3. Haldon 2002, is a persuasive demonstration that the transformations of the seventh century have their origins in the pressures of the later sixth century.

60. Procopius *Bell.* P. 166, 399, 404 (7.10.1–2), tr. Kaldellis.

61. For the rise of institutions of public credit, see Edling 2003. "Was always late": Procopius *Anek.* p. 82, tr. Kaldellis. "Began openly cheating": Agathias, *Hist.* 5.14.2, tr. Frendo. See also John Malalas, *Chron.* 18.132. Malalas offers a generally favorable account of Justinian's reign, so that his inclusion of the struggle to pay soldiers appropriately is entirely credible. Border units: Treadgold 1995, 150. In general, Treadgold 1995, 159–66.

62. Justinian, *Novellae* 147, tr. Blume.

63. "When the plague": Procopius, *Anek.* 23.20–21, tr. Kaldellis. Aphrodito: Zuckerman 2004, 120 and esp. 215. As Barnish, Lee, and Whitby 2000, 185, rightly note, "the evidence is complex and contentious." A more detailed study of sixth century fiscality, exploring the

possibility of enormous demographic contraction, would be welcome. I concur with Van Minnen 2006, 165–66, about late antique taxes rising.

64. Justin II, *Novellae* 148, tr. Blume. On these remissions, Haldon 2016, 182.

65. Recruitment under Tiberius II: Evagrius, *Hist. Eccl.* 5.14; Theophylact Simocatta, *Hist.* 3.12.3–4. Maurice: Michael the Syrian, *Chron.* 11.21 (362); John of Ephesus, *Hist. Eccl.* III.6.14. With Whitby 1995, 81. Still put large armies in the field: Whitby 1995, 100. But "finance was the key problem throughout Maurice's reign": Whitby 2000a, 99. Reduction in pay: Theophylact Simocatta, *Hist.* 3.1.2 and 7.1.2–9; Evagrius, *Hist. Eccl.* 6.4. In the words of A. H. M. Jones, these were "dangerous economies": Jones 1964, 678. The best treatment of Roman military recruitment in this period is Whitby 1995, although I would emphasize the extent of crisis and its simultaneous demographic and fiscal dimensions.

66. Booth 2013, 44–45.

67. John Moschus, *Prat. Spir.* 131, tr. Wortley.

68. Nea Church: Procopius, *De Aedif.* 5.6.1. See esp. Graham 2008; Tsafrir 2000.

69. Cameron 1978. On late antique ideas of Mary, Pentcheva 2006 is especially compelling.

70. Antioch: Evagrius Scholasticus, *Hist. eccl.* 4.29 (177). Sinai: Anastasius of Sinai, *Ques. Resp.* 28.9 and 66. Islamic examples: Conrad 1992, 92–95.

71. Year 500: Magdalino 1993, 4–5.

72. See esp. Meier 2003. "Went into ecstasy": John Malalas, *Chron.* 18.90, tr. Jeffreys. "According to the ancient oracles": Agathias, *Hist.* 5.10.5, tr. Frendo. For communal responses to the Black Death, see Dols 1974. Norse myth: Gräslund and Price 2012, noting the exceptional declines in settlement in the sixth century. China: Barrett 2007.

73. "The people": John of Ephesus, in *Chronicle of Zuqnin*, tr. Witakowski 1996, 87. "Love for mankind": Justinian, *Novellae* 122 (AD 544). See Demacopoulos 2015, 93; Kaldellis 2007, 7.

74. Clermont: Gregory of Tours, *Hist. Franc.* 4.5. East: Ebied and Young 1972, no. XXX. Little 2007b, 26–27.

75. Gregory of Tours, *Hist. Franc.* 10.1, tr. Thorpe.

76. Brown 2012. See Kaldellis 2007, 9.

77. Egypt: MacCoull 2004–5. Building: Di Segni 2009 and 1999. See also Gatier 2011. Petra: Frösén et al. 2002, 181–87; with Benovitz 2014, 498. Ravenna: see the insightful discussion of Deliyannis 2010, 252–53. Mary: e.g., a church at al-Rouhhweyb in the Jabal Hass: Trombley 2004, 77; Mouterde and Poidebard, no. 17. Piccirillo 1981, 58, for a votive restoration on behalf of the Archangel Michael at Um el-Jimal. Piccirillo 1981, 84, at Rihab, a church for Mary, begging, "Have mercy on the world and help us who give." Ovadiah 1970, 28–29, a basilica at Beit Sha'ar for St. Zechariah, late 6C, referring to deliverance (SEG 8, no. 238; see TIR, 77). Ovadiah 1970, 54–55, at 'Ein el-Jadida, a late 6C church in thanks for deliverance. Ovadiah 1970, 172–73, inscriptions asking for help for the donors (SEG 8, no. 21). At Nessana, see Colt 1962, no. 92 (AD 601/2) dedication to Holy Mary the God Bearer, begging "Help and have pity." See also no. 72 (an unknown building, a dedication in AD 605 for the salvation of certain donors) and no. 94 (AD 601, for the salvation of certain benefactors) and no. 95. Donceel-Voûte 1988, 275, a church built at Resafe in 559, for the mercy of God. Donceel-Voûte 1988, 139, at Houad, a mosaic paid for in AD 568 seeking help from St. George. See also Donceel-Voûte 1988, 356 (at Jiye) and 416 (Qabr Hiram). Further examples in Madden 2014.

78. Copy of New Testament: "Discourse on Saint Michael the Archangel by Timothy, Archbishop of Alexandria," in Budge 1915, 1028; I thank Michael Beshay for clues that put me on the trail of this magnificent reference. "Angel of God": Gregory of Tours, *Hist. Franc.* 4.5. For the spread of Michael veneration (underway well before the great pestilence), see Arnold 2013; Rohland 1977.

79. Cameron 1978, 80 and 87 (liturgy's potent influence on society). Hypapante: *ODB* 961. Allen 2011, 78. It is worth noting that, upon his installation as emperor in AD 565, Justin II

went to pray at a shrine of the Archangel Michael, while his wife went to a Church of the Virgin Mother: Corippus, *In Laud. Iust.* 2, with Cameron 1976, 149–50. Liturgy as eschatological: Magdalino 1993, 15. Domestic artifacts: Maguire 2005. *Akathistos* hymn: Pentcheva 2006.

80. Cameron 1978. See also Meier 2005. Latham 2015; Wolf 1990, esp. 131–35 on the medieval context of the legend.

81. E.g., the kind of litanies that are recorded explicitly in Constantinople in the aftermath of the earthquake of AD 554: Theophanes the Confessor, *Chron.* s.a. 6046; Kaldellis 2007, 7–8. "These scourges": recorded in Gregory of Tours, *Hist. Franc.* 10.1, tr. Thorpe. "Not entirely deserted": see Markus 1997, 63.

82. For the status of the *Apocalypse*, see now Shoemaker 2016. Commentaries on *Apoc.*: Hoskier 1928 (Oecumenius); Schmid 1955–56 (Andreas of Caesarea); Primasius, *Comm. In Apoc.* See Meier 2003, 21; though see Podskalsky 1972, for some precedents 79–80. "Although patristic theology": Magdalino 1993, 9.

83. Nicholas: *Vit. Nich. Sion.* 50–52.

84. "The Holy One": Reeves 2005, 123. In general, Himmelfarb 2017; Reeves 2005; van Bekkum 2002; Dagron and Déroche 1991. On the political position of the Jews, see now esp. Bowersock 2017.

85. Watchword: Theophylact Simocatta, *Hist.* 5.10.4. "Psychological impact": Drijvers 2002, 175.

86. Return of the true cross: recently, Zuckerman 2013, on an event that is extremely complex in its details, with earlier literature cited there. Heraclius: see the especially valuable collection of essays in Reinink and Stolte 2002; Magdalino 1993, 19.

87. Islamic conquests: Kennedy 2007b; Kaegi 1992.

88. Bowersock 2017; Robin 2012; Conrad 2000; Donner 1989. Jewish South Arabia: Bowersock 2013, 78–91. Constantinople and Muhammad: Lecker 2015 and now Bowersock 2017, 108–11.

89. Hoyland 2012; Donner 2010; Cook 2002; Bashear 1993. Casanova 1911, already delineates many of the key ideas of the apocalyptic origins of Islamic. See Al-Azmeh 2014 for the influence of late antique imperial monotheistic frameworks on the emergence of Islam, downplaying the importance of apocalyptic elements.

90. "The Coming": Shoemaker 2012, 120. "A warning": Qur'an 53:57, tr. Asad. "God's is the knowledge": Qur'an 16:77, tr. Asad.

91. Sophronius, *Sermon on the Epiphany*, tr. Hoyland 1997, 73. On the suggestion that parts of the Nea Church (which was huge and so completely disappeared that it was only found in the 1970s) were reused, see Nees 2016, 108.

Epilogue: Humanity's Triumph?

1. Malthus 1826, 257. On the intellectual context of the *Essay*, emphasizing its global perspective, see now Bashford and Chaplin 2016.

2. McNeill 2015; Livi-Bacci 2012; Klein Goldewijk, Beusen, and Janssen 2010; Maddison 2001; McEvedy and Jones 1978.

3. Pomeranz 2000 launched contemporary global-scale comparative analysis of economic development, emphasizing the equality of England and parts of China as late as 1800. More recent work tends to place the "great divergence" earlier. See Broadberry, Guan, and Li 2014; Chen and Kung 2016 for the Malthusian regime in modern China.

4. We should add that Malthus believed there were two kinds of check, the preventative and the positive. The preventative check controlled population in advance of disaster, through mechanisms that moderated fertility. The positive check controlled population via mortality. The figures derive from the sources cited in Note 2 above, with the mortality estimates presented in this book.

5. McNeill 2015, 77; Russell 2011. Technically, the International Commission on Stratigraphy of the International Union of Geological Sciences controls the nomenclature of geological epochs, but the term has broader meanings and remains the object of intense discussion. See Finney and Edwards 2016, for a recent overview.

6. One trillion species: Locey and Lennon 2016. 40 trillion cells: Sender, Fuchs, and Milo 2016. 1400 pathogens: Woolhouse and Gaunt 2007. In general, see now the remarkable overview of microbial ecology by Yong 2016.

BIBLIOGRAPHY

Primary Sources: Literary Texts

Ach. Tat. = Garnaud, J.-P. ed. 1991, *Le roman de Leucippé et Clitophon*, Paris.

Acta Acacii = Weber, J. ed. 1913, *De actis S. Acacii*, Strassburg.

Aelius Aristides = Behr, C. A. ed. 1981–86, *P. Aelius Aristides: The Complete Works*, 2 vols., Leiden.

Agapios = Vasiliev, A. ed. *Kitab al-ʾUnvan*, *Patrologia Orientalis* 5.4 (1910); 7.4 (1911); 8.3 (1912); 11.1 (1915).

Agathias = Keydell, R. ed. 1967, *Historiae*, Berlin.

Translation: Frendo, J. D. 1975, *Agathias: The Histories*, Berlin.

Ambrose = Faller, O. ed. 1968, *Epistulae*, CSEL 82, Vienna.

Translation: Liebeschuetz, J. H. W. G. 2005, *Ambrose of Milan: Political Letters and Speeches*, TTH no. 43, Liverpool.

Ammianus Marcellinus = Seyfarth, W. ed. 1978, *Res gestae*, 2 vols., Leipzig.

Translation: Rolfe, J. C. 1935–39, *Ammianus Marcellinus: Histories*, LCL 300, 315, 331, Cambridge, MA.

Anastasius of Sinai = Munitiz, J. A. and Richard, M. eds. 2006, *Anastasii Sinaitae Questiones et Responsiones*, CCSG 59, Turnhout.

Translation: Munitiz, J. A. 2011, *Questions and Answers*, Turnhout.

Appian = Mendelssohn, L. and Viereck, P. eds. 1986, *Historia Romana*, Leipzig.

Apuleius = Helm, R. ed. 1959, *Florida*, Leipzig.

Aretaeus = Hude, K. ed. 1958, *Aretaeus*, 2nd edn., Berlin.

[Aristotle] = Bekker, I. ed. 1960, *Aristotelis opera*, vol. 2, Berlin: 859a1–967b27.

Augustine = Verheijen, L. ed. 1981, *Confessiones*, CC 27, Turnhout.

= Dekkers, E. and Fraipont, J. eds. 1956, *Enarrationes in Psalmos*, CC 38-40, Turnhout.

Aurelius Victor = Dufraigne, P. ed. 1975, *De Caesaribus*, Paris.

Translation: Bird, H. W. 1994, *Liber de Caesaribus*, TTH 17, Liverpool.

Basil of Caesarea = Courtonne, Y. ed. 1935, *Saint Basile: homélies sur la richesse*, Paris.

Cassiodorus = Fridh, Å. J. ed. 1973, *Magni Aurelii Cassiodori: Variarum libri XII*, CC 96, Turnhout.

Translation: Barnish, S. J. B. 1992, *Cassiodorus: Selected* Variae, TTH 12, Liverpool.

Cassius Dio = Boissevain, U. P. ed. 1955, *Cassii Dionis Cocceiani historiarum Romanarum quae supersunt*, 3 vols., Berlin.

Celsus = Spencer, W. G. 1935–38, *Celsus*: *De medicina*, LCL 292, 304, 336, Cambridge, MA.

Choricius of Gaza = Foerster, R. and Richtsteig, E. eds. *Choricii Gazaei opera*, Leipzig.

Chronicle of 1234 = Chabot, J.-B. ed. *Anonymi auctoris Chronicon ad annum Christi 1234 pertinens*. CSCO 81, 82, 109; Syr. 36, 37, 56 (Paris, 1916, 1920, 1937).

Chronicle of Zuqnin = Chabot, J.-B. ed. *Chronicon anonymum pseudo-dionysianum vulgo dictum*, CSCO 91, 104; Syr. 43, 53 (Paris 1927, 1933).
Translation: Witakowski, W. 1996, *Chronicle: Known Also As the Chronicle of Zuqnin. Part III*, TTH 22, Liverpool.
Claudian = Hall, J. B. ed. 1985, *Carmina*, Leipzig.
Translation: Platnauer, M. 1922, *Claudian*, LCL 135–36, Cambridge, MA.
Columella = Lundström, V. ed. 1897, *Opera quae extant*, Upssala.
Consularia Caesaraugustana = De Hartmann, C. C. ed. 2001, *Tunnunensis Chronicon cum reliquiis ex Consularibus Caesaraugustanis et Iohannis Biclarensis Chronicon*, Turnhout.
Corippus = Cameron, A. ed. 1976, *In laudem Iustini*, London.
= Diggle, J. and Goodyear, F. R. D. eds., 1970, *Iohannidos seu De bellis Libycis libri VIII*, London.
Translation: Shea, G. W. 1998, *The Iohannis, or, De bellis Libycis*, Lewiston, NY.
Cosmas Indicopleustes = Wolska-Conus, W. ed. 1968-73, *Topographie chrétienne*, SC 141, 159, 197, Paris.
Cyprian = Hartel, G. ed. *S. Thasci Caecili Cypriani Opera Omnia*, 3 vols., CSEL 3.1–3, Vienna.
Pseudo-Cyprian = L. Ciccolini, ed. forthcoming, *De laude martyrii*.
Cyril of Scythopolis = Schwartz, E. 1939, *Kyrillos von Skythopolis*, TU 49.2, Leipzig.
De cerem. = Moffatt, A. and Tall, M. 2012, *Constantine Porphyrogennetos: The Book of Ceremonies, with the Greek Edition of the Corpus Scriptorum Historiae Byzantinae (Bonn, 1829)*, Canberra.
Dio Chrysostom = de Arnim, J. ed. 1896, *Orationes*, 2 vols., Berlin.
Eugippius = Régerat, P. ed. 1991, *Vie de saint Séverin*, SC 374, Paris.
Eusebius = *Historia ecclesiastica*, Bardy, G. ed. 1952–94, SC 31, 41, 55, 73, Paris.
= *Praeparatio evangelica*, Places, É. ed. 1974–91, SC 206, 215, 228, 262, 266, 292, 307, 338, 369, Paris.
= *Vita Constantini*, Winkelmann, F. ed. 1975, *Eusebius Werke*, Band 1.1: *Über das Leben des Kaisers Konstantin*, Berlin.
Eutropius = Dietsch, H. ed. 1850, *Breviarium historiae Romanae*, Leipzig.
Evagrius Scholasticus = *Historia ecclesiastica*, Bidez, J. and Parmentier, L. eds. 1898, *The Ecclesiastical History of Evagrius*, London.
Excerpta Salmasiana = Roberto, U. ed. 2005. *Ioannis Antiocheni Fragmenta ex Historia chronica*, Berlin.
Fronto = van den Hout, M. P. ed. 1988, *M. Cornelii Frontonis Epistulae*, Leipzig.
Galen =
Alim. Fac. = *De alimentorum facultatibus*. Koch, K. et al. eds. 1923, CMG 5.4.2, Leipzig.
Anat. Admin. = *De anatomicis administrationibus*. Garofalo, I. ed., 1986, *Galenus: Anatomicarum administrationum libri quae supersunt novem*, 2 vols., Naples.
Translation: Singer, C. 1956, *On Anatomical Procedures*, London.
Anim. Affect. Dign. = *De animi cuiuslibet affectuum et peccatorum dignotione et curatione*. De Boer, W. ed. 1937, CMG 5.4.1.1, Leipzig.
Atra Bile = *De atra bile*. De Boer, W. ed. 1937, CMG 5.4.1.1, Leipzig.
Translation: Grant, M. 2000, *Galen on Food and Diet*, London.

Bon. Mal. Succ. = *De bonis malisque succis*. Koch, K. et al. eds. 1923, CMG 5.4.2, Leipzig.
Hipp. Artic. = *In Hippocratis librum De articulis commentarius*. Ed. Kühn, C.G. 1829, vol. 18.1, Leipzig: 300–45, 423–767.
Hipp. Epid. 3 = *In Hippocratis librum iii epidemiarum commentarii iii*. Wenkebach, E. 1936, CMG 5.10.2.1, 1–187.
Meth. Med. = *De methodo medendi*. Johnston, I. and Horsley, G. H. R. eds. 2011, *Galen: Method of Medicine*, 3 vols., LCL 516–18, Cambridge, MA.
Morb. Temp. = *De morborum temporibus*. Wille, I. ed. 1960, *Die Schrift Galens Peri tõn en tais nósois kairõn und ihre Überlieferung*, Kiel.
Praecog. = *De praecognitione*. Nutton, V. ed. 1979, CMG 5.8.1, Berlin.
Praes. Puls. = *De praesagitione ex pulsibus*. Kühn, C.G. ed. 1825, vol. 9, Leipzig: 205–430.
Purg. Med. Fac. = *De purgantium medicamentorum facultate*. Ehlert, J. ed. 1959, *Galeni de purgantium medicamentorum facultate*, Göttingen.
Subs. Fac. Nat. = *De Substantia Facultatum Naturalium*. Kühn, C.G. ed. 1822, vol. 4, Leipzig: 757–766.
Temp. = *De temperamentis*. Helmreich, G. ed. 1904. *Galeni De temperamentis libri III*, Leipzig.
Translation: Singer, P. N. 1997, *Galen: Selected Works*, Oxford.
George Kedrenos = de Boor, C. ed. 1904, *Georgii Monachi Chronicon*, Leipzig.
Gerontius = *Vita Melaniae*, Gorce, D. ed. 1962, SC 90, Paris.
= *Vita Melaniae* (L), Laurence, P. ed. 2002, Jerusalem.
Gregory of Nazianzus = *De pauperum amore*. PG 35: 857–909.
Translation: Vinson, M. 2003, *Selected Orations*, Washington, D.C.
= *Fun. Or. in Laud. Bas*. Boulenger, F. ed. 1908, *Grégoire de Nazianze. Discours funèbres en l'honneur de son frère Césaire et de Basile de Césarée*, Paris: 58–230.
Gregory of Nyssa = *De beneficentia*. van Heck, A. ed. 1967, *Opera*, vol. 9, Leiden.
Translation: Holman 2001.
=*In illud: Quatenus uni ex his fecistis mihi fecistis*. van Heck, A. ed. 1967, *Opera*, vol. 9, Leiden.
= *De vita Gregorii Thaumaturgi*. PG 46: 893–957.
Translation: Slusser, M. 1998, *Saint Gregory Thaumaturgus*, Washington, D.C.
Gregory of Tours = *Gloria confessorum*. Krusch, B. ed. 1885, MGH SS rer. Merov. 1.2, Hannover: 294–370.
Translation: Van Dam, R. 1988, *Glory of the Confessors*, TTH 4, Liverpool.
= *Gloria martyrum*. Krusch, B. ed. 1885, MGH SS rer. Merov. 1.2, Hannover: 34–111.
Translation: Van Dam, R. 1988, *Glory of the Martyrs*, TTH 3, Liverpool.
= *Libri historiarum X*. Krusch, B. and Levison, W. eds. 1937–51, MGH SS rer. Merov. 1.1, Hannover.
Translation: Thorpe, L. 1976, *The History of the Franks*, Harmondsworth.
= *Vitae patrum*. Krusch, B. ed. 1885, MGH SS rer. Merov. 1.2, Hannover: 211–293.
Gregory the Great = *Registrum epistularum*, Norberg, D. ed. 1982, CC 140–40A, Turnhout.
= *Dialogorum libri iv*. de Vogüé, A. ed. 1978–80, SC 251, 260, 265, Paris.
= *Homiliae in evangelia*. Étaix, R. ed. 1999, CCSL 141, Turnhout.
= *Homiliae in Hiezechihelem prophetam*. Adriaen, M. ed. 1971, CCSL 142, Turnhout.

Herodian = Stavenhagen, K. ed. *Herodiani ab excessu divi Marci libri octo*, Leipzig.
Translation: Whittaker, C. 1969–70, *Herodian: History of the Empire*, LCL 454–55, Cambridge, MA.
Historia Augusta = Hohl, H. ed. 1997, 3rd edn., 2 vols., Leipzig.
Historia monachorum in Aegypto = Festugière, A.-J. 1971, Brussels.
Hydatius = Burgess, R. W. 1988, *Hydatius: A Late Roman Chronicler in Post-Roman Spain: An Historiographical Study and New Critical Edition of the Chronicle*, Oxford.
Isaac of Antioch = Moss, C. 1929–32, "Homily on the Royal City," *Zeitschrift für Semitistik und verwandte Gebiete* 7: 295–306 and 8: 61–72.
Jerome = Helm, R. ed. 1956, *Eusebii Caesariensis Chronicon: Hieronymi Chronicon*, Berlin.
John Chrysostom = De Lazaro. PG 48: 963–1054.
= *In epistulam ad Ephesios*. PG 62: 9–176 .
= *In principium Actorum*. PG 51: 65–112.
John Lydus = Wünsch, R. ed. 1898, *Ioannis Lydi liber de mensibus*, Leipzig.
= Wachsmuth, C. ed. 1897, *Ioannis Laurentii Lydi liber de ostentis et calendaria Graeca omnia*, Leipzig.
John Malalas = Thurn, I. ed. 2000, *Ioannis Malalae chronographia*, Berlin.
Translation: Jeffreys, E. et al. 1986, *The Chronicle of Malalas*, Melbourne.
John Moschus = *Pratum spirituale*. PG 87.3: 2582–3112.
= Nissen, T. 1938, "Unbekannte Erzählungen aus dem Pratum Spirituale," *Byzantinische Zeitschrift* 38: 354–372.
Translation: Wortley, J. 1992, *The Spiritual Meadow*, Kalamazoo.
John of Ephesus = Brooks, E. W. ed., 1923–5, *John of Ephesus. Lives of the Eastern Saints*, PO 17, 18, 19, Paris.
= *see also Chronicle of Zuqnin*; Michael the Syrian
John Zonaras = Dindorf, L. ed. 1870, *Ioannis Zonarae epitome historiarum*, Leipzig.
Jordanes = Mommsen, T. ed. 1882, *Getica*, MGH AA 5.1, Hannover: 53–138.
Translation: Mierow, C. C. 1960, *The Gothic History of Jordanes in English Version*, Cambridge.
Josephus = *De bello Judaico*. Niese, B. ed. 1955, *Flavii Iosephi opera*, vol. 6, Berlin: 3–628.
Translation: Whiston, W. 1961, *The Life and Works of Flavius Josephus*, New York.
Joshua the Stylite = Wright, W. ed. 1882, *The Chronicle of Joshua the Stylite Composed in Syriac A.D. 507*, Cambridge.
Lactantius = Creed, J. L. ed. 1984, *De mortibus persecutorum*, Oxford.
Libanius = Foerster, R. ed. 1903–8, *Orationes = Opera*, vols. 1–4, Leipzig.
Liber pontificalis = Mommsen, T. ed. 1898, MGH Gesta Pont. Rom. 1, Hannover.
Lucan = Shackleton Bailey, D. R. ed. *M. Annaei Lucani De bello civili libri X*, Stuttgart.
Lucian = *Alexander*. Harmon, A.M. 1925, *Lucian*, vol. 4, LCL 162, Cambridge, MA: 174–252.
Lucretius = Martin, J. ed. 1963, *De rerum natura*, Leipzig.
Ps.-Macarius = *Sermones*, Berthold, H. ed, 1973, *Makarios/Symeon Reden und Briefe*, 2 vols., Berlin.
Marcellinus Comes = Mommson, T. ed. 1894, *Chronicon ad annum DXVIII*, MGH AA 11, Hannover: 60–108.
Marcus Aurelius = Farquharson, A. S. L. ed. 1944, *The Meditations of the Emperor Marcus Aurelius*, Oxford.

Mart. Areth. = Detoraki, M. ed. 2007, *Le martyre de Saint Aréthas et de ses compagnons (BHG 166)*, Paris.

Menander = Russell, D. and Wilson, N. eds. 1981, *Division of Epideictic Speeches*, Oxford.

Michael the Syrian = Chabot, J.-B. ed. 1899–1924, *Chronique de Michel le Syrien, patriarche jacobite d'Antioche (1166–1199)*, Paris.

Olympiodorus = Blockley, R. C. ed. 1981–3, *The Fragmentary Classicising Historians of the Later Roman Empire: Eunapius, Olympiodorus, Priscus, and Malchus*, Liverpool.

Orac. Sibyll. = Potter, D. W. 1990, *Prophecy and History in the Crisis of the Roman Empire: A Historical Commentary on the Thirteenth Sibylline Oracle*, Oxford.

Oribasius = Raeder, J. ed. 1928–33, *Oribasii collectionum medicarum reliquiae*, CMG 6.1.1–6.2.2, Leipzig.

= Raeder, J. ed. 1926, *Oribasii synopsis ad Eustathium et libri ad Eunapium*, CMG 6.3, Leipzig.

Orosius = *Historia adversos paganos*. Arnaud-Lindet, M.-P. ed. 1990–1, *Histoires: contre les païens*, Paris.

Ovid = Alton, E.H., Wormell, D. E. W, and Courtney, E. eds. 1978, *P. Ovidi Nasonis Fastorum libri sex*, Leipzig.

Palladius = Rodgers, R. ed. 1975, *De insitione*, Leipzig.

Palladius = Butler, C. ed. 1904, *The Lausiac History of Palladius*, 2 vols., Hildesheim.

Ps.-Palladius = Desantis, G. ed. 1992, *Le genti dell'India e i Brahmani*, Rome.

Paul the Deacon = Waits, G. ed. 1878, *Historia Langobardorum*, SS rer. Germ. 48, Hannover.

Translation: Foulke, W. D. 1907, *History of the Langobards*, Philadelphia.

Pausanias = Spiro, F. ed. 1903, *Pausaniae Graeciae descriptio*, 3 vols., Leipzig.

Periplus Mar. Eryth. = Casson, L. 1989, *The Periplus Maris Erythraei: Text with Introduction, Translation, and Commentary*, Princeton.

Philostratus = Jones, C. P. 2006, *Philostratus: Apollonius of Tyana*, LCL 458, Cambridge, MA.

= Kayser, C.L. ed. 1871, *Flavii Philostrati opera*, vol. 2, Leipzig.

Pliny the Elder = Jahn, L., Semi, F. and Mayhoff, C. eds. 1967–80, *C. Plini Secundi Naturalis historiae libri XXXVII*, Stuttgart.

Pliny the Younger = Schuster, M. ed. 1958, *Epistularum libri novem. Epistularum ad Traianum liber. Panegyricus*, Leipzig.

Plutarch = Paton, W. R. et al. eds. 1925–67, *Plutarchi moralia*, Leipzig.

Pontius = *Vita Cypriani*. Bastiaensen, A. A. R. ed. 1975, *Vite dei santi* III Rome: 4–48.

Primasius = Adams, A. W. ed. 1985, *Commentarius in Apocalypsin*, Turnhout.

Procopius = *Anecdota*. Wirth, G. ed. 1963, *Procopii Caesariensis opera omnia*, vol. 3, Leipzig.

Translation: Kaldellis, A. 2010, *The Secret History: With Related Texts*, Indianapolis.

= *De bellis*. Wirth, G. ed. 1962–3, *Procopii Caesariensis opera omnia*, vols. 1–2, Leipzig.

Translation: Kaldellis, A. [revised and modernized version of translation by H. B. Dewing] 2014, *The Wars of Justinian*, Indianapolis.

= *De aedificiis*. Wirth, G. ed. 1964, *Procopii Caesariensis opera omnia*, vol. 4, Leipzig.

Translation: Dewing, H. B. 1971, *Procopius*, LCL 343, Cambridge, MA.

Prosper of Aquitaine = Santelia, S. ed. 2009, *Ad coniugem suam: in appendice: Liber epigrammatum*, Naples.

Ptolemy = Grasshoff, G. and Stückelberger, A. eds. 2006, *Klaudios Ptolemaios Handbuch der Geographie*, Basel.
Translation: Germanus, N. 1991, *The Geography*, Mineola.
Seneca = *Epistulae morales ad Lucilium*. Hense, O. ed. 1938, Leipzig.
= *De clementia*. Hosius, E. ed. 1914, Leipzig.
Sidonius Apollinaris = *Carmina*. Lütjohann, C. ed. 1887, MGH AA 8, Berlin: 173–264.
Sophronius = *Homilia in nativitatem Christi*. Usener, H. 1886, "Weinachtpredigt des Sophronius," *Rheinisches Museum für Philologie* 41: 501–16.
Soranus = Ilberg, J. ed. 1927, *Sorani Gynaeciorum libri iv*, CMG 4, Leipzig.
Translation: Owsei, T. 1956, *Soranus' Gynecology*, Baltimore.
Statius = *Silvae*. Marastoni, A. ed. 1970, *P. Papini Stati Silvae*, Leipzig.
Strabo = Meineke, A. ed. 1877, *Strabonis geographica*, Leipzig.
Suetonius = Ihm, M. ed. 1958, *De vita Caesarum: Libri VIII*, Stuttgart.
Sulpicius Severus = Halm, C. ed. 1866, *Dialogorum libri ii*, Vienna.
Translation: Hoare, F. R. 1954, *The Western Fathers*, New York.
Symeon the Logothete = Wahlgren, S. ed. 2006, *Symeonis Magistri et Logothetae Chronicon*, Berlin.
Symmachus = Seeck, O. ed. 1883, *Relationes*, MGH AA 6.1, Berlin: 279–317.
Translation: Barrow, R. H. 1973, *Prefect and Emperor: The Relationes of Symmachus, A.D. 384*, Oxford.
Synesius = Garzya, A. ed. 1979, *Epistulae*, Rome.
Translation: Fitzgerald, A. 1926, *The Letters of Synesius of Cyrene*, Oxford.
Tertullian = Waszink, J. H. ed. 2010, *De anima*, Leiden.
Theophanes the Confessor = de Boor, C. ed. 1883, *Theophanis chronographia*, Leipzig.
Theophylact Simocatta = de Boor, C. ed. 1887, *Theophylacti Simocattae historiae*, Leipzig.
Translation: Whitby, M. and Whitby, M. 1986, *The History of Theophylact Simocatta*, Oxford.
Victor of Tunnuna = Mommsen, T. ed. 1894, *Chronica a. CCCCXLIV–DLXVII*, MGH AA 11, Berlin: 184–206.
Vit. Ioh. Eleem. = Dawes, E. and Baynes, N. 1948, *Three Byzantine Saints: Contemporary Biographies of St. Daniel the Stylite, St. Theodore of Sykeon and St. John the Almsgiver*, London.
Vita Nich. Sion. = Sevcenko I. and Sevcenko, N. P. eds. 1984, *The Life of Saint Nicholas of Sion*, Brookline.
Vita Sym. Styl. Iun. = van den Ven, P. ed. 1962, *La vie ancienne de S. Syméon Stylite le jeune (521–592)*, Brussels.
Vita Theod. Syk. = Dawes, E. and Baynes, N. 1948, *Three Byzantine Saints: Contemporary Biographies of St. Daniel the Stylite, St. Theodore of Sykeon and St. John the Almsgiver*, London.
Pseudo-Zacharias of Mitylene = Brooks, E. W. ed. 1919–24, *Historia ecclesiastica Zachariae rhetori vulgo adscripta*, CSCO 83–84/38–39, Paris.
Translated: Greatrex, G. et al. 2011, *The Chronicle of Pseudo-Zachariah Rhetor: Church and War in Late Antiquity*, TTH 55, Liverpool.
Zosimus = Paschoud, F. ed. 1971–89, *Historia nova*, 3 vols., Paris.

Primary Sources: Inscriptions

AE = *L'Année épigraphique* (1888–)

CIL = *Corpus Inscriptionum Latinarum* (Berlin, 1863–)

I. Didyma = Rehm, A. ed. 1958, *Didyma*, vol. 2: *Die Inschriften*, Berlin.

I. Ephesos = Wankel, H. et al. eds. 1979–84, *Die Inschriften von Ephesos*, Bonn.

I. Erythrai-Klazomenai = Engelmann, H. and Merkelbach, R. eds. 1972–3, *Die Inschriften von Erythrai und Klazomenai*, Bonn.

I. Palaestina Tertia = Meimaris, Y. and Kritikakou, K. eds. 2005–, *Inscriptions from Palaestina Tertia*, Athens.

I. Priene = Hiller von Gaertringen, F. ed. 1906, *Inschriften von Priene*, Berlin.

I. Sestos = Krauss, J. ed. 1980, *Die Inschriften von Sestos und der thrakischen Chersones*, Bonn.

ICUR = Silvangi, A. Ferrua, A. et al. eds. 1922–, *Inscriptiones Christianae Urbis Romae. Nova series*, Rome.

IG = *Inscriptiones Graecae*. 1903–, Berlin.

IGLS = *Inscriptions grecques et latines de la Syrie*. 1929–, Paris.

IGRR = Cagnat, R. ed. 1906–27, *Inscriptiones Graecae ad res Romanas pertinentes*, Paris.

ILS = Dessau, H. ed. 1892–1916, *Inscriptiones Latinae Selectae*, Berlin.

ISMDA = Petsas, P et al. eds. 2000, *Inscriptions du sanctuaire de la mère des dieux autochtone de Leukopétra (Macédoine)*, Athens.

MAMA = *Monumenta Asiae Minoris Antiqua*. 1928–.

Merkelbach and Stauber = Merkelbach, R. and Stauber, J. eds. 1998–2004, *Steinepigramme aus dem griechischen Osten*, 5 vols., Munich.

OGIS = Dittenberger, W. ed. 1903–5, *Orientis Graeci Inscriptiones Selectae*. Leipzig.

SEG = *Supplementum Epigraphicum Graecum*.

Primary Sources: Legal Texts

CJ = *Codex Justinianus* = Krueger, P. ed. 1915, *Corpus iuris civilis*, vol. 2, Berlin.

CT = *Codex Theodosianus* = Mommsen, T. and Krueger, P. eds. 1905, *Theodosiani libri XVI cum constitutionibus sirmondianis* . . . Berlin.

Digest = Mommsen, T. and Krueger, P. eds. 1922, *Corpus iuris civilis*, vol. 1, Berlin.

Edictum De Pretiis Rerum Venalium = Lauffer, S. 1971, *Diokletians Preisedikt*, Berlin.

Novellae Justiniani, Kroll, W. and Schöll, R. eds. 1895, *Corpus iuris civilis*, vol. 3, Berlin.

Primary Sources: Papyri

All papyrological abbreviations follow the standard formats and editions in the "Checklist of Editions of Greek, Latin, Demotic, and Coptic Papyri, Ostraca, and Tablets," available on papyri.info.

Secondary Sources

Abbott, A. 2001, "Earliest Malaria DNA Found in Roman Baby Graveyard," *Nature* 412: 847.

Abbott, D. et al. 2014, "What Caused Terrestrial Dust Loading and Climate Downturns between A.D. 533 and 540?" *Geological Society of America Special Papers* 505: 421–38.
Abtew, W. et al. 2009, "El Niño Southern Oscillation Link to the Blue Nile River Basin Hydrology," *Hydrological Processes* 23: 3653–60.
Abulafia, D. 2011, *The Great Sea: A Human History of the Mediterranean*, Oxford.
Achtman, M. 2016, "How Old Are Bacterial Pathogens?" *Proceedings of the Royal Society B* 283: 20160990.
Al-Azmeh, A. 2014, *The Emergence of Islam in Late Antiquity*, Cambridge.
Alchermes, J. 2005, "Art and Architecture in the Age of Justinian," in M. Maas, ed., *The Cambridge Companion to the Age of Justinian*, Cambridge: 343–75.
Alchon, S. A. 2003, *A Pest in the Land: New World Epidemics in a Global Perspective*, Albuquerque.
Aldrete, G. S. 2006, *Floods of the Tiber in Ancient Rome*, Baltimore.
Alexander, P. 1985, *The Byzantine Apocalyptic Tradition*, Berkeley.
Alfani, G. 2013, "Plague in Seventeenth-Century Europe and the Decline of Italy: An Epidemiological Hypothesis," *European Review of Economic History* 17: 408–30.
Alföldy, G. 1974, "The Crisis of the Third Century as Seen by Contemporaries," *Greek Roman and Byzantine Studies* 15: 89–11.
Allen, P. 1979, "The 'Justinianic Plague,'" *Byzantion* 49: 5–20.
Allen, P. 2011, "Portrayals of Mary in Greek Homiletic Literature (6th–7th Centuries)," in L. Brubaker and M. Cunningham, eds., *The Cult of the Mother of God in Byzantium: Texts and Images*, Farnham: 68–88.
Alpert, P. et al. 2006, "Relations between Climate Variability in the Mediterranean Region and the Tropics: ENSO, South Asian and African Monsoons, Hurricanes and Saharan Dust," *Developments in Earth and Environmental Sciences* 4: 149–77.
Alston, R. 1994, "Roman Military Pay from Caesar to Diocletian," *Journal of Roman Studies* 84: 113–23.
———. 1995, *Soldier and Society in Roman Egypt: A Social History*, London.
———. 2001, "Urban Population in Late Roman Egypt and the End of the Ancient World," in W. Scheidel, ed., *Debating Roman Demography*, Leiden: 161–204.
———. 2002, *The City in Roman and Byzantine Egypt*, London.
———. 2009, "The Revolt of the Boukoloi: Geography, History and Myth," in K. Hopwood, ed., *Organized Crime in Antiquity*, London: 129–53.
Alter, G. 2004, "Height, Frailty, and the Standard of Living: Modeling the Effects of Diet and Disease on Declining Mortality and Increasing Height," *Population Studies* 58: 265–79.
Altizer, S. et al. 2006, "Seasonality and the Dynamics of Infectious Diseases," *Ecology Letters* 9: 467–84.
Ameling, W. 1993, *Karthago: Studien zu Militär, Staat und Gesellschaft*, Munich.
Ando, C. 2000, *Imperial Ideology and Provincial Loyalty in the Roman Empire*, Berkeley.
———. 2012, *Imperial Rome AD 193 to 284: The Critical Century*, Edinburgh.
———. forthcoming, "Forests: The Ancient Mediterranean," in M. Meier, ed., *A Cultural History of the Environment in the Classical Age (3500 BCE – 400 CE)*, London.

Andreau, J. 1986, "Declino e morte dei mestieri bancari nel Mediterraneo occidentale (II–IV D.C.)," in A. Giardina, ed., *Società romana e impero tardoantico*, I, Rome: 601–15, 814–18.

———. 1988, "Huit questions pour une histoire financière de l'antiquité tardive," in *Atti dell'Accademia romanistica costantiniana: XII convegno internazionale sotto l'alto patronato del Presidente della repubblica in onore di Manlio Sargenti*, Naples: 53–63.

———. 1999, *Banking and Business in the Roman World*, Cambridge.

Anisimov, A., Lindler, L., and Pier, G. 2004, "Intraspecific Diversity of *Yersinia pestis*," *Clinical Microbiology Reviews* 17: 434–64.

Arce, J. 1999, "El inventario de Roma: Curiosum y Notitia," in W. V. Harris, ed., *The Transformation of Urbs Roma in Late Antiquity*, Portsmouth: 15–22.

Ari, T. B. et al. 2011, "Plague and Climate: Scales Matter," *PLoS Pathogens* 7: e1002160.

Arjava, A. 1998, "Divorce in Later Roman Law," *Arctos* 22: 5–21.

———. 2005, "The Mystery Cloud of 536 CE in the Mediterranean Sources," *Dumbarton Oaks Papers* 59: 73–94.

Armelagos, G. J. et al. 2005, "Evolutionary, Historical and Political Economic Perspectives on Health and Disease," *Social Science & Medicine* 61: 755–65.

Arnold, J. 2014, *Theoderic and the Roman Imperial Restoration*, New York.

Arnold, J. C. 2013, *The Footprints of Michael the Archangel: The Formation and Diffusion of a Saintly Cult, c. 300-c. 800*, New York.

Atkins, E. M. and Osborne, R. eds. 2006, *Poverty in the Roman World*, Cambridge.

Audoin-Rouzeau, F. 2003, *Les chemins de la peste: le rat, la puce et l'homme*, Rennes.

Avni, G. 2014, *The Byzantine-Islamic Transition in Palestine: An Archaeological Approach*, Oxford.

Avni, G. et al. 2006, "Geomorphic Changes Leading to Natural Desertification Versus Anthropogenic Land Conservation in an Arid Environment, the Negev Highlands, Israel," *Geomorphology* 82: 177–200.

Aiewsakun, P. and Katzourakis, A. 2015, "Endogenous Viruses: Connecting Recent and Ancient Viral Evolution," *Virology* 479–80: 26–37.

Babkin, I. V. and Babkina, I. N. 2012, "A Retrospective Study of the Orthopoxvirus Molecular Evolution," *Infection, Genetics and Evolution* 12: 1597–604.

———. 2015, "The Origin of the Variola Virus," *Viruses* 7: 1100–12.

Bagnall, R. 1982, "Religious Conversion and Onomastic Change," *Bulletin of the American Society of Papyrologists* 19: 105–24.

———. 1985, *Currency and Inflation in Fourth Century Egypt*, Chico.

———. 1987a, "Church, State, and Divorce in Late Roman Egypt," in K.-L. Selig and R. Somerville, eds., *Florilegium Columbianum: Essays in Honor of Paul Oskar Kristeller*, New York: 41–61.

———. 1987b, "Conversion and Onomastics: A reply," *Zeitschrift für Papyrologie und Epigraphik* 69: 243–50.

———. 1988, "Combat ou vide: Christianisme et paganisme dans l'Égypte romaine tardive," *Ktema* 13: 285–96.

———. 1992, "Landholding in Late Roman Egypt: The Distribution of Wealth," *Journal of Roman Studies* 82: 128–49.

———. 1993, *Egypt in Late Antiquity*, Princeton.

———. 1997, "Missing Females in Roman Egypt," *Scripta Classica Israelica* 16: 121–38.

———. 2002, "The Effects of Plague: Model and Evidence," *Journal of Roman Archaeology* 15: 114–20.

Bagnall, R. S. and Frier, B. W. 1994, *The Demography of Roman Egypt*, Cambridge.

Baillie, M. G. L. 1999, *Exodus to Arthur: Catastrophic Encounters with Comets*, London.

———. 2008, "Proposed Re-dating of the European Ice Core Chronology by Seven Years Prior to the 7th century AD," *Geophysical Research Letters* 35: L15813.

Baillie, M. G. L. and McAneney, J. 2015, "Tree Ring Effects and Ice Core Acidities Clarify the Volcanic Record of the First Millennium," *Climate of the Past* 11: 105–14.

Baker, A. et al. 2015, "A Composite Annual-Resolution Stalagmite Record of North Atlantic Climate over the Last Three Millennia," *Scientific Reports* 5: 10307.

Ballais, J.-L. 2004, "Dynamiques environnementales et occupation du sol dans les Aurès pendant la période antique," *Aouras: revue annuelle* 2: 154–68.

Banaji, J. 2007, *Agrarian Change in Late Antiquity: Gold, Labour, and Aristocratic Dominance*, Oxford.

———. 2016, *Exploring the Economy of Late Antiquity: Selected Essays*, Cambridge.

Bang, P. F. 2013, "The Roman Empire II: The Monarchy," in P. F. Bang and W. Scheidel, eds., *Oxford Handbook of the State in the Ancient Near East and Mediterranean*, Oxford: 412–72.

Barker, G., Gilbertson, D. D., Mattingly, D. 2007, *Archaeology and Desertification: The Wadi Faynan Landscape Survey, Southern Jordan*, Oxford.

Barker, G., Hodges, R., and Clark, G. 1995, *A Mediterranean Valley: Landscape Archaeology and Annales History in the Biferno Valley*, London.

Barnes, T. D. 1982, *The New Empire of Diocletian and Constantine*, Cambridge, MA.

———. 2011, *Constantine: Dynasty, Religion and Power in the Later Roman Empire*, Chichester.

Barnish, S. 1985, "The Wealth of Julius Argentarius: Late Antique Banking and the Mediterranean Economy," *Byzantion* 55: 5–38.

Barnish, S., Lee, A. D., and Whitby, M. 2000, "Government and Administration," in A. Cameron, B. Ward-Perkins, and M. Whitby, eds., *The Cambridge Ancient History*, Vol. 14: *Late Antiquity: Empire and Successors*, Cambridge: 164–206.

Barrett, R. and Armelagos, G. 2013, *An Unnatural History of Emerging Infections*, Oxford.

Barrett, T. H. 2007, "Climate Change and Religious Response: The Case of Early Medieval China," *Journal of the Royal Asiatic Society of Great Britain & Ireland* 17: 139–56.

Barry, J. 2004, *The Great Influenza: The Epic Story of the Deadliest Plague in History*, New York.

Bartsch, S. 2006, *The Mirror of the Self: Sexuality, Self-knowledge, and the Gaze in the Early Roman Empire*, Chicago.

Bashear, S. 1993, "Muslim Apocalypses and the Hour: A Case-Study in Traditional Reinterpretation," in J. L. Kraemer, ed., *Israel Oriental Studies XIII*, Leiden: 75–99.

Bashford, A. and Chaplin, J. 2016, *The New Worlds of Thomas Robert Malthus: Rereading the Principle of Population*, Princeton.

Bastien, P. 1988, *Monnaie et donativa au Bas-Empire*, Wetteren.

Bastien, P. and Metzger, C. 1977, *Le trésor de Beaurains, dit d'Arras*, Wetteren.

Bazin-Tacchela, S., Quéruel, D., and Samama, E. eds. 2001, *Air, miasmes et contagion: les épidémies dans l'Antiquité et au Moyen Age*, Prez-sur-Marne.
Beard, M. 2015, *SPQR: A History of Ancient Rome*, New York.
Becker, M. J. 1992, "Analysis of the Human Skeletal Remains from Osteria dell'Osa," in A. M. Bietti Sestieri, ed., *The Iron Age Community of Osteria dell'Osa: A Study of Socio-Political Development in Central Tyrrhenian Italy*, Cambridge: 53–191.
———. 1993, "Human Skeletons from Tarquinia: A Preliminary Analysis of the 1989 Cimitero Site Excavations with Implications for the Evolution of Etruscan Social Classes," *Studi etruschi* 58: 211–48.
———. 1999, "Calculating Stature from in situ Measurements of Skeletons and from Long Bone Lengths: An Historical Perspective Leading to a Test of Formicola's Hypothesis at 5th Century BCE Satricum, Lazio, Italy," *Rivista di Antropologia* 77: 225–47.
Beer J. et al. 2006, "Solar Variability Over the Past Several Millennia," *Space Science Reviews* 125: 67–79.
Beer, J., Mende, W., and Stellmacher, R. 2000, "The Role of the Sun in Climate Forcing," *Quaternary Science Reviews* 19: 403–15.
Behr, C. 1968, *Aelius Aristides and the Sacred Tales*, Amsterdam.
Belcastro, M. G. and Guisberti, G. 1997, "La necropoli romano-imperiale di Casalecchio di Reno (Bologna, II–IV sec. d.C.): analisi morfometrica sincronica e diacronica," *Rivista di antropologia* 75: 129–44.
Belich, J., Darwin, J., and Wickham, C. 2016, "Introduction: The Prospect of Global History," in J. Belich et al., eds., *The Prospect of Global History*, Oxford: 3–22.
Beloch, J. 1886, *Die Bevölkerung der griechisch-römischen Welt*, Leipzig.
Belyi, V. et al. 2010, "Unexpected Inheritance: Multiple Integrations of Ancient Bornavirus and Ebolavirus/Marburgvirus Sequences in Vertebrate Genomes," *PLoS Pathogens* 6: 2010–17.
Benedictow, O. 1992, *Plague in the Medieval Nordic Countries: Epidemiological Studies*, Oslo.
———. 2004, *The Black Death, 1346–1353: The Complete History*, Woodbridge.
Benito, G. et al. 2015a, "Recurring Flood Distribution Patterns Related to Short-Term Holocene Climatic Variability," *Scientific Reports* 5: 16398.
Benito, G. et al. 2015b, "Holocene Flooding and Climate Change in the Mediterranean," *Catena* 130: 13–33.
Benovitz, N. 2014, "The Justinianic Plague: Evidence from the Dated Greek Epitaphs of Byzantine Palestine and Arabia," *Journal of Roman Archaeology* 27: 487–98.
Bi, Y. 2016, "Immunology of *Yersinia pestis* Infection," in R. Yang and A. Anisimov, eds., *Yersinia pestis: Retrospective and Perspective*, Dordrecht: 273–92.
Bianucci, R. et al. 2015, "The Identification of Malaria in Paleopathology—An In-depth Assessment of the Strategies to Detect Malaria in Ancient Remains," *Acta Tropica* 152: 176–80.
Biraben, J.-N. 1975, *Les hommes et la peste en France et dans les pays européens et méditerranéens,* Vol. 1: *La peste dans l'histoire*, Paris.
Biraben, J.-N. and Le Goff, J. 1969, "La peste dans le Haut Moyen Age," *Annales ESC* 24: 1484–510.
Birley, A. R. 1987, *Marcus Aurelius: A Biography*, New Haven.
———. 1988, *Septimius Severus: The African Emperor*, New Haven.

———. 1997, *Hadrian: The Restless Emperor*, London.

———. 2000, "Hadrian to the Antonines," in P. Garnsey, D. Rathbone, and A. K. Bowman, eds., *The Cambridge Ancient History*, Vol. 11: *The High Empire, A.D. 70–192*, Cambridge: 132–94.

Bivar, A. D. H. 1970, "Hāritī and the Chronology of the Kuṣāṇas," *Bulletin of the School of Oriental and African Studies* 33: 10–21.

Bjornlie, S. 2013, *Politics and Tradition between Rome, Ravenna and Constantinople: A Study of Cassiodorus and the Variae 527–554*, Cambridge.

Bkhairi, A. and Karray, M. R. 2008, "Les terrasses historiques du basin de Kasserine (Tunisie central)," *Géomorphologie: relief, processus, environment* 14: 201–13.

Black, E. 2012, "The Influence of the North Atlantic Oscillation and European Circulation Regimes on the Daily to Interannual Variability of Winter Precipitation in Israel," *International Journal of Climatology* 32: 1654–64.

Bland, R. 1997, "The Changing Pattern of Hoards of Precious-Metal Coins in the Late Empire," *Antiquité tardive* 5: 29–55.

Bleckmann, B. 1992, *Die Reichskrise des III. Jahrhunderts in der spätantiken und byzantinischen Geschichtsschreibung: Untersuchungen zu den nachdionischen Quellen der Chronik des Johannes Zonaras*, Munich.

———. 2006, "Zu den Motiven der Christenverfolgung des Decius," in K.-P. Johne, T. Gerhardt, and U. Hartmann, eds., *Deleto paene imperio Romano: Transformationsprozesse des Römischen Reiches im 3. Jahrhundert und ihre Rezeption in der Neuzeit*, Stuttgart: 57–71.

Blois, L. de 1976, *The Policy of the Emperor Gallienus*, Leiden.

Blouin, K. 2014, *Triangular Landscapes: Environment, Society, and the State in the Nile Delta under Roman Rule*, Oxford.

Boak, A. E. R. 1955, *Manpower Shortage and the Fall of the Roman Empire in the West*, Ann Arbor.

Boatwright, M. 2000, *Hadrian and the Cities of the Roman Empire*, Princeton.

Boch, R. and Spötl, C. 2011, "Reconstructing Palaeoprecipitation from an Active Cave Flowstone," *Journal of Quaternary Science* 26: 675–87.

Bodel, J. 2008, "From Columbaria to Catacombs: Collective Burials in Pagan and Christian Rome," in L. Brink and D. Green, eds., *Commemorating the Dead: Texts and Artifacts in Context: Studies of Roman, Jewish and Christian Burials*, Berlin: 177–242.

Bogaert, R. 1973, "Changeurs et banquiers chez les Pères de l'Église," *Ancient Society* 4: 239–70.

———. 1994, *Trapezitica Aegyptiaca: recueil de recherches sur la banque en Egypte gréco-romaine*, Florence.

Bond, G. et al. 2001, "Persistent Solar Influence on North Atlantic Climate During the Holocene," *Science* 294: 2130–36.

Bonfiglioli, B., Brasili, P., and Belcastro, M. G. 2003, "Dento-Alveolar Lesions and Nutritional Habits of a Roman Imperial Age Population (1st–4th c. AD): Quadrella (Molise, Italy)," *Homo* 54: 36–56.

Bonneau, D. 1964, *La crue du Nil, divinité égyptienne, à travers mille ans d'histoire (332 av.–641 ap. J.-C.) d'après les auteurs grecs et latins, et les documents des époques ptolémaïque, romaine et byzantine*, Paris.

———. 1971, *Le fisc et le Nil: incidences des irrégularités de la crue du Nil sur la fiscalité foncière dans l'Égypte grecque et romaine*, Paris.

Bonsall, L. A. 2013, *Variations in the Health Status of Urban Populations in Roman Britain: A Comparison of Skeletal Samples from Major and Minor Towns*, diss., University of Edinburgh.

Bookman, R. et al. 2004, "Late Holocene Lake Levels of the Dead Sea," *Geological Society of America Bulletin* 116: 555–71.

Booth, P. 2013, *Crisis of Empire: Doctrine and Dissent at the End of Late Antiquity*, Berkeley.

Borgognini Tarli, S. M. and La Gioia, C. 1977, "Studio antropologico di un gruppo di scheletri di età romana (I a.C – I d.C.) rinvenuti nella necropoli di Collelongo (L'Aquila, Abruzzo)," *Atti della Società toscana di scienze naturali, Memorie. Serie B* 84: 193–226.

Borsch, S. 2005, *The Black Death in Egypt and England: A Comparative Study*, Austin.

———. 2014, "Plague Depopulation and Irrigation Decay in Medieval Egypt," *The Medieval Globe* 1: 125–56.

Bos, K. I. et al. 2011, "A Draft Genome of *Yersinia pestis* from Victims of the Black Death," *Nature* 478: 506–10.

———. 2014, "Pre-Columbian Mycobacterial Genomes Reveal Seals as a Source of New World Human Tuberculosis," *Nature* 514: 494–97.

———. 2016, "Eighteenth Century *Yersinia pestis* Genomes Reveal the Long-Term Persistence of an Historical Plague Focus," *eLIFE* 5: e12994.

Boswell, J. 1988, *The Kindness of Strangers: The Abandonment of Children in Western Europe from Late Antiquity to the Renaissance*, New York.

Boudon, V. 2001, "Galien face à la 'peste antonine' ou comment penser l'invisible," in S. Bazin-Tacchela, D. Quéruel, and E. Samama, eds., *Air, miasmes et contagion: les épidémies dans l'Antiquité et au Moyen Age*, Prez-sur-Marne: 29–54.

Bowden, W., Hodges, R., and Cerova, Y. 2011, *Butrint 3: Excavations at the Triconch Palace*, Oxford.

Bowden, W., Lavan, L., and Machado, C. eds. 2003, *Recent Research on the Late Antique Countryside*, Leiden.

Bowersock, G. W. 1969, *Greek Sophists in the Roman Empire*, Oxford.

———. 2001, "Lucius Verus in the Near East," in C. Evers and A. Tsingarida, eds., *Rome et ses provinces: Hommages à Jean Charles Balty*, Brussels: 73–77.

———. 2009, *From Gibbon to Auden: Essays on the Classical Tradition*, Oxford.

———. 2012, *Empires in Collision in Late Antiquity*, Waltham.

———. 2013, *The Throne of Adulis: Red Sea Wars on the Eve of Islam*, Oxford.

———. 2017, *The Crucible of Islam*, Cambridge.

Bowman, A. K. 1985, "Landholding in the Hermopolite Nome in the Fourth Century A.D.," *Journal of Roman Studies* 75: 137–63.

———. 2005, "Diocletian and the First Tetrarchy, A.D. 294–305," in A. K. Bowman, P. Garnsey, and A. Cameron, eds., *The Cambridge Ancient History*, Vol. 12: *The Crisis of Empire, A.D. 193–337*, Cambridge: 67–89.

———. 2011, "Ptolemaic and Roman Egypt: Population and Settlement," in A. Bowman and A. Wilson, eds., *Settlement, Urbanization, and Population*, Oxford: 317–58.

Bowman, A. and Wilson, A. eds. 2009, *Quantifying the Roman Economy: Methods and Problems*, Oxford.

Brandes, W. 1999, "Byzantine Cities in the Seventh and Eighth Centuries: Different Sources, Different Histories," in G. P. Brogiolo and B. Ward-Perkins, eds., *The*

Idea and Ideal of the Town between Late Antiquity and the Early Middle Ages, Leiden: 25–57.

Bransbourg, G. 2015, "The Later Roman Empire," in A. Monson and W. Scheidel, eds., *Fiscal Regimes and the Political Economy of Premodern States*, Cambridge: 258–81.

Brasili Gualandi, P. 1989, "I reperti scheletrici della necropolis di Monte Bibele (IV–II sec. a.C.): Nota Preliminare," in V. Morrone, ed., *Guida al Museo 'L. Fantini' di Monterenzio e all'area archeologica di Monte Bibele*, Bologna: 52–5.

Braudel, F. 1972–73, *The Mediterranean and the Mediterranean World in the Age of Philip II* [2nd rev. ed., orig. 1949], New York.

Brenot, C. and Loriot, X. eds. 1992, *Trouvailles de monnaies d'or dans l'Occident romain: actes de la Table Ronde tenue à Paris les 4 et 5 décembre 1987*, Paris.

Brent, A. 2010, *Cyprian and Roman Carthage*, Cambridge.

Broadberry, S., Guan, H., and Li, D. D. 2014, "China, Europe, and the Great Divergence: A Study in Historical National Accounting, 980–1850," Economic History Department Paper, London School of Economics.

Brock, S. 1979–80, "Syriac Historical Writing: A Survey of the Main Sources," *Journal of the Iraqi Academy (Syriac Corporation)* 5: 296–326.

Brogiolo, G. and Chavarría Arnau, A. 2005, *Aristocrazie e campagne nell'Occidente da Costantino a Carlo Magno*, Florence.

Brogiolo, G., Gauthier, N., and Christie, N. eds. 2000, *Towns and Their Territories between Late Antiquity and the Early Middle Ages*, Leiden.

Broodbank, C. 2013, *The Making of the Middle Sea: A History of the Mediterranean from the Beginning to the Emergence of the Classical World*, London.

Brooke, J. 2014, *Climate Change and the Course of Global History: A Rough Journey*, New York.

———. 2016, "Malthus and the North Atlantic Oscillation: A Reply to Kyle Harper," *Journal of Interdisciplinary History* 46: 563–78.

Brooks, F. J. 1993, "Revising the Conquest of Mexico: Smallpox, Sources, and Populations," *Journal of Interdisciplinary History* 24: 1–29.

Brosch, R. et al. 2002, "A New Evolutionary Scenario for the Mycobacterium Tuberculosis Complex," *Proceedings of the National Academy of Sciences* 99: 3684–89.

Brown, P. R. L. 1971, *The World of Late Antiquity: From Marcus Aurelius to Muhammad*, London.

———. 2002, *Poverty and Leadership in the Later Roman Empire*, Hanover.

———. 2012, *Through the Eye of a Needle: Wealth, the Fall of Rome, and the Making of Christianity in the West, 350–550 AD*, Princeton.

———. 2016, *Treasure in Heaven: The Holy Poor in Early Christianity*, Charlottesville.

Brun, A. 1992, "Pollens dans les séries marines du Golfe de Gabès et du plateau des Kerkennah (Tunisie): signaux climatiques et anthropiques," *Quaternaire* 3: 31–39.

Brunetti, M. et al. 2002, "Droughts and Extreme Events in Regional Daily Italian Precipitation Series," *International Journal of Climatology* 22: 543–58.

Brunt, P. A. 1987, *Italian Manpower, 225 B.C.–A.D. 14*, Oxford.

Bruun, C. 2003, "The Antonine Plague in Rome and Ostia," *Journal of Roman Archaeology* 16: 426–34.

———. 2007, "The Antonine Plague and the 'Third-Century Crisis,'" in O. Hekster, G. de Kleijn, and D. Slootjes, eds., *Crises and the Roman Empire: Proceedings of the Seventh Workshop of the International Network Impact of Empire, Nijmegen, June 20–24, 2006*, Leiden: 201–18.

———. 2012, "La mancanza di prove di un effetto catastrofico della "peste antonina" (dal 166 d.C. in poi)," in E. Lo Cascio, ed., *L'impatto della "peste antonina"*, Bari: 123–65.

———. 2016, "Tracing Familial Mobility: Female and Child Migrants in the Roman West," in L. de Ligt and L. E. Tacoma, eds., *Migration and Mobility in the Early Roman Empire*, Leiden: 176–204.

Buccellato, A. et al. 2003, "Il comprensorio della necropoli di Via Basiliano (Roma): un' indagine multidisciplinare," *Mélanges de l'École Française de Rome Antiquité* 115: 311–76.

———. 2008, "La nécropole de Collatina," *Les dossiers d'archéologie* 330: 22–31.

Buell, D.K. 2005, *Why This New Race: Ethnic Reasoning in Early Christianity*, New York.

Büntgen, U. et al. 2011, "2500 Years of European Climate Variability and Human Susceptibility," *Science* 311: 578–82.

Büntgen, U. et al. 2016, "Cooling and Societal Change during the Late Antique Little Ice Age from 536 to around 660 AD," *Nature Geoscience* 9: 231–36.

Buraselis, K. 1989, *Theia dōrea: das göttlich-kaiserliche Geschenk: Studien zur Politik der Severer und zur Constitutio Antoniniana*, Athens.

Burroughs, W. J. 2005, *Climate Change in Prehistory: The End of the Reign of Chaos*, Cambridge.

Butcher, K. 2004, *Coinage in Roman Syria: Northern Syria, 64 BC–AD 253*, London.

Butcher, K. and Ponting, M. 2012, "The Beginning of the End? The Denarius in the Second Century," *Numismatic Chronicle* 172: 63–83.

Butler, T. et al. 1982, "Experimental *Yersinia pestis* Infection in Rodents after Intragastric Inoculation and Ingestion of Bacteria," *Infection and Immunity* 36: 1160–67.

Buttrey, T. V. 2007, "Domitian, the Rhinoceros, and the Date of Martial's 'Liber De Spectaculis,'" *Journal of Roman Studies* 97: 101–12.

Butzer, K. W. 2012, "Collapse, Environment, and Society," *Proceedings of the National Academy of Sciences* 109: 3632–39.

Caldwell, J. C. 2004, "Fertility Control in the Classical World: Was There an Ancient Fertility Transition?" *Journal of Population Research* 21: 1–17.

Callu, J.-P. 1969, *La politique monétaire des empereurs romains, de 238 à 311*, Paris.

Callu, J.-P. and Loriot, X. eds. 1990, *La dispersion des aurei en Gaule romaine sous l'Empire*, Paris.

Cameron, A. 1970, *Claudian: Poetry and Propaganda at the Court of Honorius*, Oxford.

Cameron, A. ed. 1976, *In laudem Iustini Augusti minoris*, London.

———. 1978, "The Theotokos in Sixth-Century Constantinople: A City Finds Its Symbol," *Journal of Theological Studies* 29: 79–108.

———. 1985, *Procopius and the Sixth Century*, Berkeley.

———. 2000a, "Justin I and Justinian," in A. Cameron, B. Ward-Perkins, and M. Whitby, eds., *The Cambridge Ancient History*, Vol. 14: *Late Antiquity: Empire and Successors*, Cambridge: 63–85.

———. 2000b, "Vandal and Byzantine Africa," in A. Cameron, B. Ward-Perkins, and M. Whitby, eds., *The Cambridge Ancient History*, Vol. 14: *Late Antiquity: Empire and Successors*, Cambridge: 552–69.

———. 2005, "The Reign of Constantine, A.D. 306–337," in A. K. Bowman, P. Garnsey, and A. Cameron, eds., *The Cambridge Ancient History,* Vol. 12: *The Crisis of Empire, A.D. 193–337*, Cambridge: 90–109.

Cameron, C. M., Kelton, P., and Swedlund, A.C. eds. 2015, *Beyond Germs: Native Depopulation in North America*, Tucson.

Camilli, L. and Sorda, S. eds. 1993, *L'"inflazione" nel quarto secolo d.C.: atti dell'incontro di studio, Roma 1988*, Rome.

Campbell, B. 2005a, "The Severan Dynasty," in A. K. Bowman, P. Garnsey, and A. Cameron, eds., *The Cambridge Ancient History,* Vol. 12: *The Crisis of Empire, A.D. 193–337*, Cambridge: 1–27.

———. 2005b, "The Army," in A. K. Bowman, P. Garnsey, and A. Cameron, eds., *The Cambridge Ancient History,* Vol. 12: *The Crisis of Empire, A.D. 193–337*, Cambridge: 110–30.

———. 2010, "Nature as Historical Protagonist: Environment and Society in Pre-Industrial England," *Economic History Review* 63: 281–314.

———. 2016, *The Great Transition: Climate, Disease and Society in the Late-Medieval World*, Cambridge.

Camuffo, D. and Enzi, S. 1996, "The Analysis of Two Bi-Millennial Series: Tiber and Po River Floods," in P. D. Jones, R. S. Bradley, and J. Jouzel, eds., *Climactic Variations and Forcing Mechanisms of the Last 2000 Years*, I: 433–50.

Cantor, N. 2001, *In the Wake of the Plague: The Black Death and the World It Made*, New York.

Capasso, L. 2001, *I fuggiaschi di Ercolano: paleobiologia delle vittime dell'eruzione vesuviana del 79 d.C.*, Rome.

Capitanio, M. 1974, "La necropolis di Potenzi (Macerata), di epoca romana: notizie antropologiche," *Archivio per l'antropologia e la etnologia* 104: 179–209.

———. 1981, "Anthropologische Bemerkungen über die spätrömischer Bestatteten von Pfatten-Laimburg (Vadena) bei Bozen," *Der Schlern* 55: 189–96.

———. 1985, "Gli scheletri umani di epoca barbarica rinvenuti al Dossello di Offanengo (Cremona)," *Insula fulcheria: rassegna di studi e documentazioni di Crema e del cremasco a cura del Museo civico di Crema*, Crema: 59–79.

———. 1986–87, "Esame antropologica degli inumati di Valeggio sul Mincio (Verona) d'epoca romana (I sec a.C. – I sec. d.C.)," *Atti e memorie dell'Accademia d'agricoltura scienze e lettere di Verona*: 159–98.

Cappers, R. 2006, *Roman Foodprints at Berenike: Archaeobotanical Evidence of Subsistence and Trade in the Eastern Desert of Egypt*, Los Angeles.

Carandini, A. 2011, *Rome: Day One*, Princeton.

Carmichael, A. G. 2006, "Infectious Disease and Human Agency: An Historical Overview," *Scripta Varia* 106: 3–46.

———. 2014, "Plague Persistence in Western Europe: A Hypothesis," *The Medieval Globe* 1: 157–91.

Carmichael, A. and Silverstein, A. 1987, "Smallpox in Europe before the Seventeenth Century: Virulent Killer or Benign Disease?" *Journal of the History of Medicine and Allied Sciences* 42: 147–68.

Carra, R. M. B. 1995, *Agrigento: la necropoli paleocristiana sub divo*, Rome.

Carrié, J.-M. 1994, "Dioclétien et la fiscalité," *Antiquité tardive* 2: 34–64.

———. 2003, "Solidus et crédit: qu'est-ce que l'or a pu changer?" in E. Lo Cascio, ed., *Credito e moneta nel mondo romano*, Bari: 265–79.

———. 2005, "Developments in Provincial and Local Administration," in A. K. Bowman, P. Garnsey, and A. Cameron, eds., *The Cambridge Ancient History*, Vol. 12: *The Crisis of Empire, A.D. 193–337*, Cambridge: 269–312.

———. 2007, "Les crises monétaires de l'Empire romain tardif," in B. Théret, ed., *La monnaie dévoilée par ses crises*, Vol. 1. Paris: 131–63.

Carrié, J.-M. and Rousselle, A. 1999, *L'Empire romain en mutation: des Sévères à Constantin, 192–337*, Paris.

Carter, J. C. ed. 1998, *The Chora of Metaponto: The Necropoleis*, Austin.

Casana, J. 2008, "Mediterranean Valleys Revisited: Linking Soil Erosion, Land Use and Climate Variability in the Northern Levant," *Geomorphology* 101: 429–42.

———. 2014, "The Late Roman Landscape of the Northern Levant: A View from Tell Qarqur and the Lower Orontes River Valley," *Oxford Journal of Archaeology* 33: 193–219.

Casanova, G. 1984, "Epidemie e fame nella documentazione greca d'Egitto," *Aegyptus* 64: 163–201.

Casanova, P. 1911, *Mohammed et la fin du monde: étude critique sur l'Islam primitive*, Paris.

Casson, L. 1989, *The Periplus Maris Erythraei: Text with Introduction, Translation, and Commentary*, Princeton.

———. 1990, "New Light on Maritime Loans: P. Vindob. G 40822," *Zeitschrift für Papyrologie und Epigraphik* 84: 195–206.

Catalano, P. et al. 2001a, "Le necropolis di Roma—Die Nekropolen Roms: il contributo dell'antropologia—der Beitrag der Anthropologie," *Mitteilungen des deutschen archeologischen Instituts* 108: 353–81.

———. 2001b, "Le necropoli romane di età imperiale: un contributo all'interpretazione del popolamento e della qualità della vita nell'antica Roma," L. Quilici and S. Quilici Gigli, eds., *Urbanizzazione delle campagne nell'Italia antica*, Rome: 127–37.

Cavanaugh, D. C. 1971, "Specific Effect of Temperature upon Transmission of the Plague Bacillus by the Oriental Rat Flea, *Xenopsylla cheopis*," *American Journal of Tropical Medicine and Hygiene* 20: 264–73.

Cavanaugh, D. C. and Marshall, J. D. 1972, "The Influence of Climate on the Seasonal Prevalence of Plague in the Republic of Vietnam," *Journal of Wildlife Diseases* 8: 85–94.

Cerati, A. 1975, *Caractère annonaire et assiette de l'impôt foncier au Bas-Empire*, Paris.

Chabal, L. 2001, "Les potiers, le bois et la forêt à l'époque romaine, à Sallèles d'Aude (Ier–IIIe s. ap. J.-C.)," in F. Laubenheimer, ed., *20 ans de recherches à Sallèles d'Aude. Colloque des 27–28 septembre 1996 (Sallèles d'Aude)*, Besançon: 93–110.

Chain, P. S. G. et al. 2004, "Insights into the Evolution of *Yersinia pestis* through Whole-Genome Comparison with *Yersinia pseudotuberculosis*," *Proceedings of the National Academy of Sciences* 101: 13826–31.

Chavarría, A. and Lewit, T. 2004, "Archaeological Research on the Late Antique Countryside: A Bibliographical Essay," in W. Bowden, L. Lavan, and C. Machado, eds., *Recent Research on the Late Antique Countryside*, Leiden: 3–51.

Chen, F.-H. et al. 2010, "Moisture Changes over the Last Millennium in Arid Central Asia: A Review, Synthesis and Comparison with Monsoon Region," *Quaternary Science Reviews* 29: 1055–68.

Chen, S. and Kung, J. K. 2016, "Of Maize and Men: The Effect of a New World Crop on Population and Economic Growth in China," *Journal of Economic Growth* 21: 71–99.

Cherian, P. J. 2011, *Pattanam Excavations: Fifth Season Field Report*, Trivandrum.

Choat, M. 2007, "The Archive of Apa Johannes: Notes on a Proposed New Edition," in J. Frösén, T. Purola, and E. Salmenkivi, eds., *Proceedings of the 24th International Congress of Papyrology, Helsinki, 1–7 August, 2004*, Helsinki: 175–83.

Christie, N. 2006, *From Constantine to Charlemagne: An Archaeology of Italy, AD 300–800*, Aldershot.

———. 2011, *The Fall of the Western Roman Empire: An Archaeological and Historical Perspective*, London.

Christol, M. 1982, *Les reformes de Gallien et la carriere senatoriale*, Rome.

———. 1986. *Essai sur l'évolution des carrières sénatoriales dans la seconde moitié du IIIe siècle ap. J.C.*, Paris.

Clark, E. 1984, *The Life of Melania the Younger: Introduction, Translation, Commentary*, New York.

Cline, E. 2014, *1177 B.C.: The Year Civilization Collapsed*, Princeton.

Colt, H. D. 1962, *Excavations at Nessana*, London.

Comas, I. et al. 2013, "Out-of-Africa Migration and Neolithic Coexpansion of Mycobacterium Tuberculosis with Modern Humans," *Nature Genetics* 45: 1176–82.

Conheeney, J. 1997, "The Human Bone," in T. W. Potter and A. C. King, eds., *Excavations at the Mola di Monte Gelato: A Roman and Medieval Settlement in South Etruria*, Rome: 119–70.

Conrad, L. I. 1981, *The Plague in the Early Medieval Near East*, diss., University of Princeton.

———. 1992, "Epidemic Disease in Formal and Popular Thought in Early Islamic Society," in T. Ranger and P. Slack, eds., *Epidemics and Ideas: Essays on the Historical Perception of Pestilence*, Cambridge: 77–99.

———. 1994, "Epidemic Disease in Central Syria in the Late Sixth Century: Some New Insights from the Verse of Hassān ibn Thābit," *Byzantine and Modern Greek Studies* 18: 12–59.

———. 2000, "The Arabs," in A. Cameron, B. Ward-Perkins and M. Whitby, eds., *The Cambridge Ancient History*, Vol. 14: *Late Antiquity: Empire and Successors*, Cambridge: 678–700.

Cook, D. 2002, *Studies in Muslim Apocalyptic*, Princeton.

Cook, E. 2013, "Megadroughts, ENSO, and the Invasion of Late-Roman Europe by the Huns and Avars," in W. V. Harris, ed., *The Ancient Mediterranean Environment between Science and History*, Leiden: 89–102.

Cooper, D. and Kiple, K. F. 1993, "Yellow Fever," in K. F. Kiple, ed., *The Cambridge World History of Human Disease*, New York: 1100–7.

Coppa, A. et al. 1987, "Gli inumati dell'Età del Ferro di Campovalano (Abruzzo, area Medio-Adriatica)," *Rivista di antropologia* 65: 105–38.

Corbier, M. 2001, "Child Exposure and Abandonment," in S. Dixon, ed., *Childhood, Class and Kin in the Roman World*, London: 52–73.

———. 2005a, "Coinage and Taxation: The State's Point of View, A.D. 193–337," in A. K. Bowman, P. Garnsey, and A. Cameron, eds., *The Cambridge Ancient History*, Vol. 12: *The Crisis of Empire, A.D. 193–337*, Cambridge: 327–92.

———. 2005b, "Coinage, Society and Economy," in A. K. Bowman, P. Garnsey, and A. Cameron, eds., *The Cambridge Ancient History*, Vol. 12: *The Crisis of Empire, A.D. 193–337*, Cambridge: 393–439.

Corcoran, S. 2000, *The Empire of the Tetrarchs: Imperial Pronouncements and Government, AD 284–324*, rev. ed., Oxford.

Cornelis, G. R. and Wolf-Watz, H. 1997, "The Yersinia Yop Virulon: A Bacterial System for Subverting Eukaryotic Cells," *Molecular Microbiology* 23: 861–67.

Corrain, C. 1971, "Dati osteometrici su resti umani antichi del territoria Atestino (Padova)," in *Oblatio: Raccolta di studi di antichità ed arte in onore di Aristide Calderini*, Como: 247–86.

Corrain, C. and Capitanio, M. 1967, "I resti scheletrici umani del 'Dos dell'Arca,' Valcamonica," *Bollettino del Centro camuno di studi preistorici* 3: 149–73.

———. 1969, "I resti scheletrici umani della necropoli di Sirolo (Numana) nelle Marche," in *Scritti sul quaternario in onore de Angelo Pasa*, Verona: 207–27.

———. 1972, "I resti scheltrici della necropoli di Fermo, nelle Marche," *Homo 23*: 19–36.

———. 1988, "I resti scheletrici della necropolis tardo-romana e alto-medievale di Mont Blanc (Aosta)," *Quaderni di scienze antropologiche* 14: 79–235.

Corrain, C., Capitanio, M., and Erspamer, G. 1972, "I resti scheletrici della necropoli di Salapia (Cerignola), secoli IX–III a.C.," *Atti e memorie dell'Academia Patavina di Scienze, Lettere ed Arti* 84: 75–103.

———. 1977, "I resti scheltrici della necropoli picena di Camerano, nelle Marche (secoli VI–III a.C.)," *Archivio per l'antropologia e la etnologia* 107: 81–153.

———. 1982, "Alcune necropoli romane delle Marche," *Archivio per l'antropologia e la etnologia* 112: 151–225.

———. 1986, "I resti scheletrici umani della necropolis tardo-romana e alto-medievale di Mont-Blanc (Aosta): Nota riassuntiva," *Archivio per l'antropologia e la etnologia* 116: 215–19.

Corrain, C. and Nalin, G. 1965, "Resti scheletrici umani della necropoli protostorica di Monte Saraceno presso Mattinata (Gargano)," *Atti della X. Riunione scientifica In memoria di Francesco Zorzi: Verona, 21–23 novembre 1965*, Verona: 309–38.

Cosme, P. 2007, "À propos de l'Édit de Gallien," in O. Hekster, G. de Kleijn, and D. Slootjes, eds., *Crises and the Roman Empire*, Leiden: 97–109.

Cotton, H. 1993, "The Guardianship of Jesus Son of Babatha: Roman and Local Law in the Province of Arabia," *Journal of Roman Studies* 83: 94–113.

Craig, O. E. et al. 2009, "Stable Isotopic Evidence for Diet at the Imperial Roman Coastal Site of Velia (1st and 2nd Centuries AD) in Southern Italy," *American Journal of Physical Anthropology* 139: 572–83.

Crawford, D. H. 2007, *Deadly Companions: How Microbes Shaped Our History*, Oxford.

Cremaschi, M. et al. 2006, "Cupressus dupreziana: a Dendroclimatic Record for the Middle-Late Holocene in the Central Sahara," *The Holocene* 16: 292–303.

Cremaschi, M., Marchetti, M., and Ravazzi, C. 1994, "Geomorphological Evidence for Land Surfaces Cleared from Forest in the Central Po Plain (Northern Italy) during the Roman Period," in B. Frenzel, ed., *Evaluation of Land Surfaces Cleared from Forests in the Mediterranean Region during the Time of the Roman Empire*, Stuttgart: 119–32.

Cremaschi, M. and Zerboni, A. 2010, "Human Communities in a Drying Landscape: Holocene Climate Change and Cultural Response in the Central Sahara," in I. P. Martini and W. Chesworth, eds., *Landscapes and Societies: Selected Cases*, Dordrecht: 67–89.

Cremonini, S., Labate, D., and Curina, R. 2013, "The Late-antiquity Environmental Crisis in Emilia Region (Po River Plain, Northern Italy): Geoarchaeological Evidence and Paleoclimatic Considerations," *Quaternary International* 316: 162–78.

Croke, B. 1981, "Two Early Byzantine Earthquakes and Their Liturgical Commemoration," *Byzantion* 51: 122–47.

———. 2005, "Justinian's Constantinople," in M. Maas, ed., *The Cambridge Companion to the Age of Justinian*, Cambridge: 60–86.

Cronon, W. 1983, *Changes in the Land: Indians, Colonists, and the Ecology of New England*, New York.

Crosby, A. W. 1986, *Ecological Imperialism: The Biological Expansion of Europe, 900–1900*, Cambridge.

Crow, J. 2012, "Water and Late Antique Constantinople: 'It Would Be Abominable for the Inhabitants of This Beautiful City to Be Compelled to Purchase Water,'" in L. Grig and G. Kelly, eds., *Two Romes: Rome and Constantinople in Late Antiquity*, Oxford: 116–35.

Cucina, A. et al. 2006, "The Necropolis of Vallerano (Rome, 2nd–3rd Century AD): An Anthropological Perspective on the Ancient Romans in the *Suburbium*," *International Journal of Osteoarchaeology* 16: 104–117.

Cui, Y. et al. 2013, "Historical Variations in Mutation Rate in an Epidemic Pathogen, *Yersinia pestis*," *Proceedings of the National Academy of Sciences* 110: 577–82.

Cummings, C. 2009, "Meat Consumption in Roman Britain: The Evidence from Stable Isotopes," in M. Driessen, et al., eds., *TRAC 2008: Proceedings of the Eighteenth Annual Theoretical Roman Archaeology Conference, Amsterdam 2008*, Oxford: 73–83.

Curran, J. 1998, "From Jovian to Theodosius," in A. Cameron and P. Garnsey, eds., *The Cambridge Ancient History*, Vol. 13: *The Late Empire, A.D. 337–425*, Cambridge: 78–110.

Currás, A. et al. 2012, "Climate Change and Human Impact in Central Spain During Roman Times: High-Resolution Multi-Proxy Analysis of a Tufa Lake Record (Somolinos, 1280 m asl)," *Catena* 89: 31–53.

Cutler, A. 1985, *The Craft of Ivory: Sources, Techniques, and Uses in the Mediterranean World, A.D. 200–1400*, Washington, D.C.

Dagens, C. 1970, "La fin des temps et l'Église selon Saint Grégoire le Grand," *Recherches de science religieuse* 58: 273–88.

Dagron, G. 1984, *Naissance d'une capitale: Constantinople et ses institutions de 330 à 451*, Paris.

Dagron, G. and Déroche, V. 1991, "Juifs et chrétiens dans l'Orient du VIIe siècle," *Travaux et Mémoires* 11: 17–46.

Dal Santo, M. 2013, "Gregory the Great, the Empire and the Emperor," in B. Neil and M. Dal Santo, eds., *A Companion to Gregory the Great*, Cambridge: 57–82.

Darley, R. 2013, *Indo-Byzantine Exchange, 4th to 7th Centuries: A Global History*, diss., University of Birmingham.

———. 2015, "Self, Other and the Use and Appropriation of Late Roman Coins in South India and Sri Lanka (4th–7th centuries A.D.)," in H. Ray, ed., *Negotiating Cultural Identity: Landscapes in Early Medieval South Asian History*, London: 60–84.
Dark, P. 2000, *The Environment of Britain in the First Millennium AD*, London.
Daube, D. 1966–67, "The Marriage of Justinian and Theodora: Legal and Theological Reflections," *Catholic University Law Review* 16: 380–99.
Davies, T. J. et al. 2011, "The Influence of Past and Present Climate on the Biogeography of Modern Mammal Diversity," *Philosophical Transactions of the Royal Society, B* 366: 2526–35.
de Blois, L. 1976, *The Policy of the Emperor Gallienus*, Leiden.
De Callataÿ, F. 2005, "The Graeco-Roman Economy in the Super Long-Run: Lead, Copper, and Shipwrecks," *Journal of Roman Archaeology* 18: 361–72.
De Greef, G. 2002, "Roman Coin Hoards and Germanic Invasions AD 253–269: A Study of the Western Hoards from the Reigns of Valerian, Gallienus and Postumus," *Revue belge de numismatique et de sigillographie* 148: 41–100.
de la Vaissière, E. 2003, "Is There a 'Nationality of the Hephtalites'?" *Bulletin of the Asia Institute* 17: 119–32.
———. 2005a, *Sodgian Traders: A History*, Leiden.
———. 2005b, "Huns et Xiongnu," *Central Asiatic Journal* 49: 3–26.
———. 2015, "The Steppe World and the Rise of the Huns," in M. Maas, ed., *The Cambridge Companion to the Age of Attila*, Cambridge: 175–92.
De Ligt, L. 2012, *Peasants, Citizens and Soldiers: Studies in the Demographic History of Roman Italy 225 BC–AD 100*, Cambridge.
De Ligt, L. and Tacoma, R. eds. 2016, *Migration and Mobility in the Early Roman Empire*, Leiden.
De Putter, T. et al. 1998, "Decadal Periodicities of Nile River Historical Discharge (A.D. 622–1470) and Climatic Implications," *Geophysical Research Letters* 25: 3193–96.
De Romanis F. and Tchernia, A. eds. 1997, *Crossings: Early Mediterranean Contacts with India*, New Delhi.
———. 2015, "Comparative Perspectives on the Pepper Trade," in F. De Romanis and M. Maiuro, eds., *Across the Ocean: Nine Essays on Indo-Roman Trade*, Leiden: 127–50.
De Ste. Croix, G. E. M. 1981, *The Class Struggle in the Ancient Greek World*, Ithaca.
Dean, J. et al. 2013, "Palaeo-seasonality of the Last Two Millennia Reconstructed from the Oxygen Isotope Composition of Carbonates and Diatom Silica from Nar Gölü, Central Turkey," *Quaternary Science Reviews* 66: 35–44.
Decker, M. 2009, *Tilling the Hateful Earth: Agricultural Production and Trade in the Late Antique East*, Oxford.
———. 2016, *The Byzantine Dark Ages*, London.
Deliyannis, D. M. 2010, *Ravenna in Late Antiquity*, Cambridge.
Demacopoulos, G. 2015, *Gregory the Great: Ascetic, Pastor, and First Man of Rome*, Notre Dame.
Demandt, A. 1984, *Der Fall Roms. Die Auflösung des römischen Reiches im Urteil der Nachwelt*, Munich [2nd edition, 2014].
Deng, K. 2004, "Unveiling China's True Population Statistics for the Pre-Modern Era with Official Census Data," *Population Review* 43: 1–38.

Depauw, M. and Clarysse, W. 2013, "How Christian Was Fourth Century Egypt? Onomastic Perspectives on Conversion," *Vigiliae Christianae* 67: 407–35.

Dermody, B. J. et al. 2011, "Revisiting the Humid Roman Hypothesis: Novel Analyses Depict Oscillating Patterns," *Climate of the Past Discussions* 7: 2355–89.

Dermody, B. J. et al. 2014, "A Virtual Water Network of the Roman World," *Hydrology and Earth System Sciences* 18: 5025–40.

Detriche, S. et al. 2008, "Late Holocene Palaeohydrology of Lake Afourgagh (Middle-Atlas, Morocco) from Deposit Geometry and Facies," *Bulletin de la Société Géologique de France* 179: 41–50.

DeWitte, S. and Hughes-Morey, G. 2012, "Stature and Frailty during the Black Death: The Effect of Stature on Risks of Epidemic Mortality in London, A.D. 1348–1350," *Journal of Archaeological Science* 39: 1412–19.

DeWitte, S. 2014, "The Anthropology of Plague: Insights from Bioarchaeological Analyses of Epidemic Cemeteries," *The Medieval Globe* 1: 97–123.

Di Cosmo, N. 2002, *Ancient China and Its Enemies: The Rise of Nomadic Power in East Asian History*, Cambridge.

Di Segni, L. 1999, "Epigraphic Documentation on Building in the Provinces of Palaestina and Arabia, 4th–7th c.," in J. H. Humphrey, ed., *The Roman and Byzantine Near East II*, Portsmouth: 149–78.

———. 2009, "Greek Inscriptions in Transition from the Byzantine to the Early Islamic Period," in H. Cotton et al., eds., *From Hellenism to Islam: Cultural and Linguistic Change in the Roman Near East*, Cambridge: 352–73.

Di Stefano, G. 2009, "Nuove ricerche sulle cisterne de La malga," in M. al-Waṭanī lil-Turāth, ed., *Contrôle et distribution de l'eau dans le Maghreb antique et medieval*, Rome: 1000–22.

Diamond, J. 1997, *Guns, Germs, and Steel: The Fates of Human Societies*, New York.

Dols, M. W. 1974, "The Comparative Communal Responses to the Black Death in Muslim and Christian Societies," *Viator* 5: 269–88.

Donceel-Voûte, P. 1988, *Les pavements des églises byzantines de Syrie et du Liban: décor, archéologie et liturgie*, Louvain-la-Neuve.

Donner, F. M. 1989, "The Role of Nomads in the Near East in Late Antiquity (400–800 C.E.)," in F. M. Clover and R. S. Humphreys, eds., *Tradition and Innovation in Late Antiquity*, Madison: 73–85.

———. 2010, *Muhammad and the Believers: At the Origins of Islam*, Cambridge, MA.

Donoghue, H. D. et al. 2015, "A Migration-Driven Model for the Historical Spread of Leprosy in Medieval Eastern and Central Europe," *Infection, Genetics and Evolution* 31: 250–56.

Downie, J. 2013, *At the Limits of Art: A Literary Study of Aelius Aristides' Hieroi Logoi*, Oxford.

Dragoni, W. 1998, "Some Considerations on Climatic Changes, Water Resources and Water Needs in the Italian Region South of 43°N," in A. S. Issar and N. Brown, eds., *Water, Environment and Society in Times of Climate Change*, Dordrecht: 241–71.

Drancourt, M. et al. 1998, "Detection of 400-year Old *Yersinia pestis* DNA in Human Dental Pulp: An Approach to the Diagnosis of Ancient Septicemia," *Proceedings of the National Academy of Sciences* 95: 12637–40.

Drinkwater, J. 1987, *The Gallic Empire: Separatism and Continuity in the North-Western Provinces of the Roman Empire, A.D. 260–274*, Wiesbaden.

———. 2005, "Maximinus to Diocletian and the 'Crisis,'" in A. K. Bowman, P. Garnsey, and A. Cameron, eds., *The Cambridge Ancient History*, Vol. 12: *The Crisis of Empire, A.D. 193–337*, Cambridge: 28–66.

Drijvers, J. W. 2002, "Heraclius and the Restitutio Crucis: Notes on Symbolism and Ideology," in G. J. Reinink and H. Stolte, eds., *The Reign of Heraclius (610–641): Crisis and Confrontation*, Groningen: 175–90.

Duggan, A. et al. 2016, "17th Century Variola Virus Reveals the Recent History of Smallpox," *Current Biology* 26: 3407–12.

Duncan-Jones, R. P. 1982, *The Economy of the Roman Empire: Quantitative Studies*, 2nd ed., Cambridge.

———. 1990, *Structure and Scale in the Roman Economy*, Cambridge.

———. 1996, "The Impact of the Antonine Plague," *Journal of Roman Archaeology* 9: 108–93.

———. 2004, "Economic Change and the Transition to Late Antiquity," in S. Swain and M. Edwards, eds., *Approaching Late Antiquity: The Transformation from Early to Late Empire*, Oxford: 20–52.

Dunn, A. 2004, "Continuity and Change in the Macedonian Countryside from Gallienus to Justinian," in W. Bowden, L. Lavan, and C. Machado, eds., *Recent Research on the Late Antique Countryside*, Leiden: 535–86.

Dunn, R. et al. 2010, "Global Drivers of Human Pathogen Richness and Prevalence," *Proceedings of the Royal Society B* 277: 2587–95.

DuPont, H. L. 1993, "Diarrheal Diseases," in K. Kiple, ed., *The Cambridge World History of Human Disease*, Cambridge: 676–80.

Durliat, J. 1989, "La peste du VIe siècle. Pour un nouvel examen des sources byzantines," in J. Lefort and J. Morrisson, eds., *Hommes et richesses dans l'empire byzantin*, Vol. 1: *IVe–VIIe siècle*, Paris: 107–19.

———. 1990, *De la ville antique à la ville byzantine: le problème des subsistances*, Rome.

Ebied, R. Y. and Young, M. J. L. 1972, "A Treatise in Arabic on the Nestorian Patriarchs," *Le Muséon* 87: 87–113.

Eck, W. 2000a, "Emperor, Senate, and Magistrates," in P. Garnsey, D. Rathbone, and A. K. Bowman, eds., *The Cambridge Ancient History*, Vol. 11: *The High Empire, A.D. 70–192*, Cambridge: 214–37.

———. 2000b, "The Growth of Administrative Posts," in P. Garnsey, D. Rathbone, and A. K. Bowman, eds., *The Cambridge Ancient History*, Vol. 11: *The High Empire, A.D. 70–192*, Cambridge: 238–65.

———. 2012, "Die Seuche unter Mark Aurel: ihre Auswirkungen auf das Heer," in E. Lo Cascio, ed., *L'impatto della "peste antonina,"* Bari: 63–77.

Eckstein, A. 2006, *Mediterranean Anarchy, Interstate War, and the Rise of Rome*, Berkeley.

Eddy, J. J. 2015, "The Ancient City of Rome, Its Empire, and the Spread of Tuberculosis in Europe," *Tuberculosis* 95: 23–28.

Edling, M. 2003, *A Revolution in Favor of Government: Origins of the U.S. Constitution and the Making of the American State*, New York.

Eger, A. A. 2015, *The Islamic-Byzantine Frontier: Interaction and Exchange among Muslim and Christian Communities*, London.

Eisen, R., Dennis, D., and Gage, K. 2015, "The Role of Early-Phase Transmission in the Spread of *Yersinia pestis*," *Journal of Medical Entomology* 52: 1183–92.

Eisen, R. and Gage, K. 2009, "Adaptive Strategies of *Yersinia pestis* to Persist during Inter-epizootic and Epizootic Periods," *Veterinary Research* 40: 01.

———. 2012, "Transmission of Flea-Borne Zoonotic Agents," *Annual Review of Entomology* 57: 61–82.

Elliott, C. 2016, "The Antonine Plague, Climate Change and Local Violence in Roman Egypt," *Past & Present* 231: 3–31.

Eltahir, E. 1996, "El Niño and the Natural Variability in the Flow of the Nile River," *Water Resources Research* 32: 131–37.

Elton, H. 1996, *Warfare in Roman Europe, AD 350–425*, Oxford.

———. 2015, "Military Developments in the Fifth Century," in M. Maas, ed., *The Cambridge Companion to the Age of Attila*, Cambridge, 125–39.

Enscore, R. E. et al. 2002, "Modeling Relationships between Climate and the Frequency of Human Plague Cases in the Southwestern United States, 1960–1997," *American Journal of Tropical Medicine and Hygiene* 2: 186–96.

Erdkamp, P. 2005, *The Grain Market in the Roman Empire: A Social, Political and Economic Study*, Cambridge.

Erspamer, G. 1985, "Analisi antropologica sui resti scheletrici di otto tombe di epoca tardo-romana (IV sec. dC) rinvenute in area Sacripanti a Civitanova marche (Macerata)–Scavo del 1977," *Quaderni di scienze antropologiche* 11: 12–22.

Essefi, E. et al. 2013, "Record of the Climatic Variability and the Sedimentary Dynamics during the Last Two Millennia at Sebkha Dkhila, Eastern Tunisia," *ISRN Geology* 2013: 936198.

Estiot, S. 1996, "Le troisième siècle et la monnaie: crises et mutations," in J.-L. Fiches, ed., *Le IIIe siècle en Gaule Narbonnaise. Données régionales sur la crise de l'Empire*, Sophia Antipolis: 33–70.

Evans Grubbs, J. 1995, *Law and Family in Late Antiquity: The Emperor Constantine's Marriage Legislation*, Oxford.

Facchini, F. 1968, "I resti scheletrici del sepolcreto gallico di S. Martino in Gattara (Ravenna)," *Studi etruschi* 36: 73–90.

Facchini, F. and Evangelisti, M. C. 1975, "Scheletri etruschi della Certosa di Bologna," *Studi etruschi* 41: 161–95.

Facchini, F. and Brasili Gualandi, P. 1977–79a, "I reperti scheletrici di età arcaica della necropolis di Castiglione (Ragusa), VII–VI sec. a.C.," *Rivista di antropologia* 60: 113–42.

———. 1977–79b, "Reperti antropologici di epoca romana provenienti dalla necropolis di 'Le Palazzette' (Ravenna) (I–III sec. d.C.)," *Rivista di antropologia* 60: 159–69.

———. 1980, "Reperti scheletrici della necropolis arcaica di Monte Cassaia (Ragusa) (VII–VI secolo a.C.)," *Studi etruschi* 48: 253–76.

Facchini, F. and Stella Guerra, M. 1969, "Scheletri della necropolis romana di Bagnacavallo (Ravenna)," *Archivio per l'antropologia e l'etnologia* 99: 25–54.

Farah, K. O. et al. 2004, "The Somali and the Camel: Ecology, Management and Economics," *Anthropologist* 6: 45–55.

Faraone, C. A. 1992, *Talismans and Trojan Horses: Guardian Statues in Ancient Greek Myth and Ritual*, New York.

Fareh, H. 2007, "L'Afrique face aux catastrophes naturelles: l'apport de la documentation," in A. Mrabet and J. Rodríguez, eds., *In Africa et in Hispania: études sur l'huile africaine*, Barcelona: 145–66.

Faust, D. et al. 2004, "High-resolution Fluvial Record of Late Holocene Geomorphic Change in Northern Tunisia: Climatic or Human Impact?" *Quaternary Science Reviews* 23: 1757–75.

Faust, C. and Dobson, A. P. 2015, "Primate Malarias: Diversity, Distribution and Insights for Zoonotic Plasmodium," *One Health* 1: 66–75.

Feissel, D. 1995, "Aspects de l'immigration à Constantinople d'après les épitaphes protobyzantines," in C. Mango and G. Dagron, eds., *Constantinople and Its Hinterland: Papers from the Twenty-Seventh Spring Symposium of Byzantine Studies, Oxford, April 1993*, Aldershot: 367–77.

———. 2006, *Chroniques d'épigraphie byzantine: 1987–2004*, Paris.

Feldman, M. et al. 2016, "A High-Coverage *Yersinia pestis* Genome from a 6th-century Justinianic Plague Victim," *Molecular Biology and Evolution*: 2911–23.

Feldmann, H. and Geisbert, T. W. 2011, "Ebola Haemorrhagic Fever," *The Lancet* 377: 849–62.

Fenn, E. A. 2001, *Pox Americana: The Great Smallpox Epidemic of 1775–82*, New York.

Fenner, F. 1988, *Smallpox and Its Eradication*, Geneva.

Fentress, L. et al. 2004, "Accounting for ARS: Fineware and Sites in Sicily and Africa," in S. Alcock and J. Cherry, eds., *Side-by-Side Survey: Comparative Regional Studies in the Mediterranean World*, Oxford: 147–62.

Fentress, L. and Wilson, A. I. 2016, "The Saharan Berber Diaspora and the Southern Frontiers of Byzantine North Africa," in S. Stevens and J. Conant, eds., *North Africa under Byzantium and Early Islam*, Washington D.C.: 41–63.

Ferguson, J. and Keynes, M. 1978, "China and Rome," *Aufstieg und Niedergang der römischen Welt* 2.9.2: 581–603.

Ferrari, G. and Livi Bacci, M. 1985, "Sulle relazioni fra temperatura e mortalità nell'Italia unita, 1861–1914," in Società Italiana di Demografia Storica, *La Popolazione italiana nell'Ottocento: continuità e mutamenti: relazioni e comunicazioni presentate al convegno tenuto ad Assisi nei giorni 26–28 aprile 1983*, Bologna: 273–98.

Ferrill, A. 1986, *The Fall of the Roman Empire: The Military Explanation*, New York.

Ferrua, A. 1978, "L'epigrafia cristiana prima di Costantino," *Atti del IX congresso internazionale di archeologia cristiana*, Vatican City: 583–613.

Fine, P. 2015, "Ecological and Evolutionary Drivers of Geographic Variation in Species Diversity," *Annual Review of Ecology, Evolution, and Systematics* 46: 369–92.

Finné, M. et al. 2011, "Climate in the Eastern Mediterranean, and Adjacent Regions, During the Past 6000 Years—A Review," *Journal of Archaeological Science* 38: 3153–73.

Finné, M. et al. 2014, "Speleothem Evidence for Late Holocene Climate Variability and Floods in Southern Greece," *Quaternary Research* 81: 213–27.

Finney, S. and Edwards, L. 2016, "The 'Anthropocene' Epoch: Scientific Decision or Political Statement?" *GSA Today* 26: 4–10.

Fiocchi Nicolai, V. and Guyon, J. 2006, *Origine delle catacombe romane: atti della giornata tematica dei Seminari di archeologia cristiana (Roma, 21 marzo 2005)*, Vatican City.

Fleitmann, D. et al. 2009, "Timing and Climate Impact of Greenland Interstadials Recorded in Stalagmites from Northern Turkey," *Geophysical Research Letters* 36: L19707.

Floud, R. et al. 2011, *The Changing Body: Health, Nutrition, and Human Development in the Western World Since 1700*, Cambridge.

Folke, C. 2006, "Resilience: The Emergence of a Perspective for Social–Ecological Systems Analyses," *Global Environment Change* 16: 253–67.

Foss, C. 1997, "Syria in Transition, AD 550–750: An Archaeological Approach," *Dumbarton Oaks Papers* 51: 189–269.

Fowden, G. 2005, "Polytheist Religion and Philosophy," in A. Cameron and P. Garnsey, eds., *The Cambridge Ancient History*, Vol. 13: *The Late Empire, A.D. 337–425*, Cambridge: 538–60.

Foxhall, L. 1990, "The Dependent Tenant: Land Leasing and Labour in Italy and Greece," *Journal of Roman Studies* 80: 97–114.

Frankfurter, D. 2014, "Onomastic Statistics and the Christianization of Egypt: A Response to Depauw and Clarysse," *Vigiliae Christianae* 68: 284–89.

Frankopan, P. 2015, *The Silk Roads: A New History of the World*, New York.

Fraser, P. M. 1951, "A Syriac 'Notitia Urbis Alexandrinae,'" *Journal of Egyptian Archaeology* 37: 103–8.

Fredouille, J.-C. 2003, *Cyprien de Carthage: A Démétrien*, Paris.

Frier, B. W. 1982, "Roman Life Expectancy: Ulpian's Evidence," *Harvard Studies in Classical Philology* 86: 213–51.

———. 1983, "Roman Life Expectancy: The Pannonian Evidence," *Phoenix* 37: 328–44.

———. 1994, "Natural Fertility and Family Limitation in Roman Marriage," *Classical Philology* 89: 318–33.

———. 2000, "Demography," in P. Garnsey, D. Rathbone, and A. K. Bowman, eds., *The Cambridge Ancient History*, Vol. 11: *The High Empire, A.D. 70–192*, Cambridge: 787–816.

———. 2001, "More Is Worse: Some Observations on the Population of the Roman Empire," in W. Scheidel, ed., *Debating Roman Demography*, Leiden: 139–59.

Frier, B. W. and Kehoe, D. P. 2007, "Law and Economic Institutions," in W. Scheidel, I. Morris, and R. P. Saller, eds., *The Cambridge Economic History of the Greco-Roman World*, Cambridge: 113–43.

Fries, H. 1994, *Historische Inschriften zur römischen Kaiserzeit: von Augustus bis Konstantin*, Darmstadt.

Frisia, S. et al. 2005, "Climate Variability in the SE Alps of Italy Over the Past 17,000 Years Reconstructed from a Stalagmite Record," *Boreas* 34: 445–55.

Frösén, J. et al. 2002, *Petra: A City Forgotten and Rediscovered*, Helsinki.

Frye, R. 2005, "The Sasanians," in A. K. Bowman, P. Garnsey, and A. Cameron, eds., *The Cambridge Ancient History*, Vol. 12: *The Crisis of Empire, A.D. 193–337*, Cambridge: 461–80.

Gage, K. et al. 2008, "Climate and Vectorborne Diseases," *American Journal of Preventive Medicine* 35: 436–50.

Gage, K. and Kosoy, M. 2005, "Natural History of Plague: Perspectives from More Than a Century of Research," *Annual Review of Entomology* 50: 505–28.

Gaertner, M. A. et al. 2001, "The Impact of Deforestation on the Hydrological Cycle in the Western Mediterranean: An Ensemble Study with Two Regional Climate Models," *Climate Dynamics* 17: 857–73.

Galloway, P. R. 1986, "Long-Term Fluctuations in Climate and Population in the Preindustrial Era," *Population and Development Review* 12: 1–24.

Garnsey, P. 1988, *Famine and Food Supply in the Graeco-Roman World: Responses to Risk and Crisis*, Cambridge.

———. 1998, *Cities, Peasants, and Food in Classical Antiquity: Essays in Social and Economic History*, Cambridge.

———. 1999, *Food and Society in Classical Antiquity*, Cambridge.

———. 2004, "Roman Citizenship and Roman Law in the Late Empire," in S. Swain and M. Edwards, eds., *Approaching Late Antiquity: The Transformation from Early to Late Empire*, Oxford: 133–55.

Garrett, L. 1994, *The Coming Plague: Newly Emerging Diseases in a World Out of Balance*, New York.

Gates, L. D. and Ließ, S. 2001, "Impacts of Deforestation and Afforestation in the Mediterranean Region as Simulated by the MPI Atmospheric GCM," *Global and Planetary Change* 30: 309–28.

Gatier, P.-L. 2011, "Inscriptions grecques, mosaïques et églises des débuts de l'époque islamique au Proche-Orient (VIIe–VIIIe s.)," in A. Borrut, ed., *Le Proche-Orient de Justinien aux abbassides: peuplement et dynamiques spatiales: actes du Colloque "Continuités de l'occupation entre les périodes byzantine et abbasside au Proche-Orient, VIIe–IXe siècles," Paris, 18–20 octobre 2007*, Turnhout: 7–28.

Giannecchini, M. and Moggi-Cecchi, J. 2008, "Stature in Archeological Samples from Central Italy: Methodological Issues and Diachronic Changes," *American Journal of Physical Anthropology* 135: 284–92.

Gibbon, E. 1776–89. *The History of the Decline and Fall of the Roman Empire*, 6 vols. London.

Gilbertson, D. 1996, "Explanations: Environment as Agency," in G. Barker et al., eds., *Farming the Desert: The UNESCO Libyan Valleys Archaeological Survey*, Vol. 1: *Synthesis*, Paris: 291–318.

Gilliam, J. F. 1961, "The Plague Under Marcus Aurelius," *American Journal of Philology* 94: 225–51.

Ginkel, J. J. van. 1995, *John of Ephesus: A Monophysite Historian in Sixth-century Byzantium*, diss. University of Groningen.

Gitler, H. 1990, "Numismatic Evidence on the Visit of Marcus Aurelius to the East," *Israel Numismatic Journal* 11: 36–51.

Gitler, H. and Ponting, M. 2003, *The Silver Coinage of Septimius Severus and His Family, 193–211 A.D.: A Study of the Chemical Composition of the Roman and Eastern Issues*, Milan.

Göktürk, O. M. 2011, *Climate in the Eastern Mediterranean through the Holocene Inferred from Turkish Stalagmites*, diss., University of Bern.

Goldstone, J. A. 2002, "Efflorescences and Economic Growth in World History: Rethinking the 'Rise of the West' and the Industrial Revolution," *Journal of World History* 13: 323–89.

Goldstone, J. A., and Haldon, J. F. 2009, "Ancient States, Empires, and Exploitation: Problems and Perspectives," in I. Morris and W. Scheidel, eds., *The Dynamics of Ancient Empires: State Power from Assyria to Byzantium*, Oxford: 3–29.

Goldsworthy, A. 2003, *The Complete Roman Army*, New York.

Goudsmit, J. 2004, *Viral Fitness: The Next SARS and West Nile in the Making*, Oxford.

Gourevitch, D. 2005, "The Galenic Plague: A Breakdown of the Imperial Pathocoenosis and *Longue Durée*," *History and Philosophy of the Life Sciences* 27: 57–69.

Gowland, R. and Garnsey, P. 2010, "Skeletal Evidence for Health, Nutrition and Malaria in Rome and the Empire," in H. Eckardt, ed., *Roman Diasporas: Archaeological Approaches to Mobility and Diversity in the Roman Empire*, Portsmouth: 131–56.

Gowland R. and Walther, L. forthcoming, "Tall Stories: The Bioarchaeological Study of Growth and Stature in the Roman Empire," in W. Scheidel, ed., *The Science of Roman History*, Princeton.

Graf, F. 1992, "An Oracle against Pestilence from a Western Anatolian Town," *Zeitschrift für Papyrologie und Epigraphik* 92: 267–79.

Graham, S. L. 2008, "Justinian and the Politics of Space," in J. L. Berquist and C. V. Camp, eds., *Constructions of Space II: The Biblical City and Other Imagined Spaces*, New York: 53–77.

Gräslund, B. and Price, N. 2012, "Twilight of the Gods? The 'Dust Veil Event' of AD 536 in Critical Perspective," *Antiquity* 86: 428–43.

Grassly, N. and Fraser, C. 2006, "Seasonal Infectious Disease Epidemiology," *Proceedings of the Royal Society of London B: Biological Sciences* 273: 2541–50.

Grauel, A.-L. et al. 2013, "What Do SST Proxies Really Tell Us? A High-resolution Multiproxy ($U^{K'}_{37}$, TEX^{H}_{86} and Foraminifera $\delta^{18}O$) Study in the Gulf of Taranto, Central Mediterranean Sea," *Quaternary Science Reviews* 73: 115–31.

Gray, L. J. et al. 2010, "Solar Influences on Climate," *Review of Geophysics* 48: 1–53.

Greatrex, G. 2005, "Byzantium and the East in the Sixth Century," in M. Maas, ed., *The Cambridge Companion to the Age of Justinian*, Cambridge: 477–509.

Green, M. 2014a, "Taking 'Pandemic' Seriously: Making the Black Death Global," *The Medieval Globe* 1: 27–61.

———. ed. 2014b, *Pandemic Disease in the Medieval World: Rethinking the Black Death*, Kalamazoo, MI.

———. 2017, "The Globalizations of Disease," in N. Boivin et al., eds., *Human Dispersals and Species Movement*, Cambridge: 494–520.

Greenberg, J. 2003, "Plagued by Doubt: Reconsidering the Impact of a Mortality Crisis in the 2nd c. A.D.," *Journal of Roman Archaeology* 16: 413–25.

Greene, K. 2000, "Industry and Technology," in A. K. Bowman, P. Garnsey, and D. W. Rathbone, eds., *The Cambridge Ancient History,* Vol. 11: *The High Empire, A.D. 70–192*, Cambridge: 741–68.

Griffin, M. 2000, "Nerva to Hadrian," in A. K. Bowman, P. Garnsey, and D. Rathbone, eds., *The Cambridge Ancient History,* Vol. 11: *The High Empire, A.D. 70–192*, Cambridge: 84–131.

Grig, L. and Kelly, G. eds. 2012, *Two Romes: Rome and Constantinople in Late Antiquity*, Oxford.

Grout-Gerletti, D. 1995, "Le vocabulaire de la contagion chez l'évêque Cyprien de Carthage (249–258): de l'idée à l'utilisation," in C. Deroux, ed., *Maladie et maladies dans les textes latins antiques et médiévaux*, Brussels: 228–46.

Grove, A. T., and Rackham, O. 2001, *The Nature of Mediterranean Europe: An Ecological History*, New Haven.

Gruppioni, G. 1980, "Prime osservazioni sui resti scheletrici del sepolcreto di Monte Bibele (Bologna) (IV–II sec. a.C.)," *Atti della societa dei naturalisti e matematici di Modena. Serie VI* 111: 1–17.

Guasti, L. 2007, "Animali per Roma," in E. Papi, ed., *Supplying Rome and the Empire: The Proceedings of an International Seminar Held in Siena–Certosa di Pontig-*

nano on May 2–4, 2004, on Rome, the Provinces, Production and Distribution, Portsmouth, RI: 138–152.

Guernier, V. et al. 2004, "Ecology Drives the Worldwide Distribution of Human Diseases," *PLoS Biology* 2: 740-6.

Gunn, J. ed. 2000, *The Years without Summer: Tracing A.D. 536 and Its Aftermath*, Oxford.

Gutsmiedl, D. 2005, "Die justinianische Pest nördlich der Alpen? Zum Doppelgrab 166/167 aus dem frühmittelalterlichen Reihengräberfeld von Aschheim-Bajuwarenring," in B. Päffgen, E. Pohl, and M. Schmauder, eds., *Cum grano salis. Beiträge zur europäischen Vor- und Frühgeschichte. Festschrift für Volker Bierbrauer zum 65. Geburtstag*, Friedberg: 199–208.

Gutsmiedl-Schümann, D. 2010, *Das frühmittelalterliche Gräberfeld Aschheim-Bajuwarenring*, Kallmünz.

Haas, C. 1997, *Alexandria in Late Antiquity: Topography and Social Conflict*, Baltimore.

Hadas, G. 1993, "Where Was the Harbour of 'Ein Gedi Situated?" *Israel Exploration Journal* 43: 45–49.

Haeberli, W. et al. 1999, "On Rates and Acceleration Trends of Global Glacier Mass Changes," *Geografiska Annaler:* Series A, *Physical Geography* 81: 585–91.

Haensch, S. et al. 2010, "Distinct Clones of *Yersinia pestis* Caused the Black Death," *PLoS Pathogens* 6: e1001134.

Hahn, W. R. O. 2000, *Money of the Incipient Byzantine Empire: (Anastastius I–Justinian I, 491–565)*, Vienna.

Haines, M. R., Craig, L. E., and Weiss, T. 2003, "The Short and the Dead: Nutrition, Mortality, and the 'Antebellum Puzzle' in the United States," *Journal of Economic History* 63: 382–413.

Haklai-Rotenberg, M. 2011, "Aurelian's Monetary Reform: Between Debasement and Public Trust," *Chiron* 41: 1–39.

Haldon, J. 2002, "The Reign of Heraclius: A Context for Change?" in G. J. Reinink and H. Stolte, eds., *The Reign of Heraclius (610–641): Crisis and Confrontation*, Groningen: 1–16.

———. 2005, "Economy and Administration: How Did the Empire Work?" in M. Maas, ed., *The Cambridge Companion to the Age of Justinian*, Cambridge: 28–59.

———. 2016, *The Empire That Would Not Die: The Paradox of Eastern Roman Survival, 640–740*, Cambridge, MA.

Haldon, J. et al. 2014, "The Climate and Environment of Byzantine Anatolia: Integrating Science, History, and Archaeology," *Journal of Interdisciplinary History* 45: 113–61.

Halfmann, H. 1986, *Itinera principum: Geschichte und Typologie der Kaiserreisen im römischen Reich*, Stuttgart.

Han, B. A., Kramer, A. M., and Drake, J. M. 2016, "Global Patterns of Zoonotic Disease in Mammals," *Trends in Parasitology* 32: 565–77.

Hansen, I., Hodges, R., and Leppard, S. 2013, *Butrint 4: The Archaeology and Histories of an Ionian Town*, Oxford.

Hanson, C. et al. 2012, "Beyond Biogeographic Patterns: Processes Shaping the Microbial Landscape," *Nature Reviews: Microbiology* 10: 497–506.

Hanson, J. W. 2016, *An Urban Geography of the Roman World, 100 BC to AD 300*, Oxford.

Harbeck, M. et al. 2013, "*Yersinia pestis* DNA from Skeletal Remains from the 6th Century AD Reveals Insights into Justinianic Plague," *PLoS Pathogens* 9: e1003349.

Harkins, K. M. and Stone, A. C. 2015, "Ancient Pathogen Genomics: Insights into Timing and Adaptation," *Journal of Human Evolution* 79: 137–49.

Harper, K. 2011, *Slavery in the Late Roman World, AD 275–425*, Cambridge.

———. 2012, "Marriage and Family in Late Antiquity," in S. F. Johnson, ed., *The Oxford Handbook of Late Antiquity*, Oxford: 667–714.

———. 2013a, *From Shame to Sin: The Christian Transformation of Sexual Morality in Late Antiquity*, Cambridge.

———. 2013b, "L'ordine sociale costantiniano: schiavitù, economia e aristocrazia," in *Costantino I: Enciclopedia Costantiniana sulla figura e l'immagine dell'imperatore del cosidetto editto di Milano, 313–2013*, vol. 1, Rome: 369–86.

———. 2015a, "Pandemics and Passages to Late Antiquity: Rethinking the Plague of c. 249–270 Described by Cyprian," *Journal of Roman Archaeology* 28: 223–60.

———. 2015b, "Landed Wealth in the Long Term: Patterns, Possibilities, Evidence," in P. Erdkamp, K. Verboven, and A. Zuiderhoek, eds., *Ownership and Exploitation of Land and Natural Resources in the Roman World, Oxford*: 43–61.

———. 2015c, "A Time to Die: Preliminary Notes on Seasonal Mortality in Late Antique Rome," in C. Laes, K. Mustakallio, and V. Vuolanto, eds., *Children and Family in Late Antiquity. Life, Death and Interaction*, Leuven: 15–34.

———. 2016a, "People, Plagues, and Prices in the Roman World: The Evidence from Egypt," *Journal of Economic History* 76: 803–39.

———. 2016b, "The Environmental Fall of the Roman Empire," *Daedalus* 145 (2): 101–11.

———. 2016c, "Another Eye-witness to the Plague Described by Cyprian and Notes on the 'Persecution of Decius,'" *Journal of Roman Archaeology* 29: 473–76.

———. forthcoming, "Invisible Environmental History: Infectious Disease in Late Antiquity," in A. Izdebski and M. Mulryan, eds., *Environment and Society in the Long Late Antiquity*.

Harper, K., and McCormick, M. forthcoming, "Reconstructing the Roman Climate," in W. Scheidel, ed., *The Science of Roman History*, Princeton.

Harper, K. N. and Armelagos, G. J. 2013, "Genomics, the Origins of Agriculture, and Our Changing Microbe-scape: Time to Revisit Some Old Tales and Tell Some New Ones," *American Journal of Physical Anthropology* 152: 135–152.

Harries, J. 2012, *Imperial Rome AD 284 to 363: The New Empire*, Edinburgh.

Harris, W. V. 1985, *War and Imperialism in Republican Rome, 327–70 B.C.*, Oxford.

———. 1994, "Child-Exposure in the Roman Empire," *Journal of Roman Studies* 84: 1–22.

———. 1999a, "Demography, Geography, and the Sources of Roman Slaves," *Journal of Roman Studies* 89: 62–75.

———. ed. 1999b, *The Transformations of Urbs Roma in Late Antiquity*, Portsmouth.

———. 2000, "Trade," in A. K. Bowman, P. Garnsey, and D. W. Rathbone, eds., *The Cambridge Ancient History*, Vol. 11: *The High Empire, A.D. 70–192*, Cambridge: 710–40.

———. ed. 2005, *The Spread of Christianity in the First Four Centuries: Essays in Explanation*, Leiden.

———. 2006, "A Revisionist View of Roman Money," *Journal of Roman Studies* 96: 1–24.
———. ed. 2008, *The Monetary Systems of the Greeks and Romans*, Oxford.
———. 2011, "Bois et déboisement dans la Méditerranée antique," *Annales. Histoire, Sciences Sociales* 66: 105–40.
———. 2012, "The Great Pestilence and the Complexities of the Antonine-Severan Economy," in E. Lo Cascio, ed., *L'impatto della "peste antonina"*, Bari: 331–338.
———. ed. 2013a, *The Ancient Mediterranean Environment between Science and History*, Leiden.
———. 2013b, "Defining and Detecting Mediterranean Deforestation, 800 BCE to 700 CE," in W. V. Harris, ed., *The Ancient Mediterranean Environment between Science and History*, Leiden: 173–94.
———. 2016, *Roman Power: A Thousand Years of Empire*, New York.
Harris, W. V. and Holmes, B. eds. 2008, *Aelius Aristides between Greece, Rome, and the Gods*, Leiden.
Harvey, Susan Ashbrook. 1990, *Asceticism and Society in Crisis: John of Ephesus and the Lives of the Eastern Saints*, Berkeley.
Hassall, M. 2000, "The Army," in P. Garnsey, D. Rathbone, and A. K. Bowman, eds., *The Cambridge Ancient History,* Vol. 11: *The High Empire, A.D. 70–192*, Cambridge: 320–43.
Hassan, F. 2007, "Extreme Nile Floods and Famines in Medieval Egypt (AD 930–1500) and Their Climatic Implications," *Quaternary International* 173–74: 101–12.
Hatcher, J. 2003, "Understanding the Population History of England 1450–1750," *Past & Present* 180: 83–130.
Hays, J. N. 1998, *The Burdens of Disease: Epidemics and Human Response in Western History*, New Brunswick.
Heather, P. 1995, "The Huns and the End of the Roman Empire in Western Europe," *English Historical Review* 110: 4–41.
———. 1998a, "Goths and Huns, *c.* 320–425," in A. Cameron and P. Garnsey, eds., *The Cambridge Ancient History,* Vol. 13: *The Late Empire, A.D. 337–425*, Cambridge: 487–515.
———. 1998b, "Senators and Senates," in A. Cameron and P. Garnsey, eds., *The Cambridge Ancient History,* Vol. 13: *The Late Empire, A.D. 337–425*, Cambridge: 184–210.
———. 2006, *The Fall of the Roman Empire: A New History of Rome and the Barbarians*, Oxford.
———. 2010, *Empires and Barbarians: The Fall of Rome and the Birth of Europe*, New York.
———. 2015, "The Huns and Barbarian Europe," in M. Maas, ed., *The Cambridge Companion to the Age of Attila*, Cambridge: 209–229.
Heide, A. 1997, *Das Wetter und Klima in der römischen Antike im Westen des Reiches*, diss., University of Mainz.
Hekster, O., de Kleijn, G., and Slootjes, D. eds. 2007, *Crises and the Roman Empire*, Leiden.
Henneberg, M. and Henneberg, R. 2002, "Reconstructing Medical Knowledge in Ancient Pompeii from the Hard Evidence of Bones and Teeth," in J. Renn

and G. Castagnetti, eds., *Homo Faber: Studies on Nature, Technology, and Science at the Time of Pompeii*, Rome: 169–87.

Hermansen, G. 1978, "The Population of Imperial Rome: The Regionaries," *Historia* 27: 129–68.

Hickey, T. 2012, *Wine, Wealth, and the State in Late Antique Egypt: The House of Apion at Oxyrhynchus*, Ann Arbor.

Himmelfarb, M. 2017, *Jewish Messiahs in a Christian Empire: A History of the Book of Zerubbabel*, Cambridge, MA.

Hin, S. 2013, *The Demography of Roman Italy: Population Dynamics in an Ancient Conquest Society, 201 BCE–14 CE*, Cambridge.

Hinnebusch, B. J. et al. 2002, "Role of Yersinia Murine Toxin in Survival of *Yersinia pestis* in the Midgut of the Flea Vector," *Science* 296: 733–35.

Hinnebusch, B. J. et. al. 2017, "Comparative Ability of *Oropsylla montana* and *Xenopsylla cheopis* Fleas to Transmit *Yersinia pestis* by Two Different Mechanisms," *PLOS Neglected Tropical Diseases* 11: e0005276.

Hinnebusch, B. J., Chouikha, I., and Sun, Y.-C. 2016, "Ecological Opportunity, Evolution, and the Emergence of Flea-Borne Plague," *Infection and Immunity* 84: 1932–40.

Hirschfeld, Y. 2006, "The Crisis of the Sixth Century: Climatic Change, Natural Disasters and the Plague," *Mediterranean Archaeology and Archaeometry* 6: 19–32.

Hitchner, B. 1988, "The Kasserine Archaeological Survey, 1982–86," *Antiquités africaines* 24: 7–41.

———. 1989, "The Organization of Rural Settlement in the Cillium-Thelepte Region (Kasserine, Central Tunisia)," *L'Africa Romana* 6: 387–402.

———. 1990, "The Kasserine Archaeological Survey: 1987," *Antiquités africaines* 26: 231–60.

Hobson, B. 2009, *Latrinae et Foricae: Toilets in the Roman World*, London.

Hobson, D. W. 1984, "P. VINDOB. GR. 24951 + 24556: New Evidence for Tax-Exempt Status in Roman Egypt," *Atti del XVII Congresso internazionale di papirologia*, 847–64.

Hoelzle, M. et al. 2003, "Secular Glacier Mass Balances Derived from Cumulative Glacier Length Changes," *Global and Planetary Change* 36: 295–306.

Hoffman, D. 1969–70, *Das Spätrömische Bewegungsheer und die Notitia dignitatum*, Düsseldorf.

Holman, S. R. 2001, *The Hungry Are Dying: Beggars and Bishops in Roman Cappadocia*, Oxford.

———. ed. 2008, *Wealth and Poverty in Early Church and Society*, Grand Rapids.

Holleran, C. and Pudsey, A. eds. 2011, *Demography and the Graeco-Roman World: New Insights and Approaches*, Cambridge.

Holloway, K. L. et al. 2011, "Evolution of Human Tuberculosis: A Systematic Review and Meta-Analysis of Paleopathological Evidence," *Homo* 62: 402–58.

Holloway, R. R. 1994, *The Archaeology of Early Rome and Latium*, London.

Holum, K. 2005, "The Classical City in the Sixth Century: Survival and Transformation," in M. Maas, ed., *The Cambridge Companion to the Age of Justinian*, Cambridge: 87–112.

Holzhauser, H. et al. 2005, "Glacier and Lake-Level Variations in West-Central Europe Over the Last 3500 Years," *Holocene* 15: 789–801.

Honoré, T. 1978, *Tribonian*, Ithaca.

———. 2002, *Ulpian: Pioneer of Human Rights*, Oxford.

———. 2010, *Justinian's Digest: Character and Compilation*, Oxford.

Hopkins, C. 1972, *Topography and Architecture of Seleucia on the Tigris*, Ann Arbor.

Hopkins, D. R. 2002, *The Greatest Killer: Smallpox in History, with a New Introduction*, Chicago.

Hopkins, K. 1980, "Taxes and Trade in the Roman Empire (200 B.C.–A.D. 400)," *Journal of Roman Studies* 70: 101–25.

———. 1998, "Christian Number and Its Implications," *Journal of Early Christian Studies* 6: 185–226.

———. 2009a, "The Political Economy of the Roman Empire," in I. Morris and W. Scheidel, eds., *The Dynamics of Ancient Empires: State Power from Assyria to Byzantium*, Oxford: 178–204.

———. 2009b, *A World Full of Gods: The Strange Triumph of Christianity*, New York.

Horden, P. 2005, "Mediterranean Plague in the Age of Justinian," in M. Maas, ed., *The Cambridge Companion to the Age of Justinian*, Cambridge: 134–60.

Horden, P. and Purcell, N. 2000, *The Corrupting Sea: A Study of Mediterranean History*, Oxford.

Hoskier, H. C. 1928, *The Complete Commentary of Oecumenius on the Apocalypse: Now Printed for the First Time from Manuscripts at Messina, Rome, Salonika, and Athos*, Ann Arbor.

Howgego, C., Butcher, K., Ponting, M. et al. 2010, "Coinage and the Roman Economy in the Antonine Period: The View from Egypt." *Oxford Roman Economy Project: Working Papers*.

Hoyland, R. G. 1997, *Seeing Islam as Others Saw It: A Survey and Evaluation of Christian, Jewish, and Zoroastrian Writings on Early Islam*, Princeton.

———. 2012, "Early Islam as a Late Antique Religion," in S. F. Johnson, ed., *The Oxford Handbook of Late Antiquity*, Oxford: 1053–77.

Hughes, D. J. 1994, *Pan's Travail: Environmental Problems of the Ancient Greeks and Romans*, Baltimore.

———. 2011, "Ancient Deforestation Revisited," *Journal of the History of Biology* 44: 43–57.

Humfress, C. 2005, "Law and Legal Practice in the Age of Justinian," in M. Maas, ed., *The Cambridge Companion to the Age of Justinian*, Cambridge: 161–84.

Humphries, M. 2000, "Italy, A.D. 425–605," in A. Cameron, B. Ward-Perkins, and M. Whitby, eds., *The Cambridge Ancient History*, Vol. 14: *Late Antiquity: Empire and Successors*, Cambridge: 525–51.

———. 2007, "From Emperor to Pope? Ceremonial, Space, and Authority at Rome from Constantine to Gregory the Great," in K. Cooper and J. Hillner, eds., *Religion, Dynasty, and Patronage in Early Christian Rome, 300–900*, Cambridge: 21–58.

Huntington, E. 1917, "Climatic Change and Agricultural Exhaustion as Elements in the Fall of Rome," *Quarterly Journal of Economics* 31: 173–208.

Hurrell, J. W. et al. 2003, "An Overview of the North Atlantic Oscillation," in J. W. Hurrell, ed., *The North Atlantic Oscillation: Climatic Significance and Environmental Impact*, Washington, D.C.: 1–35.

Hyams, E. 1952, *Soil and Civilization*, London.

Ibbetson, D. 2005, "High Classical Law," in A. K. Bowman, P. Garnsey, and A. Cameron, eds., *The Cambridge Ancient History*, Vol. 12, *The Crisis of Empire, A.D. 193–337*, Cambridge: 184–99.

Ieraci Bio, A. M. ed. 1981, *De bonis malisque sucis*, Naples.

Inskip, S. A. et al. 2015, "Osteological, Biomolecular and Geochemical Examination of an Early Anglo-Saxon Case of Lepromatous Leprosy," *PLoS One* 10: 1–22.

Isaac, B. 1992, *The Limits of the Empire: The Roman Army in the East*, rev. ed., Oxford.

Israelowich, I. 2012, *Society, Medicine and Religion in the Sacred Tales of Aelius Aristides*, Leiden.

Issar, A. and Zohar, M. 2004, *Climate Change: Environment and Civilization in the Middle East*, Berlin.

Ivleva, T. 2016, "Peasants into Soldiers: Recruitment and Military Mobility in the Early Roman Empire," in L. de Ligt and L. E. Tacoma, eds., *Migration and Mobility in the Early Roman Empire*, Leiden: 158–75.

Izdebski, A. 2013, *A Rural Economy in Transition: Asia Minor from Late Antiquity into the Early Middle Ages*, Warsaw.

Izdebski, A. et al. 2015, "Realising Consilience: How Better Communication Between Archaeologists, Historians and Natural Scientists Can Transform the Study of Past Climate Change in the Mediterranean," *Quaternary Science Reviews* 30: 1–18.

Izdebski, A. et al. 2016, "The Environmental, Archaeological and Historical Evidence for Regional Climatic Changes and Their Societal Impacts in the Eastern Mediterranean in Late Antiquity," *Quaternary Science Reviews* 136: 189–208.

Jablonski, D. et al. 2017, "Shaping the Latitudinal Diversity Gradient: New Perspectives from a Synthesis of Paleobiology and Biogeography," *American Naturalist* 189: 1–12.

Jaouadi, S. et al. 2016, "Environmental Changes, Climate, and Anthropogenic Impact in South-east Tunisia during the Last 8 kyr," *Climate of the Past* 12: 1339–59.

Jaritz, H. and Carrez-Maratray, J.-Y. 1996, *Pelusium: prospection archéologique et topographique de la région de Kana'is: 1993 et 1994*, Stuttgart.

Jenkins, C. et al. 2013, "Global Patterns of Terrestrial Vertebrate Diversity and Conservation," *Proceedings of the National Academy of Sciences* 110: E2602–E2610.

Jennison, G. 1937, *Animals for Show and Pleasure in Ancient Rome*, Manchester.

Jiang, J. et al. 2002, "Coherency Detection of Multiscale Abrupt Changes in Historic Nile Flood Levels," *Geophysical Research Letters* 29: 1271.

Johnson, S. F. 2016, *Literary Territories: Cartographical Thinking in Late Antiquity*, Oxford.

Johnston, D. 2005, "Epiclassical Law," in A. K. Bowman, P. Garnsey, and A. Cameron, eds., *The Cambridge Ancient History*, Vol. 12: *The Crisis of Empire, A.D. 193–337*, Cambridge: 200–11.

Jones, A. H. M. 1957, "Capitatio and Iugatio," *Journal of Roman Studies* 47: 88–94.

———. 1964, *The Later Roman Empire, 284–602: A Social, Economic, and Administrative Survey*, Norman.

Jones, C. P. 1971, "A New Letter of Marcus Aurelius to the Athenians," *Zeitschrift für Papyrologie und Epigraphik* 8: 161–63.

———. 1972, "Aelius Aristides, ΕΙΣ ΒΑΣΙΛΕΑ," *Journal of Roman Studies* 62: 134–52.

———. 2005, "Ten Dedications 'To the Gods and Goddesses' and the Antonine Plague," *Journal of Roman Archaeology* 18: 293–301.

———. 2006, "Addendum to JRA 18 (2005): Cosa and the Antonine Plague?" *Journal of Roman Archaeology* 19: 368–69.

———. 2007, "Procopius of Gaza and the Water of the Holy City," *Greek, Roman, and Byzantine Studies* 47: 455–67.

———. 2008, "Aristides' First Admirer," in W. V. Harris and B. Holmes, eds., *Aelius Aristides between Greece, Rome and the Gods*, Leiden: 253–62.

———. 2011, "The Historian Philostratus of Athens," *Classical Quarterly* 61: 320–22.

———. 2012a, "Galen's Travels," *Chiron* 42: 399–419.

———. 2012b, "Recruitment in Time of Plague: The Case of Thespiae," in E. Lo Cascio, ed., *L'impatto della "peste Antonina"*, Bari: 79–85.

———. 2013, "Elio Aristide e i primi anni di Antonino Pio," in P. Desideri and F. Fontanella, eds., *Elio Aristide e la legittimazione greca dell'impero di Roma*, Bologna: 39–67.

———. 2014, *Between Pagan and Christian*, Cambridge, MA.

———. 2016, "An Amulet from London and Events Surrounding the Antonine Plague," *Journal of Roman Archaeology* 29: 469–72.

Jones, D. S. 2003, "Virgin Soils Revisited," *William and Mary Quarterly* 60: 703–42.

Jones, K. E. et al. 2008, "Global Trends in Emerging Infectious Diseases," *Nature* 451: 990–994.

Jongman, W. M. 2007, "The Early Roman Empire: Consumption," in W. Scheidel, I. Morris, and R. P. Saller, eds., *The Cambridge Economic History of the Greco-Roman World*, Cambridge: 592–618.

———. 2012, "Roman Economic Change and the Antonine Plague: Endogenous, Exogenous, or What?" in E. Lo Cascio, ed., *L'impatto della "peste Antonina"*, Bari: 253–63.

Juliano, A. and Lerner, J. 2001, *Monks and Merchants: Silk Road Treasures from Northwest China*, New York.

Kaegi, W. 1992, *Byzantium and the Early Islamic Conquests*, Cambridge.

Kaldellis, A. 2004, *Procopius of Caesarea: Tyranny, History, and Philosophy at the End of Antiquity*, Philadelphia.

———. 2007, "The Literature of Plague and the Anxieties of Piety in Sixth-Century Byzantium," in F. Mormando and T. Worcester, eds., *Piety and Plague: From Byzantium to the Baroque*, Kirksville: 1–22.

Kamash, Z. 2012, "Irrigation Technology, Society and Environment in the Roman Near East," *Journal of Arid Environments* 86: 65–74.

Kaniewski, D. et al. 2007, "A High-Resolution Late Holocene Landscape Ecological History Inferred from an Intramontane Basin in the Western Taurus Mountains, Turkey," *Quaternary Science Reviews* 26: 2201–18.

Karlen, A. 1995, *Man and Microbes: Disease and Plagues in History and Modern Times*, New York.

Kausrud, K. et al. 2010, "Modeling the Epidemiological History of Plague in Central Asia: Palaeoclimatic Forcing on a Disease System over the Past Millennium," *BioMed Central Biology* 8: 112.

Keenan, J. G. 1973, "The Names Flavius and Aurelius as Status Designations in Later Roman Egypt," *Zeitschrift für Papyrologie und Epigraphik* 11: 33–63.

———. 1974, "The Names Flavius and Aurelius as Status Designations in Later Roman Egypt," *Zeitschrift für Papyrologie und Epigraphik* 13: 283–304.

———. 2003, "Deserted Villages: From the Ancient to the Medieval Fayyūm," *Bulletin of the American Society of Papyrologists* 40: 119–39.

Kehoe, D. P. 1988, *The Economics of Agriculture on Roman Imperial Estates in North Africa*, Göttingen.

———. 2007, *Law and the Rural Economy in the Roman Empire*, Ann Arbor.

Kelly, C. 1998, "Emperors, Government, and Bureaucracy," in A. Cameron and P. Garnsey, eds., *The Cambridge Ancient History,* Vol. 13: *The Late Empire, A.D. 337–425*, Cambridge: 138–83.

———. 2006, "Bureaucracy and Government," in N. Lenski, ed., *The Cambridge Companion to the Age of Constantine*, Cambridge: 183–204.

———. 2015, "Neither Conquest Nor Settlement: Attila's Empire and Its Impact," in M. Maas, ed., *The Cambridge Companion to the Age of Attila*, Cambridge: 193–208.

Kennedy, H. N. 2000, "Syria, Palestine and Mesopotamia," in A. Cameron, B. Ward-Perkins, and M. Whitby, eds., *The Cambridge Ancient History,* Vol. 14: *Late Antiquity: Empire and Successors*, Cambridge: 588–611.

———. 2007a, "Justinianic Plague in Syria and the Archaeological Evidence," in L. K. Little, ed., *Plague and the End of Antiquity: The Pandemic of 541–750*, New York: 87–95.

———. 2007b, *The Great Arab Conquests: How the Spread of Islam Changed the World We Live In*, Philadelphia.

Keys, D. 2000, *Catastrophe: An Investigation into the Origins of the Modern World*, New York.

Killgrove, K. 2010a, *Migration and Mobility in Imperial Rome*, diss., University of North Carolina.

———. 2010b, "Response to C. Bruun's Water, Oxygen Isotopes and Immigration to Ostia-Portus," *Journal of Roman Archaeology* 23: 133–36.

———. 2014. "Bioarchaeology in the Roman Empire," in C. Smith, ed., *Encyclopedia of Global Archaeology*, New York: 876–82.

King, A. 1999, "Diet in the Roman World: A Regional Inter-Site Comparison of the Mammal Bones," *Journal of Roman Archaeology* 12: 168–202.

Kirbihler, F. 2006, "Les émissions de monnaies d'homonoia et les crises alimentaires en asie sous Marc-Aurèle," *Revue des études anciennes* 108: 613–40.

Kisić, R. 2011, *Patria caelestis: die eschatologische Dimension der Theologie Gregors des Grossen*, Tübingen.

Klein Goldewijk, G. and Jacobs, J. 2013, "The Relation Between Stature and Long Bone Length in the Roman Empire," Research Institute SOM, Faculty of Economic and Business, University of Groningen.

Klein Goldewijk, K., Beusen, A. and Janssen, P. 2010, "Long-term Dynamic Modeling of Global Population and Built-Up Area in a Spatially Explicit Way: HYDE 3.1," *The Holocene* 20: 565–73.

Knapp, A. B. and Manning, S. 2016, "Crisis in Context: The End of the Late Bronze Age in the Eastern Mediterranean," *American Journal of Archaeology* 120: 99–149.

Koder, J. 1995, "Ein inschriftlicher Beleg zur 'justinianischen' Pest in Zora (Azra'a)," *Byzantinoslavica* 56: 12–18.

Koepke, N. and Baten, J. 2005, "The Biological Standard of Living in Europe During the Last Two Millennia," *European Review of Economic History* 9: 61–95.

Kolb, F. 1977, "Der Aufstand der Provinz Africa Proconsularis im Jahr 238 n. Chr.: Die wirtschaftlichen und sozialen Hintergründe," *Historia: Zeitschrift für Alte Geschichte* 26: 440–78.

Koloski-Ostrow, A. O. 2015, *The Archaeology of Sanitation in Roman Italy: Toilets, Sewers, and Water Systems*, Chapel Hill.

Kominko, M. 2013, *The World of Kosmas: Illustrated Byzantine Codices of the Christian Topography*, Cambridge.

Komlos, J. 2012. "A Three-Decade History of the Antebellum Puzzle: Explaining the Shrinking of the U.S. Population at the Onset of Modern Economic Growth," *Journal of the Historical Society* 12: 395–445.

Körner, C. 2002, *Philippus Arabs: Ein Soldatenkaiser in der Tradition des Antoninisch-Severischen Prinzipats*, Berlin.

Kostick, C. and Ludlow, F. 2015, "The Dating of Volcanic Events and Their Impact upon European Society, 400–800 CE," *Post-Classical Archaeologies* 5: 7–30.

Kouki, P. 2013, "Problems of Relating Environmental History to Human Settlement in the Classical and Late Classical Periods—The Example of Southern Jordan," in W. V. Harris, ed., *The Ancient Mediterranean Environment between Science and History*, Leiden: 197–211.

Krause, J. and Pääbo, S. 2016, "Genetic Time Travel," *Genetics* 203: 9–12.

Krause, J.-U. 1994, *Witwen und Waisen im römischen Reich*, Stuttgart.

Krause, J.-U. and Witschel, C. eds. 2006, *Die Stadt in der Spätantike: Niedergang oder Wandel?: Akten des internationalen Kolloquiums in München am 30. und 31. Mai 2003*, Stuttgart.

Krebs, C. 2013, *Population Fluctuations in Rodents*, Chicago.

Krishnamurthy, R. 2007, *Late Roman Copper Coins from South India: Karur, Madurai and Tirukkoilur*, Chennai.

Krom, M. D. et al. 2002, "Nile River Sediment Fluctuations over the Past 7000 yr and Their Key Role in Sapropel Development," *Geology* 30: 71–74.

Kron, G. 2005, "Anthropometry, Physical Anthropology, and the Reconstruction of Ancient Health, Nutrition, and Living Standards," *Historia: Zeitschrift für alte Geschichte* 54: 68–83.

———. 2012, "Nutrition, Hygiene and Mortality. Setting Parameters for Roman Health and Life Expectancy Consistent with Our Comparative Evidence," in E. Lo Cascio, ed., *L'impatto della "peste Antonina"*, Bari: 193–252.

Kulikowski, M. 2004, *Late Roman Spain and Its Cities*, Baltimore.

———. 2006, "The Late Roman City in Spain," in J.-U. Krause and C. Witschel, eds., *Die Stadt in der Spätantike: Niedergang oder Wandel?: Akten des internationalen Kolloquiums in München am 30. und 31. Mai 2003*, Stuttgart: 129–49.

———. 2007, "Plague in Spanish Late Antiquity," in L. K. Little, ed., *Plague and the End of Antiquity: The Pandemic of 541–750*, New York: 150–70.

———. 2012, "The Western Kingdoms," in S. F. Johnson, ed., *The Oxford Handbook of Late Antiquity*, Oxford: 31–59.

———. 2016, *The Triumph of Empire: The Roman World from Hadrian to Constantine*, Cambridge.

Kutiel, H. and Benaroch, Y. 2002, "North Sea-Caspian Pattern (NCP)—An Upper Level Atmospheric Teleconnection Affecting the Eastern Mediterranean: Identification and Definition," *Theoretical and Applied Climatology* 71: 17–28.

Kutiel, H. and Türkeş, M. 2005, "New Evidence for the Role of the North Sea — Caspian Pattern on the Temperature and Precipitation Regimes in Continental Central Turkey," *Geografiska Annaler: Series A, Physical Geography* 87: 501–13.

Kuzucuoğlu, C. et al. 2011, "Mid- to Late-Holocene Climate Change in Central Turkey: The Tecer Lake Record," *The Holocene* 21: 173–88.

Lafferty, K. 2009, "Calling for an Ecological Approach to Studying Climate Change and Infectious Diseases," *Ecology* 90: 932–33.

Lamb, H. H. 1982, *Climate, History, and the Modern World*, London.

Landers, J. 1993, *Death and the Metropolis: Studies in the Demographic History of London, 1670–1830*, Cambridge.

Lane Fox, R. 1987, *Pagans and Christians*, New York.

Larsen, C. S. 2015, *Bioarchaeology: Interpreting Behavior from the Human Skeleton*, Cambridge.

Latham, J. 2015, "Inventing Gregory 'the Great': Memory, Authority, and the Afterlives of the Letania Septiformis," *Church History* 84: 1–31.

Launaro, A. 2011, *Peasants and Slaves: The Rural Population of Roman Italy (200 BC to AD 100)*, Cambridge.

Lavan, L. and Bowden, W. eds. 2001, *Recent Research in Late-Antique Urbanism*, Portsmouth.

Lavan, M. 2016, "The Spread of Roman Citizenship, 14–212 CE: Quantification in the Face of High Uncertainty," *Past & Present* 230: 3–46.

Lazer, E. 2009, *Resurrecting Pompeii*, London.

Le Bohec, Y. 1994, *The Imperial Roman Army*, London.

Le Roy, M. et al. 2015, "Calendar-Dated Glacier Variations in the Western European Alps During the Neoglacial: The Mer de Glace Record, Mont Blanc Massif," *Quaternary Science Reviews* 108: 1–22.

Le Roy Ladurie, E. 1973, "Un concept: L'unification microbienne du monde (XIVe–XVIIe siècles)," *Schweizerische Zeitschrift für Geschichte* 23: 627–96.

Lechat, M. 2002, "The Paleoepidemiology of Leprosy: An Overview," in C. Roberts, M. Lewis, and K. Manchester, eds., *Past and Present of Leprosy: Archaeological, Historical, Paleopathological, and Clinical Approaches: Proceedings of the International Congress on the Evolution and Palaeoepidemiology of the Infectious Diseases 3 (ICEPID), University of Bradford, 26th–31st July 1999*, Oxford: 460–70.

Lecker, M. 2015, "Were the Ghassānids and the Byzantines behind Muḥammad's *hijra*?" in D. Genequand and C. J. Robin, eds., *Les Jafnides. Des rois arabes au service de Byzance (VIe siècle de l'ère chrétienne)*, Paris: 277–93.

Lee, A. D. 1998, "The Army," in A. Cameron and P. Garnsey, eds., *The Cambridge Ancient History*, Vol. 13: *The Late Empire, A.D. 337–425*, Cambridge: 211–37.

———. 2007, *War in Late Antiquity. A Social History*, Malden.

Lee H. F., Fok, L., and Zhang, D. 2008, "Climatic Change and Chinese Population Growth Dynamics over the Last Millennium," *Climatic Change* 88: 131–56.

Lehoux, D. 2007, *Astronomy, Weather, and Calendars in the Ancient World: Parapegmata and Related Texts in Classical and Near-Eastern Studies*, Cambridge.

Lenski, N. ed. 2006, *The Cambridge Companion to the Age of Constantine*, Cambridge.

———. 2016, *Constantine and the Cities: Imperial Authority and Civic Politics*, Philadelphia.

Leone, A. 2012, "Water Management in Late Antique North Africa: Agricultural Irrigation," *Water History* 4: 119–33.

Lepelley, C. 2006, "La cité africaine tardive, de l'apogée du IVe siècle à l'effondrement du VIIe siècle," in J.-U. Krause and C. Witschel, eds., *Die Stadt in der Spätantike: Niedergang oder Wandel?: Akten des internationalen Kolloquiums in München am 30. und 31. Mai 2003*, Stuttgart: 13–32.

Leroy, S. A. G. 2010, "Pollen Analysis of Core DS7-1SC (Dead Sea) Showing Intertwined Effects of Climatic Change and Human Activities in the Late Holocene," *Journal of Archaeological Science* 37: 306–16.

Leveau, P. 2014, "Évolution climatique et construction des ouvrages hydrauliques en Afrique romaine," in F. Baratte, C. J. Robin, and E. Rocca, eds., *Regards croisés d'Orient et d'Occident les barrages dans l'Antiquité tardive*, Paris: 125–38.

Levick, B. 2000a, "Greece and Asia Minor," in A. K. Bowman, P. Garnsey, and D. Rathbone, eds., *The Cambridge Ancient History*, Vol. 11: *The High Empire, A.D. 70–192*, Cambridge: 604–34.

———. 2000b, *The Government of the Roman Empire: A Sourcebook*, 2nd ed., London.

———. 2014, *Faustina I and II: Imperial Women of the Golden Age*, Oxford.

Lewis, M. E. 2007, *The Early Chinese Empires: Qin and Han*, Cambridge.

Lewis-Rogers, N. and Crandall, K. A. 2010, "Evolution of Picornaviridae: An Examination of Phylogenetic Relationships and Cophylogeny," *Molecular Phylogenetics and Evolution* 54: 995–1005.

Lewit, T. 2004, *Villas, Farms and the Late Roman Rural Economy, Third to Fifth Centuries AD*, Oxford.

Li, B. et al. 2012, "Humoral and Cellular Immune Responses to *Yersinia pestis* Infection in Long-Term Recovered Plague Patients," *Clinical and Vaccine Immunology* 19: 228–34.

Li, Y. et al. 2007. "On the Origin of Smallpox: Correlating Variola Phylogenetics with Historical Smallpox Records," *Proceedings of the National Academy of Sciences* 104: 15787–92.

Lieberman, V. 2003, *Strange Parallels: Southeast Asia in Global Context*, New York.

Liebeschuetz, J. H. W. G. 2000, "Administration and Politics in the Cities of the Fifth to the Mid Seventh Century: 425–640," in A. Cameron, B. Ward-Perkins, and M. Whitby, eds., *The Cambridge Ancient History*, Vol. 14: *Late Antiquity: Empire and Successors*, Cambridge: 207–37.

———. 2001, *The Decline and Fall of the Roman City*, Oxford.

Lintott, A. 1999, *The Constitution of the Roman Republic*, Oxford.

Lionello, P. ed. 2012, *The Climate of the Mediterranean Region: From the Past to the Future*, London.

Little, L. K. ed. 2007a, *Plague and the End of Antiquity: The Pandemic of 541–750*, New York.

———. 2007b, "Life and Afterlife of the First Plague Pandemic," in L. K. Little, ed., *Plague and the End of Antiquity: The Pandemic of 541–750*, New York: 3–32.

Littman, R. J. 2009, "The Plague of Athens: Epidemiology and Paleopathology," *Mount Sinai Journal of Medicine* 76: 456–67.

Littman, R. J. and Littman M. L. 1973, "Galen and the Antonine Plague," *American Journal of Philology* 94: 243–55.

Livi Bacci, M. 2006, "The Depopulation of Hispanic America after the Conquest," *Population and Development Review* 32: 199–232.

———. 2012, *A Concise History of World Population*, 5th ed., Oxford.

Lo Cascio, E. 1986, "Teoria e politica monetaria a Roma tra III e IV d.C.," in A. Giardina, ed., *Società romana e impero tardoantico*, I, Rome: 535–57, 779–801.

———. 1991, "Fra equilibrio e crisi," in A. Schiavone, ed., *Storia di Roma,* vol. 2.2, Turin: 701–31.

———. 1993, "Prezzo dell'oro e prezzi delle merci," in L. Camilli, and S. Sorda, eds., *L'"inflazione" nel quarto secolo d.C.: atti dell'incontro di studio, Roma 1988*, Rome: 155–88.

———. 1994, "The Size of the Roman Population: Beloch and the Meaning of the Augustan Census Figures," *Journal of Roman Studies* 84: 23–40.

———. 1995, "Aspetti della politica monetaria nel IV secolo," in *Atti dell'Accademia romanistica costantiniana: X Convegno internazionale in onore di Arnaldo Biscardi*, Naples: 481–502.

———. 1998, "Considerazioni su circolazione monetaria, prezzi e fiscalità nel IV secolo," in *Atti dell'Accademia romanistica costantiniana: XII convegno internazionale sotto l'alto patronato del Presidente della repubblica in onore di Manlio Sargenti*, Naples: 121–36.

———. 2005a, "The New State of Diocletian and Constantine: From the Tetrarchy to the Reunification of the Empire," in A. K. Bowman, P. Garnsey, and A. Cameron, eds., *The Cambridge Ancient History,* Vol. 12: *The Crisis of Empire, A.D. 193–337*, Cambridge: 170–83.

———. 2005b, "General Developments," in A. K. Bowman, P. Garnsey, and A. Cameron, eds., *The Cambridge Ancient History,* Vol. 12: *The Crisis of Empire, A.D. 193–337*, Cambridge: 131–36.

———. 2005c, "The Government and Administration of the Empire in the Central Decades of the Third Century," in A. K. Bowman, P. Garnsey, and A. Cameron, eds., *The Cambridge Ancient History,* Vol. 12: *The Crisis of Empire, A.D. 193–337*, Cambridge: 156–69.

———. 2006, "Did the Population of Imperial Rome Reproduce Itself?" in G. Storey, ed., *Urbanism in the Preindustrial World: Cross-Cultural Approaches*, Tuscaloosa: 52–68.

———. 2009, *Crescita e declino: studi di storia dell'economia romana*, Roma.

———. 2012, ed. *L'impatto della "peste Antonina"*, Bari.

———. 2016, "The Impact of Migration on the Demographic Profile of the City of Rome: A Reassessment," in L. de Ligt and L. E. Tacoma, eds., *Migration and Mobility in the Early Roman Empire*, Leiden: 23–32.

Lo Cascio, E. and Malanima, P. 2005, "Cycles and Stability. Italian Population Before the Demographic Transition (225 B.C.–A.D.1900)," *Rivista di storia economica* 21: 5–40.

Locey, K. and Lennon, J. 2016, "Scaling Laws Predict Global Microbial Diversity," *Proceedings of the National Academy of Sciences* 113: 5970–5.

Loveluck, C. 2013, *Northwest Europe in the Early Middle Ages, c. AD 600–1150: A Comparative Archaeology*, Cambridge.

Loy, D. E. et al. 2016, "Out of Africa: Origins and Evolution of the Human Malaria Parasites *Plasmodium falciparum* and *Plasmodium vivax*," *International Journal for Parasitology* 47: 87–97.

Luijendijk, A. 2008, *Greetings in the Lord: Early Christians and the Oxyrhynchus papyri*, Cambridge, MA.

Lusnia, S. 2014, *Creating Severan Rome*, Brussels.

Luterbacher, J. et al. 2013, "A Review of 2000 Years of Paleoclimatic Evidence in the Mediterranean," in P. Lionello, ed., *The Climate of the Mediterranean Region from the Past to the Future*, London: 87–185.

Luttwak, E. 2009, *The Grand Strategy of the Byzantine Empire*, Cambridge, MA.

———. 2016, *The Grand Strategy of the Roman Empire: From the First Century CE to the Third*, rev. ed. (orig. 1976), Baltimore.

Maas, M. 1992. *John Lydus and the Roman Past: Antiquarianism and Politics in the Age of Justinian*, London.

———. 2005, "Roman Questions, Byzantine Answers: Contours of the Age of Justinian," in M. Maas, ed., *The Cambridge Companion to the Age of Justinian*, Cambridge: 3–27.

———. 2015, "Reversals of Fortune: An Overview of the Age of Attila," in M. Maas, ed., *The Cambridge Companion to the Age of Attila*, Cambridge: 3–25.

Maat, G. J. R. 2005, "Two Millennia of Male Stature Development and Population Health and Wealth in the Low Countries," *International Journal of Osteoarchaeology* 15: 276–90.

MacCoull, L. 2004–5, "The Antaiopolite Estate of Count Ammonios: Managing for This World and the Next in a Time of Plague," *Analecta Papyrologica* 16–17: 109–116.

MacKinnon, M. 2006, "Supplying Exotic Animals for the Roman Amphitheatre Games: New Reconstructions Combining Archaeological, Ancient Textual, Historical and Ethnographic Data," *Mouseion* 6: 1–25.

———. 2007, "Osteological Research in Classical Archaeology," *American Journal of Archaeology* 111: 473–504.

Macklin, M. G. et al. 2015, "A New Model of River Dynamics, Hydroclimatic Change and Human Settlement in the Nile Valley Derived from Meta-Analysis of the Holocene Fluvial Archive," *Quaternary Science Reviews* 130: 109–23.

MacMullen, R. 1976, *Roman Government's Response to Crisis, A.D. 235–337*, New Haven.

———. 1982, "The Epigraphic Habit in the Roman Empire," *American Journal of Philology* 103: 233–46.

———. 1984, *Christianizing the Roman Empire (A.D. 100–400)*, New Haven.

Madden, A. M. 2014, *Corpus of Byzantine Church Mosaic Pavements from Israel and the Palestinian Territories*, Leuven.

Maddicott, J. 2007, "Plague in Seventh-Century England," in L. K. Little, ed., *Plague and the End of Antiquity: The Pandemic of 541–750*, New York: 171–214.

Maddison, A. 2001, *The World Economy: A Millennial Perspective*, Paris.

Maenchen-Helfen, O. 1973, *The World of the Huns: Studies in their History and Culture*, Berkeley.

Magdalino, P. 1993, "The History of the Future and Its Uses: Prophecy, Policy and Propaganda," R. Beaton and C. Roueché, eds., *The Making of Byzantine History: Studies Dedicated to Donald M. Nicol*, Aldershot: 3–34.

Magny, M. et al. 2012a, "Contrasting Patterns of Precipitation Seasonality During the Holocene in the South- and North-Central Mediterranean," *Journal of Quaternary Science* 27: 290–96.

———. 2012b, "Holocene Palaeohydrological Changes in the Northern Mediterranean Borderlands as Reflected by the Lake-Level Record of Lake Ledro, Northeastern Italy," *Quaternary Research* 77: 382–96.
Magny, M. et al. 2007, "Holocene Climate Changes in the Central Mediterranean as Recorded by Lake-Level Fluctuations at Lake Accesa (Tuscany, Italy)," *Quaternary Science Reviews* 26: 13–14.
Maguire, H. 2005, "Byzantine Domestic Art as Evidence for the Early Cult of the Virgin," in M. Vassilaki, ed., *Images of the Mother of God: Perceptions of the Theotokos in Byzantium*, Aldershot: 183–94.
Malanima, P. 2013, "Energy Consumption and Energy Crisis in the Roman World," in W. V. Harris, ed., *The Ancient Mediterranean Environment between Science and History*, Leiden: 13–36.
Mallegni, F., Fornaciari, G., and Tarabella, N. 1979, "Studio antropologico dei resti scheletrici della necropoli dei Monterozzi (Tarquinia)," *Atti della Societa toscana di scienze naturali, Memorie. Serie B* 86: 185–221.
Mallegni, F. et al. 1998, "Su alcuni gruppi umani del territorio piemontese dal IV al XVIII secolo: aspetti di paleobiologia," *Archeologia in Piemonte* 3: 233–61.
Malthus, T. R. 1826, *An Essay on the Principle of Population, or, A View of Its Past and Present Effects on Human Happiness*, 6th ed. [orig. 1798], London.
Manders, E. 2011, "Communicating Messages through Coins: A New Approach to the Emperor Decius," *Jaarboek Munt- en Penningkunde* 98: 1–22.
———. 2012, *Coining Images of Power: Patterns in the Representation of Roman Emperors on Imperial Coinage, A.D. 193–284*, Leiden.
Mangini, A., Spötl, C., and Verdes, P. 2005, "Reconstruction of Temperature in the Central Alps During the Past 2000 yr from a δ18O Stalagmite Record," *Earth and Planetary Science Letters* 235: 741–51.
Mango, C. 1986, "The Development of Constantinople as an Urban Centre," in *The 17th International Byzantine Congress: Major Papers*, New Rochelle: 117–36.
Manning, S. W. 2013, "The Roman World and Climate: Context, Relevance of Climate Change, and Some Issues," in W. V. Harris, ed., *The Ancient Mediterranean Environment between Science and History*, Leiden: 103–70.
Mansvelt Beck, B. J. 1986, "The Fall of Han," in D. Twitchett and M. Loewe, eds., *The Cambridge History of China,* Vol. 1: *The Ch'in and Han Empires, 221 BC–AD 220*, Cambridge: 317–76.
Manzi, G. 1999, "Discontinuity of Life Conditions at the Transition from the Roman Imperial Age to the Early Middle Ages: Example from Central Italy Evaluated by Pathological Dento-Alveolar Lesions," *American Journal of Human Biology* 11: 327–41.
Marciniak, S. et al. 2016, "*Plasmodium falciparum* Malaria in 1st–2nd Century CE Southern Italy," *Current Biology* 26: R1205–25.
Marcone, A. 2002, "La peste antonina: Testimonianze e interpretazioni," *Rivista storica italiana* 114: 803–19.
Marcozzi, V. and Cesare, B. M., 1969, "Le ossa lunghe della città di Spina," *Archivio per l'antropologia e l'etnologia* 99: 1–24.
Marino, A. 2012, "Una rilettura delle fonti storico-letterarie sulla peste di età antonina," in E. Lo Cascio, ed., *L'impatto della "peste antonina"*, Bari: 29–62.
Mariotti, A. et al. 2005, "Decadal Climate Variability in the Mediterranean Region: Roles of Large-Scale Forcings and Regional Processes," *Climate Dynamics* 38: 1129–45.

Mark, S. 2002, "Alexander the Great, Seafaring, and the Spread of Leprosy," *Journal of the History of Medicine and Allied Sciences* 57: 285–311.

Marks, R. 2012, *China: Its Environment and History*, Lanham.

Markus, R. A. 2007, *Gregory the Great and His World*, Cambridge.

Marquer, B. 2008, *Les romans de la Salpêtrière: réception d'une scénographie clinique: Jean-Martin Charcot dans l'imaginaire fin-de-siècle*, Geneva.

Marriner, N. et al. 2012, "ITCZ and ENSO-like Pacing of Nile Delta Hydro-Geomorphology during the Holocene," *Quaternary Science Reviews* 45: 73–84.

———. 2013, "Tracking Nile Delta Vulnerability to Holocene Change," *PLOS ONE* 8: e69195.

Marshall, J. et al. 2001, "North Atlantic Climate Variability: Phenomena, Impacts and Mechanisms," *International Journal of Climatology* 21: 1863–98.

Martín-Chivelet, J. et al. 2011, "Land Surface Temperature Changes in Northern Iberia Since 4000 yr BP, Based on $\delta^{13}C$ of Speleothems," *Global and Planetary Change* 77: 1–12.

Martín-Puertas, C. et al. 2009, "The Iberian–Roman Humid Period (2600–1600 cal yr BP) in the Zoñar Lake Varve Record (Andalucía, Southern Spain)," *Quaternary Research* 71: 108–20.

Martiny, J. et al. 2006, "Microbial Biogeography: Putting Microorganisms on the Map," *Nature Reviews: Microbiology* 4: 102–12.

Martuzzi Veronesi, F. and Malacarne, G. 1968, "Note antropologiche su reperti romani e medioevali del territorio di Classe (Ravenna)," *Archivio per l'antropologia e l'etnologia* 98: 147–64.

Marty, A. M. et al. 2006, "Viral Hemorrhagic Fevers," *Clinics in Laboratory Medicine* 2: 345–86.

Mattern, S. P. 1999, *Rome and the Enemy: Imperial Strategy in the Principate*, Berkeley.

———. 2013, *The Prince of Medicine: Galen in the Roman Empire*, Oxford.

Matthews, J. 1975, *Western Aristocracies and Imperial Court, A.D. 364–425*, Oxford.

———. 2010, *Roman Perspectives: Studies in the Social, Political and Cultural History of the First to Fifth Centuries*, Swansea.

Mattingly, D. 1994, *Tripolitania*, Ann Arbor.

———. 2006, *An Imperial Possession: Britain in the Roman Empire, 54 BC–AD 409*, London.

———. ed. 2003–13, *The Archaeology of the Fazzān*, London.

Mayerson, P. 1993, "A Confusion of Indias: Asian India and African India in the Byzantine Sources," *Journal of the American Oriental Society* 113: 169–74.

Mayewski, P. A. et al. 2004, "Holocene Climate Variability," *Quaternary Research* 62: 243–55.

Mazza, M. 2001, *L'Archivio degli Apioni: terra, lavoro e proprietà senatoria nell'Egitto tardoantico*, Bari.

McAnany, P. A. and Yoffee, N. eds. 2010, *Questioning Collapse: Human Resilience, Ecological Vulnerability, and the Aftermath of Empire*, Cambridge.

McCormick, M. 1986, *Eternal Victory: Triumphal Rulership in Late Antiquity, Byzantium, and the Early Medieval West*, Cambridge.

———. 1998, "Bateaux de vie, bateaux de mort. Maladie, commerce, transports annonaires et le passage économique du bas-empire au moyen âge," *Settimane di studio—Centro Italiano di studi alto medioevo* 45: 35–118.

———. 2001, *Origins of the European Economy: Communications and Commerce, A.D. 300–900*, Cambridge.

———. 2003, "Rats, Communications, and Plague: Toward an Ecological History," *Journal of Interdisciplinary History* 34: 1–25.

———. 2007, "Toward a Molecular History of the Justinianic Pandemic," in L. K. Little, ed., *Plague and the End of Antiquity: The Pandemic of 541–750*, New York: 290–312.

———. 2011, "History's Changing Climate: Climate Science, Genomics, and the Emerging Consilient Approach to Interdisciplinary History," *Journal of Interdisciplinary History* 42: 251–73.

———. 2012, "Movements and Markets in the First Millennium: Information, Containers, and Shipwrecks," in C. Morrisson, ed., *Trade and Markets in Byzantium*, Washington, D.C.: 51–98.

———. 2013a, "Coins and the Economic History of Post-Roman Gaul: Testing the Standard Model in the Moselle, ca. 400–750," in J. Jarnut and J. Strothman, eds., *Die Merowingischen Monetarmünzen als Quelle zum Verständnis des 7. Jahrhunderts in Gallien*, Paderborn: 337–76.

———. 2013b, "What Climate Science, Ausonius, Nile Floods, Rye, and Thatch Tell Us about the Environmental History of the Roman Empire," in W. V. Harris, ed., *The Ancient Mediterranean Environment between Science and History*, Leiden: 61–88.

———. 2015, "Tracking Mass Death during the Fall of Rome's Empire (I)," *Journal of Roman Archaeology* 28: 325–57.

———. 2016, "Tracking Mass Death during the Fall of Rome's Empire (II): A First Inventory of Mass Graves," *Journal of Roman Archaeology* 29: 1004-46

McCormick, M. et al. 2012, "Climate Change During and After the Roman Empire: Reconstructing the Past from Scientific and Historical Evidence," *Journal of Interdisciplinary History* 43: 169–220.

McCormick, M., Harper, K., More, A.F., and Gibson, K. 2012, "Historical Evidence on Roman and Post-Roman Climate, 100 BC to 800 AD," DARMC Scholarly Data Series 2012-1. darmc.harvard.edu

McDermott, F. 2004, "Palaeo-Climate Reconstruction from Stable Isotope Variations in Speleothems: A Review," *Quaternary Science Reviews* 23: 901–18.

McDermott, F. et al. 2011, "A First Evaluation of the Spatial Gradients in δ18O Recorded by European Holocene Speleothems," *Global and Planetary Change* 79: 275–87.

McEvedy, C. and Jones, R. 1978, *Atlas of World Population History*, Harmondsworth.

McGinn, T. A. J. 1999, "The Social Policy of Emperor Constantine in Codex Theodosianus 4, 6, 3," *Legal History Review* 67: 57–73.

McKeown, T. 1988, *The Origins of Human Disease*, Oxford.

McLaughlin, R. 2010, *Rome and the Distant East: Trade Routes to the Ancient Lands of Arabia, India and China*, London.

McMichael, A. J. 2010, "Paleoclimate and Bubonic Plague: A Forewarning of Future Risk?" *BioMed Central: Biology* 8: 108.

———. 2015, "Extreme Weather Events and Infectious Disease Outbreaks," *Virulence* 6: 543–47.

McNally, A. et al., 2016, "'Add, Stir and Reduce': Yersinia spp. as Model Bacteria for Pathogen Evolution," *Nature Reviews Microbiology* 14: 177–90.

McNeill, J. 2010, *Mosquito Empires: Ecology and War in the Greater Caribbean, 1620–1914*, New York.

———. 2015, "Energy, Population, and Environmental Change since 1750: Entering the Anthropocene," in J. R. McNeill and K. Pomeranz, eds., *The Cambridge World History*, Vol. 7: *Production, Destruction, and Connection, 1750–Present, Part 1: Structures, Spaces, and Boundary Making*, Cambridge: 51–82.

McNeill, W. H. 1976, *Plagues and Peoples*, Garden City.

Mee, C. and Forbes, H. 1997, *A Rough and Rocky Place: The Landscape and Settlement History of the Methana Peninsula, Greece: Results of the Methana Survey Project*, Liverpool.

Meier, M. 2003, *Das andere Zeitalter Justinians: Kontingenzerfahrung und Kontingenzbewältigung im 6. Jahrhundert n. Chr.*, Göttingen.

———. 2005, "'Hinzu kam auch noch die Pest . . .' Die sogenannte Justinianische Pest und ihre Folgen," in M. Meier, ed., *Pest—Die Geschichte eines Menschheitstraumas*, Stuttgart, 86–107, 396–400.

———. 2016, "The 'Justinianic Plague': The Economic Consequences of the Pandemic in the Eastern Roman Empire and Its Cultural and Religious Effects," *Early Medieval Europe* 24: 267–92.

Meiggs, R. 1982, *Trees and Timber in the Ancient Mediterranean World*, Oxford.

Memmer, M. 2000, "Die Ehescheidung im 4. und 5. Jahrhundert n. Chr," in M. Schermaier, et al., eds., *Iurisprudentia universalis: Festschrift für Theo Mayer—Maly zum 70. Geburtstag*, Cologne: 489–510.

Mezzabotta, M. R. 2000, "Aspects of Multiculturalism in the *Mulomedicina* of Vegetius," *Akroterion* 45: 52–64.

Miarinjara, A. et al. 2016, "*Xenopsylla brasiliensis* Fleas in Plague Focus Areas, Madagascar," *Emerging Infectious Diseases* 22: 2207–8.

Migowski, C. et al. 2006, "Holocene Climate Variability and Cultural Evolution in the Near East from the Dead Sea Sedimentary Record," *Quaternary Research* 66: 421–31.

Mikhail, A. ed. 2012, *Water on Sand: Environmental Histories of the Middle East and North Africa*, New York.

Millar, F. 2004, *Rome, the Greek World, and the East*, Vol. 2: *Government, Society, and Culture in the Roman Empire*, H. Cotton and G. M. Rogers, eds., Chapel Hill.

Miller, S. S. 1992, "R. Hanina bar Hama at Sepphoris," in L. Levine, ed., *The Galilee in Late Antiquity*, New York: 175–200.

Mitchell, J. 1992, *Up in the Old Hotel and Other Stories*, New York.

Mitchell, P. D. 2017, "Human Parasites in the Roman World: Health Consequences of Conquering an Empire," *Parasitology* 144: 48–58.

Mitchell, S. 2015, *A History of the Later Roman Empire, AD 284–641*, 2nd ed., Malden.

Modrzejewski, J. 1970, "La règle de droit dans l'Égypte romaine," in *Proceedings of the xiith International Congress of Papyrology*, Toronto: 317–77.

Monot, M. et al. 2005, "On the Origin of Leprosy," *Science* 308: 1040–42.

Morens, D. M. and Littman, R. J. 1992, "Epidemiology of the Plague of Athens," *Transactions of the American Philological Association* 122: 271–304.

Morley, N. 1996, *Metropolis and Hinterland: The City of Rome and the Italian Economy, 200 B.C.–A.D. 200*, Cambridge.

———. 2007, *Trade in Classical Antiquity*, Cambridge.

———. 2011, "Population Size: Evidence and Estimates," in P. Erdkamp, ed., *The Cambridge Companion to Ancient Rome*, Cambridge: 29–44.

Morony, M. G. 2007, "'For Whom Does the Writer Write?': The First Bubonic Plague Pandemic according to Syriac Sources," in L. K. Little, ed., *Plague and the End of Antiquity: The Pandemic of 541–750*, New York: 58–86.

Morris, I. 2010, *Why the West Rules—For Now: The Patterns of History, and What They Reveal About the Future*, London.

———. 2013, *The Measure of Civilization: How Social Development Decides the Fate of Nations*, Princeton.

Morrison, C. and Sodini, J.-P. 2002, "The Sixth-Century Economy," in A. Laiou, ed., *The Economic History of Byzantium: From the Seventh through the Fifteenth Century*, Washington, D.C.: 171–220.

Morrisson, C. et al. 1985, *L'or Monnayé, Purification et altération de Rome à Byzance*, Paris.

Mossner, E. C. 1980, *The Life of David Hume*, Oxford.

Mouritsen, H. 2013, "The Roman Empire I: The Republic," in P. F. Bang and W. Scheidel, eds., *Oxford Handbook of the State in the Ancient Near East and Mediterranean*, Oxford: 383–411.

Mouterde, R. and Poidebard, A. 1945, *Le limes de Chalcis: organisation de la steppe en haute Syrie romaine*, Paris.

Moy, C. M. et al. 2002, "Variability of El Niño/Southern Oscillation Activity at Millennial Timescales during the Holocene Epoch," *Nature* 420: 162–65.

Muldner, G. and Richards, M. P. 2007, "Stable Isotope Evidence for 1500 Years of Human Diet at the City of York, UK," *American Journal of Physical Anthropology* 133: 682–97.

Mulligan, B. 2007, "The Poet from Egypt? Reconsidering Claudian's Eastern Origin," *Philologus* 151: 285–310.

Müller, R. et al. 2014, "Genotyping of Ancient *Mycobacterium tuberculosis* Strains Reveals Historic Genetic Diversity," *Proceedings of the Royal Society, B* 281: 20133236.

Musurillo, H. 1972, *The Acts of the Christian Martyrs*, Oxford.

Nappo, D. 2015, "Roman Policy on the Red Sea in the Second Century CE," in F. De Romanis and M. Maiuro, eds., *Across the Ocean: Nine Essays on Indo-Roman Trade*, Leiden: 55–72.

Needham, J. et al. 2000, *Science and Civilisation in China,* Vol. 6: *Biology and Biological Technology,* Part VI: *Medicine*, Cambridge.

Nees, L. 2016, *Perspectives on Early Islamic Art in Jerusalem*, Leiden.

Neri, V. 1998, *I marginali nell'Occidente tardoantico: poveri, "infames" e criminali nella nascente società cristiana*, Bari.

Neumann, F. et al. 2010, "Vegetation History and Climate Fluctuations on a Transect along the Dead Sea West Shore and Their Impact on Past Societies over the Last 3500 Years," *Journal of Arid Environments* 74: 756–64.

Newfield, T. 2015, "Human-Bovine Plagues in the Early Middle Ages," *Journal of Interdisciplinary History* 46: 1–38.

———. 2016, "The Global Cooling Event of the Sixth Century: Mystery No Longer?" Historical Climatology Blog: www.historicalclimatology.com/blog/something-cooled-the-world-in-the-sixth-century-what-was-it. Accessed August 8, 2016.

Nicholson, S. and Kim, J. 1997, "The Relationship of the El Nino–Southern Oscillation to African Rainfall," *International Journal of Climatology* 17: 117–35.

Nordh, A. 1949, *Libellus de Regionibus Urbis Romae*, Lund.

Noreña, C. 2011, *Imperial Ideals in the Roman West*, Cambridge.

Nutton, V. 1973, "The Chronology of Galen's Early Career," *Classical Quarterly* 23: 158–71.

Ober, J. 2015, *The Rise and Fall of Classical Greece*, Princeton.

Oberhänsli, H. et al. 2007, "Climate Variability during the Past 2,000 years and Past Economic and Irrigation Activities in the Aral Sea Basin," *Irrigation and Drainage Systems* 21: 167–83.

———. 2011, "Variability in Precipitation, Temperature and River Runoff in W. Central Asia during the Past ~2000 yrs," *Global and Planetary Change* 76: 95–104.

O'Donnell, J. 2008, *The Ruin of the Roman Empire*, New York.

Oesterheld, C. 2008, *Göttliche Botschaften für zweifelnde Menschen: Pragmatik und Orientierungsleistung der Apollon-Orakel von Klaros und Didyma in hellenistisch-römischer Zeit*, Göttingen.

Oldstone, M. B. A. 2010, *Viruses, Plagues, and History: Past, Present, and Future*, Oxford.

Olsen, J. et al. 2012, "Variability of the North Atlantic Oscillation over the Past 5,200 Years," *Nature Geoscience* 5: 808–12.

Orland, I. J. et al. 2009, "Climate Deterioration in the Eastern Mediterranean as Revealed by Ion Microprobe Analysis of a Speleothem That Grew from 2.2 to 0.9 ka in Soreq Cave, Israel," *Quaternary Research* 71: 27–35.

O'Sullivan, L. et al. 2008, "Deforestation, Mosquitoes, and Ancient Rome: Lessons for Today," *BioScience* 58: 756–60.

Ovadiah, A. 1970, *Corpus of Byzantine Churches in the Holy Land*, Bonn.

Paine R. R. and Storey, G. R. 2012, "The Alps as a Barrier to Epidemic Disease during the Republican Period: Implications for the Dynamic of Disease in Rome," in E. Lo Cascio, ed., *L'impatto della "peste antonina"*, Bari: 179–91.

Pannekeet, C. G. J. 2008, *Vier eeuwen keizers/munten*, Slootdorp.

Panzac, D. 1985, *La peste dans l'Empire Ottoman, 1700–1850*, Leuven.

Pardini, E. et al. 1982, "Gli inumati di Pontecagnano (Salerno) (V–IV secolo a.C.)," *Archivio per l'antropologia e la etnologia* 112: 281–329.

Pardini, E. and Manucci, P. 1981, "Gli Etruschi di Selvaccia (Siena): studio antropologico," *Studi etruschi* 49: 203–15.

Parke, H. W. 1985, *The Oracles of Apollo in Asia Minor*, London.

Parker, G. 2008, *The Making of Roman India*, Cambridge.

Parker, G. 2013, *Global Crisis: War, Climate Change and Catastrophe in the Seventeenth Century*, New Haven.

Parkin, T. G. 1992, *Demography and Roman Society*, Baltimore.

Parkin, T. G. and Pomeroy, A. 2007, *Roman Social History: A Sourcebook*, London.

Patlagean, E. 1977, *Pauvreté économique et pauvreté sociale à Byzance, 4e–7e siècles*, Paris.

Peachin, M. 1991, "Philip's Progress: From Mesopotamia to Rome in A.D. 244," *Historia: Zeitschrift für Alte Geschichte* 40: 331–42.

Pearce-Duvet, J. M. C. 2006, "The Origin of Human Pathogens: Evaluating the Role of Agriculture and Domestic Animals in the Evolution of Human Disease," *Biological Review* 81: 369–82.

Pechous, R. D. 2016, "Pneumonic Plague: The Darker Side of Yersinia pestis," *Trends in Microbiology* 24: 190–7.

Peck, J. J. 2009, *The Biological Impact of Culture Contact: A Bioarchaeological Study of Roman Colonialism in Britain*, diss., Ohio State University.

Pentcheva, B. V. 2006, *Icons and Power: The Mother of God in Byzantium*, University Park.

Percoco, M. 2013, "The Fight against Disease: Malaria and Economic Development in Italian Regions," *Economic Geography* 89: 105–25.

Perdrizet, P. 1903, "Une inscription d'Antioche qui reproduit un oracle d'Alexandre d'Abonotichos," *Comptes rendus des séances de l'Académie des Inscriptions et Belles-Lettres* 47: 62–66.

Pérez-Sanz, A. et al. 2013, "Holocene Climate Variability, Vegetation Dynamics and Fire Regime in the Central Pyrenees: The Basa de la Mora Sequence (NE Spain)," *Quaternary Science Reviews* 73: 149–69.

Pergola, P. 1998, *Le catacombe romane: storia e topografia*, Rome.

Perry, R. T. and Halsey, N. A., 2004, "The Clinical Significance of Measles: A Review," *Journal of Infectious Diseases* 189: 4–16.

Petit, P. 1957, "Les sénateurs de Constantinople dans l'oeuvre de Libanius," *L'Antiquité classique* 26: 347–82.

Petrucci, A. 1998, "Persistenza di negozi bancari nelle fonti giuridiche tra la fine del III e i primi decenni del V secolo D.C.," in *Atti dell'Accademia romanistica costantiniana: XII convegno internazionale sotto l'alto patronato del Presidente della repubblica in onore di Manlio Sargenti*, Naples: 223–50.

Pettegrew, D. K. 2007, "The Busy Countryside of Late Roman Corinth: Interpreting Ceramic Data Produced by Regional Archaeological Surveys," *Hesperia* 76: 743–84.

Pflaum, H.-G. 1976, "Zur Reform des Kaisers Gallienus," *Historia: Zeitschrift für Alte Geschichte* 25: 109–17.

Phillips, C., Villeneuve, F., and Facey, W. 2004, "A Latin Inscription from South Arabia," *Proceedings of the Seminar for Arabian Studies* 34: 239–50.

Piccirillo, M. 1981, *Chiese e mosaici della Giordania settentrionale*, Jerusalem.

Pieri, D. 2005, *Le commerce du vin oriental à l'époque byzantine, Ve–VIIe siècles: le témoignage des amphores en Gaule*, Beirut.

Pighi, G. 1967, *De ludis saecularibus populi romani quiritium*, Chicago.

Pinault, J. 1992, *Hippocratic Lives and Legends*, Leiden.

Piso, I. 2014, "Zur Reform des Gallienus anläßlich zweier neuer Inschriften aus den Lagerthermen von Potaissa," *Tyche: Beiträge zur Alten Geschichte, Papyrologie und Epigraphik* 29: 125–46.

Pitts, M. and Versluys, M. J. eds. 2015, *Globalisation and the Roman World: World History, Connectivity and Material Culture*, Oxford.

Podskalsky, G. 1972, *Byzantinische Reichseschatologie: die Periodisierung der Weltgeschichte in den vier Grossreichen (Daniel 2 und 7) und dem tausendjährigen Friedensreiche (Apok. 20) Eine motivgeschichtliche Untersuchung*, Munich.

Pomeranz, K. 2000, *The Great Divergence: China, Europe, and the Making of the Modern World Economy*, Princeton.

Popper, W. 1951, *The Cairo Nilometer: Studies in Ibn Taghrî Birdî's Chronicles of Egypt: I*, Berkeley.

Potter, D. S. 1990, *Prophecy and History in the Crisis of the Roman Empire: A Historical Commentary on the Thirteenth Sibylline Oracle*, Oxford.

———. 2004, *The Roman Empire at Bay: AD 180–395*, London.

———. 2013, *Constantine the Emperor*, Oxford.
———. 2015, *Theodora: Actress, Empress, Saint*, New York.
Power, T. 2012, *The Red Sea from Byzantium to the Caliphate: AD 500–1000*, Cairo.
Prowse, T. 2016, "Isotopes and Mobility in the Ancient Roman World," in L. de Ligt and L. E. Tacoma, eds., *Migration and Mobility in the Early Roman Empire*, Leiden: 205–33.
Prowse, T. et al. 2004, "Isotopic Paleodiet Studies of Skeletons from the Imperial Roman-Age Cemetery of Isola Sacra, Rome, Italy," *Journal of Archaeological Science* 31: 259–72.
———. 2007, "Isotopic Evidence for Age-Related Immigration to Imperial Rome," *American Journal of Physical Anthropology* 132: 510–19.
———. 2008, "Isotopic and Dental Evidence for Infant and Young Child Feeding Practices in an Imperial Roman Skeletal Sample," *American Journal of Physical Anthropology* 137: 294–308.
Purcell, N. 1985, "Wine and Wealth in Ancient Italy," *Journal of Roman Studies* 75: 1–19.
———. 2000, "Rome and Italy," in P. Garnsey, D. Rathbone, and A. K. Bowman, eds., *The Cambridge Ancient History,* Vol. 11: *The High Empire, A.D. 70–192*, Cambridge: 405–43.
———. 2016, "Unnecessary Dependences: Illustrating Circulation in Pre-Modern Large-Scale History," in J. Belich et al., eds., *The Prospect of Global History*, Oxford: 65–79.
Quammen, D. 2012. *Spillover: Animal Infections and the Next Human Pandemic*, New York.
———. 2014, *Ebola: The Natural and Human History of a Deadly Virus*, New York.
Rambeau, C. and Black, S. 2011, "Palaeoenvironments of the Southern Levant 5,000 BP to Present: Linking the Geological and Archaeological Records," in S. Mithen and E. Black, eds., *Water, Life and Civilisation: Climate, Environment and Society in the Jordan Valley*, Cambridge: 94–104.
Raoult, D. et al. 2000, "Molecular Identification by 'Suicide PCR' of Yersinia pestis as the Agent of Medieval Black Death," *Proceedings of the National Academy of Sciences* 97: 12800–3.
Raschke, M. G. 1978, "New Studies in Roman Commerce with the East," *Aufstieg und Niedergang der römischen Welt* 2.9.2: 604–1361.
Rasmussen, S. et al. 2015, "Early Divergent Strains of Yersinia pestis in Eurasia 5,000 Years Ago," *Cell* 163: 571–582.
Rathbone, D. W. 1990, "Villages, Land and Population in Graeco-Roman Egypt," *Proceedings of the Cambridge Philological Society* 36: 103–42.
———. 1991, *Economic Rationalism and Rural Society in Third-Century A.D. Egypt: The Heroninos Archive and the Appianus Estate*, Cambridge.
———. 1996, "Monetisation, Not Price-inflation, in Third-century AD Egypt?," in C. E. King and D. G. Wigg, eds., *Coin Finds and Coin Use in the Roman World: The Thirteenth Oxford Symposium on Coinage and Monetary History, 25.–27.3.1993*, Berlin: 321–39.
———. 1997, "Prices and Price Formation in Roman Egypt," *Economie antique. Prix et formation des prix dans les economies antiques*, Saint-Bertrand-de-Comminges: 183–244.

———. 2000, "The 'Muziris' Papyrus (SB XVIII 13167): Financing Roman Trade with India," in M. El-Abbadi et al., eds., *Alexandrian Studies II in Honour of Mostafa El Abbadi*, Alexandria: 39–50.

———. 2007, "Roman Egypt," in W. Scheidel, I. Morris, and R. Saller, eds., *The Cambridge Economic History of the Greco-Roman World*, Cambridge: 698–719.

Rathbone, D. W. and Temin, P. 2008, "Financial Intermediation in 1st-century AD Rome and 18th-century England," in K. Verboven, K. Vandorpe, and V. Chankowski, eds., *Pistoi dia tèn technèn: Bankers, Loans, and Archives in the Ancient World: Studies in Honour of Raymond Bogaert*, Leuven: 183–244.

Rathbone, D. W. and von Reden, S. 2015, "Mediterranean Grain Prices in Classical Antiquity," in R. J. Van der Spek, J. Luiten, and B. van Zanden, eds., *A History of Market Performance: From Ancient Babylonia to the Modern World*, London: 149–235.

Ratovonjato, J. et al. 2014, "*Yersinia pestis* in *Pulex irritans* Fleas during Plague Outbreak, Madagascar," *Emerging Infectious Diseases* 20: 1414–15.

Rea, J. R. 1997, "Letter of a Recruit: P. Lond. III 982 Revised," *Zeitschrift für Papyrologie und Epigraphik* 115: 189–93.

Reale, O. and Dirmeyer, P. 2000, "Modeling the Effects of Vegetation on Mediterranean Climate During the Roman Classical Period, Part I: Climate History and Model Sensitivity," *Global and Planetary Change* 25: 163–84.

Reale, O. and Shukla, J. 2000, "Modeling the Effects of Vegetation on Mediterranean Climate During the Roman Classical Period, Part II: Model Simulation," *Global and Planetary Change* 25: 185–214.

Rebillard, É. 2009, *The Care of the Dead in Late Antiquity*, Ithaca.

Redfern, R. C. et al. 2015, "Urban–Rural Differences in Roman Dorset, England: A Bioarchaeological Perspective on Roman Settlements," *American Journal of Physical Anthropology* 157: 107–20.

Redfern, R. C. and DeWitte, S. N. 2011a, "A New Approach to the Study of Romanization in Britain: A Regional Perspective of Cultural Change in Late Iron Age and Roman Dorset Using the Siler and Gompertz–Makeham Models of Mortality," *American Journal of Physical Anthropology* 144: 269–85.

Redfern, R. C. and DeWitte, S. N. 2011b, "Status and Health in Roman Dorset: The Effect of Status on Risk of Mortality in Post–Conquest Populations," *American Journal of Physical Anthropology* 146: 197–208.

Reeves, J. C. 2005, *Trajectories in Near Eastern Apocalyptic: A Postrabbinic Jewish Apocalypse Reader*, Atlanta.

Reff, D. 2005, *Plagues, Priests, and Demons: Sacred Narratives and the Rise of Christianity in the Old World and New*, Cambridge.

Reilly, K. 2010, "The Black Rat," in T. O'Connor and N. Sykes, eds., *Extinctions and Invasions: A Social History of British Fauna*, Oxford: 134–45.

Reinink, G. J. 2002, "Heraclius, the New Alexander: Apocalyptic Prophecies during the Reign of Heraclius," in G. J. Reinink and H. Stolte, eds., *The Reign of Heraclius (610–641): Crisis and Confrontation*, Groningen: 81–94.

Reinink, G. J. and Stolte, H. eds. 2002, *The Reign of Heraclius (610–641): Crisis and Confrontation*, Groningen.

Rey, E. and Sormani, G. 1878, "Statistica delle cause di morte," in *Monografia della città di Roma e della campagna romana*, Rome: 121–48.

Reynolds, D. W. 1996, *Forma Urbis Romae: The Severan Marble Plan and the Urban Form of Ancient Rome*, diss., University of Michigan.

Rich, J. ed. 1992, *The City in Late Antiquity*, London.

Rickman, G. 1980, *The Corn Supply of Ancient Rome*, Oxford.

Riley, J. C. 2010, "Smallpox and American Indians Revisited," *Journal of the History of Medicine and Allied Sciences* 65: 445–77.

Ritterling, E. 1904, "Epigraphische Beiträge zur römischen Geschichte," *Rheinisches Museum für Philologie* 59: 186–99.

Ritti, T., Şimşek, C., and Yıldız, H. 2000, "Dediche e KATAGRAPHAI dal Santuario Frigio di Apollo Lairbenos," *Epigraphica Anatolica* 32: 1–87.

Rives, J. B. 1999, "The Decree of Decius and the Religion of Empire," *Journal of Roman Studies* 89: 135–54.

Roberts, C. A. 2002, "The Antiquity of Leprosy in Britain: The Skeletal Evidence," in C. A. Roberts, M. Lewis, and K. Manchester, eds., *Past and Present of Leprosy: Archaeological, Historical, Paleopathological, and Clinical Approaches: Proceedings of the International Congress on the Evolution and Palaeoepidemiology of the Infectious Diseases 3 (ICEPID), University of Bradford, 26th–31st July 1999*, Oxford: 213–22.

———. 2015, "Old World Tuberculosis: Evidence from Human Remains with a Review of Current Research and Future Prospects," *Tuberculosis* 95: 117–21.

Roberts, C. A. and Buikstra, J. 2003, *The Bioarchaeology of Tuberculosis: A Global View on a Reemerging Disease*, Gainesville.

Roberts, C. A. and Cox, M. 2003, *Health & Disease in Britain: From Prehistory to the Present Day*, Stroud.

Roberts, C. A., Lewis, M., and Manchester, K. eds. 2002, *Past and Present of Leprosy: Archaeological, Historical, Paleopathological, and Clinical Approaches: Proceedings of the International Congress on the Evolution and Palaeoepidemiology of the Infectious Diseases 3 (ICEPID), University of Bradford, 26th–31st July 1999*, Oxford.

Roberts, M. 1992, "Barbarians in Gaul: The Response of the Poets," in J. Drinkwater and H. Elton, eds., *Fifth-Century Gaul: A Crisis of Identity?*, Cambridge: 97–106.

Roberts, N. et al. 2012, "Paleolimnological Evidence for an East-West Climate See-Saw in the Mediterranean since AD 900," *Global and Planetary Change* 84–5: 23–34.

Robin, C. 1992, "Guerre et épidémie dans les royaumes d'Arabie du Sud, d'après une inscription datée (IIe s. de l'ère chrétienne)," *Comptes rendus des séances de l'Académie des Inscriptions et Belles-Lettres* 136: 215–34.

———. 2012, "Arabia and Ethiopia," in S. F. Johnson, ed., *The Oxford Handbook of Late Antiquity*, Oxford: 247–332.

Rogers, G. 1991, *The Sacred Identity of Ephesus: Foundation Myths of a Roman City*, London.

Rohland, J. 1977, *Der Erzengel Michael, Arzt und Feldherr: zwei Aspekte des vor- und frühbyzantinischen Michaelskultes*, Leiden.

Roselaar, S. T. 2016, "State-Organized Mobility in the Roman Empire: Legionaries and Auxiliaries," in L. de Ligt and L. E. Tacoma, eds., *Migration and Mobility in the Early Roman Empire*, Leiden: 138–57.

Rosen, W. 2007, *Justinian's Flea: Plague, Empire, and the Birth of Europe*, New York.

Rossignol, B. 2012, "Le climat, les famines et la guerre: éléments du contexte de la peste antonine," in E. Lo Cascio, ed., *L'impatto della "peste antonina"*, Bari: 87–122.

Rossignol, B. and Durost, S. 2007, "Volcanisme global et variations climatiques de courte durée dans l'histoire romaine (Ier s. av. J.-C.–IVème s. ap. J.-C.): leçons d'une archive glaciaire (GISP2)," *Jahrbuch des römisch-germanischen Zentral-museums Mainz* 54: 395–438.

Røstvig, M. S. 1962, *The Happy Man: Studies in the Metamorphoses of a Classical Ideal*, Oslo.

Roucaute et al. 2014, "Analysis of the Causes of Spawning of Large-Scale, Severe Malarial Epidemics and Their Rapid Total Extinction in Western Provence, Historically a Highly Endemic Region of France (1745–1850)," *Malaria Journal* 13: 1–42.

Roueché, C. and Reynolds, J. 1989, *Aphrodisias in Late Antiquity: The Late Roman and Byzantine Inscriptions Including Texts from the Excavations at Aphrodisias Conducted by Kenan T. Erim*, London.

Rowlandson, J. 1996, *Landowners and Tenants in Roman Egypt: The Social Relations of Agriculture in the Oxyrhynchite Nome*, Oxford.

Rubin, R. 1989, "The Debate over Climate Changes in the Negev, Fourth–Seventh Centuries C.E.," *Palestine Exploration Quarterly* 121: 71–78.

Rubini, M. et al. 2014, "Paleopathological and Molecular Study on Two Cases of Ancient Childhood Leprosy from the Roman and Byzantine Empires," *International Journal of Osteoarchaeology* 24: 570–82.

Ruddiman, W. F. 2001, *Earth's Climate: Past and Future*, New York.

———. 2005, *Plows, Plagues, and Petroleum: How Humans Took Control of Climate*, Princeton.

Russell, E. 2011, *Evolutionary History: Uniting History and Biology to Understand Life on Earth*, Cambridge.

Rutgers, L. V. et al 2009, "Stable Isotope Data from the Early Christian Catacombs of Ancient Rome: New Insights into the Dietary Habits of Rome's Early Christians," *Journal of Archaeological Science* 36: 1127–34.

Sadao, N. 1986, "The Economic and Social History of Former Han," in D. Twitchett and M. Loewe, eds., *The Cambridge History of China,* Vol. 1: *The Ch'in and Han Empires, 221 BC–AD 220*, Cambridge: 545–607.

Sadori, L. et al. 2016, "Climate, Environment and Society in Southern Italy during the Last 2000 Years: A Review of the Environmental, Historical and Archaeological Evidence," *Quaternary Science Reviews* 136: 173–88.

Sage, M. M. 1975, *Cyprian*, Cambridge, MA.

Said, R. 1993, *The River Nile: Geology, Hydrology, and Utilization*, Oxford.

Sallares, R. 1991, *The Ecology of the Ancient Greek World*, New York.

———. 2002, *Malaria and Rome: A History of Malaria in Ancient Italy*, Oxford.

———. 2004, "The Spread of Malaria to Southern Europe in Antiquity: New Approaches to Old Problems," *Medical History* 48: 311–28.

———. 2007a, "Ecology," in W. Scheidel, I. Morris, and R. Saller, eds., *The Cambridge Economic History of the Greco-Roman World*, Cambridge: 15–37.

———. 2007b, "Ecology, Evolution, and Epidemiology of Plague," in L. K. Little, ed., *Plague and the End of Antiquity: The Pandemic of 541–750*, New York: 231–89.

Sallares, R. et al. 2003, "Identification of a Malaria Epidemic in Antiquity Using Ancient DNA," in K. R. Brown, ed., *Archaeological Sciences 1999. Proceedings of the Archaeological Sciences Conference, University of Bristol, 1999*, Oxford, 120–5.
Saller, R. P. 1982, *Personal Patronage under the Early Empire*, Cambridge.
———. 1994, *Patriarchy, Property and Death in the Roman Family*, Cambridge.
———. 2000, "Status and Patronage," in P. Garnsey, D. Rathbone, and A. K. Bowman, eds., *The Cambridge Ancient History*, Vol. 11: *The High Empire, A.D. 70–192*, Cambridge: 817–54.
Santelia, S. ed. 2009, *Ad coniugem suam: in appendice: Liber epigrammatum*, Naples.
Sarris, P. 2002, "The Justinianic Plague: Origins and Effects," *Continuity and Change* 17: 169–82.
———. 2006, *Economy and Society in the Age of Justinian*, Cambridge.
Scattarella, V. and De Lucia, A. 1982, "Esame antropologico dei resti scheletrici della necropoli classica di Purgatorio presso Rutigliano (Bari)," *Taras* 2: 137–47.
Scheffer, M. 2009, *Critical Transitions in Nature and Society*, Princeton.
Scheidel, W. 1996, *Measuring Sex, Age and Death in the Roman Empire*, Ann Arbor.
———. 1999, "Emperors, Aristocrats, and the Grim Reaper: Towards a Demographic Profile of the Roman Élite," *Classical Quarterly* 49: 254–81.
———. 2001a, *Death on the Nile: Disease and the Demography of Roman Egypt*, Leiden.
———. ed. 2001b, *Debating Roman Demography*, Leiden.
———. 2001c, "Roman Age Structure: Evidence and Models," *Journal of Roman Studies* 91: 1–26.
———. 2002, "A Model of Demographic and Economic Change in Roman Egypt after the Antonine Plague," *Journal of Roman Archaeology* 15: 97–114.
———. 2003, "Germs for Rome," in C. Edwards and G. Woolf, eds., *Rome the Cosmopolis*, Cambridge: 158–76.
———. 2007, "Marriage, Families, and Survival: Demographic Aspects," in P. Erdkamp, ed., *A Companion to the Roman Army*, Oxford: 417–34.
———. 2009, "In Search of Roman Economic Growth," *Journal of Roman Archaeology* 22: 46–70.
———. 2012, *The Cambridge Companion to the Roman Economy*, Cambridge.
———. 2013, "The First Fall of the Roman Empire," Ronald Syme Lecture, Wolfson College, University of Oxford, October 31, 2013.
———. 2014, "The Shape of the Roman World: Modelling Imperial Connectivity," *Journal of Roman Archaeology* 27: 7–32.
———. ed. 2015a, *State Power in Ancient China and Rome*, Oxford.
———. 2015b, "State Revenue and Expenditure in the Han and Roman Empire," in W. Scheidel, ed., *State Power in Ancient China and Rome*, Oxford: 150–80.
———. 2017, *The Great Leveler: Violence and the History of Inequality from the Stone Age to the Twenty-First Century*, Princeton.
———. ed. forthcoming, *The Science of Roman History*, Princeton.
Scheidel, W. and Friesen, S. J. 2009, "The Size of the Economy and the Distribution of Income in the Roman Empire," *Journal of Roman Studies* 99: 61–91.
Scheidel W. and Monson, A. eds. 2015, *Fiscal Regimes and the Political Economy of Premodern States*, Cambridge.
Schlange-Schöningen, H. 2003, *Die römische Gesellschaft bei Galen: Biographie und Sozialgeschichte*, Berlin.

Schmid, J. 1955–56, *Studien zur Geschichte des griechischen Apokalypse-Textes*, Munich.

Schneider, H. 2007, "Technology," in W. Scheidel, I. Morris, and R. P. Saller, eds., *The Cambridge Economic History of the Greco-Roman World*, Cambridge: 144–174.

Schor, A. M. 2009, "Conversion by the Numbers: Benefits and Pitfalls of Quantitative Modelling in the Study of Early Christian Growth," *Journal of Religious History* 33: 472–98.

———. 2011. *Theodoret's People: Social Networks and Religious Conflict in Late Roman Syria*, Berkeley.

Schuenemann, V. J. et al. 2011, "Targeted Enrichment of Ancient Pathogens Yielding the pPCP1 Plasmid of Yersinia pestis from Victims of the Black Death," *Proceedings of the National Academy of Sciences* 108: E746-52.

Schuenemann, V. J. et al. 2013, "Genome-Wide Comparison of Medieval and Modern *Mycobacterium leprae*," *Science* 341: 179–83.

Scobie, A. 1986, "Slums, Sanitation, and Mortality in the Roman World," *Klio* 68: 399–433.

Scranton, R. 1957, *Mediaeval Architecture in the Central Area of Corinth*, Princeton.

Sebbane, F. et al. 2006, "Role of the Yersinia pestis Plasminogen Activator in the Incidence of Distinct Septicemic and Bubonic Forms of Flea-Borne Plague," *Proceedings of the National Academy of Sciences* 103: 5526–30.

Seifert, L. et al. 2016, "Genotyping Yersinia pestis in Historical Plague: Evidence for Long-Term Persistence of Y. pestis in Europe from the 14th to the 17th Century," *PLoS ONE* 11: e0145194.

Seland, E. H. 2007, "Ports, Ptolemy, Periplus and Poetry — Romans in Tamil South India and on the Bay of Bengal," in E. H. Seland, ed., *The Indian Ocean in the Ancient Period: Definite Places, Translocal Exchange*, Oxford: 69–82.

———. 2012, "The Liber Pontificalis and Red Sea Trade of the Early to Mid 4th Century AD," in D. Agius, et al., eds., *Navigated Spaces, Connected Places: Proceedings of the Fifth International Conference on the People of the Red Sea, Exeter 2010*, Oxford: 117–26.

———. 2014, "Archaeology of Trade in the Western Indian Ocean, 300 BC–AD 700," *Journal of Archaeological Research* 22: 367–402.

Selinger, R. 2002, *The Mid-third Century Persecutions of Decius and Valerian*, Frankfurt.

Sender, R., Fuchs, S., and Milo, R. 2016, "Revised Estimates for the Number of Human and Bacteria Cells in the Body," *PLOS Biology* 14(8): e1002533.

Shah, S. 2010, *The Fever: How Malaria Has Ruled Humankind for 500,000 Years*, New York.

———. 2016, *Pandemic: Tracking Contagions, from Cholera to Ebola and Beyond*, New York.

Shanzer, D. 2002, "*Avulsa a Latere Meo*: Augustine's Spare Rib—Confessions 6.15.25," *Journal of Roman Studies* 92: 157–76.

Sharpe, P. 2012, "Explaining the Short Stature of the Poor: Chronic Childhood Disease and Growth in Nineteenth-Century England," *Economic History Review* 65: 1475–94.

Shaw, B. D. 1987, "The Age of Roman Girls at Marriage: Some Reconsiderations," *Journal of Roman Studies* 77: 30–46.

———. 1995, *Environment and Society in Roman North Africa*, Aldershot.

———. 1996, "Seasons of Death: Aspects of Mortality in Imperial Rome," *Journal of Roman Studies* 86: 100–38.

———. 2000, "Rebels and Outsiders," in P. Garnsey, D. Rathbone, and A. K. Bowman, eds., *The Cambridge Ancient History*, Vol. 11: *The High Empire, A.D. 70–192*, Cambridge: 361–403.

———. 2015, *Bringing in the Sheaves: Economy and Metaphor in the Roman World*, Toronto.

Shchelkunov, S. 2009, "How Long Ago Did Smallpox Virus Emerge?" *Archives of Virology* 154: 1865–71.

Shelton, J. 2015, "Creating a Malaria Test for Ancient Human Remains," *YaleNews*, news.yale.edu, Accessed August 8, 2016.

Sheridan, M. 2015, "John of Lykopolis," in G. Gabra and H. Takla, eds., *Christianity and Monasticism in Middle Egypt*, Cairo: 123–32.

Shindell, D. T. 2001, "Solar Forcing of Regional Climate Change During the Maunder Minimum," *Science* 294: 2149–52.

Shindell, D. T. et al. 2003, "Volcanic and Solar Forcing of Climate Change during the Preindustrial Era," *Journal of Climate* 16: 4094–107.

Shoemaker, S. 2012, *Death of a Prophet: The End of Muhammad's Life and the Beginnings of Islam*, Philadelphia.

———. 2016, "The Afterlife of the Apocalypse of John in Byzantium," in D. Krueger and R. Nelson, eds., *The New Testament in Byzantium*, Washington D.C.: 301–16.

Sidebotham, S. E. 2011, *Berenike and the Ancient Maritime Spice Route*, Berkeley.

Sigl, M. et al. 2015, "Timing and Climate Forcing of Volcanic Eruptions for the Past 2,500 Years," *Nature* 523: 543–62.

Singh, P. et al. 2015, "Insight into the Evolution and Origin of Leprosy Bacilli from the Genome Sequence of *Mycobacterium lepromatosis*," *Proceedings of the National Academy of Sciences* 112: 4459–64.

Sirks, A. J. B. 1991, *Food for Rome: The Legal Structure of the Transportation and Processing of Supplies for the Imperial Distributions in Rome and Constantinople*, Amsterdam.

Six, D. and Vincent, C. 2014, "Sensitivity of Mass Balance and Equilibrium-Line Altitude to Climate Change in the French Alps," *Journal of Glaciology* 60: 867–78.

Slack, P. 2012, *Plague: A Very Short Introduction*, Oxford.

Slim, H. 2004, *Le littoral de la Tunisie: étude géoarchéologique et historique*, Paris.

Sodini, J.-P. et al. 1980, "Déhès (Syrie du nord): campagnes I–III (1976–1978)," *Syria* 57: 1–308.

Soren, D. and Soren, N. 1999, *A Roman Villa and a Late Roman Infant Cemetery: Excavation at Poggio Gramignano, Lugnano in Teverina*, Rome.

Sorrel, P. et al. 2007, "Climate Variability in the Aral Sea Basin (Central Asia) during the Late Holocene Based on Vegetation Changes," *Quaternary Research* 67: 357–70.

Southern, P. 2006, *The Roman Army: A Social and Institutional History*, Santa Barbara.

Speidel, M. A. 2007, "Ausserhalb Des Reiches? Zu neuen römischen Inschriften aus Saudi Arabien und zur Ausdehnung der römischen Herrschaft am Roten Meer," *Zeitschrift für Papyrologie und Epigraphik* 163: 296–306.

———. 2014, "Roman Army Pay Scales Revisited: Responses and Answers," in M. Reddé, ed., *De l'or pour les braves!: soldes, armées et circulation monétaire dans le monde romain: actes de la table ronde organisée par l'UMR 8210 (AnHiMa) à l'Institut national d'histoire de l'art (12–13 septembre 2013)*, Bordeaux: 53–62.

Spera, L. 2003, "The Christianization of Space along the Via Appia: Changing Landscape in the Suburbs of Rome," *American Journal of Archaeology* 107: 23–43.

Sperber, D. 1974, "Drought, Famine and Pestilence in Amoraic Palestine," *Journal of the Economic and Social History of the Orient* 17: 272–98.

Spurr, M. 1986, *Arable Cultivation in Roman Italy, c. 200 B.C.–c. A.D. 100*, London.

Squatriti, P. 1998, *Water and Society in Early Medieval Italy: AD 400–1000*, Cambridge.

———. 2010, "The Floods of 589 and Climate Change at the Beginning of the Middle Ages: An Italian Microhistory," *Speculum* 85: 799–826.

Stark, R. 1996, *The Rise of Christianity: A Sociologist Reconsiders History*, Princeton.

Staskiewicz, A. 2007, "The Early Medieval Cemetery at Aschheim-Bajuwarenring—A Merovingian Population under the Influence of Pestilence?" in G. Grupe and J. Peters, eds., *Skeletal Series and Their Socio-Economic Context*, Rahden: 35–56.

Stathakopoulos, D. 2000, "The Justinianic Plague Revisited," *Byzantine and Modern Greek Studies* 24: 256–76.

———. 2004, *Famine and Pestilence in the Late Roman and Early Byzantine Empire: A Systematic Survey of Subsistence Crises and Epidemics*, Burlington.

Steckel, R. H. 2013, "Biological Measures of Economic History," *Annual Review of Economics* 5: 401–23.

Steger, F. 2016, *Asklepios: Medizin und Kult*, Stuttgart.

Stein, E. 1968, *Histoire du Bas-Empire*, Amsterdam.

Steinhilber, F., Beer, J., and Fröhlich, C. 2009, "Total Solar Irradiance during the Holocene," *Geophysical Research Letters* 36(19).

Steinhilber, F. et al. 2012, "9,400 Years of Cosmic Radiation and Solar Activity from Ice Cores and Tree Rings," *Proceedings of the National Academy of Sciences* 109: 5967–71.

Stephens, P. et al. 2016, "The Macroecology of Infectious Diseases: A New Perspective on Global-Scale Drivers of Pathogen Distributions and Impacts," *Ecology Letters* 19: 1159–71.

Stevenson, A. C. et al. 1993, "The Palaeosalinity and Vegetational History of Garaet el Ichkeul, Northwest Tunisia," *Holocene* 3: 201–10.

Stiros, S. 2001, "The AD 365 Crete Earthquake and Possible Seismic Clustering during the Fourth to Sixth Centuries AD in the Eastern Mediterranean: A Review of Historical and Archaeological Data," *Journal of Structural Geology* 23: 545–62.

Stone, A. C. et al. 2009, "Tuberculosis and Leprosy in Perspective," *Yearbook of Physical Anthropology* 52: 66–94.

Stothers, R. B. and Rampino, M. R. 1983, "Volcanic Eruptions in the Mediterranean before A.D. 630 from Written and Archaeological Sources," *Journal of Geophysical Research: Solid Earth* 88: 6357–71.

Strauch, I. ed. 2012, *Foreign Sailors on Socotra: The Inscriptions and Drawings from the Cave Hoq*, Bremen.

Straw, C. 1988, *Gregory the Great: Perfection in Imperfection*, Berkeley.

Strobel, K. 1993, *Das Imperium Romanum im "3. Jahrhundert": Modell einer historischen Krise?: Zur Frage mentaler Strukturen breiterer Bevölkerungsschichten in der Zeit von Marc Aurel bis zum Ausgang des 3. Jh. n. Chr.*, Stuttgart.

Swain, S. 2007, "Introduction," in S. Swain, S. Harrison, and J. Elsner, eds., *Severan Culture*, Cambridge: 1–28.
Swain, S. and Edwards, M. eds. 2004, *Approaching Late Antiquity: The Transformation from Early to Late Empire*, Oxford.
Swain, S., Harrison, S., and Elsner, J. eds. 2007, *Severan Culture*, Cambridge.
Syme, R. 1971, *Emperors and Biography: Studies in the 'Historia Augusta'*, Oxford.
———. 1983, *Historia Augusta Papers*, Oxford.
———. 1984, *Roman Papers III*, Oxford.
Tacoma, L. E. 2006, *Fragile Hierarchies: The Urban Elites of Third-Century Roman Egypt*, Leiden.
———. 2016, *Moving Romans: Migration to Rome in the Principate*, Oxford.
Tate, G. 1992, *Les campagnes de la Syrie du Nord du II au VII siècle: un exemple d'expansion démographique et économique à la fin de l'antiquité*, Paris.
Taylor, D. J. et al. 2010, "Filoviruses Are Ancient and Integrated into Mammalian Genomes," *BMC Evolutionary Biology* 10: 193.
Taylor, G. M., Young, D. B., and Mays, S. A. 2005, "Genotypic Analysis of the Earliest Known Prehistoric Case of Tuberculosis in Britain," *Journal of Clinical Microbiology* 2005: 2236–40.
Taylor, K. C. et al. 1993, "The 'Flickering Switch' of Late Pleistocene Climate Change," *Nature* 7: 432–36.
Tchalenko, G. 1953–58, *Village antiques de la Syrie du Nord, I–III*, Paris.
Tchernia, A. 1986, *Le vin de l'Italie romaine*, Rome.
Telelēs, I. 2004, *Meteōrologika phainomena kai klima sto Byzantio: symboles stēn ereuna tēs hellēnikēs kai latinikēs grammateias*, Athens.
Temin, P. 2004, "Financial Intermediation in the Early Roman Empire," *Journal of Economic History* 64: 705–33.
———. 2013, *The Roman Market Economy*, Princeton.
Thompson, E. A. 1958, "Early Germanic Warfare," *Past & Present* 14: 2–29.
———. 1996, *The Huns*, Oxford.
Thurmond, D. L. 1994, "Some Roman Slave Collars in CIL," *Athenaeum* 82: 459–93.
Tiradritti, F. 2014, "Of Kilns and Corpses: Theban Plague Victims," *Egyptian Archaeology* 44: 15–18.
Todd, M. 2005, "The Germanic Peoples and Germanic Society," in A. K. Bowman, P. Garnsey, and A. Cameron, eds., *The Cambridge Ancient History*, Vol. 12: *The Crisis of Empire, A.D. 193–337*, Cambridge: 440–60.
Tomber, R. 2008, *Indo-Roman Trade: From Pots to Pepper*, London.
———. 2012, "From the Roman Red Sea to Beyond the Empire: Egyptian Ports and Their Trading Partners," *British Museum Studies in Ancient Egypt and Sudan* 18: 201–15.
Tomlin, R. S. O. 2014, "'Drive Away the Cloud of Plague': A Greek Amulet from Roman London," in R. Collins and F. McIntosh, eds., *Life in the Limes: Studies of the People and Objects of the Roman Frontiers Presented to Lindsay Allason-Jones on the Occasion of Her Birthday and Retirement*, Oxford: 197–205.
Toner, J. P. 2014, *The Day Commodus Killed a Rhino: Understanding the Roman Games*, Baltimore.
Toohey, M. et al. 2016, "Climatic and Societal Impacts of a Volcanic Double Event at the Dawn of the Middle Ages," *Climatic Change* 136: 401–12.

Toubert, P. 2016, "La Peste Noire (1348), entre Histoire et biologie moléculaire," *Journal des savants*: 17–31.

Touchan, R. et al. 2016, "Dendroclimatology and Wheat Production in Algeria," *Journal of Arid Environments* 124: 102–10.

Treadgold, W. 1995, *Byzantium and Its Army, 284–1081*, Stanford.

Treggiari, S. 1991, *Roman Marriage: Iusti Coniuges from the Time of Cicero to the Time of Ulpian*, Oxford.

Treme, J. and Craig, L. A. 2013, "Urbanization, Health and Human Stature," *Bulletin of Economic Research* 65: 130–41.

Trombley, F. R. 2004, "Epigraphic Data on Village Culture and Social Institutions: An Interregional Comparison (Syria, Phoenice Libanensis, and Arabia)," *Late Antique Archaeology* 2: 73–101.

Trouet, V. et al. 2009, "Persistent Positive North Atlantic Oscillation Mode Dominated the Medieval Climate Anomaly," *Science* 324: 78–80.

Trueba, G. 2014, "The Origin of Human Pathogens," in A. Yamada et al., eds., *Confronting Emerging Zoonoses: The One Health Paradigm*, Tokyo: 3–11.

Tsafrir, Y. 2000, "Justinian and the Nea Church," *Antiquité tardive* 8: 149–64.

Tsiamis, C., Poulakou-Rebelakou, E., and Petridou, E. 2009, "The Red Sea and the Port of Clysma: A Possible Gate of Justinian's Plague," *Gesnerus* 66: 209–17.

Turner, P. 1989, *Roman Coins from India*, London.

Usoskin, I. G. et al. 2016, "Solar Activity During the Holocene: The Hallstatt Cycle and Its Consequence for Grand Minima and Maxima," *Astronomy and Astrophysics* 27295: 1–10.

Usoskin, I. G. and Kromer, B. 2005, "Reconstruction of the ^{14}C Production Rate from Measured Relative Abundance," *Radiocarbon* 47: 31–37.

Valtuena, A. A. et al., forthcoming. "The Stone Age Plague: 1000 Years of Persistence in Eurasia," bioRxiv 094243.

van Bekkum, W. J. 2002, "Jewish Messianic Expectations in the Age of Heraclius," in G. J. Reinink and H. Stolte, eds., *The Reign of Heraclius (610–641): Crisis and Confrontation*, Groningen: 95–112.

Van Dam, R. 1982, "Hagiography and History: The Life of Gregory Thaumaturgus," *Classical Antiquity* 1: 272–308.

———. 2007, *The Roman Revolution of Constantine*, New York.

———. 2010, *Rome and Constantinople: Rewriting Roman History during Late Antiquity*, Waco.

van der Vliet, J. 2015, "Snippets from the Past: Two Ancient Sites in the Asyut Region: Dayr al-Gabrawi and Dayr al-ʿIzam," in G. Gabra and H. Takla, eds., *Christianity and Monasticism in Middle Egypt*, Cairo: 161–88.

van Minnen, P. 1995, "Deserted Villages: Two Late Antique Town Sites in Egypt," *Bulletin of the American Society of Papyrologists* 32: 41–56.

———. 2001, "P. Oxy. LXVI 4527 and the Antonine Plague in the Fayyum," *Zeitschrift für Papyrologie und Epigraphik* 135: 175–177.

———. 2006, "The Changing World of Cities in Later Roman Egypt," in J.-U. Krause and C. Witschel, eds., *Die Stadt in der Spätantike: Niedergang oder Wandel?: Akten des internationalen Kolloquiums in München am 30. und 31. Mai 2003*, Stuttgart: 153–79.

Van Tilburg, C. 2015, *Streets and Streams: Health Conditions and City Planning in the Graeco-Roman World*, Leiden.

Vanhaverbeke, H. et al. 2009, "What Happened after the 7th Century AD? A Different Perspective on Post-Roman Anatolia," in T. Vorderstrasse and J. Roodenberg, eds., *Archaeology of the Countryside in Medieval Anatolia*, Leiden: 177–90.
Varlık, N. 2014, "New Science and Old Sources: Why the Ottoman Experience of Plague Matters," *The Medieval Globe* 1: 193–227.
———. 2015, *Plague and Empire in the Early Modern Mediterranean World: The Ottoman Experience, 1347–1600*, New York.
Verjbitski, D. T., Bannerman, W. B., and Kápadiâ, R. T. 1908, "Reports on Plague Investigations in India," *Journal of Hygiene* 8: 161–308.
Vionis, A. K., Poblome, J., and Waelkens, M. 2009, "Ceramic Continuity and Daily Life in Medieval Sagalassos, SW Anatolia (ca. 650–1250 AD)," in T. Vorderstrasse and J. Roodenberg, eds., *Archaeology of the Countryside in Medieval Anatolia*, Leiden: 191–213.
Visbeck, M. H. et al. 2001, "The North Atlantic Oscillation: Past, Present, and Future," *Proceedings of the National Academy of Sciences* 98: 12876–77.
Vollweiler, N. et al. 2006, "A Precisely Dated Climate Record for the Last 9 Kyr from Three High Alpine Stalagmites, Spannagel Cave, Austria," *Geophysical Research Letters* 33: L20703.
Vonmoos, M., Beer, J., and Muscheler, R. 2006, "Large Variations in Holocene Solar Activity: Constraints from ^{10}Be in the Greenland Ice Core Project Ice Core," *Journal of Geophysical Research* 111: 1–14.
Waelkens, M. et al. 1999, "Man and Environment in the Territory of Sagalassos, a Classical City in SW Turkey," *Quaternary Science Reviews* 18: 697–709.
———. 2006, "The Late Antique to Early Byzantine City in Southwest Anatolia. Sagalassos and Its Territory: A Case Study," in J.-U. Krause and C. Witschel, eds., *Die Stadt in der Spätantike: Niedergang oder Wandel?: Akten des internationalen Kolloquiums in München am 30. und 31. Mai 2003*, Stuttgart: 199–255.
Wagner, D. M. et al. 2014, "Yersinia pestis and the Plague of Justinian 541–543 AD: A Genomic Analysis," *Lancet Infectious Diseases* 14: 319–26.
Walburg, R. 2008, *Coins and Tokens from Ancient Ceylon*, Wiesbaden.
Walker, D.R. 1976, *The Metrology of the Roman Silver Coinage*, Oxford.
Walker, M. J. C. et al. 2012, "Formal Subdivision of the Holocene Series/Epoch: A Discussion Paper by a Working Group of INTIMATE (Integration of Ice-Core, Marine and Terrestrial Records) and the Subcommission on Quaternary Stratigraphy (International Commission on Stratigraphy)," *Journal of Quaternary Science* 27: 649–59.
Walmsley, A. 2007, *Early Islamic Syria: An Archaeological Assessment*, London.
Wanner, H. et al. 2008, "Mid- to Late Holocene Climate Change: An Overview," *Quaternary Science Reviews* 27: 1791–828.
Ward-Perkins, B. 2000a, "Specialized Production and Exchange," in A. Cameron, B. Ward-Perkins, and M. Whitby, eds., *The Cambridge Ancient History*, Vol. 14: *Late Antiquity: Empire and Successors*, Cambridge: 346–91.
———. 2000b, "Land, Labour and Settlement," in A. Cameron, B. Ward-Perkins, and M. Whitby, eds., *The Cambridge Ancient History*, Vol. 14: *Late Antiquity: Empire and Successors*, Cambridge: 315–45.
———. 2005, *The Fall of the Rome: And the End of Civilization*, Oxford.
Ware, C. 2012, *Claudian and the Roman Epic Tradition*, Cambridge.
Watts, E. 2015, *The Final Pagan Generation*, Oakland.

Wertheim, J. O. and Pond, S. L. K. 2011, "Purifying Selection Can Obscure the Ancient Age of Viral Lineages," *Molecular Biology and Evolution* 28: 3355–65.

Weinreich, O. 1913, "Heros Propylaios und Apollo Propylaios," *Mitteilungen des deutschen archäologischen Instituts, athenische Abteilung* 38: 62–72.

Whitby, M. 1985, "Justinian's Bridge over the Sangarius and the Date of Procopius' de Aedificiis," *Journal of Hellenic Studies* 105: 129–48.

———. 1995, "Recruitment in Roman Armies from Justinian to Heraclius (ca. 565–615)," in A. Cameron, ed., *The Byzantine and Early Islamic Near East III: States, Resources and Armies*, Princeton: 61–124.

———. 2000a, "The Successors of Justinian," in A. Cameron, B. Ward-Perkins, and M. Whitby, eds., *The Cambridge Ancient History*, Vol. 14: *Late Antiquity: Empire and Successors*, Cambridge: 86–111.

———. 2000b, "The Army, c. 420–602," in A. Cameron, B. Ward-Perkins, and M. Whitby, eds., *The Cambridge Ancient History*, Vol. 14: *Late Antiquity: Empire and Successors*, Cambridge: 288–314.

———. 2004, "Emperors and Armies, AD 235–395," in S. Swain and M. Edwards, eds., *Approaching Late Antiquity: The Transformation from Early to Late Empire*, Oxford: 156–86.

White S. 2011, *The Climate of Rebellion in the Early Modern Ottoman Empire*, New York.

White, T. C. R. 2008, "The Role of Food, Weather and Climate in Limiting the Abundance of Animals," *Biological Reviews* 83: 227–48.

Whitewright, J. 2009, "The Mediterranean Lateen Sail in Late Antiquity," *International Journal of Nautical Archaeology* 38: 97–104.

Whittaker, C. R. 1994, *Frontiers of the Roman Empire: A Social and Economic Study*, Baltimore.

Wickham, C. 2005, *Framing the Early Middle Ages: Europe and the Mediterranean 400–800*, Oxford.

———. 2016, *Medieval Europe*, New Haven.

Wiechmann, I. and Grupe, G. 2005, "Detection of Yersinia pestis DNA in Two Early Medieval Skeletal Finds from Aschheim (Upper Bavaria, 6th Century AD)," *American Journal of Physical Anthropology* 126: 48–55.

Wilkes, J. 1996, *The Illyrians*, Oxford.

———. 2005a, "Provinces and Frontiers," in A. K. Bowman, P. Garnsey, and A. Cameron, eds., *The Cambridge Ancient History*, Vol. 12: *The Crisis of Empire, A.D. 193–337*, Cambridge: 212–68.

———. 2005b, "The Roman Danube: An Archaeological Survey," *Journal of Roman Studies* 95: 124–225.

Wilson, A. I. 1998, "Water-supply in Ancient Carthage," in J. J. Rossiter, et al., eds., *Carthage Papers: The Early Colony's Economy, Water Supply, a Private Bath and the Mobilization of State Olive Oil*, Portsmouth: 65–102.

———. 2002, "Machines, Power and the Ancient Economy," *Journal of Roman Studies* 92: 1–32.

———. 2007, "Urban Development in the Severan Empire," in S. Swain, S. Harrison, and J. Elsner, eds., *Severan Culture*, Cambridge: 290–326.

———. 2009, "Indicators for Roman Economic Growth: A Response to Walter Scheidel," *Journal of Roman Archaeology* 22: 71–82.

———. 2011, "City Sizes and Urbanization in the Roman Empire," in A. Bowman and A. Wilson, eds., *Settlement, Urbanization, and Population*, Oxford: 161–95.

———. 2012, "Saharan Trade in the Roman Period: Short-, Medium- and Long-Distance Trade Networks," *Azania: Archaeological Research in Africa* 47: 409–49.

———. 2013, "The Mediterranean Environment in Ancient History: Perspectives and Prospects," in W. V. Harris, ed., *The Ancient Mediterranean Environment between Science and History*, Leiden: 259–76.

———. 2015, "Red Sea Trade and the State," in F. De Romanis and M. Maiuro, eds., *Across the Ocean: Nine Essays on Indo-Mediterranean Trade*, Leiden: 13–32.

Wilson, E. O. 1998, *Consilience: The Unity of Knowledge*, New York.

Wipszycka, E. 1986, "La valeur de l'onomastique pour l'histoire de la christianisation de l'Egypte. A propos d'une étude de RS Bagnall," *Zeitschrift für Papyrologie und Epigraphik* 62: 173–81.

———. 1988. "La christianisation de l'Égypte aux IV e–VI e siècles. Aspects sociaux et ethniques," *Aegyptus* 68: 117–65.

Witakowski, W. tr. 1996, *Pseudo-Dionysius of Tel-Mahre, Chronicle: Known Also as the Chronicle of Zuqnin. Part III*, Liverpool.

———. 2010, "Why Are the So-called Dead Cities of Northern Syria Dead?" in P. Sinclair et al., eds., *The Urban Mind: Cultural and Environmental Dynamics*, Uppsala: 295–309.

Witcher, R. E. 2011, "Missing Persons? Models of Mediterranean Regional Survey and Ancient Populations," in A. Bowman and A. I. Wilson, eds., *Settlement, Urbanization and Population*, Oxford, 36–75.

Witschel, C. 1999, *Krise, Rezession, Stagnation?: der Westen des römischen Reiches im 3. Jahrhundert n. Chr.*, Frankfurt am Main.

Wolf, G. 1990, *Salus populi Romani: die Geschichte römischer Kultbilder im Mittelalter*, Weinheim.

Wolfe, N. D., Dunavan, C. P., and Diamond, J. 2007, "Origins of Major Human Infectious Diseases," *Nature* 447: 279–283.

Wolska-Conus, W. 1962, *La Topographie chrétienne de Cosmas Indicopleustès. Théologie et science au VIe siècle*, Paris.

———. ed. 1968–73, *Topographie chrétienne*, Paris.

Woodbridge, J. and Roberts, N. 2011, "Late Holocene Climate of the Eastern Mediterranean Inferred from Diatom Analysis of Annually-Laminated Lake Sediments," *Quaternary Science Reviews* 30: 3381–92.

Woolf, G. 1998, *Becoming Roman: The Origins of Provincial Civilization in Gaul*, Cambridge.

Woolhouse, M. and Gaunt, E. 2007, "Ecological Origins of Novel Human Pathogens," *Critical Reviews in Microbiology* 33: 231–42.

Wrigley, E. A. 1988, *Continuity, Chance and Change: The Character of the Industrial Revolution in England*, Cambridge.

Xoplaki, E. 2002, *Climate Variability Over the Mediterranean*, diss., University of Bern.

Xu, L. et al. 2014, "Wet Climate and Transportation Routes Accelerate Spread of Human Plague," *Proceedings of the Royal Society of London B: Biological Sciences* 281: 20133159.

———. 2015, "The Trophic Responses of Two Different Rodent–Vector–Plague Systems to Climate Change," *Proceedings of the Royal Society of London B: Biological Sciences* 282: 20141846.

Ying-Shih, Y. 1986, "Han Foreign Relations," in D. Twitchett and M. Loewe, eds., *The Cambridge History of China*, Vol. 1: *The Ch'in and Han Empires, 221 BC–AD 220*, Cambridge: 377–462.

Yong, E. 2016, *I Contain Multitudes: The Microbes within Us and a Grander View of Life*, New York.

Zanchetta, G. et al. 2012, "Multiproxy Record for the Last 4500 Years from Lake Shkodra (Albania/Montenegro)," *Journal of Quaternary Science* 27: 780–9.

Zanchettin, D., Traverso, P., and Tomasino, M. 2008, "Po River Discharges: A Preliminary Analysis of a 200-Year Time Series," *Climate Change* 89: 411–33.

Zehetmayer, M. 2011, "The Continuation of the Antebellum Puzzle: Stature in the US, 1847–1894," *European Review of Economic History* 15: 313–27.

Zelener, Y. 2003, *Smallpox and the Disintegration of the Roman Economy after 165 AD*, diss., Columbia University.

———. 2012, "Genetic Evidence, Density Dependence and Epidemiological Models of the 'Antonine Plague,'" in E. Lo Cascio, ed., *L'impatto della "peste antonina,"* Bari: 167–78.

Zhang, Z. et al. 2007, "Relationship between Increase Rate of Human Plague in China and Global Climate Index as Revealed by Cross-Spectral and Cross-Wavelet Analyses," *Integrative Zoology* 2: 144–53.

Zimbler, D. L. et al. 2015, "Early Emergence of Yersinia pestis as a Severe Respiratory Pathogen," *Nature Communications* 6: 1–10.

Zocca, E. 1995, "La '*senectus mundi*': Significato, fonti, e fortuna di un tema ciprianeo," *Augustinianum* 35: 641–77.

Zuckerman, C. 1995, "The Hapless Recruit Psois and the Mighty Anchorite, Apa John," *Bulletin of the American Society of Papyrologists* 32: 183–94.

———. 2004, *Du village à l'empire: autour du registre fiscal d'Aphroditô, 525–526*, Paris.

———. 2013, "Heraclius and the Return of the Holy Cross," *Travaux et Mémoires* 17: 197–218.

INDEX